To Julia, whose copy editing was even more brilliant than usual.

FIFTH EDITION

Business Data Networks and Telecommunications

Raymond R. Panko
Ray@Panko.com

University of Hawai`i

PEARSON

Prentice Hall

Pearson Education
International

Executive Editor: Robert Horan
Publisher: Natalie E. Anderson
Project Editor: Lori Cerreto
Editorial Assistant: Robyn Goldenberg
Senior Media Project Manager: Joan Waxman
Marketing Assistant: Danielle Torio
Managing Editor: John Roberts
Production Editor: Suzanne Grappi
Production Assistant: Joe DeProspero
Manufacturing Buyer: Michelle Klein
Design Manager: Maria Lange
Cover Design: Jayne Conte
Manager, Print Production: Christy Mahon
Composition/Full-Service Project Management: *The GTS Companies, Inc.*
Printer/Binder: Phoenix Book Tech

Microsoft® and Windows® are registered trademarks of the Microsoft Corporation in the U.S.A. and other countries. Screen shots and icons reprinted with permission from the Microsoft Corporation. This book is not sponsored or endorsed by or affiliated with the Microsoft Corporation.

If you purchased this book within the United States or Canada you should be aware that it has been wrongfully imported without the approval of the Publisher or the Author.

Pearson Education LTD.
Pearson Education Singapore, Pte. Ltd
Pearson Education, Canada, Ltd
Pearson Education—Japan
Pearson Education Australia PTY, Limited

Pearson Education North Asia Ltd
Pearson Educación de Mexico, S.A. de C.V.
Pearson Education Malaysia, Pte. Ltd
Pearson Education Upper Saddle River, New Jersey

10 9 8 7 6 5 4 3 2 1
ISBN 0-13-127315-9

Brief Contents

Contents

Preface for Teachers

THREE QUESTIONS

Teachers who are considering this book typically have three questions.
- Why select this book?
- How can I teach with it?
- What's new since the last edition?

WHY SELECT THIS BOOK?

Relevant Information

The most important reason to select a textbook is the information it covers. This is especially important today because it is much harder for students to find jobs now than it was three or four years ago. Students need to achieve a higher level of content mastery.

Up-to-Date Content
Of course, the content needs to be up to date. The fifth edition has a full chapter on wireless LANs, a strong section on 3G cellular technologies, and information on many other important new areas. The fifth edition is not simply a minor update. Most chapters have been completely or heavily rewritten to reflect changes since the fourth edition.

Market-Driven Content
Even more importantly, the book's content is strongly market driven. Too many textbooks try to cover every technology that ever existed, even when its use today is almost nonexistent. This leaves far too little time for today's critical technologies and emerging technologies. Students become historians, not market-ready graduates. In fact, some books seem to ignore market data. One recent textbook even called Frame Relay a new technology despite the fact that Frame Relay revenues are almost equal to those of private line revenues in the WAN market. Here are some examples of the book's market-driven content.

> **Wireless LANs.** The fifth edition has a full chapter on wireless LANs (Chapter 5). The chapter looks at WLAN security and at the details of trade-offs between different WLAN standards.

> **Ethernet.** Ethernet has won the LAN wars, but it is no longer a simple install-and-forget technology. The book has a full chapter on Ethernet (Chapter 4).

> **Quality of Service (QoS).** It is no longer enough for networks to function fairly well. Today, they must work *very well*. Quality of service is introduced in the first chapter, and it is reinforced throughout the book.

➤ **Security.** As you would expect from the author, whose security textbook is energizing security teaching in IS, security is pervasive throughout the text. One of the final chapters (Chapter 9) anchors the security information.

➤ **Frame Relay.** In the WAN market, Frame Relay is almost as large as private lines today. The fifth edition has more on Frame Relay, including how to design a Frame Relay network and mesh versus hub-and-spoke design (Chapter 7). Of course, metropolitan Ethernet, VPNs, and ATM are covered, but they are all very small today.

➤ **TCP/IP.** Above the physical and data link layers used in LANs and WANs, TCP/IP is consolidating its hold on upper-layer technologies. In the fifth edition, TCP/IP is pervasive, with Chapter 8 as its anchor chapter.

Job-Ready Detail

In the past, students were often asked in job interviews to name the OSI layers and nothing else. Today, however, even interviews for non-networking jobs grill applicants on the details of networking. The days of feel-good "network appreciation" textbooks with far too little detail for job applicants should be over.

For example, in the Ethernet chapter (Chapter 4), the book goes well beyond basic topology and switch operation to look at VLANs, link aggregation, the Spanning Tree Protocol and the Rapid Spanning Tree Protocol, overprovisioning versus priority, and switch purchasing considerations. Detail has been beefed up in other areas as well. The TCP/IP chapter (Chapter 8) takes a detailed look at how routers operate, while many other books cover this critical topic poorly or even incorrectly.

Great Teacher Support

Teaching networking is very difficult, so textbooks must provide strong teacher support.

Detailed PowerPoint Lectures

The book has full PowerPoint lectures created by the author—not just "a few selected slides." The PowerPoint lectures include builds for more complex figures. They also include information gathered since the book went to press. Students can download the lectures in full PowerPoint format. They can also download handouts with six slides per page in PDF format for faster downloading. Teachers can obtain annotated versions of the PowerPoint presentations to help them prepare and present lectures.

The PowerPoint slides are keyed directly to figures in the book. This is no accident. The book was designed so that almost all key points are covered in figures—including "study figures" that summarize key points in more complex sections. The PowerPoint lectures are created and updated by the author.

Website

The book's website, http://www.prenhall.com/panko, is rich in teacher resources. This is where teachers can download answer keys, test item file questions, TestGen, and the latest versions of the PowerPoint lectures (which are updated once or twice a year). The website also is created and updated by the author.

Flexibility: An 11-Chapter Core

The book has 11 core chapters. In a three-credit, one-semester course, this leaves one to two weeks for other material. The 11 core chapters form a complete course, so the additional time can be spent in enrichment activities. These may include hands-on activities (discussed below), additional TCP/IP material (or other material in the advanced modules), term projects, or whatever the teacher wishes to cover. In addition, many teachers cover only the 11 core chapters to ease the learning burden on students.

Answer Keys and Test Item File

The chapters have Test Your Understanding questions roughly once per page, so that students can do a brain check on what they have just read. In addition, end-of-chapter thought questions, design questions, and troubleshooting questions help the student attain higher-level mastery of the material. Answer keys for all questions are available to teachers.

Multiple-choice test item file questions are keyed to specific chapter questions. This allows teachers who wish to be selective to specify those questions students should master and then develop tests that reflect those selected questions. The test questions are provided in Word for Windows format.

Materials for Your Online Course

Prentice Hall supports adopters using online courses by providing files ready for upload into both WebCT and Blackboard course management systems for your testing, quizzing, and other supplements. Please contact your local PH representative or mis_service@prenhall.com for further information on your particular course.

Mailing List

There is a low-volume mailing list that is used a few times per year to update adopters on new developments—most commonly new material at the website. The mailing list is also used to solicit adopter feedback on the text.

Pedagogy

Learning networking is difficult. Many students find that networking is the most conceptually difficult course in IS programs. Networking books need to have very strong pedagogy.

Clear Writing

All editions of this book have received accolades for clear writing—especially their ability to teach difficult and complex topics. Every chapter is classroom-tested, and each chapter is proofread by two separate proofreaders, one of which has considerable networking knowledge.

Hands-On Opportunities

Students want opportunities to do things hands on. With the fifth edition, they can.

> ➤ OPNET IT Guru and ACE. OPNET Technologies, Inc., has kindly made the student versions of their IT Guru and ACE programs available to adopters—together with a series of lab exercises to reinforce key networking concepts. IT Guru is a powerful network simulation tool, while ACE focuses on application-level performance. (These exercises require 32-bit versions of Windows, so not everyone can run them.)

➤ End-of-Chapter Hands On Questions. Chapters 1, 4 and 10 have Hands On questions to reinforce concepts.

➤ Chapter 3a discusses how to cut and connectorize UTP. Teaching this material requires an investment of about $200, but undergraduate students love it.

➤ Windows XP Networking and Security. Windows XP is rapidly becoming the main client version of Windows, and it is not due to be replaced for some time. Chapter 1a shows students how to set up a Windows XP client for networking. If you have an understanding lab manager, students can go into your lab, set up a connection, set up a workgroup, and then undo what they have done. In turn, Chapter 9a shows students how to set up Windows XP security. Again, the chapter has material that you can have students do in your lab.

➤ TCP/IP Play. Layering is difficult for students. The TCP/IP Play in Chapter 2a allows teams of three students to go through a simple HTTP download to develop a skin understanding of how TCP works.

Running Case and Case Studies

Students like real-world examples because they help make complex concepts more concrete. The book has a running case—the First Bank of Paradise—which is a composite of several actual banks (much of the information presented is too sensitive for bank identification).

Chapter 1 describes how a bank vice president set up a home PC network (a topic that most students will find interesting). In addition, several of the "a" chapters following the main chapters are case studies. Chapter 2b is a case study in the development of a small SOHO PC network. Chapter 4a, in turn, is a LAN design exercise for a larger building. Chapter 7a is a case study on the First Bank of Paradise's wide area networks.

Chapter Questions

The book gives students many opportunities to check their knowledge. Approximately once per page, there are Test Your Understanding questions to help students see whether they have understood the material just read.

End-of-chapter questions help the student integrate the material in the chapter. Thought questions help the student think more deeply about the material. Troubleshooting questions and Design questions also help students develop important troubleshooting and design skills, which are critical in networking. Some chapters also have hands-on questions for your students to do at home.

As noted earlier, test item file questions are keyed to specific Test Your Understanding and end-of-chapter questions.

Up Through the Layers/Familiar to the Unfamiliar

Like most books, the fifth edition takes an up-through-the-layers approach. However, this approach is significantly modified, because most books that take this approach teach one layer at a time in isolation. Only at the end of the book does the student get the whole picture. During the process, they have only a cursory framework within which to integrate chapter knowledge.

➤ The book begins, in Chapters 1 through 3, with a strong framework to help students understand networking broadly so that when new knowledge appears, they understand its place. The difficult concept of layered network architectures is introduced early and is reinforced throughout the book.

➤ Chapters 4 through 7 deal with LAN, telephone, and WAN technologies. Every LAN and WAN technology is a mixture of Layer 1 (physical) and Layer 2 (data link) technologies. For this reason, this book covers Layer 1 and 2 technologies within the context of specific LAN and WAN technologies rather than individually (although Chapter 3 introduces specific physical layer information).

➤ Chapter 8 deals with internetworking, especially TCP/IP internetworking at Layer 3 (internet) and Layer 4 (transport). Once the student understands LAN and WAN technologies, they can appreciate the need to interconnect them.

➤ Chapters 9 and 10 cover material that cuts throughout the layers—security and network management. These topics are introduced early, but a full discussion has to wait until students have a solid understanding of layer technologies.

➤ Chapter 11 covers the application layer (in OSI, application layers). It might seem better to cover this information after Chapter 8, but many schools cover applications in a separate course.

Synopsis Sections

Each chapter ends in a synopsis section that summarizes key points. In classroom testing of the fifth edition chapters, these synopsis sections were very popular with students.

TEACHING WITH THIS BOOK

As noted earlier, this book has 11 core chapters. These form a complete course.

Freshman and Sophomore Courses

For freshman and sophomore courses, it is good practice to stay with the 11 core chapters, going over chapter questions in class. If you want to do hands-on material, it is advisable to cut some material from the core chapters.

Junior and Senior Courses

With courses for juniors and seniors, covering the 11 core chapters (including "a" chapters that are case studies) will probably leave you with one or two semester weeks "free." As noted earlier, this leaves time for hands-on activities (discussed above), additional TCP/IP material (or other material in the advanced modules), a term project, or whatever content you wish to cover. However, the entire book, including all hands-on material, should not be covered front-to-back in a semester.

Graduate Courses

Graduate courses tend to look a lot like junior and senior level courses but with greater depth. More focus can be placed on end-of-chapter questions and novel hands-on exercises, such as OPNET simulations. It is also typical to have a term project.

Changes Since the Fourth Edition

The fifth edition generally follows the same basic flow as the fourth. The following table lists some specific changes. Speaking broadly, the fifth edition offers a running case study, many more hands-on opportunities, and substantially stronger treatments of wireless LANs, Ethernet, TCP/IP, security, and network management. Systems (server) administration coverage is reduced and is covered primarily in Chapter 1b; this reflects feedback from adopters who wish to focus more on core networking concerns.

Fifth Edition	Remarks Relative to the Fourth Edition (*4e*)
Chapter 1. Introduction	Similar to Chapter 1 in *4e* but with the addition of the book's running case study, a discussion of how to build a small home PC network, and hands-on questions.
Chapter 1a. Windows XP Home Networking	Similar to Chapters 4a and 11a in *4e* but focuses on XP. Details on how to set up a Windows XP home network for Internet access and peer-to-peer service. My test students really liked this chapter.
Chapter 1b. Design Exercise: XTR Consulting: A SOHO Network with Dedicated Servers	Draws material from *4e*'s Chapter 4, 5a, and 10. In Chapter 1 of the new edition, the student learns how to create a home network. In Chapter 1b, they learn how to design a somewhat large SOHO PC network with a dedicated server. Cover 1b if you wish to cover the basics of systems administration.
Chapter 2. Layered Standards Architectures	Similar to Chapter 2 in *4e* but has a more stream-lined structure and presents more detail on Ethernet and TCP/IP.
Chapter 3. Physical Layer Propagation	Similar to Chapter 3 in *4e*. However, better treatment of optical fiber, moves UTP categories up from Chapter 4, and moves building wiring up from Chapter 6. Overall, avoids the spreading of physical layer material found in the fourth edition.
Chapter 3a. Hands On: Cutting and Connectorizing UTP	Essentially unchanged from the fourth edition.
Chapter 4. Ethernet LANs	In *4e*, Ethernet was spread over Chapters 4 and 5. In the fifth edition, it is introduced in Chapter 1 but is covered mostly in Chapter 4. Beefed-up presentation reflects the fact that Ethernet today has pushed legacy technologies almost completely out of the LAN arena but has grown more complex itself.
Chapter 4a. Case Study: Rewiring a College Building	Essentially Chapter 6a in the fourth edition but somewhat updated.
Chapter 5. Wireless LANs	The most comprehensive change in the book. This full chapter reflects the explosive growth of wireless LAN technologies. It covers both general technology and 802.11 security.

Fifth Edition	Remarks Relative to the Fourth Edition (*4e*)
Chapter 6. The Public Switched Telephone Network (PSTN)	Similar to Chapter 6 in *4e*, but quite heavily rewritten. Better coverage of PSTN technology, including its growing ATM transport core. Transport versus signaling. Stronger coverage of VoIP and 3G cellular service. Moves private lines (called leased lines in *4e*) up from Chapter 7 in *4e*.
Chapter 7. Wide Area Networks (WANs)	Similar to Chapter 7 in *4e*. Private lines and Frame Relay now account for almost 90% of the WLAN market. So more information on Frame Relay, including Frame Relay network design. Introduces mesh, hub-and-spoke, and combined private line network design. Drops ISDN coverage which is now negligible in corporations. ATM coverage is moved here, but ATM is primarily used in PSTN carrier cores today, not in corporations. Improved VPN section.
Chapter 7a. Case Study: First Bank of Paradise's Wide Area Networks	Basically Chapter 7a in *4e* but with an update—the bank's intended switch back from Frame Relay to private lines to connect its branches.
Chapter 8. TCP/IP Internetworking	Reworked Chapter 8 from the fourth edition. Covers a bit more because Chapter 2 in the fifth edition introduces more TCP/IP concepts earlier in the book.
Chapter 8a. Hands On: Packet Capture and Analysis with WinDUMP and TCPdump	New. How to use WinDUMP to capture a stream of packets and display them. By looking at the details of TCP/IP sessions, students can solidify their understanding of how TCP/IP really works.
Chapter 9. Security	Completely rewritten since the fourth edition based on the author's perspective after writing his security textbook. A bit less on encryption details (which are still available in Module D), significantly more comprehensive coverage.
Chapter 9a. Hands-On: Windows XP Home Security	New. How to make a Windows XP client secure. Hands-on exercises solidify student security knowledge.
Chapter 10. Network Management	Much more detail on management software tools than in the fourth edition. Discussion of simulation tied to hands-on OPNET IT Guru and ACE hands-on exercises. New discussion of tools to check network connections and troubleshoot problems. Managing IP networks, including subnet planning and managing DHCP, DNS, and WINS servers. Hands-on exercises.

Fifth Edition	Remarks Relative to the Fourth Edition (*4e*)
Chapter 11. Networked Applications	Largely the same as in the fourth edition. E-mail section has added section on viruses, Trojan horses, and spam. More on instant messaging in the peer-to-peer (P2P) section.
Module A. More on TCP/IP	Largely unchanged from *4e*, although MPLS is now in Chapter 8 to reflect its actual use and growing importance. Keyed to Chapter 8.
Module B. More on Modulation	Although modem modulation is no longer a critical topic, many teachers still wish to cover it. Keyed to Chapter 7.
Module C. Telephone Service	Specific voice services offered by PSTN carriers. Good to cover with Chapter 6. This material was in Module D in *4e*.
Module D. Cryptographic Processes	If you have an extra class and really want to get into how cryptographic systems work, this is good material to cover after Chapter 9.

REVIEWERS

We wish to thank the following faculty for their participation in reviews for this edition:

Sylvia Bembry, Winston Salem State University
John Carson, The George Washington University
Todd Edwards, Wake Forest University
Qing Hu, Florida Atlantic University
Harry Reif, James Madison University

Preface for Students

PERSPECTIVE

Initially, information systems (IS) graduates had only a single career track: programmer –analyst–database administrator–manager. Today, however, many IS graduates are going into the networking career track—often to their surprise. This course is an introduction to the networking track.

However, even programmers need a strong understanding of networking. In the past, programmers wrote stand-alone programs that ran on a single computer. Today, however, most programmers write networked applications that work cooperatively with other programs on other computers.

LEARNING NETWORKING

Networking Is Difficult

Networking is an exciting topic. It is also a difficult topic. In programming, the focus is on creating running programs. In networking, the critical skills are design, product selection, and troubleshooting. These rather abstract skills require a broad and deep knowledge of many concepts. Many IS students have a difficult time adjusting to these more cerebral skill requirements.

Employers Are Growing More Demanding

In the past, many teachers tried to deal with the complexity of networking by selecting what was in essence a "network appreciation" book—a feel-good book that lacked the detailed knowledge needed for actual networking jobs.

Today, however, employers demand—and get—much strong job readiness from new graduates. If you want to get a job in the IS field, you will need to have a competitive level of knowledge in every IS field you study. Even applicants for database jobs are grilled in networking knowledge (and networking applicants are grilled in database and other areas).

How to Study the Book

There are several keys to studying this book.

- ➤ Reading chapters once will not be enough. You will need to really study the chapters.
- ➤ Slow down for the tough parts. Some sections will be fairly easy, others difficult. Too many students study the harder stuff at the same speed they use to study the easier stuff.
- ➤ When you finish studying a section, do the Test Your Understanding questions immediately. If you don't get one of the questions, go back over the text.

Networking is strongly cumulative, and if you skim over one section, you will have problems with other sections later. Multiple-choice questions in the test item file are taken entirely from the Test Your Understanding questions and the end of chapter questions.

➤ Later, in groups, go over the Test Your Understanding questions to see if you got the correct answer.

➤ Study the figures. Nearly every key point in the chapter is covered in the figures. If there is something in a figure you don't understand, you need to study that section.

➤ If several concepts, for instance different network technologies, are presented in a section or chapter, do not just study them individually. You need to know which one to use in a particular situation, and that requires compare/contrast knowledge. Study figures that compare concepts, and make your own if the book does not have them.

➤ Study the Synopsis at the end of the chapter. The Synopsis summarizes the core concepts in the chapter. Be very sure you know them well. You might even study them before the chapter to get a broad understanding of the material.

Hands On

One way to make networking less abstract is to do as many hands-on activities as possible. If you have a Windows XP computer, do the work in Chapters 1a and 9a. Also be sure to do the hands-on exercises in Chapters 1, 4, and 10. If you can, do the OPNET exercises at the website, http://www.prenhall.com/panko. To really understand TCP/IP, download WinDUMP and play with it.

A NETWORKING CAREER

If you like the networking course and think you want a networking career, there are a number of steps you should take before graduation, even if your school does not have advanced networking courses.

➤ Most importantly, do a networking internship. Employers really want job experience—often preferring it to an absurd degree over academic preparation.

➤ Learn systems administration (the management of servers). Learn the essentials of Unix and Windows Server. You can download a server version of Linux and install it on your home computer in order to play with Unix commands and network management functions.

➤ Learn about security. Security and networking are now inextricably intertwined.

➤ Think about getting one or more industry certifications. In networking, the low-level CompTIA Network+ certification should be obtainable with just a bit more study after taking your core networking course. Cisco's CCNA (Cisco Certified Network Associate) certification, which focuses on switching and routing, will require substantially more study. Microsoft server certification is also valuable. Employers like applicants who are job-ready.

About the Author

Ray Panko is a professor of IT management at the University of Hawai`i's College of Business Administration. Before coming to the university, he was a research physicist at Boeing, where he flew on an early flight test of the 747 prototype, and was a project manager at Stanford Research Institute (now SRI International), where he worked for Doug Englebart (the inventor of the mouse). He received his B.S. in physics and his MBA from Seattle University. He received his doctorate from Stanford University, where his dissertation was conducted under contract to the Office of the President of the United States. In his spare time, he collects die-cast models and races in six-seat Hawai`ian outrigger canoes.

INTRODUCTION

Learning Objectives:

By the end of this chapter, you should be able to:

- Discuss the First Bank of Paradise (FBP), our running case study for this book.
- Discuss the major types of networked applications.
- List the eight elements of networks.
- Explain the major types of networks in businesses: LANs, WANs, internets, intranets, and extranets.
- Discuss major technical concerns for network managers: network architecture, standards, security, wireless networking, efficiency, and quality of service (QoS).
- Explain the elements and operation of a small home PC network using a LAN.
- Use some key hands-on network management tools, including bandwidth measurement services, ping, ping 127.0.0.1, tracert, ipconfig, winipconfig, nslookup, and the use of Windows Calculator to compute dotted decimal notation IP addresses.

FIRST
BANK
OF
PARADISE

INTRODUCTION

Bill Hannagan, CEO of Hawai`i's First Bank of Paradise,[1] recently reflected on the importance of networking in his firm: "When I joined FBP thirty years ago, computers were big mysterious boxes that lived in glass enclosures. Today computers are everywhere, and they all talk to each other. Networking has allowed our employees to work together in ways that were impossible before. It has also given us new products to offer to customers and new ways of talking to our customers."

Applications

We will see a little later in this chapter that networking at the First Bank of Paradise involves complex systems of hardware, software, and transmission components. However, network users do not care about that. They are only concerned with **networked applications** that help the First Bank of Paradise's customers do banking and that help FBP's internal staff do its work.

Networked Applications
>   Applications made possible by networking
>   Users only care about applications; the rest is mere details

E-Commerce
>   Buying and selling on the Internet
>   Users interact with databases
>   Customers will soon be able to talk to customer representatives

Transaction Processing
>   Simple, highly-structured, and high-volume interactions such as check processing
>   Built around databases
>   External settlement networks
>   Back-office transaction processing applications
>>   Accounting, payroll, purchasing, human resources, etc.
>>   Functional databases in individual departments

Office Applications
>   Word processing, spreadsheeting, etc.
>   E-mail, instant messaging (IM), and Web access

Figure 1-1   Networked Applications at the First Bank of Paradise (Study Figure)

---

[1]   The First Bank of Paradise is a composite of several banks in Hawai`i. For security reasons, a single bank cannot divulge identifiable information about itself publicly. However, the banks in this composite were generous with information that could be provided without appropriate disguise.

FBP is a mid-sized bank, but it is not a small company. The bank has annual revenues of $4 billion. It has 50 branches and 350 ATMs. It has more than 500 switches, 400 routers, 2,000 desktop and notebook PCs, 200 Windows servers, 30 Unix servers, and 10 obsolete Novell NetWare file servers. Its information systems staff has 150 employees.

E-Commerce
The networked application that is most visible to the bank's customers is the bank's **e-commerce** website. Today, users interact with the bank's databases. Users can see information about their own accounts and make appropriate changes.

FBP customers surfing the e-commerce website can already send e-mail to bank representatives. Soon customers will be able to use instant messaging to type messages back and forth with customer representatives in real time. In about two years, if their computers have microphones and speakers, customers will be able to talk to a bank representative while viewing the same screen the representative sees.

Transaction Processing
Transaction processing involves simple, highly structured, and high-volume interactions. In banking, the classic transaction processing application is check processing. Even a medium-size bank like the First Bank of Paradise processes five to seven million checks per day. Transaction processing, although not glamorous, dominates network traffic volume.

Databases   Almost all transaction-processing applications are linked to highly structured databases, such as the database of customer accounts. When FBP processes a check from one of its customers to another of its customers, the bank will debit one customer's account record and credit the other customer's account record.

External Settlement Networks   Of course, many transactions involve a customer of another bank. Banks have extensive settlement networks that allow them to handle cross-institution transactions.

Back-Office Business Operations   In addition to money-handling operations for customers, FBP also has extensive **"back-office"** transaction processing applications for its internal needs, such as accounting, payroll, purchasing, human resources, and billing. These often involve transaction processing and functional databases in individual departments.

Office Applications
In addition to stand-alone office applications, such as word processing and spreadsheeting, the bank's general office workers—from clerks to senior managers—use many networked applications. E-mail, instant messaging (IM), and Web access to internal webservers are just a few examples.

TEST YOUR UNDERSTANDING
1.   a) What are the major networked applications at the bank? b) What is transaction processing? c) What transaction processing applications used by the bank were mentioned in the text?

Networks at the Bank
To serve its communication needs, the First Bank of Paradise uses three types of networks: local area networks (LANs), wide area networks (WANs), and internets. We will look at each of these types of networks.

The Eight Elements of a Network

Before we look at specific networks, we should note that most networks share common elements, as Figure 1-2 illustrates. In general, networks have eight elements: application programs, client stations, server stations, switches, routers, access lines, trunk lines, and messages.

A network, then, is a transmission system that allows any application on any station in the network to communicate with any application on any other station in the network.

A network is a transmission system that allows any application on any station in the network to communicate with any application on any other station in the network.

Stations and Applications   The computers that communicate over networks are called **stations.** They come in two versions: client stations and server stations. As the name suggests, **server stations** provide service to **client stations.** Typically, these two types of computers are referred to simply as clients and servers.

Clients and servers run **application programs.** For example, in webservice, the client runs a browser, while the webserver runs a server application program. To users, of course, only the applications are important; everything else is mere details.

Messages   When applications need to communicate, stations send messages to one another. In single networks, these **messages** are called **frames.**

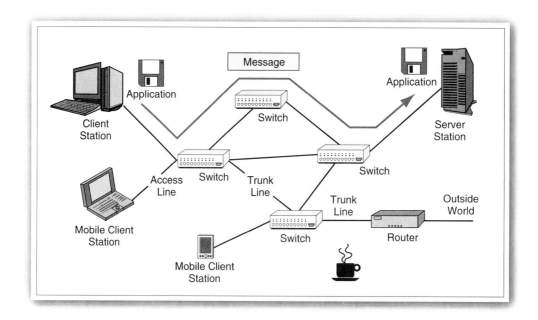

Figure 1-2   Elements of a Network

Switches and Routers   Frames are forwarded within a network by devices called **switches.** When a station transmits a frame, the frame goes to the switch to which the station connects. That switch forwards the frame to a switch closer to the destination station. This continues at intermediate switches, until the frame is delivered from the final switch to the destination station.

Again, switches only forward messages within a single network. In turn, **routers** forward messages outside of a single network, to another single network.

Transmission Lines: Access Lines and Trunk Lines   In turn, there are two types of transmission lines. **Access lines** connect stations to a switch. **Trunk lines,** in turn, link switches to each other, routers to each other, or a router to a switch. Trunk lines carry the traffic of many station-to-station conversations, so they need to have higher speed.

Message Transmission: Packet Switching   In modern networks, transmissions are broken into short messages (typically a few hundred bits long) that are sent individually. The breaking of transmissions into short messages is known as **packet switching,** even when the message itself is called a frame.

Multiplexing to Reduce Trunk Line Costs   Figure 1-3 shows why packet switching is done. It allows the messages of many conversations to share trunk lines. This is called **multiplexing.** Multiplexing reduces trunk line costs compared to traditional telephone transmission, in which each call has dedicated (reserved) capacity on each trunk line. This means that packet switching reduces overall costs, despite the fact that it increases some non–trunk line costs, for instance by making switches more complex and therefore more expensive.

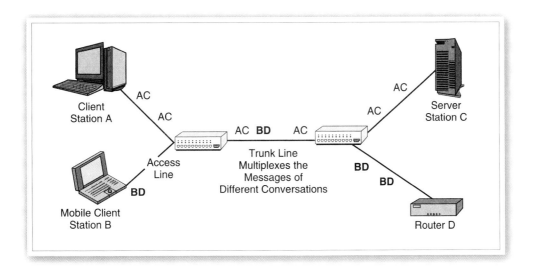

Figure 1-3   Multiplexing in a Packet-Switched Network

How to Talk about Network Costs   Note that when you discuss costs, you should mention which specific element of a network is involved: stations, switches, routers, access lines, or trunk lines. Packet switching reduces trunk line costs. It does nothing for access line costs. In fact, it somewhat increases switching costs, but cost savings from trunk line multiplexing more than make up for increased switching costs.

TEST YOUR UNDERSTANDING

2. a) List the eight elements of a network. b) Which element do users care about? c) Distinguish between client and server stations. d) What are messages called in single networks? e) Distinguish between switches and routers. f) Distinguish between the two types of transmission lines. g) Which type of line has higher capacity? Why? h) What is packet switching? i) What is the benefit of packet switching?

The Bank's Major Buildings

Figure 1-4 shows the bank's major sites.

Headquarters   The Headquarters building houses FBP's main business offices. About a quarter of the bank's employees work in this building.

Operations   Another building, called Operations, houses the bank's check-processing computers and other large servers that provide services to many customer and employee computers.

North Shore   The third major building, North Shore, is the bank's application development center. The bank develops all of its new software there. North Shore also has backup facilities for check processing and other major services so that if Operations fails, North Shore can take over the workload in a matter of minutes.

Local Branches and One Distant Branch   In addition, FBP has more than fifty bank branches across the state. These are the buildings that customers see the most. The bank also has communication links to a number of other companies, including an FBP affiliate on Da Kine Island and a credit card processing company.

Wide Area Networks (WANs)

Networks that link different sites together are called **wide area networks (WANs).** Figure 1-4 shows that the bank uses a number of wide area networks and that these networks use different technologies. We will see these technologies in Chapter 7. They include a point-to-point private line network, two Frame Relay networks, and several other WANs.

A wide area network (WAN) is a network that links different sites together.

TEST YOUR UNDERSTANDING

3. a) List the bank's sites. b) Explain the functions of the bank's three main buildings. c) What is a WAN? d) Does FBP use a single WAN or several?

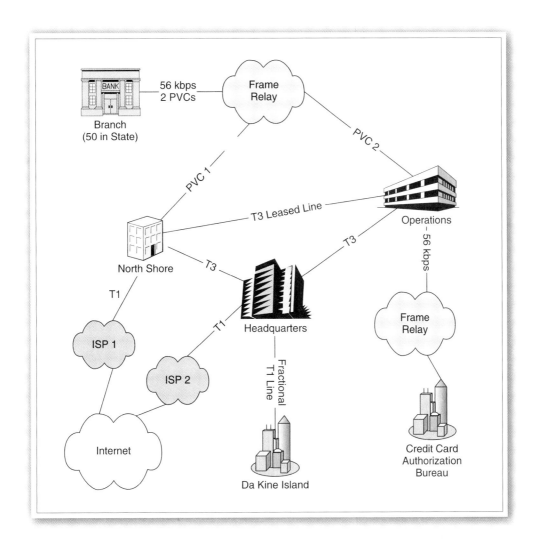

Figure 1-4 The First Bank of Paradise's Wide Area Networks (WANs)

Local Area Networks (LANs)

In turn, networks within sites are **local area networks (LANs).** The bank has a LAN within each of its buildings. We will see later that some of its managers also have LANs within their individual homes. Figure 1-5 shows a typical local area network in one of the bank's main buildings.

Local area networks (LANs) are networks within sites.

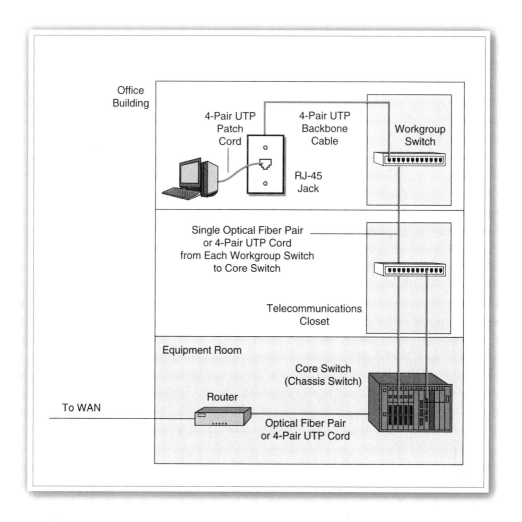

Figure 1-5   Local Area Network (LAN) in a Large Building

Devices called **switches** interconnect the computers in the building. A workgroup switch (see Figure 1-6) on each floor connects to all computers on that floor. A central core switch (see Figure 1-7) in the basement interconnects the workgroup switches so that any computer on any floor can talk to any computer on any other floor. The central core switch also connects to a router in the basement, which connects the building to the Internet and to other sites within the firm.

TEST YOUR UNDERSTANDING

4.   a) What are LANs? b) Distinguish between core and workgroup switches.

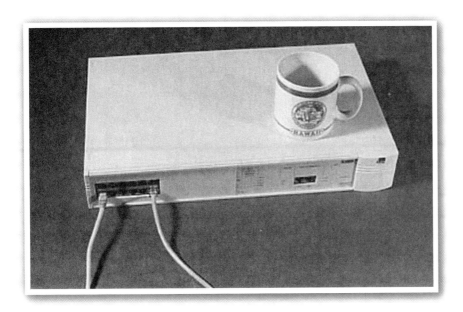

Figure 1-6   Workgroup Switch (19 inches/58 cm Wide)

Internets

Initially, computers could only communicate within individual LANs and WANs. However, Vint Cerf envisioned the linking of multiple networks into larger systems called "internets." Figure 1-8 shows a simple internet with three networks: Network X, Network Y, and Network Z.

These networks would be connected via devices called routers. An **internet,** then, is a collection of networks connected by routers so that any application on any host computer on any network can send messages called **packets** to any other application on any host computer on any other network in the internet.

An internet, then, is a collection of networks connected by routers so that any application on any host computer on any network can send messages called packets to any other application on any host computer on any other network in the internet.

Hosts

In an internet, computers are called **hosts.** In single networks, computers are usually called stations.

IP Addresses

In an internet, each host has an Internet Protocol address or **IP address.** This is a string of 32 bits (ones and zeros). It is usually written in **dotted decimal notation** in which there are four numbers separated by dots. An example would be 10.239.22.112.

Figure 1-7   Core Switch (19 inches/58 cm Wide)

Switches Versus Routers

If you are confused by the difference between switches and routers, switches provide connections *within* networks, while routers provide connections *between* networks in an internet. Routers are, in effect, super switches.

> Switches provide connections *within* networks, while routers provide connections *between* networks in an internet.

Packets and Frames

Within single networks, messages are called frames. Within internets, messages going from the source host to the destination host are called packets. As Figure 1-8 shows, this is not just a semantic distinction.

The packet, as just noted, goes all the way from the source host to the destination host. However, frames only travel through a single network. More specifically, the packet is carried in (encapsulated in, to use a technical term) a frame in each network.

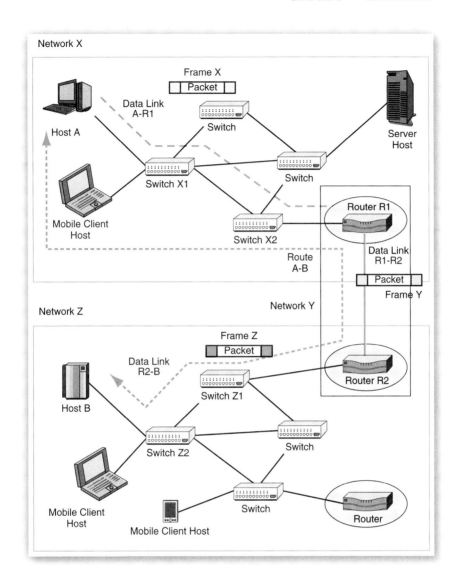

Figure 1-8   Internet with Three Networks

➤ In Network X, the source host places the packet in Frame X and sends the frame to Router R1.

➤ Router R1 takes the packet out of Frame X, places it in a new frame, Frame Y, then sends Frame Y to Router R2 in Network Y.

➤ Router R2 takes the packet out of Frame Y, places it in Frame Z, and sends Frame Z to the destination host in Network Z.

➤ The destination host takes the packet out of Frame Z. The packet has now traveled from the source host to the destination host.

Figure 1-9   Large Router (19 inches/58 cm Wide)

The Internet

When we spell internet with a lower-case *i*, we are using the term generically for any internet. However, when we spell Internet with an upper-case *I*, we mean the global **Internet,** which connects hundreds of millions of host computers around the world.

Figure 1-4 shows that the bank connects to the outside world via two connections to the Internet. Figure 1-9 shows a large router used to connect the FBP to the Internet. Figure 1-10 shows a smaller router used to connect a small business to the Internet. The First Bank of Paradise also uses these small routers to connect their branch office LANs to the corporate WANs.

Host Computers   Figure 1-11 shows the essentials of how the Internet operates. First, all computers attached to the Internet (and any individual internet) are called host computers. This includes large **servers** that provide services to residential and corporate users around the world. Webservers and mail servers are just two examples of servers.

Figure 1-10   Small Router for a Branch Office (19 inches/58 cm Wide)

Household PCs and desktop computers in organizations that use the Internet also are hosts. Even PDAs, cellphones, and other emerging small devices are hosts.

Internet Service Providers (ISPs)   To connect to the Internet, an organization needs to pay an Internet service provider (ISP). An ISP serves two basic functions.

➤   First, the ISP is an organization's on-ramp to the Internet. Companies (and individuals) cannot use the Internet without an ISP.

➤   Second, the ISP provides transmission service for an organization's Internet data. If you are communicating with a host that uses the same ISP that you do, the ISP will handle all of the transmission itself, even if the other host is in another state or even in another country.

Network Access Points   What if the other host you are trying to reach uses a different ISP? This is quite likely because there are many ISPs. The answer is that ISPs interconnect at sites called **network access points (NAPs).** ISPs exchange traffic at these NAPs so that any host on any ISP can reach any other host on any other ISP. The collection of all ISPs forms the **Internet backbone.**

TCP/IP   ISPs can work together because all ISPs (and all hosts) transmit according to the **TCP/IP** standards that we will see in Chapter 2.

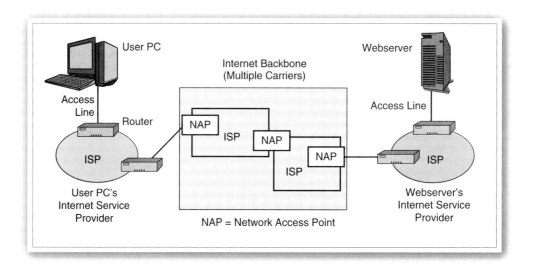

Figure 1-11   The Internet

ISPs at the First Bank of Paradise   In Figure 1-4, note that FBP uses two ISPs. This provides **redundancy** (duplication of service), so that if one ISP fails, the bank will still be connected to the Internet.

TEST YOUR UNDERSTANDING

5. a) What is an internet? b) In an internet, what device connects networks together? c) In an internet, what are hosts? d) Distinguish between frames and packets. e) In an internet, the source and destination hosts are separated by five networks (including their own networks). When the source host transmits, how many packets will travel through the internet? f) How many frames? g) Distinguish between internets and the Internet. h) Is a home PC that is connected to the Internet a host? i) Is the Internet a single network? j) What are the roles of ISPs on the Internet? k) What are NAPs? l) Why are NAPs crucial to universal connectivity on the Internet? m) What standards allow ISPs and hosts to communicate with one another? n) What is redundancy? o) Why does FBP use two ISPs?

Intranets and Extranets

Intranets   An **intranet** is an internet for internal transmission *within* firms; intranets use the TCP/IP transmission standards that govern transmission over the Internet. The First Bank of Paradise has a very widely used intranet for its internal traffic.

An intranet is an internet for internal transmission *within* firms; intranets use the TCP/IP transmission standards that govern transmission over the Internet.

---

internets

    An internet is a collection of networks connected by routers so that any host on any single network can communicate with any application on any other host on any other network in the internet

Internet components

    Computers are called hosts

    Hosts usually have IP addresses

        IP addresses are 32-bit strings of ones and zeros

        For humans, they are written in dotted decimal notation, such as 10.34.128.17

    Networks are connected by devices called routers

    Packets are carried within frames

        One packet is transmitted from the source host to the destination host

        In each network, the packet is carried in (encapsulated in) a frame

        If there are N networks between the source and destination hosts, there will be a packet and N frames for a transmission

internets versus the Internet

    internets (lower-case i): any internet that connects multiple networks

    Internet (upper-case I): the worldwide Internet

    Messages in internets are called packets

Intranet

    Internal internet for use within an organization

    Based on the TCP/IP standards created for the Internet

Extranets

    To connect multiple firms

        Only some computers from each firm are on the extranet

    Use TCP/IP standards

Intranets, Extranets, and the Internet

    Confusingly, both intranets and extranets can use the Internet for some of their transmission

Figure 1-12   The Internet, internets, Intranets, and Extranets (Study Figure)

Extranets   The bank also uses **extranets,** which use TCP/IP Internet standards to link several firms together but that are not accessible to people outside these firms. Even within the firms of the extranet, only some of each firm's computers have access to the extranet.

Extranets use TCP/IP Internet standards to link several firms together but do not allow access by people outside these firms.

Intranets, Extranets, and the Internet   Confusingly, both intranet and extranet transmission may use the global Internet for some or all of their transmission needs. However, when intranet or extranet traffic flows over the Internet, the sender encrypts it so that attackers lurking on the Internet cannot read or change messages. In addition, at connection points between corporate sites and the Internet, devices called firewalls prevent any improper traffic from entering the intranet or extranet.

TEST YOUR UNDERSTANDING

6.   a) What is an intranet? b) What is an extranet? c) Can intranets and extranets use the Internet for some of their transmission?

TECHNICAL NETWORK CONCERNS

Yvonne Champion, First Bank of Paradise's network manager, is responsible for making sure that FBP's networks meet the bank's exploding needs. Yvonne's boss, Warren Chun, is the bank's chief information officer (CIO)—the boss of the bank's information technology (IT) staff. Warren recently asked Yvonne to prepare a list of her top technical concerns as network manager. Figure 1-13 shows Yvonne's list. Most network managers would have similar lists.

Network Architecture

What Is a Network Architecture?

Yvonne's first technical priority is to develop a network architecture for FBP. A **network architecture** is a broad plan for how the firm will connect all of its computers within buildings (local area networks), between sites (wide area networks), and to the Internet. It also includes security devices and services.

Undisciplined Growth in the Past

In the past, the bank's networks grew haphazardly. Individual network projects were created with no clear overall plan. In addition, the bank's current network is an amalgam of the bank's original network and the network of another bank that the First Bank of Paradise absorbed in an acquisition.

Living with Legacy Networks

Yvonne wants a strong plan for where the network will be in five years. Then she wants to develop a plan for getting there. This will require the retiring of a number of old **legacy networks,** which use obsolete technologies that do not fit the long-term architecture.

However, these legacy networks will take a great deal of money to replace, so the First Bank of Paradise will have to be able to live with some of them for a few years to come. In fact, the bank still uses some legacy networks that date back to the 1960s.

Scalability

One concern in designing the architecture is **scalability**—the ability of selected technologies to be able to handle growth efficiently. They must not "top out" at some level of growth so that growth will be limited, and they must not jump sharply in cost

Network Architecture

  A broad plan for how the firm will connect all of its computers within buildings (local area networks), between sites (wide area networks), and to the Internet

  Undisciplined growth in the past

    No overall plan

  Legacy networks

    Use obsolete technologies that do not fit the long-term architecture

    Many exist in the bank

    Too expensive to replace quickly; must live with many for a while

Standards

  Standards govern message interactions between pairs of entities

    HTTP request and response messages for WWW access

    Request message asks for a file

    Response message delivers the requested file

  Standards create competition

    This reduces costs

    It also stimulates the development of new features

    Protects the business if the main vendors go out of business

  Competing standards organizations create incompatible standards

  LAN standards: will focus on Ethernet (some legacy technologies exist)

  WAN standards: will have fewer but still two to four

  Internetworking: will focus on TCP/IP

Security

  A Major Problem

    Many attacks

    Growing trend toward criminal attackers

  Firewalls (Figure 1-15)

    Firewalls drop and log packets identified as attack packets

  Virtual Private Networks (VPNs) (Figure 1-16)

    Provide secure communication over the Internet

    Cryptographic protection for confidentiality (eavesdroppers cannot read)

    Cryptographic authentication (confirms sender's identity)

Wireless Communication

  To improve mobility

  Drive-by hackers can eavesdrop on internal communication

Figure 1-13   Major Technical Network Concerns (Study Figure)

Efficiency (Figure 1-17)
      User demand is growing rapidly
      Budgets are growing slowly if at all
      For projects, need burning justification
      Still add new services by squeezing maximum payback from each dollar
Quality of Service
      Numerical objectives for performance that must be met by the network staff
      Transmission speed
            bits per second (bps)
                  In increasing factors of 1,000 (not 1,024)
                  kilobits per second (kbps)—lowercase k
                  megabits per second (Mbps)
                  gigabits per second (Gbps)
                  terabits per second (Tbps)
            LANs: 100 Mbps to each desktop
            WANs: most site-to-site links are 56 kbps to a few megabits per second
      Congestion, Throughput, Latency, and Response Time
            Congestion: when there is too much traffic for the network's capacity
            Throughput: the speed users actually see (often much less than rated speed)
                  Individual throughput is less than aggregate throughput on shared-speed links
            Latency: delay (usually measured in milliseconds (ms))
                  Within corporations, typically under 60 ms 90% of the time
                  On the internet, typically 30 ms to 150 ms
            Response time: the time to get a response after a user issues a command
                  0.25 second or less is good
      Availability
            Availability is the percentage of time a network can be used
            Downtime: when the user cannot use the network
            Want 24/7 availability
            Telephone network gives 99.999% availability
            Typical networks reach 98% today
      Error Rate
            Measured as the percentage of lost or damaged messages
            Substantial error rates can disrupt applications
            Substantial error rates generate more network traffic

Figure 1-13   *(Continued)*

as volume grows. For instance, the bank initially used manual processing to keep track of its inventory of network equipment. However, as the network grew larger, the manual approach became unwieldy and extremely expensive. The bank switched to an automated inventory management system, which should scale nicely as the network continues to grow.

TEST YOUR UNDERSTANDING

7. a) What is a network architecture? b) Why is it important? c) What are legacy networks? d) Why is it usually impossible to replace all legacy networks immediately? e) What is scalability? f) Why is scalability important?

Standards

Standards Govern Message Exchanges

A major part of Yvonne's network architecture will be selecting network standards[2] for various networks and applications. **Standards** are rules of operation that allow two hardware or software processes to work together. As Figure 1-14 shows, standards normally govern the exchange of messages between two entities.

Standards are rules of operation that allow two hardware or software processes to work together.

Standards normally govern the exchange of messages between two entities.

In the figure, the entities are a browser and a webserver application program on a webserver. The browser sends a request message governed by the Hypertext Transfer Protocol (HTTP). This message asks for a file. The response message, which also is governed by HTTP, contains the requested file.

Standards Create Competition

Standards are important because they allow products from different vendors to work together. The resulting competition keeps costs low, stimulates the development of

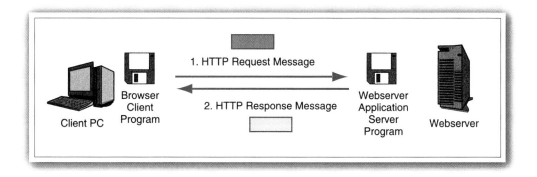

Figure 1-14   Standards Govern Interactions

---

[2] Research by Hacket Benchmarking & Research found that companies that standardize can cut their per-user annual cost by $2,000. This figure, however, includes standardizing on hardware and software as well as network standards. However, it indicates the importance of standardization on cost containment in general. Kim S. Nash, "Study: Use of Standards in IT Saves Money," Computerworld.com, http://www.computerworld.com/cwi/stories/0,1199,NAV47-68-84-88-93_STO58407,00.html.

new features (because vendors want to differentiate themselves), and protects corporations if one of their main vendors goes out of business.

Competing Incompatible Standards

Unfortunately, there are several different **official standards organizations,** and they produce standards that are incompatible with those of other standards organizations. There are also proprietary "standards" that are created by individual vendors for their own products.

FIRST BANK OF PARADISE

Today, the bank is virtually a museum of standards. Yvonne once tried to count the standards currently in use at the bank. She quit after two dozen. As she put it, "The Tower of Babel had nothing on us."

Local Area Network (LAN) Standards

In the future, Yvonne wants to slash the number of standards the bank uses. For the bank's LANs, she wants to use nothing except Ethernet technology. This is a safe decision because Ethernet almost completely dominates LAN technology today. The bank still has a few legacy LANs using non-Ethernet technology, but these will be gone within about a year.

Wide Area Network (WAN) Standards

The WAN standards situation is more complex, so Yvonne doubts that she can settle upon a single standard for WANs. However, she wants to reduce the number of major standards used in the bank's WANs to between two and four.

Internetworking

For internetworking—interconnecting the bank's LANs and WANs—Yvonne also wants to settle upon a single set of standards, namely the TCP/IP standards that govern communication over the Internet. Unfortunately, the bank currently uses four internetworking standards.

TEST YOUR UNDERSTANDING

8.   a) What are standards? b) Why are standards important? c) Why is it bad that there are multiple standards organizations? d) What standard will Yvonne select for the bank's LANs? e) What about for WANs? f) What standards will Yvonne use for internetworking?

Security

Another major concern for Yvonne is **network security.** The bank has long been a target of hackers attempting to break into the bank's computers. The bank receives approximately one attack packet per second. Most of these packets are simple probes to look for vulnerabilities or are part of unsophisticated attacks. However, a growing number of attack packet transmissions are part of serious break-in attempts.

Disturbingly, organized crime has been developing the ability to hack into banks. Why risk gunfights for a few thousand dollars when hacking thefts can bring vastly more money with little physical risk to the thief? Security will be a major component of the network architecture that Yvonne is designing for the bank.

Firewalls

For instance, Yvonne already has a firewall (see Figure 1-15) at every building's electronic connection points to the outside world. A **firewall** examines all incoming and outgoing traffic to look for dangerous content. When it finds provably dangerous packets (as it does thousands of times each day), it drops them and records them in a log file.

The firewall shown in the figure is a border firewall sitting at the border between the internal corporate network or internet and the global Internet. In the future, Yvonne will also use firewalls to divide building LANs into multiple areas that should be separated from one another. For instance, accounting servers should be protected by firewalls that only allow a few select people in other departments to reach these servers.

Virtual Private Networks (VPNs)

Another aspect of the network architecture will be the increasing use of virtual private networks. As Figure 1-16 shows, **virtual private networks (VPNs)** provide secure communication over the Internet or other public networks. VPNs give the same protection a company would have if it built its own private network, hence the name.

VPNs apply cryptographic protections to all messages flowing over them. These messages are encrypted for confidentiality, so that even if attackers intercept VPN packets, they will not be able to read them. In addition, each packet is given a digital signature to authenticate the sender, that is, prove the sender's identity.

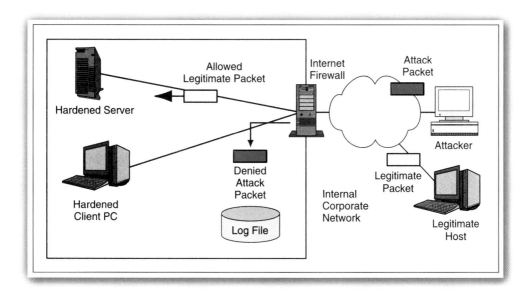

Figure 1-15   Firewalls and Intrusion Detection Systems (IDSs)

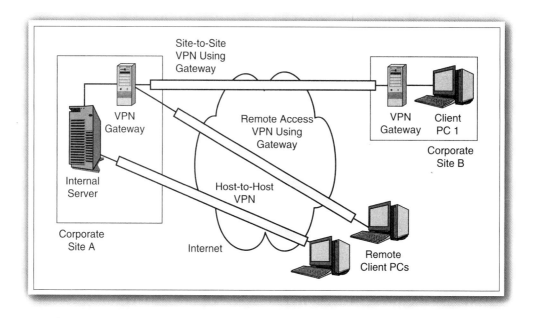

Figure 1-16 Virtual Private Networks (VPNs)

TEST YOUR UNDERSTANDING

9.   a) Why is the bank especially concerned about security? b) What is the function of a firewall? c) What is a border firewall? d) What are VPNs? e) What protections do VPNs provide?

Wireless Networking

Also on Yvonne's list is **wireless networking,** which uses radio transmission to connect devices. On the one hand, wireless networking can produce many benefits for the bank's staff members who have notebook computers, PDAs, and other mobile devices. With wireless networking, the bank's staff will no longer have to stop and find a wall jack to plug into when they need to use the bank's network or the Internet.

On the other hand, the radio signals used by wireless LANs can travel outside the bank's walls, allowing "drive-by hackers" to listen to the bank's internal communication.

The first wireless LAN standards had serious security problems. Although good standards for wireless LAN security are now becoming available, the bank is implementing wireless LANs cautiously because older products do not implement these new security standards and so can leave the bank open to attack.

TEST YOUR UNDERSTANDING

10.   a) Why is wireless networking desirable? b) Why is the bank moving cautiously on wireless LANs?

Efficiency

Figure 1-17 shows that internal and customer demands for networking service are exploding. Even if the bank's network budget were increasing rapidly, keeping up with exploding user demand would be difficult. In fact, Yvonne's network budget has never grown more than 5 percent per year and has been completely stagnant recently. This situation of exploding network demand and stagnant network budgets is almost universal in corporations. It must be considered in every networking decision in every firm.

"I tell my people not to even think about bringing me a project that doesn't have hard—really hard—cost numbers and a burning business justification." Despite this hardheaded attitude, Yvonne and her team have been able to add services by squeezing the maximum payback out of every dollar spent.

TEST YOUR UNDERSTANDING

11.   a) Compare trends in user demand and networking budgets. b) What are the implications of this comparison?

Quality of Service (QoS)

Last, but certainly not least, Yvonne is developing quantitative measures and objectives for network functioning. Now that networking is becoming critical to the corporation, it is not enough for networks to work. They must work *well*. **Quality of service (QoS)** objectives are numerical service targets that must be met by the networking staff.

> Quality of service (QoS) objectives are numerical service targets that must be met by the networking staff.

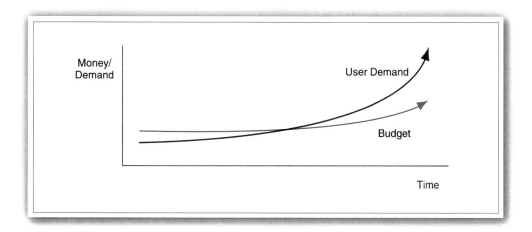

Figure 1-17 Demand Versus Budget Trends

Transmission Speed

Most obviously, users want the network to be fast. Transmission speed is measured in **bits per second (bps).** A bit is either a one or a zero. Obviously, a single bit cannot convey much information. In increasing factors of 1,000 (not 1024), we have kilobits per second (kbps[3]), megabits per second (Mbps), gigabits per second (Gbps), and terabits per second (Tbps).

Within buildings, the bank's local area networks typically bring 100 Mbps to each user's desktop. Speeds between corporate sites are much slower because long-distance transmission is expensive. Consequently, companies are more parsimonious with WAN speeds. At the First Bank of Paradise, for instance, almost all site-to-site WAN links operate at 128 kbps, although a few operate at megabit speeds. Most WAN site-to-site links in general operate at 56 kbps to a few megabits per second and are shared by multiple users.

LANs typically bring 100 Mbps to each desktop.

Most WAN site-to-site links operate at 56 kbps to a few megabits per second and are shared by multiple users.

Congestion, Throughput, Latency, and Response Time

Even on high-speed links, traffic congestion can slow the speed that users receive to a trickle if there is too much traffic for the network's capacity. (Think of a major freeway at rush hour.)

Throughput   When congestion occurs, the **throughput**—the speed that users actually see—falls dramatically. To give an analogy, when there is congestion on the freeway, your actual speed is far below the freeway's rated speed.

Aggregate and Individual Throughput   In most cases, a network serves many users at the same time. The **individual throughput** a single user receives is much less than **aggregate throughput** on shared links.

Latency   During periods of congestion, packets may be delayed by several **milliseconds (ms).** For some applications, such as e-mail, this makes no difference to users. For other applications, such as voice communication, this delay (called **latency**) can produce serious problems.

Yvonne notes that latency within her corporate network is less than 40 ms 90 percent of the time. This latency is so small that users almost never notice delays, even in highly interactive applications. On the Internet, however, latency typically is 30 ms to 150 ms. At the high end, Internet latency is high enough to bother the users of interactive applications such as voice communication.

Response Time   The opposite of latency (delay) is **response time.** This is the time it takes the user to get a response when he or she gives a command. An excellent

---

[3] In the metric system, kilo is abbreviated with a lowercase *k*. (Capital *K* is for Kelvin—a unit of temperature.) Networking people generally know this. Computer science and database people generally do not.

response time is less than a quarter of a second. Longer response times tend to be very annoying to users.

Availability
Users also expect service without disruptions. **Availability** is the percentage of time a network can be used. **Downtime** (a period of network unavailability) is simply intolerable if it goes on for more than a few minutes. Yvonne notes that networks need to be available 24/7. "The telephone network provides 99.999 percent availability. We aren't close to that now, but our availability is over 98 percent today."

Error Rate
The last major QoS measure is the transmission's system **error rate.** In networks, this usually is measured by the percentage of messages that are lost or damaged. A substantial error rate can harm application operation. Even if errors are caught and lost or damaged messages are retransmitted, this will result in latency while the receiver has to wait for the retransmission. These retransmissions, furthermore, increase network traffic.

TEST YOUR UNDERSTANDING

12.   a) What are QoS objectives? b) In what units is transmission speed measured? c) Give the names and abbreviations for speeds in increasing factors of 1,000. d) Is speed measured as bits per second or bytes per second? e) Distinguish between speed and throughput. f) Distinguish between individual and aggregate throughput. g) What is latency? h) What causes latency? i) In what units is latency measured? j) What is response time? k) What is availability? l) What is downtime? m) In what units is the error rate usually measured?

PAT LEE'S HOME LAN

We have seen that FBP has a large number of internal networks. In this first chapter, we will look at only one of these networks in detail—the LAN in the home of one of the bank's vice presidents, Pat Lee.

Microcosm
Starting with an employee's home network may seem like an odd choice. Why begin with such a small LAN, in fact one that is outside the corporation itself? The answer is that it is good pedagogy to begin with the familiar, and a growing number of students already have such LANs in their homes or are likely to install them during this course. Students interested in installing their own LANs should read Chapter 1a, which looks at PC configuration for LAN use.

More fundamentally, we will see that even networks in employee homes are fairly complex. In later chapters, when we look at larger LANs in the bank's main buildings and branches, we will see that although such networks are much larger, they primarily operate in the same ways. The home network we will see is a microcosm of the bank's larger networks. Starting with the simplest network is another example of good pedagogy. In addition, if students work through Chapter 1a, they will get hands-on experience in client configuration.

Pat Lee's Home Network

Two PCs
Pat Lee is the bank's vice president for international marketing. Like many people today, she has a computer at home. In fact, like a growing number of people, she has more than one PC at home. She has one desktop PC in her family's study. The other is a desktop PC upstairs in her daughter's room. Pat's daughter, Emily, uses her computer to do homework and to communicate with her friends via e-mail and instant messaging.

Access Router and Cable Modem
Last year, Pat networked the two computers together, as shown in Figure 1-18. From a small device called an **access router,** she ran wiring to each of the two PCs and to her cable modem, which provides high-speed access to Pat's local cable television carrier. As we will see, the access router performs several functions beyond routing.

TEST YOUR UNDERSTANDING

13.   a) List the physical elements of Pat Lee's home network.

Applications

Internet Applications
How does Pat Lee's family use her network? Most obviously, they use the Internet. They surf webpages, send and receive e-mail, chat with friends using instant messaging, and do a variety of other things.

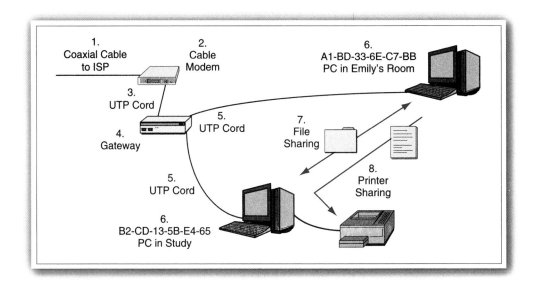

Figure 1-18   Pat Lee's Home Network

File Sharing

In addition, thanks to networking, they can now do **file sharing** between the two home PCs. The downstairs machine in the study is their main PC. As shown in Figure 1-18, Emily, working on the computer in her room, can retrieve files stored on the study computer's hard drive and can write files to the computer. Before this, Emily had to transfer files by floppy disk. Often, she had no idea whether the copy of a document on the computer in the study or on her bedroom computer was the latest version. Now, there is only one version—the one on the study PC.

Printer Sharing

Emily also does **printer sharing,** which Figure 1-18 also illustrates. The downstairs machine has a high-speed laser printer. When Emily gives the print command on her computer, the network sends her print job to the downstairs computer, which prints it on the laser printer. Before, when Emily needed to print, she would have had to interrupt the person working at the study computer to get her file on a floppy disk printed.

TEST YOUR UNDERSTANDING

14.   a) How does the family use file sharing? b) How does it use printer sharing?

The Personal Computers

Pat did not need special PCs to set up her LAN. The only hardware requirement was for each computer to have a **network interface card (NIC).** Figure 1-19 shows an internal NIC for a desktop PC and a PC Card NIC for a notebook PC. Like most desktop computers sold today, both PCs in Pat's house were sold with internal NICs, so there was nothing else to buy. However, if Pat had a notebook computer, she probably would need to buy a PC Card NIC, although some notebooks now come with built-in NICs.

Nor did Pat have to add any software to her computers. Both computers run Microsoft Windows operating systems. Emily's computer runs the older Windows ME

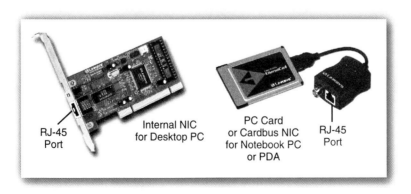

Figure 1-19   Network Interface Cards (NICs)

version, while the computer in the study runs the XP home version. Both are fully capable of handling file sharing, print service, and Internet access.

TEST YOUR UNDERSTANDING

15. a) Did Pat need special PCs to be able to network them? b) What hardware component might she have had to buy for her PCs? c) Did she need to add software to her PCs to be able to network them? d) Why or why not?

Pat's Cable Modem

To get to the Internet, Pat needed an Internet service provider (ISP), which provides access to the Internet for a monthly fee. Initially, Pat connected to her ISP with a telephone modem. Its slow speed drove her family crazy. In addition, when someone was on the Internet, nobody could use the telephone (and vice versa).

As Figure 1-18 illustrates, Pat's family upgraded to **cable modem** Internet access. A coaxial cable comes into her home from the cable television company. This is similar to the coaxial cable you use to connect a VCR to a television. The cable connection terminates in a cable modem. Wiring connects this cable modem to the access router. With telephone modems, she averaged WWW download speeds of only about 40 kbps. With her cable modem, download throughput has been averaging about 350 kbps.

TEST YOUR UNDERSTANDING

16. a) What are the advantages of cable modems over telephone modems for Internet access? b) What speeds does Pat's family receive? c) What hardware is required in Pat's network?

Wires, Connectors, and Jacks

Figure 1-20 shows the type of cabling used in Pat's home network.[4] This is **unshielded twisted pair (UTP)** wiring. We will look at UTP in more detail in Chapter 3. UTP wiring is about as thick as a pen and is rugged and flexible.

A length of UTP is a **cord**. As Figure 1-21 shows, the cord terminates at both ends in **RJ-45 connectors,** which plug into **RJ-45 jacks** (plugs). RJ-45 connectors and jacks are similar to the RJ-11 connectors and jacks used by home telephones, but RJ-45 connectors and plugs are a little wider. RJ-45 connectors simply snap into RJ-45 jacks.

To connect Pat's computers to the access router's built-in Ethernet switch and to connect the access router to the cable modem, Pat needed UTP cords of the right lengths. Fortunately, she was able to buy 4-pair UTP patch cords at her computer store. **Patch cords** are UTP cords that come pre-cut in a variety of lengths, with connectors attached. Although it is possible to buy bulk UTP cable and cut it to specific lengths, this takes special tools, including a tester (handmade cords often work poorly if at all).

TEST YOUR UNDERSTANDING

17. a) What is a length of UTP wiring called? b) Compare RJ-45 connectors and jacks to home telephone connectors and jacks. c) What are UTP patch cords? d) Why are they attractive?

---

[4] Very large Ethernet networks also use another transmission medium, optical fiber, which sends signals as light pulses. Optical fiber can carry signals much farther than UTP.

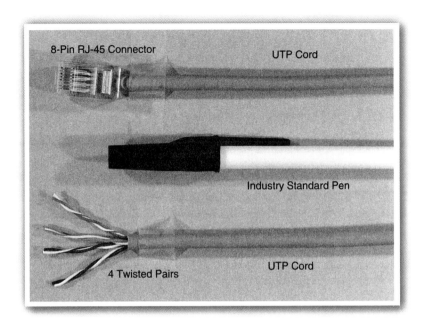

Figure 1-20   Unshielded Twisted Pair (UTP) Cord with RJ-45 Connector

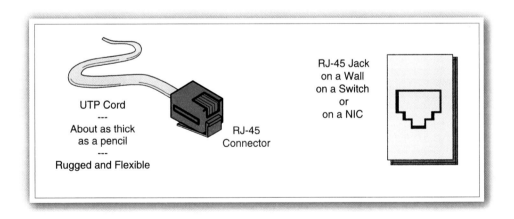

Figure 1-21   UTP Cord with RJ-45 Connector and Jack

Figure 1-22   Home Network Access Router

The Access Router

The heart of Pat's home network is her access router. This inexpensive device performs a number of important functions, as Figure 1-23 illustrates.

Ethernet Switch

First, the access router contains a 4-port **Ethernet switch.** This switch allows the two
computers to send frames to each other. It also allows each PC to send frames to and
from the internal access router function.

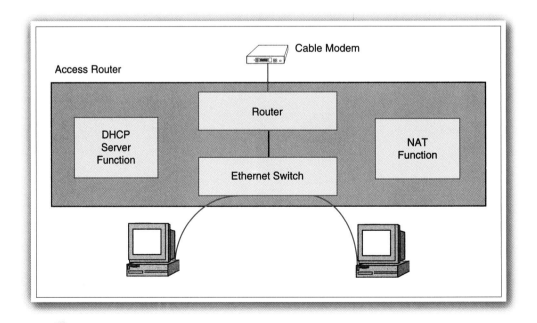

Figure 1-23   Logical Functions of the Access Router

Each computer's NIC has an Ethernet address, as Figure 1-24 illustrates. (This figure shows four computers instead of Pat's two to emphasize the generality of switching. Also, Pat's access router has a four-port Ethernet switch, so she can add another two if she wishes to do so.)

A typical Ethernet address is something like "A1-BD-33-6E-C7-BB." If one computer wishes to send a frame to another computer, it places the receiver's Ethernet address in the frame's destination address field. (Frames are divided into sections called fields, much as e-mail messages have several header fields such as "To" and "From," plus a body field.)

When the switch receives a frame addressed to one of these PCs, the switch looks up the destination address in its switching table, which associates destination addresses with port numbers. It then sends the frame out of the port indicated for the destination address.

TEST YOUR UNDERSTANDING

18.   a) What kind of address is B2-EF-43-7C-D3-BD? b) How does the switch decide which port to use to send the frame back out?

Router Functions

Pat's home network is a single Ethernet network. The telephone company's cable modem transmission line is another network that uses a different transmission standard (DOCIS). A router's job is to send packets (router messages are called **packets**) across multiple networks. Pat's router module in her access router performs that essential router function, linking her home network to the Internet via her cable connection.

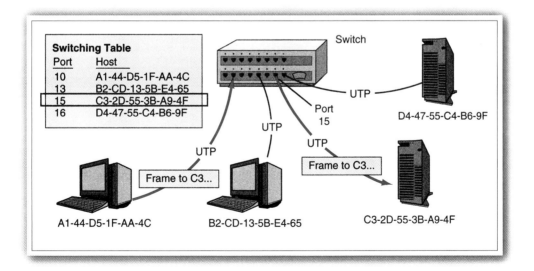

Figure 1-24   Ethernet Switch Operation

TEST YOUR UNDERSTANDING

19. What is the job of the router function in Pat's access router?

Frames and Packets

Earlier, in Figure 1-8, we saw that packets are always carried inside frames and that a router takes a packet out of a frame arriving from one network and places the packet in another frame so that it can travel through the next network.

Figure 1-25 shows that when a host computer in Pat's home transmits a packet, it places the packet in an Ethernet frame. The access router takes the packet out of the Ethernet frame, encapsulates it inside a DOCIS frame, and sends it out.

A packet traveling over the Internet is likely to travel through ten to twenty individual networks connected by routers. This means that each packet is likely to travel within ten to twenty different frames. Even if a router connects two networks of the same type, it will put the packet in a new frame when it sends the packet from one network to another.

TEST YOUR UNDERSTANDING

20. a) In Pat's home network, what frames are involved (list them in order) when a host computer transmits a packet? b) When a host computer receives a packet? c) How would things change if Pat's home network had an ATM connection to the ISP instead of a DOCIS connection? (ATM is another network technology).

DHCP Server

Ethernet Addresses are Assigned to NICs at the Factory   How does a computer get an Ethernet address? The answer is that every NIC that is sold has a unique

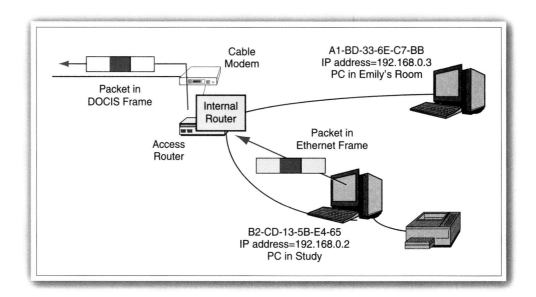

Figure 1-25   Frames and Packets

Ethernet address assigned to it at the factory. Although this Ethernet address can be changed, users rarely do so.

IP Addresses   Just as computers on individual Ethernet networks need Ethernet addresses, every computer on the Internet needs an **IP address,** where IP is the **Internet Protocol**—the standard that governs packet delivery across a series of routers. A typical IP address is 60.33.127.45. It contains four numbers separated by dots.

   Just as an Ethernet address is placed in the destination address of an Ethernet frame, an IP address is placed in the destination address field of a packet to be delivered across the Internet.

DHCP for Client IP Addresses   How do hosts receive IP addresses? It depends on whether the host is a client or a server. A client (home and office desktop) PC usually is given a *temporary IP address* each time it starts to use the Internet. As Figure 1-26 shows, the user's ISP has a **Dynamic Host Configuration Protocol (DHCP)** server that provides each user PC with a temporary IP address to use during that Internet session. DHCP response messages contain other configuration information as well. This other configuration information is based on the content covered in Chapter 8.

DHCP Service within Pat's Network   Pat's ISP's DHCP server only gives her a single IP address, say 60.47.112.6. Pat's access router, however, contains an internal DHCP server, which gives IP addresses to each of her internal PCs. As Figure 1-26 shows, Pat's access router's internal DHCP server gives her two PCs the internal addresses 192.168.0.2 and 192.168.0.3.

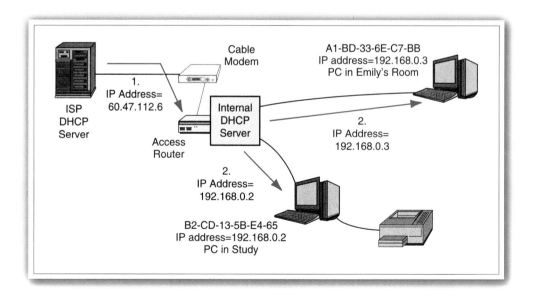

Figure 1-26   Dynamic Host Configuration Protocol (DHCP)

A Different IP Address Each Time   As a consequence of how DHCP works, a PC may get a different IP address each time it uses the Internet. This creates challenges for peer-to-peer applications in which PCs communicate directly. Ways to discover one another's IP addresses must be found for these applications to work.

Servers Receive Permanent IP Addresses   Servers, in contrast, are given **permanent IP addresses** so that you can reach them easily. (Imagine trying to shop at a store that moved to a different street address each day!) You just send packets to the server's permanent IP address, and routers along the way will get them to the server.

TEST YOUR UNDERSTANDING

21.   a) Who gives an Ethernet NIC its Ethernet address? b) What kind of address is 128.171.17.13—an Ethernet address or an IP address? c) How do client PCs normally get IP addresses? d) How does Pat's home network get its IP address? e) How do Pat's individual PCs get IP addresses? f) Does a client PC get the same IP address each time from its DHCP server? g) What kind of IP addresses do servers normally receive? h) Why?

Network Address Translation (NAT)

Pat's ISP only gives her a single IP address, but her access router's DHCP server assigns each of her internal PCs an IP address—neither of which is the IP address assigned by the ISP.

Figure 1-27 shows that the access router solves this problem by doing **network address translation (NAT).** When host 192.168.0.2 sends a packet to a webserver, the access router's NAT module replaces 192.168.0.2 in the packet's source IP address

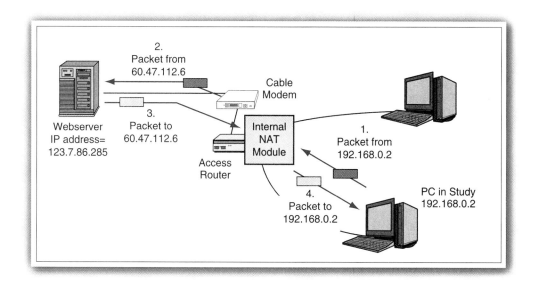

Figure 1-27   Network Address Translation (NAT)

with 60.47.112.6—the IP address assigned by the ISP. The ISP will only accept packets from this source address.

When the server sends back an IP packet addressed to 60.47.112.6, the access router's NAT module replaces this destination IP address in the packet with 192.168.0.2 and sends the packet on to the destination host. In Chapter 9 we will see more complex aspects of NAT.

TEST YOUR UNDERSTANDING

22.   a) Why is NAT needed in Pat's home network? b) How does NAT work?

Dealing with Host Names

A host's IP address, which contains 32 ones and zeros, is the host's official address on the Internet. Many hosts also have **host names,** which are easier to remember, for instance, voyager.cba.hawaii.edu. Host names are not official addresses, but users frequently only know a destination host's host name.[5]

Fortunately, the Internet has an automatic system for finding the IP addresses of named hosts. If the user types a host name in a URL or e-mail address, the DNS resolver program on the user's computer sends a DNS request message to a **domain name system (DNS)** server, as Figure 1-28 indicates. This request message gives the host name (Voyager.cba.hawaii.edu) and asks for the corresponding IP address. The DNS server

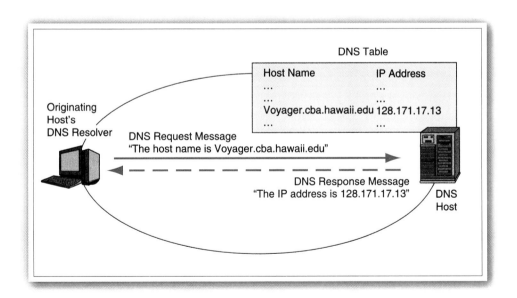

Figure 1-28   The Domain Name System (DNS)

---

[5]   In networking technology, a name refers to a particular entity, while an address gives the entity's physical location. Just as you have a name and an address, so do many servers on the Internet.

finds the IP address (128.171.17.13) either in its own files or by passing the request on to another DNS server. The DNS response message contains the IP address of the host.

The DNS server that receives DNS requests from Pat Lee's home network is located at her ISP. In addition to these ISP DNS servers, large firms have their own DNS servers. Due to the functions of these DNS servers, they often are called **name servers.**

TEST YOUR UNDERSTANDING

23.   a) Which is a host's official address—its IP address or its host name? b) What information does the user PC send to the DNS server? c) What information does the DNS server send back? d) Why is the information that the DNS server sends back important? e) What is a name server?

Dedicated Servers

Although Pat Lee's small home network is in most ways a microcosm of larger corporate networks, it is very atypical in one way. This is its use of peer-to-peer service, in which client PCs act as servers for one another. Larger networks almost always use **dedicated servers,** which act only as servers and not also as user PCs.

Some dedicated servers are ordinary personal computers, although dedicated **PC servers** run server-specific operating systems, called **network operating systems.** The most popular NOSs are Microsoft Windows Server, Linux, and Novell NetWare.

Other dedicated servers are more powerful computers that are the sizes of small to large refrigerators. (With computers, physical size is not a good indicator of processing power.) The most popular large dedicated servers are **workstation servers,** which run the Unix operating systems. Some companies use even larger server computers, called **mainframes.** Chapter 1b looks at dedicated servers in some depth.

TEST YOUR UNDERSTANDING

24.   a) What is a dedicated server? b) What type of operating system do dedicated PC servers use? c) What are the three most popular network operating systems? d) List the three types of dedicated servers in order of increasing processing power. e) What operating system do workstation servers use?

SYNOPSIS

This chapter introduced you to a number of basic networking concepts and issues that we will be seeing throughout this course.

First, we saw the First Bank of Paradise (FBP), our running case study for this book. FBP is a composite medium-sized firm operating primarily in Hawai`i. Although banks have some industry-specific needs, such as higher-than-average security requirements, the issues and technologies that the bank is facing are similar to those in most firms.

The only things that users care about are the networked applications that data communications provides to them. The First Bank of Paradise has a normal mixture of networked applications—including e-commerce, transaction processing (both check processing and standard back room applications)—and common office applications, including e-mail, Web access and instant messaging.

The bank has many single networks. Local area networks (LANs) operate entirely within the bank's premises, usually in a single building. Wide area networks (WANs) connect sites together and connect the bank to the Internet. The eight elements of single networks are: stations (hosts); messages called frames; switches to forward frames to the destination station; access lines to connect switches to switches, routers to routers, and switches to routers; and routers to connect single networks to the outside world and to other networks within the bank.

Single networks are no longer isolated islands of communication. Routers link them together into internets. Although the worldwide Internet is the dominant internet, most organizations have internal internets (called intranets if they use the TCP/IP standards created for the Internet). In addition, routers can use TCP/IP technology to link groups of companies into closed extranets for buying and selling. Messages in internets are called packets, and computers are called hosts.

FBP's network manager, Yvonne Champion, has a number of important technical concerns.

> ➤ It is important for network managers to create a strong network architecture that will guide individual projects so they contribute to a unified network environment.

> ➤ Creating a network architecture requires deciding upon which standards to use.

> ➤ The network architecture also needs a strong security component because of the growing number of internal and external attackers. Security is a special problem in wireless networking, which is both attractive and problematic.

> ➤ Security is particularly important in the bank because of major theft and fraud concerns. The bank has border firewalls at every site to drop and log attack packets attempting to get into the site. The bank also uses virtual private networks (VPNs), which provide secure communication over the Internet.

> ➤ Wireless communication is attractive to mobile users, but the bank is proceeding cautiously with wireless networking because of security threats.

> ➤ Networks today are critical to business operations, so they have to work well. Networks need to be efficient because user demand is growing rapidly while budgets are nearly stagnant or actually are shrinking.

> ➤ Networks also need to meet measurable quality-of-service (QoS) performance goals for worst-case conditions of throughput (the speed actually experienced by users), congestion-caused latency, response times, availability, and error rates.

In this chapter we looked at the PC network that FBP vice president Pat Lee installed in her home. This network uses an access router with a built-in Ethernet switch to connect Pat's two PCs together so they can do file sharing and printer sharing. The access router also contains a built-in router that connects Pat's home network to the Internet via an Internet service provider (ISP). The access router also contains a DHCP server to assign IP addresses to internal PCs and a NAT module to convert between internal IP addresses and the one IP address given to Pat by her cable provider. A cable modem links the access router to the ISP.

SYNOPSIS QUESTIONS

1. Name the major networked application categories at the First Bank of Paradise.
2. a) Distinguish between LANs and WANs. b) What are the eight components of individual networks? c) What are messages in individual networks called? d) Distinguish between single networks and internets. e) What are messages in internets called? f) Distinguish between internets and the Internet. g) Distinguish between internets, intranets, and extranets.
3. a) List Yvonne Champion's major technical concerns. b) What two major security protection technologies is the bank using? c) What are the major QoS goals? State each in terms of whether Yvonne should guarantee a minimum or maximum value (minimum throughput, etc.). d) Distinguish between speed and throughput. e) Distinguish between congestion and latency. f) What is response time? g) What is availability? h) What is the error rate?
4. a) What are the major hardware elements in Pat Lee's home network? b) What are the major functional components of the access router?

THOUGHT QUESTIONS

1. List at least two networked applications not mentioned in the Introduction.
2. Internet Exercise. Create a table. In the first column, list all of the hardware and software that Pat Lee had to purchase for her home network (or may have had to purchase, if you are not sure). In the second column, list a specific product for that need, including its model number. In the third column, give the item's price. Compute a total purchase price. Look up components and prices on the Internet.
3. Your DSL line has a listed speed of 500 kbps. However, when you do downloads, a speed counter tells you that you are only receiving 60 KBps. Can you explain this apparent inconsistency?

TROUBLESHOOTING QUESTIONS

1. Troubleshooting is a very important skill. A good idea is to draw a diagram of a system to see what might go wrong. Here is a sample problem for you to solve. You have been using a telephone modem to access the Internet. Its rated download speed is 56 kbps. You switch to a cable modem, which should allow you to receive at 800 kbps. In general, your download speed for webpages is faster than it was with your modem; however, your actual download rates is only about 400 kbps. List likely reasons for you not to be getting a full 800 kbps. Do NOT just come up with one or two possible explanations. Hint: Consider Figure 1-11, which shows the Internet.
2. In your browser, you enter the URL of a website you use daily. After some delay, you receive a DNS error message that the host does not exist. What may have happened?

IP Address	01111111101010110001000100001101			
Divided into Octets (Bytes)	01111111	10101011	00010001	00001101
Octets Converted to Decimal	127	171	17	13
Dotted Decimal Notation	127.17.17.13			

Figure 1-29   Converting Binary IP Addresses to Dotted Decimal Notation

HANDS-ON

Binary and Decimal Conversions

IP addresses are stored in packets as strings of 32 bits, as illustrated in Figure 1-29. To convert a binary IP address into dotted decimal notation, follow the steps in the figure.

If you have Microsoft Windows, the Calculator accessory shown as Figure 1-30 can convert between binary and dotted decimal notations. Go to the Start button, then to Programs or All Programs, then to Accessories, and then click on Calculator. The Windows Calculator will then pop up.

Binary to Decimal
To convert eight binary bits to decimal, first choose View and click on Scientific to make the Calculator a more advanced scientific calculator. Click on the Bin (binary) radio button and type in the 8-bit binary sequence you wish to convert. Then click on the Dec (decimal) radio button. The decimal value for that segment will appear.

Decimal to Binary
To convert decimal to binary, go to View and choose Scientific if you have not already done so. Click on Dec to indicate that you are entering a decimal number. Type the number. Now click on Bin to convert this number to binary.

One additional subtlety is that Calculator drops initial zeros. So if you convert 17, you get 10001. You must add three initial zeros to make this an 8-bit segment: 00010001.

Another subtlety is that you can only convert one 8-bit segment at a time.

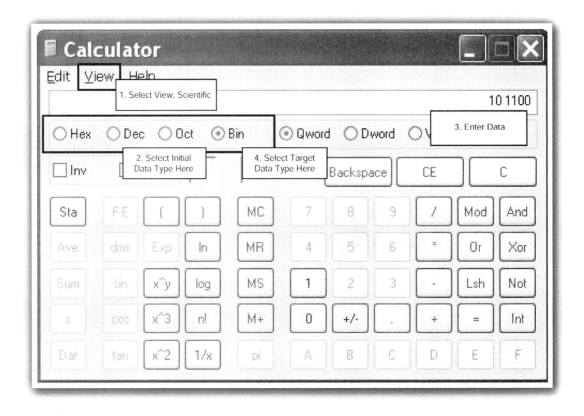

Figure 1-30   Windows Calculator

1. a) Convert 11001010 to decimal. b) Express the following IP address in binary: 128.171.17.13. *Hint:* 128 is 10000000. Put spaces between each group of eight bits. c) Convert the following address in binary to dotted decimal notation: 11110000 10101010 00001111 11100011. (Spaces are added between bytes to make reading easier.) *Hint:* 11110000 is 240 in decimal.

Test Your Download Speed

How fast is your Internet connection? Test your download speed at http://www.pcpitstop.com/internet/bandwidth.asp or http://webservices.zdnet.com/zdnet/bandwidth. If you can, test your bandwidth during periods of light and heavy use.

2. a) What kind of connection do you have (telephone modem, cable modem, LAN, etc.)? b) What was your download speed during the test?

Working with the Windows Command Line

Windows offers a number of tools from its command line prompt. Network professionals need to learn to work with these commands.

Getting to the Command Line

To get to the command line, click on the *Start* button and choose *Run*. Type either *cmd* and hit OK or *command* and then OK, depending on your version of Windows.

Command Line Rules

At the command line, you need to type everything exactly. You also need to hit Enter at the end of each line. It is also good to know that you can clear the command line screen by typing *cls[Enter]*.

Your Configuration

In Windows, you can find information about your own computer with **ipconfig** or **winipconfig.** In newer versions of windows, type the command *ipconfig/all[Enter]*. Older versions of windows have *winipconfig*. This will give you your IP address, your physical address (your Ethernet address), the IP addresses of your organization's or ISP's DNS hosts, and other information—some of which we will see in Chapter 8.

3. Use ipconfig/all or winipconfig. a) What is your computer's IP address? b) What is its Ethernet address? c) What are the IP addresses of your DNS hosts?

DNS Lookup

In this chapter we saw that if you know the host name of the host to which you want to send, then your computer's DNS resolver must look up the host's IP address. You can also do this yourself from the command line, using the **nslookup** command (DNS servers are also called name servers). For instance, type *nslookup www.google.com[Enter]* to find Google's IP address.

4. a) Do an nslookup DNS lookup on a host whose name you know and that you use frequently. What is its IP address? b) Now do an nslookup on that IP address. Do you get the host name?

Ping

To find out if you can reach a host, and to see how much latency there is when you contact a host, use the **ping** command. You ping an IP address or host name much as a submarine pings a target to see if it exists and how far it is away. To use the command, type *ping hostname[Enter]* or *ping IPaddress[Enter]*. Ping may not work if the host is behind a firewall, because firewalls typically block pings.

5. Ping a host whose name you know and that you use frequently. What is the latency? If this does not work because the host is behind a firewall, do other hosts until you succeed.

Ping 127.0.0.1 (PC, Call Home)

Ping the address 127.0.0.1. This is your computer's **loopback address.** In effect, the computer's network program sends a ping to itself. If your PC seems to be having trouble communicating over the Internet, ping 127.0.0.1. If the ping fails, you know that the problem is internal, and you need to focus on your network software's configuration. If the ping succeeds, then your computer is talking to the outside world at least.

6. Ping 127.0.0.1. Did it succeed?

Tracert

The Windows **tracert** program is like a super ping. It lists latency not only to a target host but also lists each router along the way and lists latency to that router. Actually, it shows three latencies because it tests each router three times. To use tracert, type *tracert hostname[Enter]* or *tracert IPaddress[Enter]*. Again, hosts (and routers) behind firewalls will not respond.

7. Do a tracert on a host whose name you know and that you use frequently. a) What is the destination host? b) How many routers are there between you and the destination host? If this does not work because the host is behind a firewall, do other hosts until you succeed.

To Get Your IP and Ethernet Addresses with Windows XP

Do the following if you have an XP computer. It will give you configuration information without your having to go to the command line. Choose *Start*, then *Control Panel*, then *Network and Internet Connections*, then *Network Connections*. In the window that appears, click on a *network connection* icon to select it. Choose *File, Status*. You will see the *Connection Status* dialog box for that connection. Select the *Support* tab to see your IP address. Click on *Details* while in the Support tab to see more information, including the physical (Ethernet) address of your NIC.

8. If you have a Windows XP computer, find your IP address and the physical (Ethernet) address of your NIC.

WINDOWS XP HOME NETWORKING

Learning Objectives:

By the end of this chapter, you should be able to discuss:

- The importance of Microsoft Windows XP Home.
- Setting up an Internet connection.
- Allowing peer-to-peer file sharing in the Shared Documents (SharedDocs) folder.
- Accessing shared files.
- Sharing additional directories.
- Making a printer available for peer-to-peer printing.
- Using shared printers.

INTRODUCTION

In Chapter 1, we looked at a small PC network in the home of Pat Lee, a vice president at the First Bank of Paradise. We saw that, with the major exception of using peer-to-peer service rather than dedicated servers, Pat Lee's home network is a microcosm of larger corporate networks.

In this chapter we will look at how to configure a **Microsoft Windows XP Home** computer for home networks like Pat's. XP Home is the dominant operating system today for residential PCs. It has all the functionality you need to set up a home network, including connecting to the Internet, peer-to-peer file sharing, and peer-to-peer printer sharing.

However, we will see that XP home is limited to Simple File Sharing, which is easy to implement but offers no real security. **Microsoft Windows XP Professional** offers much better security, but it can only do this when it uses dedicated servers rather than peer-to-peer networking.

SETTING UP AN INTERNET CONNECTION

Initial Steps

The first step in setting up a Windows XP Home computer for home networking is to configure it to talk to the Internet. Figure 1a-1 shows the initial steps in doing this. These are the same initial steps needed for peer-to-peer file and printer sharing.

0.
Install and Set Up Internet Connection Hardware

1.
Click on the Start Button
Choose "Control Panel"

2.
Control Panel
Click on "Network and Internet Connections"

3.
Network and Internet Connections
(Figure 1a-2)
Choices are:
"Set up or change your Internet Connection" [to set up or change an Internet connection]
"Create a connection to the network at your workplace" [for a virtual private network connection]
"Set up or change your home or small office network" [for file and printer sharing]

Figure 1a-1   Initial Network Setup Process in Windows XP Home

Hardware Installation
Before a computer can be configured for Internet communication, it needs to be physically connected to the Internet. Make sure that all hardware is connected properly, has electrical power, and is turned on.

The Start Button
Once the hardware is installed, hit the Start button. Choose "Control Panel."

Control Panel
When the Control Panel appears, click on "Network and Internet Connections."

Network and Internet Connections Dialog Box
In the Network and Internet Connections dialog box, you will have three options, as shown in Figure 1a-2.

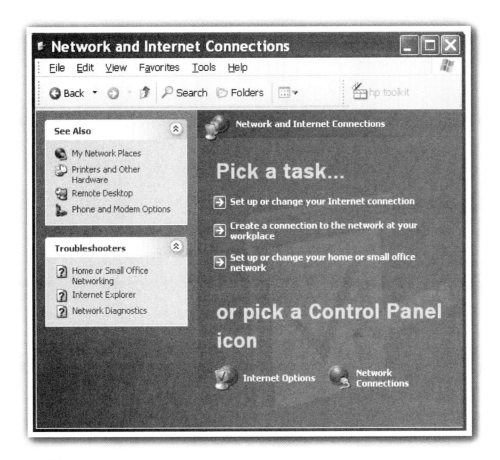

Figure 1a-2   Network and Internet Connections Dialog Box in Windows XP Home

➤ "Setup or change your Internet connection." Choose this one.

➤ "Creating a connection to the network at your workplace." This is for setting up a virtual private network (VPN). VPNs were introduced in Chapter 1. We will learn more about them in Chapter 7.

➤ "Set up or change your home or small office network." We will return to this option later, to set up peer-to-peer file sharing.

Next Steps

Figure 1a-3 shows the remaining steps needed to set up an Internet connection.

1.
Internet Properties
Select the Connections tab
Click on "Setup"

2.
New Connection Wizard
Welcome to the New Connection Wizard
Click on "Next"

3.
New Connection Wizard
Network Connection Type
Select "Connect to the Internet"

4.
New Connection Wizard
Getting Ready
Options:
Choose from a list of Internet service providers (ISPs)
Set up my connection manually [Select this option]
Use the CD I got from an ISP

5.
New Connection Wizard
Internet Connection
Options:
Connect using a dial-up modem
Connect using a broadband connection that requires a user name and password [PPPoE]
Connect using a broadband connection that is always on [Choose this option]

6.
New Connection Wizard
Completing the New Connection Wizard
Click on "Finish"

Figure 1a-3 Setting Up an Internet Connection in Windows XP Home

Internet Properties

Selecting "Set up or change your Internet connection" will take you to the Internet Properties dialog box. When this dialog box is displayed, select the Connections tab if it is not selected. Click on "Setup" to set up your connection.

New Connection Wizard: Welcome to the New Connection Wizard

Now you begin working with the New Connection Wizard. At the Welcome screen, choose "Next" to get started.

New Connection Wizard: Network Connection Type

Select "Connect to the Internet"

New Connection Wizard: Getting Ready

This dialog box gives you three options:

➤ "Choose from a list of Internet service providers." You can choose this option if you do not already have an ISP.

➤ "Set up my connection manually." *This is the option we will follow.*

➤ "Use the CD I got from an ISP." Sometimes your ISP gives you a setup disk. If it does, use it.

New Connection Wizard: Internet Connections

This dialog box asks you to specify how you connect to the Internet.

➤ "Connect using a dial-up modem." Select this option if you have a dial-up modem. You will be asked a number of modem-specific questions.

➤ "Connect using a broadband connection that requires a user name and password." As soon as some people see "broadband connection," they select this option. However, few broadband vendors use this approach. (Those that do use something called PPPoE—Point-to-Point Protocol over Ethernet.) PPPoE requires the users to log in each time they connect to the Internet.

➤ "Connect using a broadband connection that is always on." This is the normal option for broadband connections. The provider knows where you are physically, so there is no need for you to log in. *This is the option we will follow.*

New Connection Wizard: Completing the New Connection Wizard

Click on "Finish." The process of implementing your choices may take a few seconds.

ALLOWING PEER-TO-PEER DIRECTORY AND FILE SHARING

Now that the computer is connected to the Internet, the computer's owner may wish to share directories and files with others on the same SOHO (small office or home office) network. To allow peer-to-peer directory and file sharing, go through the initial steps in Figure 1a-1. In the last step, select "Set up or change your home or small office network." This initiates the Network Setup Wizard, which is illustrated in Figure 1a-4.

Network Setup Wizard: Welcome to the Network Setup Wizard

Click on "Next" to continue.

4.
Network Setup Wizard
Welcome to the New Connection Wizard
Click on "Next"

5.
Network Setup Wizard
Before you continue...
Go through the checklist, click on "Next"

6.
Network Setup Wizard
Select a connection method
(Figure 1a-5)
Options:

This computer connects directly to the Internet. The other computers on my
network connect to the Internet through this computer.

This computer connects to the Internet through another computer on my network
or through a residential gateway *[Choose this option]*

Other

7.
New Setup Wizard
Your computer has multiple connections
If this appears, choose:
Determine the appropriate connections for me (Recommended)

8.
Network Setup Wizard
Give this computer a description and name
(Figure 1a-6)
Type a name and description for the computer
This is what other computers on the network will see when they access it.

9.
Network Setup Wizard
Name your network
(Figure 1a-7)
Type a workgroup name.
All computers on the network must have the same workgroup name.
Giving computers the same workgroup name creates a network.

10.
Network Setup Wizard
Ready to apply network settings
Select "Next." This may take some time.

11.
Network Setup Wizard
Finish

Figure 1a-4   Setting Up File and Folder Sharing on a Windows XP Home Computer

Network Setup Wizard: Before You Continue

This dialog box has a checklist for your hardware. Go through it to be sure your hardware is set up correctly.

Network Setup Wizard: Select a Connection Method

This option requires you to know that there are two basic ways to connect multiple computers to the Internet. Figure 1a-5 illustrates both.

Using a Gateway (Access Router)

One option is to use a **gateway,** which is an obsolete term for "router" that Microsoft stubbornly continues to use. Many firms that sell access routers now call them gateways to be consistent with Microsoft terminology. Using a "gateway" is how Pat Lee's home network is implemented, as Figure 1a-5 shows. Access routers provide a good deal of security against attacks simply by implementing NAT.

Internet Connection Sharing (ICS)

Microsoft Windows XP also offers an **Internet connection sharing (ICS)** option, which also is illustrated in Figure 1a-5. Here one PC has two NICs. One NIC connects to a modem for Internet access. The other NIC connects to a desktop Ethernet switch, which connects to other PCs. ICS software on the PC connected directly to the Internet enables it to act as an access router.

Although ICS saves some money because access routers (gateways) are more expensive than desktop Ethernet switches, the saving is small. On the negative side, if the PC implementing ICS crashes or is turned off, all computers lose Internet access. In addition, ICS is a little complex to set up well. Without the access router for NAT security, it is important to implement a firewall on the PC implementing ICS. (Windows XP home has a built-in firewall discussed in Chapter 9a.) However, implementing a firewall on other computers can interfere with ICS operation.

ICS is especially economical if a network has only two computers. In this case, there is no need for a switch. The second NIC on the computer connected directly to the Internet can be connected directly to the NIC in the other computer with a UTP **crossover cable.** NICs transmit on Pins 1 and 2 and listen on Pins 3 and 6. Connecting the two NICs with an ordinary UTP cord would mean that both would talk but neither would listen—not the recipe for a good relationship in technology (or the human realm either). A crossover cable connects Pins 1 and 2 on one computer to Pins 3 and 6 on the other computer, allowing communication.

Network Setup Wizard: Your Computer Has Multiple Connections

You *might* see this dialog box. If you do, select "Determine the appropriate connections for me (Recommended)."

Network Setup Wizard: Give This Computer a Description and Name

This dialog box, which is shown in Figure 1a-6 is self-descriptive, but the figure is included to emphasize its importance. The name and associated description that you enter is how others on the network will see the computer.

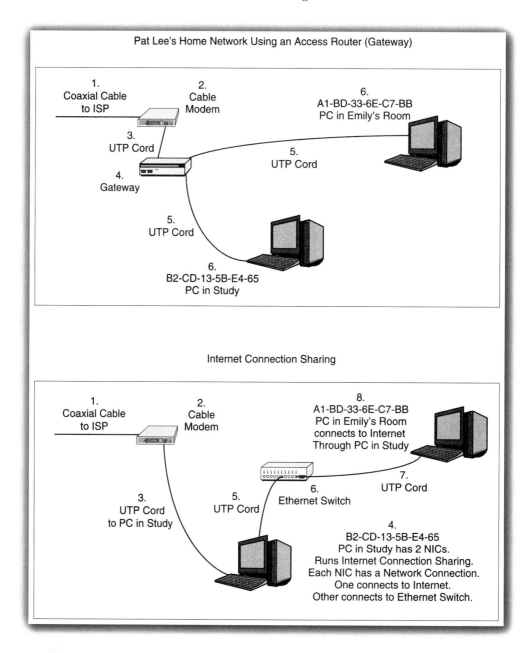

Figure 1a-5   Access Router (Gateway) Versus Internet Connection Sharing

Figure 1a-6   Give This Computer a Description and Name

New Connection Wizard: Name Your Network

This critical dialog box, which is illustrated in Figure 1a-7, is confusing to many users and is perhaps the biggest reason for failed file and directory sharing.

Although the dialog box does not make it clear, a network in Windows is a group of computers sharing the same **workgroup name.** The workgroup name is how computers find one another on a network. Giving several computers the same workgroup name automatically creates a **workgroup** (logical network).

Requiring all computers to have the same workgroup name seems silly in Pat Lee's home, where the physical network serves only a single logical network (group of users). However, in large corporate LANs, there may be several workgroups (logical groups of users) sharing the same physical network. They will be distinguished by their different workgroup names (which are case-sensitive).

Network Setup Wizard: Ready to Apply Network Settings

Now the computer is ready to apply the network settings you have just specified. As the dialog box says, this may take some time to do. Do not interrupt the process.

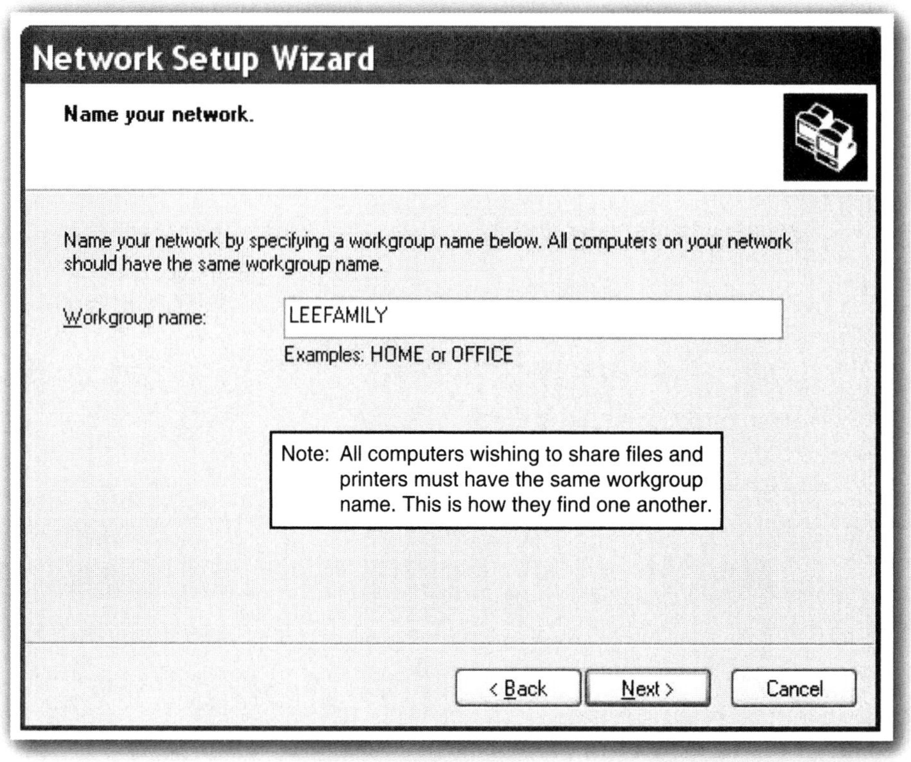

Figure 1a-7   Name Your Network

Network Setup Wizard: Finish

Your computer is set up for directory and file sharing. Select "Finish."

ACCESSING SHARED FILES

Process

Once the computers on the network are set up for file sharing, users can access files on other computers. Figure 1a-8 shows how this is done. The process is dead simple.

First, click on the Start button. Then select "My Network Places." A dialog box will appear showing which directories you can share on various computers. Figure 1a-9 shows this dialog box. It shows that **SharedDocs** (the **Shared Documents folder**) is available for sharing on both computers.

1.
Start
Choose My Network Places

2.
My Network Places
(Figure 1a-9)
Select a Directory
Initially, only the Shared Directory (SharedDocs) is shared on each computer
Files copied into the Shared Directory will be sharable over the network
Files copied from the Shared Directory to other directories will not be sharable over the network

Figure 1a-8   Using a Shared Directory via a Network

Shared Documents (SharedDocs)

By default, Windows XP only shares a single directory on each computer. This is the Shared Documents directory. (In older versions of Windows, *no* directory was shared by default.)

Making Files Available for Sharing

To share a document on the network, a user should copy or move the document from one of his or her other directories (which cannot be seen over the network) into his or her Shared Documents directory. This makes the file available to other computers in the workgroup.

Saving to the Shared Document Directories on Other Computers

Users can also copy files from any of their directories to the Shared Documents directories on other PCs in the workgroup.

Making Files Unavailable for Sharing

If someone else on the network writes a file into a user's Shared Documents directory, in turn, anyone in the workgroup can see it. To remove it from general sight, the user can move it into another directory. This makes it invisible to the network.

Weak Security in Simple File Sharing

Note that there is only very weak security on files in Shared Documents folders. There is not even a simple password. Everyone in the workgroup can read them and even modify them. The only security is that people must know the workgroup names to read and change files, but this typically is easy to do. This weak security is called **Simple File Sharing**. Although it is an option with Windows XP Professional, it is the only way to share files with Windows XP home.

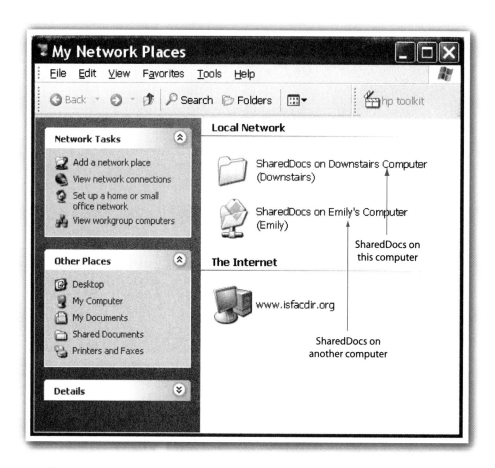

Figure 1a-9 My Network Places

SHARING ADDITIONAL DIRECTORIES

Simple File Sharing does allow additional directories besides Shared Documents to be shared. Figure 1a-10 shows how to share an additional directory besides Shared Documents.

Run the Network Setup Wizard

This process will only work after you have run the Network Setup Wizard, as described in the previous section.

Select Another Directory You Wish to Share

Go to My Computer and find the directory you wish to share. Select it.

0.
Run the Network Setup Wizard
The Shared Documents directory will be shared automatically

1.
Select Another Directory You Want to Share
Find it through My Computer
Right-click on the directory.
Select "Sharing and Security"

2.
In the Directory's Properties Dialog Box
Select the Sharing Tab.
(Figure 1a-11)
Under network sharing and security,
Click on "Share this folder on the network"

Enter a Share name

Click on "Allow network users to change my files" if desirable

Figure 1a-10   Sharing an Additional Directory in Windows XP Home

In the Directory's Properties Dialog Box

Once a directory is selected, everything else can be done from the directory's Properties dialog box. Figure 1a-11 shows the Properties dialog box for the My Music directory.

- ➤ Right-click on the directory to bring up its Properties dialog box.
- ➤ Select the Sharing tab.
- ➤ Under network sharing and security, click on "Share this folder on the network." Until this is done, the folder will not be accessible to other computers.
- ➤ Enter a "share name" by which other computers will see this directory. (In Microsoft-speak, a **share** is something that is shared—usually a directory or a printer.)
- ➤ Check the "Allow network users to change my files" box if desirable. If you do not check this box, others will have only read-only access.

Still No Security

Although this approach allows directories to be made "read only," there still is no password or other security measure. Simple File Sharing, even when it is added to directories other than Shared Documents, is a security nightmare.

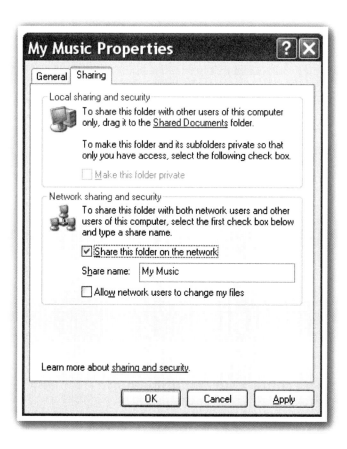

Figure 1a-11   My Music Properties Sharing Tab

SHARING PRINTERS

Making a Printer Available for Sharing

Many users would also like to share printers. In Pat Lee's home network, for instance, the downstairs printer must be shared by Emily's computer in her bedroom. Figure 1a-12 shows how to share a printer.

Install and Set Up the Printer

This step seems obvious, but most problems in printer sharing configuration occur because the computer cannot "see" a printer because it is powered off or not connected to the computer.

Click on the Start Button
Select "Printers and Faxes."

Printers and Faxes
Click on the printer on your own computer that you wish to share. In the left Printer Tasks pane, select "Share this printer." Click on "OK."

Printer Properties Dialog Box
This will take you to the printer's Properties dialog box (shown as Figure 1a-13). Click on the "Share this printer" radio button and give the computer a share name by which other computers will know it.

When you install a printer on an XP computer, you install printer driver software that is specific to Windows XP. If other computers run older versions of Windows, you may have to install additional printer drivers for them. In the printer's dialog box, click on the "Additional Drivers..." button to install drivers for older versions of Windows.

0.
Install and Set Up the Printer

1.
Click on the Start Button
Choose "Printers and Faxes"

2.
Printers and Faxes
Click on the printer you wish to share.
In the Printer Tasks pane to the left,
Choose "Share this printer."

3.
Properties
(Figure 1a-13)
Select the Sharing tab
Click the "Share this printer" radio button
If other machines run diferent versions of Windows, you may need to install additional drivers.
Click on "OK"

Figure 1a-12   Sharing a Printer in Windows XP Home

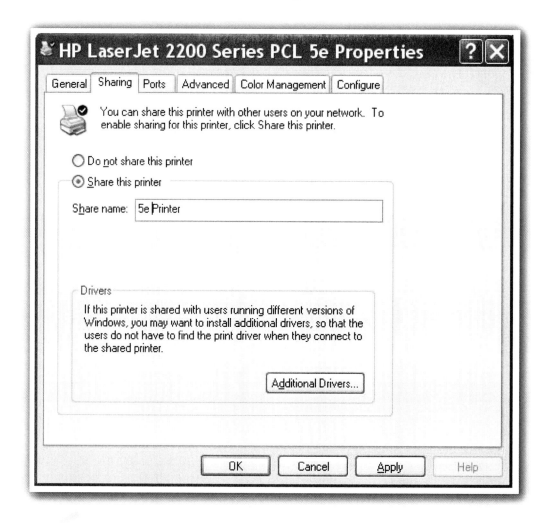

Figure 1a-13   Printer Properties Dialog Box Sharing Tab

Using a Shared Printer

Figure 1a-14 shows the steps a computer user (in the Pat Lee case study, Emily) would take to use a printer that someone else has made available for sharing.

Printers and Faxes

To begin, hit the Start button. Then select "Printers and Faxes." This will take you to the Printers and Faxes dialog box. In this dialog box, select "Add a printer." This will start the Add Printer Wizard.

Add Printer Wizard: Welcome to the Add Printer Wizard

At the initial Add Printer Wizard dialog box, select "Next"

1.
Start
Select "Printers and Faxes"

2.
Printers and Faxes
Choose "Add a Printer"

3.
Add Printer Wizard
Welcome to the Add Printer Wizard
Select "Next"

4.
Add Printer Wizard
Local or Network Printer
Click on "A network printer, or a printer attached to another computer"
Click on "Next"

5.
Add Printer Wizard
Specify a Printer
Select "Browse for a printer"
Click on "Next"

6.
Add Printer Wizard
Browse for a Printer
Select the printer you wish to use
Click on "Next"

...
Remaining steps are printer-specific.
You will be asked if you want to make the selected printer your default printer.

Figure 1a-14   Using a Shared Printer in Windows XP Home

Add Printer Wizard: Local or Network Printer
The next dialog box asks you if you will be adding a local or network printer. Select "A network printer, or a printer on a different computer." This indicates that you want to use a computer that someone else has made available for sharing.

Add Printer Wizard: Specify a Printer
At the "Specify a Printer" dialog box, select "Browse for a printer."

Add Printer Wizard: Browse for a Printer

At the next dialog box, browse for the printer you wish to use from a list of printers available on the network.

Remaining Steps

The remaining steps are printer-specific but normally ask if you wish to make the selected printer your **default printer**—the printer to which your print jobs will be sent unless you specify a different printer.

TEST YOUR UNDERSTANDING

1. a) Which offers better security—Windows XP Home or Windows XP Professional?

2. a) What link in the Control Panel is used for many networking configuration actions? b) Explain the three options in the Network and Internet Connections dialog box.

3. a) For what is "gateway" another name? b) How is Internet Connection Sharing less expensive than using an access router? c) Why is Internet Connection Sharing undesirable? d) When is a crossover UTP cord needed?

4. Explain why the workgroup name is important.

5. What is a share?

6. a) When you set up a Windows XP Home PC to allow it to share directories and files, what directory or directories is/are available for sharing by default with Windows XP Home? b) In earlier versions of Windows? c) Using the default, how can you make files available for sharing? d) Can additional directories be made available for sharing? e) Can passwords be required for directory access with Windows XP Home? f) Can directories other than Shared Documents be made read-only?

7. How does a computer user see what files are available over the network?

8. If you want to print a particular document to a printer attached to another computer, what must have already been done?

CHAPTER 1b

DESIGN EXERCISE: XTR CONSULTING: A SOHO NETWORK WITH DEDICATED SERVERS

Learning Objectives:

By the end of this chapter, you should be able to discuss:

- Dedicated server technologies (PC servers, workstation servers, mainframes, and server farms).

- Major network operating systems for PC servers: Microsoft Windows Server, Linux, and Novell NetWare.

- File servers (including file server program access) and print servers.

- Systems administration (the management of servers), including the management of access permissions.

- How to design a small office or home office (SOHO) network for a small company.

INTRODUCTION

XTR Consulting (name changed for confidentiality) is a small environmental consulting company with sixteen professionals (including managers), plus a secretary. The company has an office suite in a large downtown office building. This is called a **SOHO (small office or home office)** environment. They are planning to build a LAN. It will be quite different from the large networks we will see in later chapters.

This LAN also will be different from the small home LAN we saw in Pat Lee's house. Among these differences are the following:

➤ Most importantly, user PCs in Pat Lee's home act as servers for one another, providing both file sharing and printer sharing services. In the XTR network, however, there will be at least one **dedicated server**—a server that does not act simultaneously as a user PC. In addition to being expensive, dedicated servers require special skills to manage. In small LANs with dedicated servers, server administration, called systems administration, is far more work than the management of the rest of the network. In a small network, and in a small network only, systems administration is the biggest part of the administrator's job.

➤ In addition, the XTR network uses small electronic devices called print servers to feed printer output to shared printers.

➤ Finally, the functions of the Ethernet switch and router, which were combined in Pat Lee's inexpensive access router, are separated in the XTR network and are handled by a large Ethernet switch and by a somewhat more expensive access router than Pat Lee's. This access router will be more expensive than Lee's because the XTR access router will have a built-in firewall function.

A dedicated server is a server that does not act simultaneously as a user PC.

	Pat Lee's Home Network	XTR Consulting's Small Office Network
Number of Client PCs	2 PCs	17 user PCs (16 professionals and one secretary)
File Service	Peer-to-Peer	Dedicated Server Systems Administration
Print Service	Peer-to-Peer	Print Servers
Access Router and Ethernet Switch	Combined	Separate Boxes Access Router Has Firewall

Figure 1b-1   XTR Consulting's Network Versus Pat Lee's Network (Study Figure)

TEST YOUR UNDERSTANDING

1. a) What is a dedicated server? b) What is systems administration? c) How will XTR's network technology differ from Pat Lee's home network technology?

DEDICATED SERVERS

Peer-to-Peer Networks

In Pat Lee's home network, the client PCs provide services to one another. This is called **peer-to-peer service.**

Inexpensive Networking
Peer-to-peer service makes sense for very small networks. It would be absurd, for example, for Pat Lee to spend $1,000 to purchase a third PC as a dedicated server to serve her two client PCs.

Peer-to-Peer Networks

> Clients serving other clients
>
> Inexpensive—no need to purchase a dedicated server
>
> Operational problems for other users if a user PC is turned off or crashes
>
> Poor security: No password or shared password for shared directories

Server Technology

> PC Servers and Network Operating Systems (NOSs)
>
>> Standard PC architectures with more RAM, large and fast hard disk drives, redundant power supplies and fans, and multiple processors (multiprocessing)
>>
>> Network operating systems are server operating systems that have more functions and reliability than client operating systems
>
> Workstation Servers
>
>> Fast (and expensive) custom microprocessors for an expensive computer
>>
>> Run the Unix operating system
>>
>>> Extremely reliable
>>>
>>> Difficult to learn
>>>
>>> Not standardized
>
> Mainframe Servers
>
>> Faster, more reliable, and more expensive than workstation servers
>>
>> Require a large systems programming staff
>
> Server Farms
>
>> Group of PC servers or workstation servers
>>
>> Load-balancing router distributes the processing work

Figure 1b-2   Peer-to-Peer Service Versus Dedicated Servers (Study Figure)

Operational Problems

However, as the number of PCs rises, operational problems inevitably begin to appear in peer-to-peer networks. To see why, suppose that a user turns off his or her PC or does something that causes it to crash. Then anyone using that PC's services may have their own application crash, possibly destroying any data currently being handled. Until the computer that crashed is rebooted, furthermore, nobody can use its services, including any shared data they might need to do their work. Given such operational problems, peer-to-peer service does not scale beyond about ten client PCs, and most analysts would set the cut-off point lower.

Poor Security

In addition, peer-to-peer security is poor. Often, shared directories are not even protected by passwords, and when they are, this password usually is shared by all users. Having no password is very dangerous if the user is connected to the Internet, especially if the home has an always-on high-speed connection, as Pat Lee's home does. Shared passwords may work in homes, but in businesses, not everybody should have equal access to every resource. Of course, you can reveal a directory's password to only some employees, but employees often fail to keep shared passwords secret "because everybody knows it."

TEST YOUR UNDERSTANDING

2. a) What is peer-to-peer service? b) What operational problems does it create? c) What is the practical maximum number of client PCs for peer-to-peer networking? d) Why is peer-to-peer security usually poor?

Server Technology

Dedicated servers tend to be much more powerful than client PCs. There are three common types of dedicated servers: PC servers, workstation servers, and mainframes.

PC Servers and Network Operating Systems (NOSs)

The first type of server, the **PC server,** is an ordinary Macintosh or Windows PC. Many are designed from the ground up to be servers; they typically have a great deal of RAM, very large and fast hard disk drives, and redundant power supplies and fans. Some even have multiple microprocessors to increase their processing power. Computers with multiple microprocessors are called **multiprocessing computers.** Despite these features, they still follow the standard Windows or Macintosh hardware architecture.

Dedicated PC servers cannot use ordinary client PC operating systems, such as Microsoft Windows 98, ME, 2000 Professional, or XP. These client-based operating systems do not have the functionality or (usually) the reliability needed for dedicated servers. These dedicated PC servers use different operating systems, called network operating systems (NOSs). Popular PC server NOSs include Microsoft Windows Server, Linux, and Novell NetWare.

Workstation Servers and Unix

Specialized Multiprocessors Windows PCs use ordinary Intel or compatible mass-market microprocessors. Another type of server, the **workstation server,** uses custom-designed microprocessors. These microprocessors are faster than Pentium and compatible microprocessors but are considerably more expensive because of the need to push the state of the art in microprocessor design and the limited number that are sold. Basically, workstation servers are like racing cars in performance and price.

Unix   Workstation servers run vendor-specific versions of the **Unix** operating system. Unix is extremely reliable. In addition, Unix has a rich toolbox of management applications. Unix on a high-speed workstation server is the platform of choice for large enterprise servers such as public webservers.

However, Unix workstation servers would not make sense for XTR's small network. Most importantly, this small network will not need the high-power (or the high prices) of workstation servers.

Also, Unix is very difficult to learn and use because server administrators often have to type complex commands such as those you saw in Chapter 1 if you did the hands-on exercises that took you to the Windows command line. Although Unix vendors offer graphical user interfaces similar to that of Microsoft Windows, the functionality of these interfaces tends to be limited, so Unix administrators constantly have to drop down to the command line to give commands.

In addition, Unix is not a standardized operating system. Each vendor's version of Unix offers many proprietary features. Most application programs written for Unix will run on most versions of Unix, but vendors vary widely in the management utilities they offer. Consequently, most companies tend to use only one or two versions of Unix to minimize relearning.

Mainframe Servers

The last major dedicated server technology is the **mainframe computer,** which is much more powerful than a workstation server. In addition, mainframes are even more reliable than Unix workstation servers. However, mainframe power is more expensive than a comparable amount of workstation server power because mainframe hardware is expensive and because a mainframe requires a large staff of systems programmers to keep it functioning.

XTR Consulting

For its dedicated server or servers, XTR Consulting will go with the PC server option. A PC server or a few PC servers will be adequate for XTR's needs and will be far less expensive than a workstation server or a mainframe computer.

Server Farms

One might think that PC servers could only be used for very small applications and that workstation servers could only be used for mid-sized applications. However, as Figure 1b-3 shows, some applications can be handled by **server farms,** which are large groups of servers. A **load-balancing** router can send client requests to the first available server. Large websites, for example, typically consist of hundreds or thousands of small PC servers or workstation servers.

TEST YOUR UNDERSTANDING

3. a) List the major server technologies in terms of increasing power and price. b) What is a NOS for a PC server? c) What are the major NOSs for PC servers? d) Distinguish

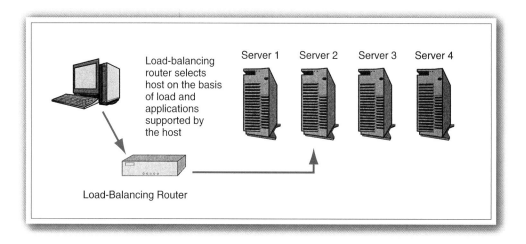

Figure 1b-3   Server Farm with Load-Balancing Router

between PC servers and workstation servers in terms of hardware and operating systems. e) Why will XTR Consulting use PC server technology? f) Why are server farms ways to provide scalability? g) What is the function of a load-balancing router?

Network Operating Systems (NOSs) for PC Servers

PC server operating systems, for historical reasons, are called **network operating systems (NOSs).** For PCs, there are three popular network operating systems: Linux, Novell Netware, and server versions of Microsoft Windows. Figure 1b-4 compares these network operating systems.

Microsoft Windows Server

For small businesses, the best choice usually is **Microsoft Windows Server,** which comes in three versions: NT, 2000, and 2003.

Ease of Learning and Operation   Windows Server is the most popular NOS because it is the simplest to learn and operate. As Figure 1b-5 shows, Windows Server 2003 looks almost exactly like client versions of Windows. In fact, server versions will execute ordinary client Windows application programs if this is required.

As the figure shows, most network management functions are located through the *Start* button, in the *Administrative Tools* program group. In addition, most Windows management tools have similar user interfaces called *Microsoft Management Consoles (MMCs).* An MMC is shown in Figure 1b-6.

Reliability   Early versions of Windows Server, through NT, had poor reliability. Newer versions have made Windows Server sufficiently reliable for departmental and small-business needs, although for central corporate servers, Unix is the gold standard for reliability.

	Microsoft Windows Server	Linux	Novell NetWare
Ease of Learning	Very Good	Poor	Good
Ease of Use	Very Good	Poor	Good
Availability of Consultants	Excellent	Modest	Moderate
Reliability	Very Good in the Most Recent Versions	Excellent	Very Good
Standardization	Excellent	Poor (Many Distributions)	Excellent
Availability of Device Drivers	Excellent	Poor	Very Good
Price	Moderate	Low or Free	Higher than Windows

Figure 1b-4   Popular PC Server Network Operating Systems (NOSs)

Linux
Linux was created as a freeware version of Unix.[1] Unlike most versions of Unix, which run on workstation servers, Linux is one of a handful of Unix variants that run on standard PCs.

Cost   A principle attraction of Linux is its cost. It can be downloaded for free, although boxed versions with instructions and phone-in support tend to cost $100 or more. Even if an expensive boxed version is purchased, the disks inside the box can be used to install Linux on multiple servers, again bringing down the cost.

Also, because the Linux kernel is highly efficient, Linux can run on a smaller computer than Windows Server can for a comparable workload. This further contains costs.

Reliability   Another benefit is that Linux has the legendary reliability of Unix. Linux rarely crashes.

Diversity and Configuration Problems   When you buy Microsoft Windows Server, you know what you are getting. However, Linux really is only an operating system kernel. What you actually buy or download for free is a **Linux distribution** consisting of

---

[1]   Linus Torvalds, who created the Linux kernel as an undergraduate student in Finland, pronounces Linux as "LEE-nucks." In English, it usually is pronounced with a short *i*, "LIH-nucks" (with a stress on the first syllable).

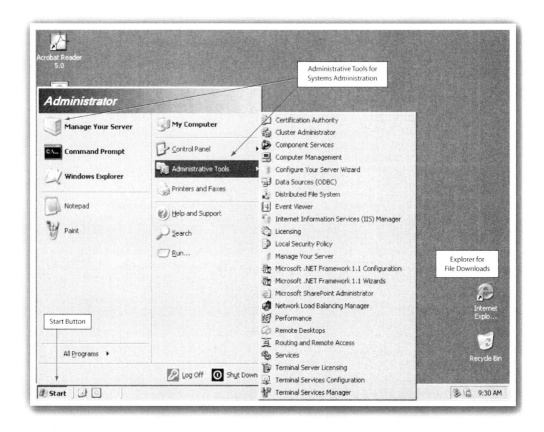

Figure 1b-5   Microsoft Windows 2003 Server User Interface

the kernel plus a collection of many other programs. These other programs usually are taken from the GNU project, which produces open-source software to run on Unix kernels. Different distributions contain different GNU and non-GNU programs, causing endless confusion. For example, Linux offers two competing graphical user interfaces as well as a command-line language.

In addition, configuring Linux to run on a server can be very difficult. Often, there is no device driver for a particular NIC, or you may have to compile a device driver from source code.

Linux Overall   For small businesses, the low cost of Linux may be a false saving because of the learning, configuration, and maintenance headaches the business will have to face. This may change as Linux matures.

Not for XTR Consulting   XTR does not have a trained systems administrator to run its dedicated server or servers. However, everyone at XTR is familiar with Microsoft

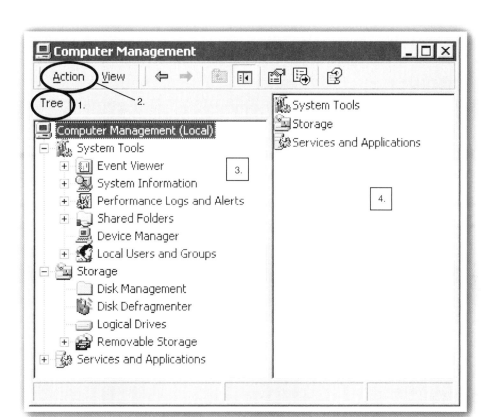

Figure 1b-6   Microsoft Management Console

Windows. Under these circumstances, it will be much easier to train someone to do day-to-day systems administration with Windows Server than with Unix. Consequently, XTR will use Windows as its NOS.

Novell NetWare

Novell NetWare once dominated the NOS market. It still has the best file and print service. However, high pricing led to a catastrophic loss of market share. Although NetWare still is an excellent product for file and print sharing, Windows Server generally is a better choice for small businesses because of its greater ease of installation and use and because there now are many more consultants who can provide help if a Windows Server has problems.

TEST YOUR UNDERSTANDING

4.   a) What is the major benefit of Microsoft Windows Server NOSs? b) What are Linux distributions? c) Why do Linux distributions cause problems? d) What problems do Linux servers have for small business? e) Why is Novell NetWare not a good choice for XTR's servers?

Systems Administration for Servers

Managing a network is called network administration. However, as noted earlier, managing a server is called **systems administration.**

Systems administration is the management of a server.

The Costs of Systems Administration

For a small network like the one that will be installed at XTR, most devices will require little installation labor or labor to manage the device after it is installed. In addition, unless unusual things occur, XTR will be able to ignore its switch, wiring, access router, and most other devices for months on end. Other than the server, only the firewall will require much upkeep over time—perhaps fifteen minutes per week on average.

However, systems administration for the firm's server or servers is another matter. Installing and configuring a server is a great deal of work and requires specialized knowledge. Even after an operating system is installed and configured, the operating system vendor will undoubtedly have issued updates since the operating system was written. Without these updates, the server will have serious security vulnerabilities and is likely to be taken over by attackers quickly after it is connected to the Internet. Downloading and installing updates from the operating system vendor's website can take longer than the initial installation and configuration.

In addition, unlike the other devices on the network, the server needs frequent attention after installation. Consequently, some member of the staff must be trained to do maintenance work such as diagnosing minor problems, doing routine software updates, and assigning access permissions (discussed next). Although this does not require as much knowledge as server installation and configuration, it still requires several days of training.

For more difficult situations, such as diagnosing complex problems or the installation of complex application programs, the company will keep a network consultant on retainer to be called as needed.

TEST YOUR UNDERSTANDING

    5. a) Why is systems administration time consuming? b) How will the functions of systems administration be handled at XTR?

Managing Access Permissions

The most time-consuming part of ongoing systems administration for a server is managing account permissions. Each user has an **account,** which is an identifiable entity that may own resources on a computer. To use a computer, the account holder types a **username** that signifies the account that he or she will be using. The account holder then types a secret **password**—a keyboard string that only the account holder should know.

Password Resets

Users frequently lose their passwords. When they do, the systems administrator must do a **password reset,** which consists of changing the password to some value known only to the systems administrator and the account owner. A typical reset password is

Accounts

      Have usernames and passwords

      Are assigned permissions in directories (what the account holder can do in the directory)

No Permissions

      Cannot even see the directory

Microsoft Windows Permissions

      List Folder Contents

            Allows the account owner to see the contents of a folder (directory).

      Read

            Allows the account owner to read files in the directory. This is read-only access. Without further permissions, the account owner cannot change the files.

      Write

            Allows the account owner to change the contents of files in the directory. For instance, if a file is a word processing document, the account owner can edit the document and then save it.

      Modify

            Gives additional permissions to act upon files, for example the permission to delete a file, which is not included in Write.

      Read and Execute

            Is a permission for executable programs. With these permissions, the account owner can run a program.

      Full Control

            Is an omnibus permission. It is equal to all of the above permissions.

Unix Permissions

      Read (only)

      Write

      Execute

Microsoft Windows Administrator Account

      Has all permissions in all directories

      Necessary for systems administration

      Dangerous in terms of security

      Comparable account in Unix is root

Assigning Permissions to Multiple Accounts in a Directory

      Windows can do this easily

      Unix can only assign permissions to the owner, one group, and all other accounts in the world

Reducing the Work of Assigning Permissions to All Accounts in All Directories

      Groups

            Place accounts with similar permissions needs in a group

            Assign permission to the group

            All accounts in the group receive the permission

      Inheritance

            If permissions are assigned in a directory

            The account inherits these permissions in all subdirectories

            Inheritance can be blocked

Figure 1b-7   Managing Access Permissions (Study Figure)

"changeme." When the user next logs in, he or she changes the password to something only he or she knows. With an organization of XTR's size, there will be an average of about one password reset per month.

Directory Permissions
Every account has a set of "permissions" in every directory on the server. These **permissions** determine what the account owner can do in that directory.

No Permissions
The simplest permission set is *no* permissions, meaning that there is nothing the account owner can do in that directory. Generally speaking, the account owner cannot even see the directory or any of its contents.

Microsoft Windows Server Permissions
In Microsoft Windows Server, there are six main permissions that the systems administrator can assign to an account in a directory:

> ➤ **List Folder Contents** allows the account owner to see the contents of a folder (directory).
> ➤ **Read** allows the account owner to read files in the directory. This is read-only access. Without further permissions, the account owner cannot change the files.
> ➤ **Write** allows the account owner to change the contents of files in the directory. For instance, if a file is a word processing document, the account owner can edit the document and then save it.
> ➤ **Modify** gives additional permissions to act upon files, for example, the permission to delete a file, which is not included in Write.
> ➤ **Read and Execute** is a permission for executable programs. With these permissions, the account owner can run a program.
> ➤ **Full Control** is an omnibus permission. It is equal to all of the above permissions.

In fact, if Windows systems administrators want and need additional flexibility, these six basic Windows permissions can be further divided into thirteen specialized permissions giving even more specific granularity in access control. In contrast, Unix only offers three permissions: Read (only), Write, and Execute.

Administrator Account
A special account on Windows servers is **Administrator.** This account, which is used by the systems administrator, automatically has full permissions in every directory on the server. This omnibus access is needed for systems administration, but it means that someone in the organization will be able to read and change every file in every directory. This is a serious security concern. In Unix, the equivalent account is called **root.**

Permissions for Multiple Accounts in a Directory
In a given directory, Windows can assign individual permissions for multiple accounts. For instance, in a task force project team's directory, only one or two member accounts may be given full control. Others may get read-only access. People not on the project team probably would get no permissions at all.

Microsoft Windows allows many different accounts to get different access permissions in a particular directory. In contrast, Unix can only specify three different sets of permissions: one set for the owner, another set for a single group, and a third set for every other account.

Groups

If a firm has many accounts, assigning permissions to every account in every directory can be extremely time consuming. One way to reduce the work of assigning permissions is the use of groups. Groups are lists of accounts. For instance, one group might be Marketing. This group would contain the account names of everyone in marketing. In some directories, the systems administrator might assign permissions to the Marketing group. Every account within the Marketing group would then have these permissions.

Inheritance

Even with groups, assigning permissions to every account in every directory is still extremely burdensome. Fortunately, permissions assigned in a directory normally are automatically **inherited** in all subdirectories.

For example, if all widely used programs are stored in a Programs directory, the systems administrator can assign Read and Execute permission in the Programs directory to a group consisting of all users. Everyone will automatically inherit permission to execute *all programs in all subdirectories of Programs.*

However, inheritance can be blocked or modified. For instance, if there is a program in a subdirectory of Programs that only the Marketing department's members should be able to use, inheritance would be blocked for everyone, and specific permissions would be assigned to the Marketing group.

Setting Permissions in Windows

Figure 1b-8 shows how directory permissions are set in Windows. This is the properties dialog box for a particular directory, My Music. Directory permissions are set on the Security tab.

Specifying Accounts and Groups   In the upper pane, there is a list of accounts and groups that have permissions set in the directory. (Other accounts and groups automatically receive no permissions.) The Add button can be used to add an account or group. The Remove button can be used to delete an account or group.

Assigning Permissions for an Account or Group   To assign permission to an account or group, the administrator first clicks on the name to select it. In the figure, the *Power Users* group is selected. Power Users is a default group that is created automatically when Windows is installed.

When an account or group is selected, its permissions appear in the lower pane. For power users, the assigned permissions currently are Read & Execute, List Folder Contents, and Read. By clicking on boxes, the administrator can add other permissions or remove those that already exist.

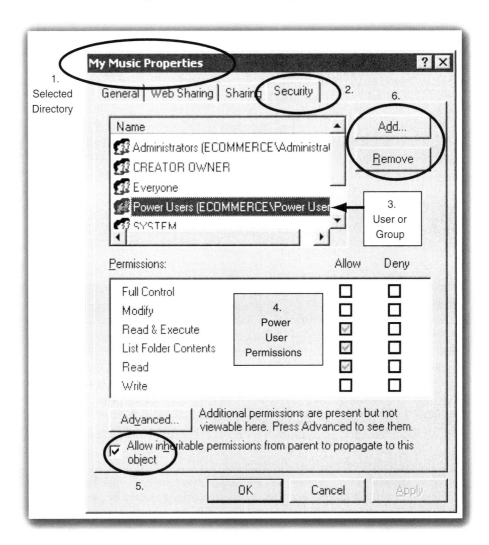

Figure 1b-8   Assigning Directory Permissions in Windows

Inheritance   At the bottom of the figure, there is a check box, "Allow inheritable permissions from parent to propagate to this object." This box is checked by default, to allow automatic inheritance. Unchecking this box prevents automatic inheritance.

Even if there is inheritance, the checked boxes in the Permissions pane take precedence. For instance, various permissions can be denied to the account or group so that they cannot be inherited in the directory.

TEST YOUR UNDERSTANDING

6. a) What are accounts, usernames, and passwords? b) What are password resets?

7. a) What are directory permissions? b) What does it mean if an account has no permissions in a directory? c) List the six Microsoft Windows Server permissions, describing each one briefly. d) How does the number of Windows Server permissions compare to the number of Unix permissions? e) What Microsoft permissions would you give to the group Everybody (which automatically includes all users) in a Public directory that contains documents that should be available to all employees? Justify your answer. f) What Microsoft permissions would you give to the group Everybody in the User directory? (The home directories of each user, in which the user should have full permission, come directly under this User directory.) Justify your answer.

8. a) Describe the Administrator account's permissions in Windows Server. b) Why does it need these permissions? c) Why are these permissions dangerous? d) What Unix account is comparable to the Windows Administrator account?

9. a) How is Windows Server more flexible than Unix in assigning permissions to multiple accounts? b) Why is assigning permissions to directories time consuming? c) How does using groups make this task easier? d) How does inheritance make this task easier?

10. How are permissions controlled in the Security tab of a directory's Properties dialog box?

XTR: THE INITIAL SITUATION

Having discussed dedicated servers, we can finally get to the XTR network design. We will begin with the initial situation in the company.

PCs

When planning for the network began, each staff member had a good personal computer. This included XTR's sixteen professionals and one secretary.

Printers

Each staff member also had a printer. Victor Chao, XTR's president, had a laser printer. So did Ann Jacobs (the secretary) and one of the consultants. The other consultants had color ink jet printers because the cost of giving each a laser printer would have been prohibitive.

Sneakernet

When a consultant with an ink jet printer needed a laser printout, he or she saved the file onto a floppy disk and gave it to Ann to print. Similarly, when consultants wanted to share a file, one had to save it to disk and walk it to the other consultant. Walking files around was jokingly called "**sneakernet.**" It wasted time and created confusion over who had the most current version of each file.

Remote Access

The consultants, who frequently are on the road, had to copy files they would need to the firm's one "loaner" notebook computer. If they needed an unexpected file on the road, there was no good way to retrieve it.

Internet Access

Although all PCs had modems, only one person could dial into the Internet at a time, and telephone access was very slow for the large maps they often had to download.

Maintenance

One of the employees, Kumiko Touchi, is very good with computers. However, there were many problems that Kumiko could not fix. Also, pulling Kumiko away from highly paid consulting work to fix a computer problem was absurd financially.

TEST YOUR UNDERSTANDING

11.   a) Describe the computer situation at XTR before the network. b) Describe the printer situation at XTR before the network. c) Describe file sharing at XTR before the network. d) Describe remote access to files at XTR before the network. e) Describe Internet access at XTR before the network. f) Describe maintenance at XTR before the network.

BROAD NETWORK DESIGN

Given this situation, the firm naturally wanted to network its computers together and to connect the internal network to the outside world. Victor Chao hired a network consultant, Robert Blanco, to create a broad design for the network. Your job will be to flesh out Blanco's broad design.

Labor Costs

XTR will hire a company to do the actual installation. The company will charge $75 per hour.

Switch

The firm will have an Ethernet switch to connect all of the elements. This switch will operate at 100 Mbps. It will need quite a few ports. This will be a very simple switch that can be plugged into a power outlet with no additional installation work beyond physical installation. Installation time will be about thirty minutes.

Wires

A UTP wire cord will run between the switch and each device. The company will run patch cords (precut UTP cords of approximately the correct length for each run). Patch cords will average $50 apiece.

The cords will run under carpets rather than neatly through false ceilings or walls. Neater but more expensive installation will not be done because the firm might be moving soon, and a neat installation would take more than four hours per UTP connection to the switch in XTR's building. Running the cords under carpets but leaving the cords otherwise exposed will only take about thirty minutes per UTP connection to the switch.

Network Interface Card (NIC)

The firm will have to install a NIC in three of the firm's current desktop PCs, now renamed client PCs. The other desktops came with adequate NICs when they were purchased. NICs, like the switch, will operate at 100 Mbps. Installation of a NIC will take about fifteen minutes. Setting up a PC with a NIC to run on the network will take an average of thirty minutes.

Dedicated Servers

In Mr. Blanco's design, the firm will purchase one or more new computers to be used as dedicated servers. They will run the Windows 2003 Server network operating system instead of the desktop versions of Windows that the client computers use.

Pricing for the server NOS depends upon the number of users who will share the server. In addition, each server will require six hours of installation time.

File Service

One server task will be to provide **file service,** meaning that the consultants can store their files on the server, as Figure 1b-9 shows. Consequently, the main server is called a **file server.** File service is good because the server is backed up nightly.

As the figure shows, one user can save a file to a directory on a file server. Other authorized users can later retrieve the file and can even change it and resave it if they are authorized to do so. However, unauthorized users cannot even see the file. Assigning and changing access permissions in various directories to the various users of a file server is a constant chore for server administrators, as noted earlier.

File Server Program Access

One important aspect of file service is **file server program access,** which is illustrated in Figure 1b-10. File server program access allows a firm to store application programs on a file server rather than on each client PC. As the figure shows, whenever a user runs a program, the program is copied temporarily from the file server to the client

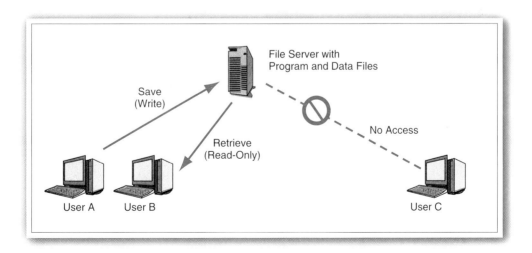

Figure 1b-9　File Service for Data Files

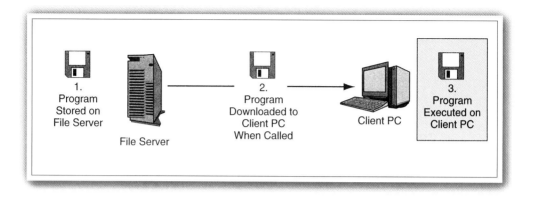

Figure 1b-10   File Server Program Access

PC, where it is executed. Note that a file server *stores* application programs but does not *execute* them. The execution is done on the client PC.

> In file server program access, the program is stored on the file server but executed on the client PC.

Users do not even realize that they are using file server program access. They merely click on a program icon and the program runs. The broad design calls for file server program access to be used for office applications, such as word processing and spreadsheet processing. It also will be used for the firm's computer-aided design applications and other software specific to the firm's professional work.

Why should an organization adopt file server program access? The answer is that this approach reduces the labor of installing and upgrading application programs. Even with seventeen client PCs, the work of installing and upgrading the firm's many applications on individual PCs is considerable. With file server program access, the installation and upgrading work for each application is done only once, on the file server. Figure 1b-11 illustrates this.

Making a single installation does not mean that the firm can buy one ordinary single-user copy of the software from a store and then have several people use it. File server program access requires the purchasing of **networked versions** of the software. These come with licenses for a certain number of users. As you would suspect, they cost much more than single-user copies.

Dedicated Print Servers

The network will replace the firm's diverse printers with the three existing high-capacity laser printers that will be spread around the office area for easy access by users. Pat Lee's network in Chapter 1 uses peer-to-peer print service. In contrast, as Figure 1b-12 shows, larger firms use dedicated **print servers.** When client users

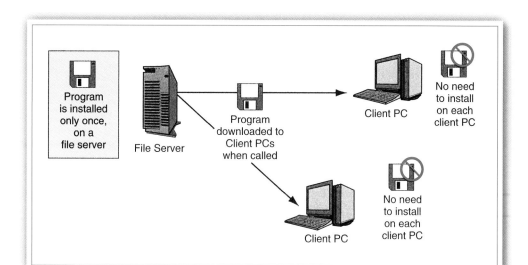

Figure 1b-11   File Server Program Access and Program Installation

print, their printout will go to a print server. The print server will then feed the print job to the attached printer.

Although print servers are called servers, they are not full computers. They are simple electronic devices that cost only $50 to $300 and have only the processing power and software needed to receive print jobs and feed them to the printer attached to them. A print server also has a built-in NIC to allow it to talk to the PCs it serves.

Each print server connects to a switch with a UTP cord and to its printer through an ordinary parallel cable or USB cord. Parallel cables and USB cords are very short— only one or two meters. UTP cords, in contrast, can be up to 100 meters long. Consequently, print servers sit right next to the printer, and there is a long UTP wiring run to the switch that connects the print server to the network.

Each print server takes about thirty minutes to set up, including configuration. To configure the print server, the installer runs an installation program on a client PC. This program on the client PC communicates across the network with the print server.

Internet Access

For Internet access, the firm will replace its slow telephone access with high-speed DSL service (see Chapter 7) from their local telephone company. Several client PCs can use this access line simultaneously. This will require purchasing an access router. One end of the access router will plug into the DSL modem via a UTP cord. The other end will plug into the switch via another UTP cord. (In Pat Lee's home network, the access

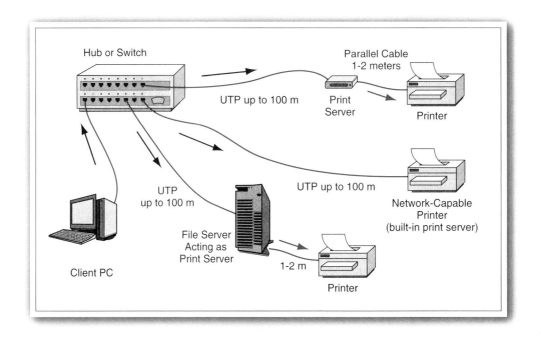

Figure 1b-12   Print Server Operation

router had an internal Ethernet switch; this is not practical for this larger network, which needs many more switch ports.)

The access router has a built-in webserver. To configure the access router, the installer runs a browser on a client PC. The installer then types the IP address of the access router to connect the browser to the built-in webserver. Physical installation and setup take about thirty minutes.

Firewall

The access router has a built-in firewall. Setting up the firewall will require about five hours. The major part of this will be working with XTR employees to determine what traffic should be blocked and what traffic should be permitted.

E-Mail

The firm will need a mail server to store incoming mail for all users. This mail server will communicate over the Internet with mail hosts in other companies. The company can either run the mail service software on the file server or purchase a separate mail server. The e-mail software will take about five hours to install and configure, including setting up mail accounts for XTR's employees. The e-mail vendor charges an annual license fee of $30 per user for the networked version of its software.

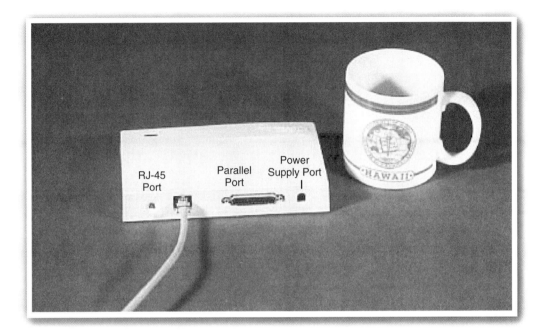

Figure 1b-13   Print Server

Remote Access Service (RAS)

One issue is whether to install **remote access service (RAS),** which would allow employees to connect to the network via the Internet. For remote access, a virtual private network (VPN) service (see Chapter 1) would have to be implemented. This would require activating and configuring VPN service on the remote access server. This would also require implementing authentication service on a file server or dedicated authentication server. Installing remote access service would take about five hours on the server.

YOUR DETAILED DESIGN

Based on this information, you will flesh out the network design for the company. To guide you, answer the following questions. (Hand them in to your teacher.)

Design Questions

1. Should the company use separate file and e-mail servers, or should it use one server for both? Justify your answer.

2. Do you think the company should purchase the server or servers you listed in the previous question; or should it save money by taking away existing PCs from employees and using them as dedicated servers?

3. How many print servers should the company use—one per printer or just one for all three printers? Justify your answer.

4. How many ports will be needed on the switch? Justify this number.

5. Switches can be purchased with 12, 24, 36, and 48 ports. Which switch should the company purchase? Explain your decision.

6. Cost out your server or servers, using an online source such as Dell.com. Provide a detailed listing of your server's or servers' features. The server should have Microsoft Windows Server with a sufficient number of licenses for XTR's employees.

7. Cost out the full network using a spreadsheet model. (Do not include remote access.) Cite specific products and sources of data.

8. What was the least expensive component of this network's cost?

9. What was the most expensive component of this network's cost?

10. Do you think the company should implement remote access? Cite reasons on both sides of the issue and give your choice, justifying it.

LAYERED STANDARDS ARCHITECTURES

Learning Objectives:

By the end of this chapter, you should be able to discuss:

- Core standards concepts: semantics, syntax, timing constraints, reliability, and connection-oriented versus connectionless service.

- Standards at Layers 1 and 2 (the physical and data link layers) and how they are implemented in Ethernet.

- Layer 3 (internet layer) standards and how IP works at this layer.

- Layer 4 (transport layer) standards and how TCP works at this layer.

- Layer 5 (application layer) standards in general and how HTTP works at this layer.

- Vertical communication among layer processes on the same node.

- Competing standards architectures and the dominance of the TCP/IP–OSI hybrid standards architecture.

INTRODUCTION

Chapter 1 introduced **standards**—rules of operation that allow two hardware or software processes to work together. In this chapter we will look at network standards in more depth. Nearly everything in networking is driven by standards, and to master networking, you must learn to think precisely about standards.

Standards are rules of operation that allow two hardware or software processes to work together.

In this chapter we often refer to standards as **protocols.** Later in the chapter, we will see that only some standards are protocols. However, all of the standards we will see in this book truly are protocols, so we will use the terms "standard" and "protocol" synonymously.

HOW STANDARDS GOVERN INTERACTIONS

As we saw in the previous chapter, standards normally govern the exchange of messages between two hardware or software entities. For instance, in Figure 2-2, the entities are a browser on a desktop PC and a webserver application program on a webserver. These messages are governed by the Hypertext Transfer Protocol (HTTP) standard.

Message Semantics (Meaning)

Standards govern the **semantics** of message exchange, that is, the meaning of each message. For instance, in Figure 2-2, the browser on the client PC transmits a message requesting a particular file on the webserver. The webserver responds with a message that either contains the requested file or gives an explanation of why the requested file could not be delivered. Although HTTP has a few other message types, there is nothing like the breadth of meanings in human communication. Computers simply do not have the flexibility of human intelligence, so protocol semantics have to be limited to only a few messages types.

Message Syntax

Standards also govern message **syntax,** that is, how messages are organized. In human languages, words in a sentence must be arranged in certain ways, although rules in human language are sometimes vague. In contrast, network standards have precise and rigid rules for structuring messages.

HTTP Request Message

For instance, in Figure 2-3, the HTTP request message consists of two lines of **plain text**(keyboard characters). The first line terminates with a carriage return/line feed [CRLF] to indicate that a new line should begin. There is rigid structure within each line.

For example, the first line must have a method (in this case, GET) that describes what the other side is to do (get a file), a single space, the path to the requested file (/reports/project1/final.htm), another single space, and what HTTP version the browser implements (1.1).

Standards Govern the Exchange of Messages
Message Semantics (Meaning)
        Only a few message types are allowed because computers to not have the
        intelligence to handle open-ended communication
        In HTTP, request and response messages
Message Syntax (Organization)
        Cannot be freely structured like human sentences
        Rigidly structured
                In HTTP, lines of text (Figure 2-3)
                Most lines are of the form "Keyword: Information"
        General Message Organization (Figure 2-4)
                Primary Components
                        Data Field (content to be delivered)
                        Header (everything before the data field)
                        Trailer (everything after the data field)
                Header and trailer are further divided into fields
                Trailers are uncommon and come mostly at the data link layer
                Some messages have only a header
Message Timing Constraints
        In client/server computing, server cannot respond unless it receives a request
        There are many more complex examples of timing constraints
        For example, see TCP later in this chapter

Figure 2-1   How Standards Govern Interactions (Study Figure)

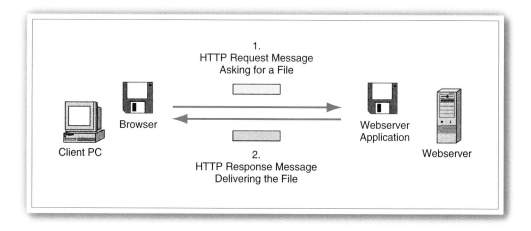

Figure 2-2   Hypertext Transfer Protocol (HTTP) Interactions

HTTP Request Message
    GET/reports/project1/final.htm HTTP/1.1[CRLF]
    Host: Voyager.cba.Hawaii.edu[CRLF]
HTTP Response Message
    HTTP/1.1 200 OK[CRLF]
    Date: Tuesday, 20-MAR-2002 18:32:15 GMT[CRLF]
    Server: name of server software[CRLF]
    MIME-version: 1.0[CRLF]
    Content-type: text/plain[CRLF]
    [CRLF]
    file to be downloaded: long string of bits

Figure 2-3   Syntax of HTTP
Request and Response Messages

Each subsequent line (there is only one subsequent line in this example) starts with a keyword, a colon (:), and the value of the keyword. The HOST keyword indicates that the colon is followed by the host name of the webserver.

HTTP Response Message

The HTTP response message, also shown in Figure 2-3, begins with an indication of which HTTP version the webserver understands. It then has a 200 code, which tells the browser that the GET request was handled successfully. The "OK," on this line says the same thing for human readers.

Afterward, we see several more lines, each beginning with a keyword, a colon, and a value for the keyword. A single blank line follows these lines. (Two carriage return/line feeds produce a blank line.) After this single blank line comes the file being delivered by the message. The file is a long string of ones and zeros.

General Message Organization

Figure 2-4 shows that messages in general may consist of three parts—a header, a data field, and a trailer. The header and trailer may be further divided into smaller units called fields.

Data Field

The **data field** is the content delivered in a message. In an HTTP response message, the data field is the file being delivered.

Header

The **header,** quite simply, is everything that comes before the data field. For the HTTP request message, the entire message is the header. There is no data field. In the HTTP response message, the header is everything that comes before the file being delivered.

Trailer

Some messages also have **trailers,** which consist of everything coming *after* the data field. HTTP messages do not have trailers. They have nothing after the data field.

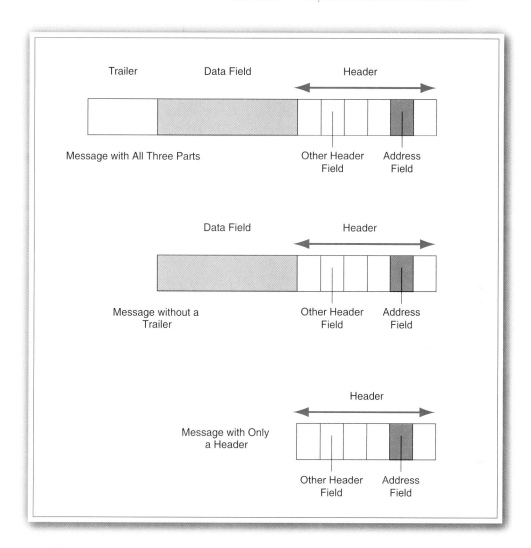

Figure 2-4   General Message Organization

Not All Messages Have All Three Parts
Only a header is present in all messages, as in the case of HTTP request messages. Data fields are not always present but are very common. Trailers are not common.

Header and Trailer Fields
The header and trailer may be subdivided into smaller sections called **fields.** In HTTP, the first line of a request message has the method, path, and version fields. The first line of a response message also is divided into fields. In both the request and response messages, each subsequent line of the header (consisting of a keyword, a colon, and content) is a field.

Message Timing

Standards also govern **message timing,** that is, when hardware or software processes may transmit. To give an analogy, you usually cannot talk any time you wish in a classroom. The teacher must call on you, or there must be some other way to recognize you.

Figure 2-2 shows a particularly simple form of interaction called **client/server interaction.** As the name implies, client/server interaction involves one entity providing service to another entity. In this case, the browser is the *client* (customer) of the webserver application program. The webserver application provides *services* to the client—the transmission of requested files.

Timing restrictions in HTTP are very simple. The client may transmit any time it wishes to do so. The webserver, however, may only send an HTTP response message after it receives an HTTP request message. This may be blindingly obvious, but we will see many more complex forms of message timing control in this book, including later in this chapter when we look at TCP (in Figure 2-6).

TEST YOUR UNDERSTANDING

1.   a) What three things do standards govern? b) Distinguish between semantics and syntax. c) What are the three general parts of messages? d) What is the purpose of the data field? e) How is the header defined? f) Is there always a data field in a message? g) Are trailers common? h) Distinguish between headers and header fields. h) In HTTP client/server transmission, what timing restrictions exist on the client application? i) In HTTP client/server transmission, what timing restrictions exist on the server application?

Connection-Oriented and Connectionless Protocols

Connection-Oriented Protocols

When you call someone on the telephone, you do not just begin speaking when the other party picks up the phone. There is at least a tacit initial negotiation that the called person is willing to speak. In addition, at the end of a conversation, it is rude simply to hang up without both sides indicating that they wish to end the call.

In networking, as Figure 2-5 illustrates, **connection-oriented** standards work this way, with explicit openings and closings.

Connectionless Protocols

In contrast, **connectionless** protocols do *not* establish a connection before transmitting. When you send someone an e-mail, for instance, you do not call him or her ahead of time to ask if you may send the e-mail message or call him or her afterward to ask if you can stop sending messages. Most of the protocols we will see in this book are connectionless.

Although connection-oriented protocols are heavyweight protocols in terms of both processing costs and network traffic (several messages may be exchanged to open or close a connection, and there will be other connection-maintenance messages as well), they have a number of advantages.

Sequence Numbers for Message Ordering

Most notably, as Figure 2-5 shows, each message during a connection can be given a sequence number. This allows the receiving process to put messages in order if they

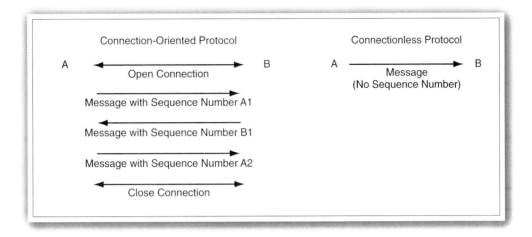

Figure 2-5   Connectionless and Connection-Oriented Protocols

arrive out of order. With connectionless protocols, there is no message sequencing. Each message is treated as an isolated event. We will next look at another advantage of connection-oriented protocols, namely reliability.

TEST YOUR UNDERSTANDING

2.   a) Distinguish between connectionless and connection-oriented services. b) Which places a heavier traffic burden on the network? c) Which places a heavier processing load on the computers that are communicating? d) Which can have sequence numbers?

Reliability

Another aspect of protocols is **reliability,** which in networking means that errors are corrected by resending lost or damaged messages. Protocols are either reliable or unreliable.

Reliable protocols correct errors by resending lost or damaged messages. Unreliable protocols do not correct errors.

Unreliable Protocols

HTTP is an **unreliable** protocol. It does *not* do error correction. This is not a problem in practice because HTTP relies on TCP to do error correction, as we will see below. TCP then gives HTTP clean data.

Reliable Protocols

Acknowledging Correctly Received Segments   Figure 2-6 shows how the Transmission Control Protocol (TCP) implements reliability. Every time a transport process receives a correct **TCP segment** (TCP messages are called TCP segments), it sends back an **acknowledgement (ACK)** segment. If the original sender receives the acknowledgement for a segment, it knows that its segment was received correctly. See segments 4 and 5 for an example of a transmission and an acknowledgement.

Not Acknowledging Incorrect or Not-Received Segments   However, the receiving transport process does not send an acknowledgement if the segment was damaged

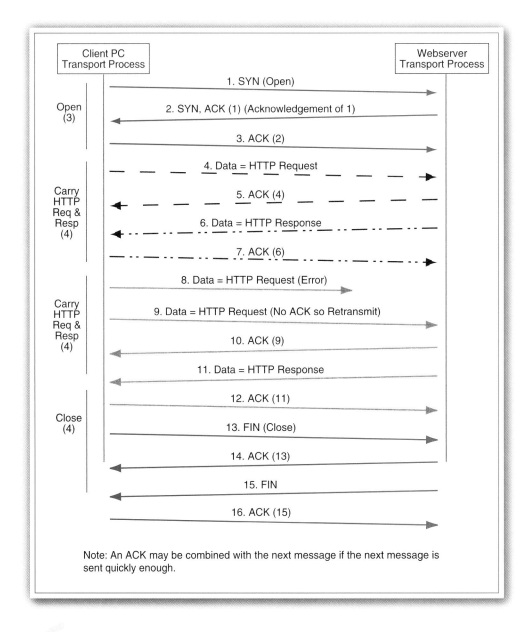

Figure 2-6   Transmission Control Protocol (TCP)

or if the segment never arrived. (See Segment 8.) If the sending transport process does *not* receive an acknowledgement for a segment promptly, it knows that the segment was not received properly. It resends the segment. (See segment 9.) This process corrects the transmission error, giving reliability. Note that the transport process that originally sends a TCP segment decides whether to resend it.

Connections and TCP   Figure 2-6 also shows how TCP, a connection-oriented protocol, implements openings and closings. One side opens a connection by sending a SYN segment to its desired communication partner. The other side sends back a single TCP segment that has a SYN, indicating that it is willing to communicate, plus an ACK (acknowledgement) of the originating side's SYN segment. The originating side sends back an ACK segment.

Closing a TCP connection is similar, with one side initiating the close with a FIN segment and the other side responding with a separate FIN/ACK segment. The side initiating the close sends back an acknowledgement segment.

Although they are not shown in Figure 2-6, TCP segments contain sequence numbers.

TEST YOUR UNDERSTANDING

3.   a) What is reliability? b) How does TCP implement reliability? c) In TCP, what is the receiver's role in reliability? d) In TCP, what is the sender's role in reliability?

The TCP/IP–OSI Architecture

Up to now, we have been talking about the characteristics of individual standards. In networking, there are dozens of standards that do different things, somewhat as the rooms of a house have different functions.

Architectures

To continue the house example, you do not design a house by building one room, then another, and then another, without any plan. Rather, you first create an **architecture**—a broad plan that specifies what rooms the house will have and how these rooms will be related in terms of traffic flow. The architecture ensures that the rooms will collectively provide everything that is needed. Only after the architecture is finished will the architect begin to design individual rooms.

TCP/IP–OSI

Similarly, data network standards are not created in isolation. A **network architecture** is a broad plan that specifies everything that must be done for two application programs on different networks on an internet to be able to work together effectively. Figure 2-7 illustrates the most popular standards architecture for networking today, the TCP/IP–OSI Architecture. Later in this chapter, we will discuss its odd name. For now, we will focus on its organization.

> A network architecture is a broad plan that specifies everything that must be done for two application programs on different networks on an internet to be able to work together effectively.

Layer	Name	Specific Function	Broad Function
5	Application	The application layer governs how two applications work with each other, even if they are from different vendors.	Interoperability of application programs
4	Transport	Transport layer standards govern aspects of end-to-end communication between two end hosts that are not handled by the internet layer. These standards also allow hosts to work together even if the two computers are from different vendors and have different internal designs.	Transmission across an internet
3	Internet	Internet layer standards govern the transmission of packets across an internet—typically by sending them through several routers along the route. Internet layer standards also govern packet organization, timing constraints, and reliability.	
2	Data Link	Data link layer standards govern the transmission of frames across a single network—typically by sending them through several switches along the data link. Data link layer standards also govern frame organization, timing constraints, and reliability.	Transmission across a single network
1	Physical	Physical layer standards govern transmission between adjacent devices connected by a transmission medium.	

Figure 2-7   TCP/IP–OSI Architecture

Standards for Single Networks

The architecture is organized as a series of layers. The bottom two layers (physical and data link) govern transmission across a single network—either a single LAN or a single WAN. Figure 2-8 compares these two layers. It shows a single network. In this network, Host A and Router R1 are separated by two switches, X1 and X2.

Physical Links   Physical layer (Layer 1) standards govern transmission between adjacent devices connected by a transmission medium. In the figure, there are three **physical links** between Host A and Router R1: from Host A to the first switch, from the first switch to the second switch, and from the second switch to Router R1.

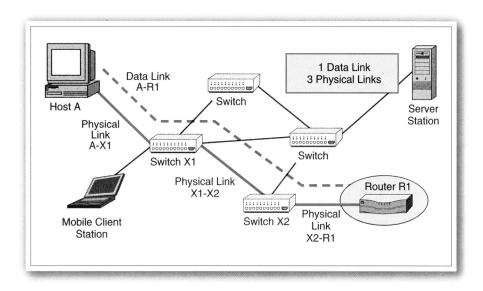

Figure 2-8   Physical and Data Link Layer Standards in a Single Network

---

Physical layer (Layer 1) standards govern transmission between adjacent devices connected by a transmission medium.

Data Links   The **data link layer (Layer 2) standards,** in turn, govern transmission all the way from Host A to Router R1. The path that a frame takes across the network is called the frame's **data link.** There is only one data link between the source and destination computers on a single network.

Data link layer standards also govern the organization of frames for that network, plus timing constraints, reliability, and other matters.

Datalink layer (Layer 2) standards govern the transmission of frames across a single network—typically by sending them through several switches along the data link. Data link layer standards also govern frame organization, timing constraints, and reliability.

There are many types of single networks, including Ethernet for LANs and Frame Relay for WANs, to name just two. Each network standard specifies both physical layer standards and data link layer standards. There are many types of single networks, so most of the network standards we will see in this book will be physical and data link layer standards.

TEST YOUR UNDERSTANDING

4. a) What devices does a physical link connect? b) What is a data link? c) Five switches separate two stations on a network. How many physical links are there between the two stations? d) How many data links are there between them? e) What do data link layer standards govern? f) Which layers govern LAN transmission? g) Which layers govern WAN transmission?

Standards for Internet Transmission

Initially, there were only standards for single networks. However, as networks began to proliferate, new standards were needed to link two or more single networks together into **internets,** which, as we saw in Chapter 1, are groups of networks connected by routers so that any application on any host on any network can communicate with any application on any other host on any other network. Standards at the internet and transport layers collectively govern transmission across an internet.

The Internet and Data Link Layers   As Figure 2-9 illustrates, there are multiple networks in an internet. Here, there are three networks. Host A is transmitting to Host B. There are three networks between the two hosts, so there are three data links (A to R1, R1 to R2, and R2 to B).

In turn, the path that the packet takes across its internet is called its **route.** There is only one route between the source and destination hosts (Host A and Host B). Note that we spell the name of the layer in lower-case (internet). The main standard for internet transmission is the Internet Protocol (IP).

Internet layer (Layer 3) standards govern the transmission of packets across an internet—typically by sending them through several routers along the route. Internet layer standards also govern packet organization, timing constraints, and other matters.

Internet layer (Layer 3) standards govern the transmission of packets across an internet—typically by sending them through several routers along the route. Internet layer standards also govern packet organization, timing constraints, and other matters.

The Internet Layer and Transport Layer   As Figure 2-10 shows, the **internet layer** governs the transmission of packets across an entire internet. It specifies what each router along the way does with packets.

The Transport Layer   In turn, **transport layer (Layer 4) standards** govern aspects of end-to-end communication between the two end hosts that are not handled by the internet layer. These standards also allow hosts to work together even if the two computers are from different vendors and have different internal designs.

Transport layer standards govern aspects of end-to-end communication between two end hosts that are not handled by the internet layer. These standards also allow hosts to work together even if the two computers are from different vendors and have different internal designs.

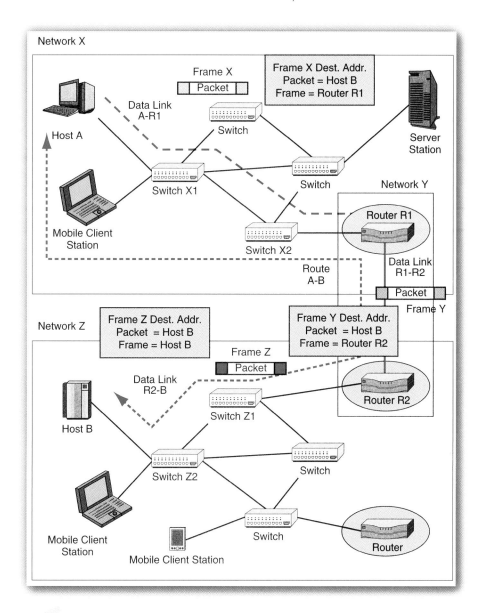

Figure 2-9   Internet and Data Link Layers in an Internet

Typically, the transport layer standard (usually TCP) is reliable, fixing any errors created at the transport layer or lower layers and delivering clean data to the application program.

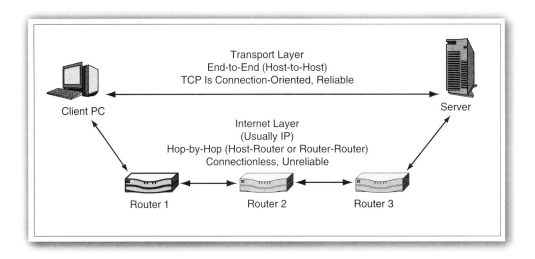

Figure 2-10   Internet and Transport Layer Standards

Standards for Applications

At the highest layer, **application layer (Layer 5) standards** govern how two applications work with each other, even if they are from different vendors. For example, a Microsoft Internet Explorer browser can work with a webserver application program built by a different vendor.

Application layer (Layer 5) standards govern how two applications work with each other, even if they are from different vendors.

In Figure 2-2 we saw the Hypertext Transfer Protocol (HTTP) application layer standard. HTTP is only one of many application protocols used on the Internet. E-mail, FTP, instant messaging, and other applications use different application protocols. There are more application standards than any other type of standard.

TEST YOUR UNDERSTANDING

5.   a) What do the internet and transport layers do collectively? b) Distinguish between what the internet and transport layer standards govern. c) What errors does the transport layer usually fix? d) What do application layer standards govern?

Why a Layered Architecture?

Breaking up Large Tasks into Smaller Tasks

Why break networking standards up into five pieces (layers)? The first answer is that whenever you have a major task, it is a good strategy to break it up into pieces and attack the pieces individually. This also works in standards development, which is a *very large* task.

Breaking Up Large Tasks into Smaller Tasks
Specialization in Standards Design
> Electrical engineers develop physical layer standards
> Specialists in an application develop application
> standards for that application

Simplification in Standards Design
> Designers have to focus only on issues relevant to
> that layer

If You Change a Standard at One Layer,
You Do Not Have to Change Standards at Other Layers

Figure 2-11   Why Layer?
(Study Figure)

Specialization in Standards Design

In addition, if a team is doing the task, team members should be assigned individual tasks that suit their skills. In networking standards development, for instance, electrical engineers usually create physical layer standards, while specialists in a particular application usually create application layer standards for that application.

Simplification in Standards Design

Another benefit of layering is that development at one layer is freed of concerns at other layers. Application standards designers can design new application standards without having to worry about how to get messages between the two hosts. Physical layer standards designers, in turn, do not have to worry about what applications will use a particular physical link.

If You Change a Standard at One Layer, You Do Not Have to Change Standards at Other Layers

A fourth reason for layering is that it allows standards to be updated or changed at various layers independently. If you switch from one LAN standard to another, say from Ethernet to an 802.11 wireless LAN, or if you upgrade from one version of Ethernet to a newer version, you do not have to change standards above the data link layer. In addition, if you switch from handling e-mail to browsing webservers (these are application layer functions), you do not need to use different lower-layer (Layer 1-4) standards.

TEST YOUR UNDERSTANDING

> 6. a) Why do standards architectures break down the standards development process into layers?

LAYERS 1 (PHYSICAL) AND 2 (DATA LINK) IN ETHERNET

Having looked at layering in general, we will begin to look at individual layers in more detail, beginning with Layers 1 and 2, which govern individual LANs and WANs.

Ethernet Physical Layer Standards

At the physical layer, we saw in Chapter 1 that Ethernet can use UTP cabling. We will see in Chapter 3 that Ethernet physical layer transmission can also be done using optical fiber cabling, which sends bits as light flashes. At the physical layer, there are no messages. Bits are sent individually as voltage changes, light flashes, radio pulses, or other discrete signaling events.

Ethernet Frames

Messages at the data link layer are called **frames**. Data link layer standards dictate the semantics, timing, and syntax of frame transmission within a single network. As we saw in Chapter 1, Ethernet switches work with Ethernet frames. Figure 2-12 illustrates the standard **Ethernet frame,** which, formally speaking, is the **802.3 MAC layer frame** for reasons we will see in Chapter 4.

Messages at the data link layer are called frames.

Octets

The Ethernet header and trailer are divided into fields. Field lengths can be measured in bits. Another common measure for field lengths in networking is the octet. An **octet** is a collection of eight bits. Isn't that a byte? Yes, exactly. "Octet" is just another name

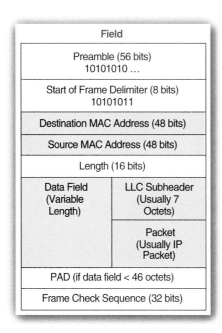

Figure 2-12   Ethernet Frame

for "byte." It is widely used in networking, however, so you need to become familiar with the term.[1]

Ethernet Addresses and Hexadecimal Notation

48-Bit Ethernet Addresses   Frames have data link layer source addresses and destination addresses that identify the sending and receiving stations, respectively. Each Ethernet network interface card (NIC) has a unique 48-bit **Ethernet address** set at the factory before the card is shipped. As Chapter 4 discusses, Ethernet addresses often are written in hexadecimal (Base 16) notation. In "hex" notation, a typical 48-bit Ethernet address is "A7-BF-23-D4-33-99."

Hexadecimal notation is only used by humans because it is easier to write and read 12 hex symbols and dashes than it is to write and read a string of 48 ones and zeros. Computers, in contrast, only work with raw 48-bit strings.

As Figure 2-13 shows, switches between the source and destination hosts read the destination Ethernet address in every arriving frame. They look up the Ethernet address in the switching table. The line containing this address contains a port number. They send the frame back out this indicated port number.

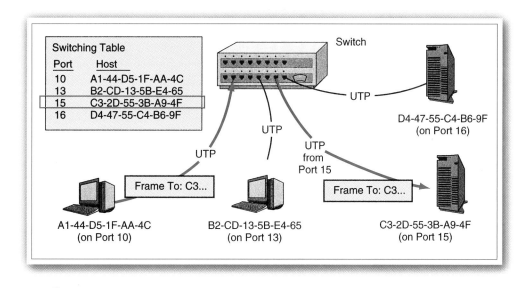

Figure 2-13   Ethernet Switching Decision

---

[1] "Octet" actually makes more sense than "byte," because "oct" means eight. We have octopuses, octagons, and octogenarians. What is the eighth month? (Careful!)

The Ethernet Data Field

In the TCP/IP–OSI architecture, an Ethernet frame's data field begins with an LLC subheader discussed in Chapter 4. More importantly, the data field usually (but not always) contains an IP packet. As we discuss later, we say that the IP packet is **encapsulated** (carried) in the Ethernet frame.

Frame Check Sequence Field: Unreliable Operation

Following the data field is a trailer with a single field, the **frame check sequence field.** This four-octet field holds a binary number that the sending NIC calculates based on the bits in other fields in the frame.

The receiving NIC recomputes the number and compares its result to the value contained in the frame check sequence field. If the two do not match, an error must have occurred during transmission.

If an error is found, the receiving NIC simply *discards* the frame. There is no request for retransmission. Having no means of error correction, the Ethernet data link layer standard is unreliable despite *detecting* errors. Error detection is not error correction because there is no automatic retransmission of a lost or damaged frame.

TEST YOUR UNDERSTANDING

7. What is an octet?
8. a) How long are Ethernet addresses? b) When are Ethernet addresses set on NICs? c) In what notation are they typically written for human reading? d) Who uses hex notation for Ethernet addresses—computers, people, or both? e) What device in a network reads the Ethernet address? f) What is its purpose in doing so?
9. What does the Ethernet data field usually contain?
10. a) How many *bits* long is the Ethernet frame check sequence field? b) What is its purpose? c) How does the receiving NIC use it? d) What happens if a receiving NIC detects an error? e) Does this error detection and discarding process make Ethernet a reliable standard?

Ethernet Characteristics

Ethernet Is Unreliable and Connectionless

In addition to being unreliable (as we have just seen), Ethernet is connectionless. Ethernet frames can be sent without first requesting permission to send messages.

Ethernet Timing

When Ethernet switches are used, Ethernet timing is extremely simple. Any NIC may transmit at any time, period. As we will see in Chapter 4, some older Ethernet networks use devices called hubs instead of switches. When hubs are used, only one NIC can transmit at a time. In that chapter we will discuss timing when Ethernet hubs are used.

TEST YOUR UNDERSTANDING

11. a) Is Ethernet connectionless or connection-oriented? Explain. b) In a switched Ethernet network, are there timing constraints on when a station may transmit?

LAYER 3: THE INTERNET PROTOCOL (IP)

Internet layer standards govern the transmission of messages called packets across an internet. As noted earlier, the Internet Protocol (IP) is the most common standard at Layer 3, the internet layer.

Layer 2 Versus Layer 3

Figure 2-9 illustrates how the data link layer and the internet layer are related. Put simply, the data link is the path a frame takes through a single network. In turn, the internet layer path, called a **route,** is the path a packet takes from the source host to the destination host across an internet.

Figure 2-9 also shows how frames and packets are related. As the packet travels through the internet, it is carried in the data field of a frame within each network. If three networks separate the source and destination hosts, there will be one packet that will be carried successively in three different frames. To give an analogy, if you mail a letter, the letter goes all the way from you to the receiver, like a packet. Along the way, however, it will be carried in a series of trucks and airplanes, which are analogous to frames.

The figure shows how addressing is done in frames and packets. The packet always has the same Layer 3 internet destination address: the IP address of Host B. However, the three frames have Layer 2 data link layer destination addresses that are the endpoints in their individual networks. For Frame X in Network X, it is the data link layer address of Router R1. For Frame Y in Network Y, it is the data link layer address of Router R2. For Frame Z in Network Z, it is the data link layer address of Host B.

TEST YOUR UNDERSTANDING

12.   a) The source and destination hosts are separated by four networks. How many packets will there be? b) How many frames will there be? c) How many routers will there be along the route? (Hint: Draw a picture showing the hosts, networks, and routers.) d) How many routes will there be? e) How many data links will there be? f) How many IP destination addresses will there be? g) How many data link layer destination addresses will there be? h) What will be the Layer 2 data link layer destination address in the frame in the first network? i) What will be the destination IP address of the packet contained in that frame?

The IP Packet

In Figure 2-12, we saw the Ethernet frame at Layer 2. We saw that an Ethernet frame typically carries an IP packet in its data field. Figure 2-14 shows the organization of an **Internet Protocol (IP)** packet.

Illustrated with 32 Bits per Line

An IP packet, like an Ethernet frame, is a long string of bits (1s and 0s). Unfortunately, drawing the packet this way would require a page several meters wide. Instead, Figure 2-14 shows an IP packet as a series of rows with 32 bits per row. In binary counting, the first bit is zero. Consequently, the first row is bits 0 through 31. The next row

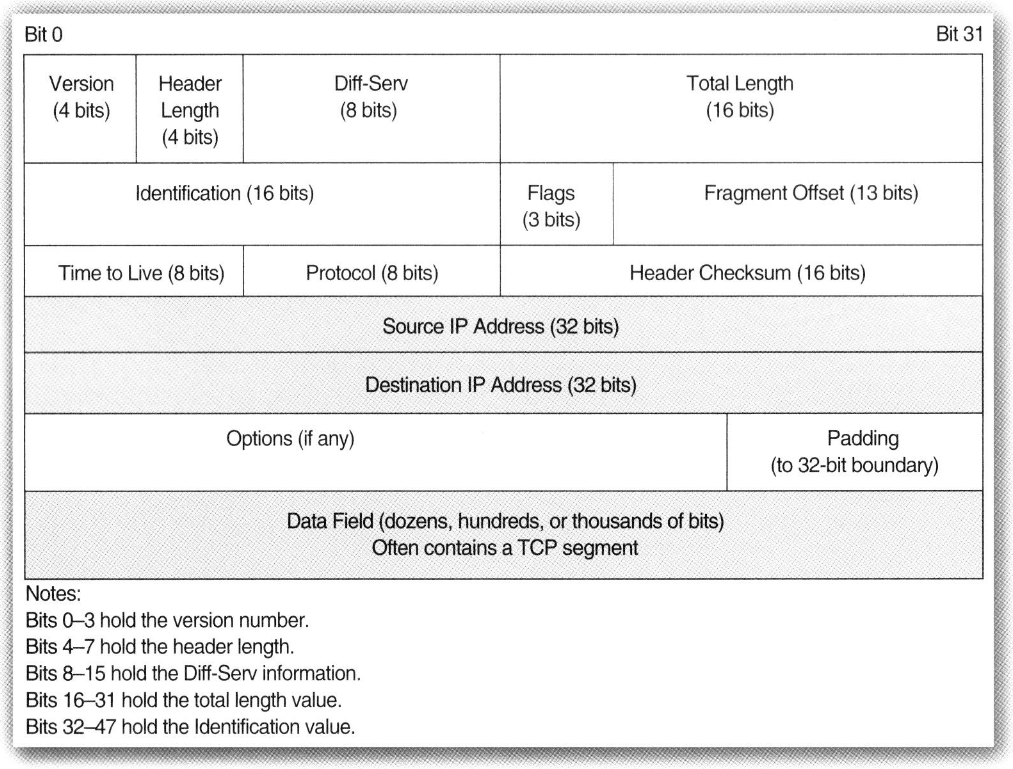

Bit 0   Bit 31

Version (4 bits)	Header Length (4 bits)	Diff-Serv (8 bits)	Total Length (16 bits)		
Identification (16 bits)			Flags (3 bits)	Fragment Offset (13 bits)	
Time to Live (8 bits)		Protocol (8 bits)	Header Checksum (16 bits)		
Source IP Address (32 bits)					
Destination IP Address (32 bits)					
Options (if any)				Padding (to 32-bit boundary)	
Data Field (dozens, hundreds, or thousands of bits) Often contains a TCP segment					

Notes:
Bits 0–3 hold the version number.
Bits 4–7 hold the header length.
Bits 8–15 hold the Diff-Serv information.
Bits 16–31 hold the total length value.
Bits 32–47 hold the Identification value.

Figure 2-14   Internet Protocol (IP) Packet

is bits 32 through 63. This is a different way of showing syntax than we saw with the Ethernet frame and with HTTP messages, but it is a common way of showing syntax, so you need to be familiar with it.

Source and Destination Addresses

32-Bit Addresses   Like an Ethernet frame, an IP packet has source and destination addresses. These **IP addresses** are 32 bits long. (In contrast, we saw earlier in this chapter that Ethernet addresses are 48 bits long.) While an Ethernet address gives a host's address on its single network, an IP address gives a host's internet layer address on an internet consisting of multiple single networks.

> While an Ethernet address gives a host's address on its single network, an IP address gives a host's internet layer address on an internet consisting of multiple single networks.

For human comprehension, it is normal to express IP addresses in **dotted decimal notation.** A typical IP address in dotted decimal notation is "128.171.17.13"—four numbers separated by dots. Like hex notation for Ethernet addresses, dotted decimal notation is only used by humans who have a difficult time remembering and writing strings of 32 ones and zeros.

Just as switches within a single network read the Ethernet address in the Ethernet header in a frame to learn how to deliver a frame, routers along the way read the IP address in the IP header to learn how to forward the IP packet to the next router or, if the destination host is attached to the router, to the destination host itself.

TEST YOUR UNDERSTANDING

13.   a) How many octets long is an IP header if there are no options? (Look at the figure.) b) What is the bit number of the first bit in the destination address? (Remember that the first bit in binary counting is Bit 0.) c) How long are IP addresses? d) You have two addresses: B7-23-DD-6F-C8-AB and 217.42.18.248. Specify what kind of address each address is. e) What device in an internet reads the IP address? f) What is its purpose in doing so?

IP Characteristics

IP Is a Connectionless Service with No Timing Constraints
Like Ethernet, IP is a connectionless protocol and has no timing constraints. A host can transmit an IP packet at any time.

IP Is an Unreliable Service
The IP packet has a **header checksum** field, which the receiver uses to check for errors in the IP header (but not its body). As in Ethernet, if the receiver detects an error, it simply discards the packet. There is error detection but no error correction, so IP is an unreliable protocol.

TEST YOUR UNDERSTANDING

14.   a) Is IP connectionless or connection-oriented? b) Does IP have timing constraints? c) Is IP reliable or unreliable?

LAYER 4: THE TRANSMISSION CONTROL PROTOCOL (TCP)

In this section we will take a closer look at the **Transmission Control Protocol (TCP).** As we saw in Chapter 1, TCP is the most common protocol at the transport layer (Layer 4).

Layers 3 and 4

Hop-by-Hop Layers

Figure 2-10 showed the relationship between the internet and transport layers. Note that the internet layer is a **hop-by-hop** layer. Although it gets packets from the source host to the destination host, it does this by governing what each router does along the way. It governs internet layer processes on the source host, on each intermittent router, and on the destination host. Similarly, data link layer standards govern hop-by-hop transmission across multiple switches.

End-to-End Layers

In contrast, the transport layer is an **end-to-end** layer. It governs communication between the transport process on the source host and the transport process on the destination host. The application layer we will see after this section also is an end-to-end layer.

TEST YOUR UNDERSTANDING

15. Two hosts are separated by ten routers. a) How many internet layer processes will be active on the two hosts and the devices between them? b) How many transport layer processes will be active? c) Which layers are hop-by-hop layers? d) Which layers are end-to-end layers?

TCP: A Reliable Protocol

Most protocols are unreliable. As we saw earlier, however, TCP, which operates at the transport layer, is reliable. As Figure 2-7 showed, TCP is the highest layer apart from the application layer. Making TCP reliable means that any errors made at the transport layer or at lower layers will be caught and corrected by TCP. This gives the application layer clean data.

Why not simply make all layers reliable? The answer is that reliability is expensive. It requires considerable processing and storage requirements on each node to hold copies of outgoing messages until correct delivery is verified and to decide what to do if there is or is not an error. In fact, error correction consumes far more processor cycles per message than any other process in switching or routing.

One reason for doing error correction only at the transport layer is that it can correct errors at all lower layers. Doing error correction at only one layer—the highest layer before the application layer—provides error-free data to the application program at minimal cost compared to doing error correction at all layers.

A second reason for doing error correction only at the transport layer is that TCP is an end-to-end layer. This means that error correction is only done once, by the two hosts. In contrast, if error correction had to be done on each hop between switches and on each hop between routers, the costs of switches and routers would be far higher. Furthermore, incurring this cost at each switch and router would be extremely wasteful because transmission errors are rare today.

TCP is a very complex protocol. We will look at it in detail in Chapter 8. However, to give you some feeling for how much traffic is generated by openings, closings, acknowledgements, and retransmissions, Figure 2-6 shows a brief TCP connection that is used to send HTTP requests and responses. Note that three segments are needed to open a connection (SYN, SYN/ACK, and ACK) and that four segments are needed to close a connection (FIN, ACK, FIN, and ACK). However, the "chattiest" aspect of TCP is all of the acknowledgements it generates in addition to the information transmissions that are the heart of the communication.

TEST YOUR UNDERSTANDING

16. a) Why are most standards unreliable? b) Why is making TCP reliable a good choice? c) In Figure 2-6 how many of the TCP segments are content information, messages (which carry application date) and how many are supervisory messages that control the transmission instead of carrying application layer data?

LAYER 5: HTTP AND OTHER APPLICATION STANDARDS

The highest layer is the application layer **(Layer 5).** Standards at this layer govern how application programs talk to one another.

The transport layer merely governs how hosts exchange messages. In contrast, application layer standards govern application-specific communication. Examples of application-specific communication are communication between a browser and a webserver, communication between an e-mail client and an e-mail server, and communication between an FTP client and an FTP server.

We looked at one application layer standard earlier. This was HTTP. Other applications have their own standards. For instance, at the application layer, e-mail uses the Simple Mail Transfer Protocol (SMTP) standard to transmit messages and the Post Office Protocol (POP) when a user downloads mail from a mail server. File transfers, as you might suspect, often use the File Transfer Protocol (FTP) standard. There are many applications on the Internet, and there are more application standards than all other types of standards combined.

TEST YOUR UNDERSTANDING

17. a) Is the application layer standard always HTTP? b) What layer has the most standards? Why? c) At what layer will you find HTTP, FTP, SMTP, and POP? d) At what layer would you expect to find standards for instant messaging? (The answer is not explicitly in this section.)

VERTICAL COMMUNICATION ON HOSTS, SWITCHES, AND ROUTERS

In this chapter so far, we have looked at horizontal communication at a single layer, between processes on different hosts, switches, and routers. Indeed, this kind of horizontal communication will be the focus of most of this book. However, communication also needs to take place *within* a single host, router, or switch. More specifically, a layer process on a device often needs to communicate vertically with the process one layer above it (Layer N+1) and the process one layer below it (Layer N−1).

Layered Communication on the Source Host

Figure 2-15 looks at vertical communication within a single host computer. Here the computer is a client PC with a browser. The browser creates an HTTP request message intended for the webserver process on a webserver destination computer.

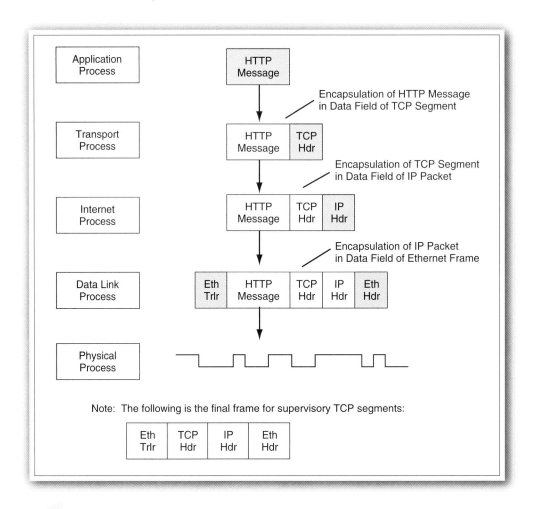

Figure 2-15   Layered Communication on the Source Host

Of course, the two application layer processes are on different computers, so they cannot communicate directly, unless they use mental telepathy. Rather, they must use processes at other layers on their own computers and on intermediate switches and routers to carry messages between the two application programs.

Figure 2-15 shows that immediately after the browser at the application layer creates the HTTP message, it passes this application message down to the next-lower layer, which is the transport layer.

The transport layer on the sending host then initiates a connection to the transport process on the destination host. (This step is not shown in the figure, but data segments cannot be sent until the connection is open.)

Once the connection is open, the transport process on the sending host then places the HTTP message in the data field of a TCP segment. A TCP segment carrying an application message, then, consists of a TCP header and the application message.

As noted earlier in the chapter, placing a message in the data field of another message is called encapsulation. The transport layer, then, encapsulates the HTTP message in the data field of a TCP segment.

Each subsequent layer on the source host (except the physical layer) passes its message down to the next-lower layer, which encapsulates the message by adding a header and (in the case of the data link layer) sometimes a trailer.

The final message, then, consists of an Ethernet header, an IP header, a TCP header, the HTTP message, and an Ethernet trailer. This is the Ethernet frame.

The physical layer converts the ones and zeros of this frame into signals and transmits them to the next device, usually a switch.

TEST YOUR UNDERSTANDING

18. a) When a layer creates a message, what does it usually do immediately afterward? b) What does the layer below it usually do after receiving the next-higher-layer message? TCP is the exception to this. c) What does TCP do before encapsulating the application layer connection and passing it down to the next-lower layer? d) What is encapsulation? e) With Web communication using HTTP, what message does IP encapsulate in packet data fields? f) What is the final frame if SMTP (an e-mail protocol that requires TCP) is used at the application layer and if Frame Relay (which has a header and a trailer) is used instead of Ethernet at the data link layer? g) What is the final frame if a transport process sends a TCP SYN message, which does not have a data field, over a Frame Relay network?

Vertical Layered Communication in End-to-End Transmission

On the Destination Host

When the signal of the last frame finally reaches the destination host, the process of **decapsulation** on the destination host is the mirror image of the encapsulation process on the source host, as Figure 2-16 shows.

➤ The physical layer turns the signals into the bits of the frame and passes the frame to the data link layer.

➤ The data link layer checks the frame for errors. If there are no errors, it decapsulates the packet and passes the packet up to the internet layer process.

➤ The internet layer checks the packet header for errors. If there are no errors, it decapsulates the TCP segment and passes the segment up to the transport layer process.

➤ The transport layer checks the TCP segment for errors. If there are no errors, it decapsulates the HTTP message and passes this message up to the application layer process.

In practice, the layer processes sometimes do other things besides checking for errors before decapsulating a message and passing it up to the next-higher layer. For instance, the TCP process on the source host opens a connection before passing the TCP segment containing the HTTP message down to the internet layer. However, these are marginal details.

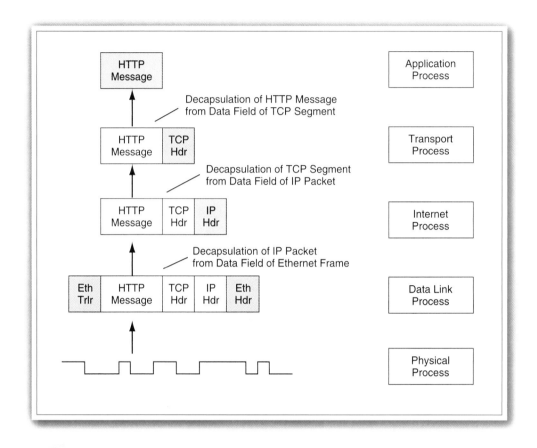

Figure 2-16 Decapsulation on the Destination Host

On Switches and Routers along the Way

So far we have looked only at layered communication on the two hosts. However, layer processing also takes place on all intermediate switches and routers, as Figure 2-17 shows.

On switches and routers, reception occurs, then transmission. This means that there is decapsulation and then encapsulation.

Also note that the highest layer on switches is the data link layer and that the highest layer on routers is the internet layer. This is why switches are called Layer 2 devices and routers are called Layer 3 devices.

TEST YOUR UNDERSTANDING

19. a) Which host decapsulates—the sending host or the receiving host? b) Describe what each layer's process does on the receiving host when an Ethernet frame containing an HTTP message is received by that host.

20. a) Why are switches called Layer 2 devices? b) Why are routers called Layer 3 devices?

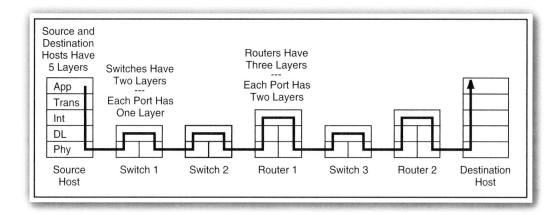

Figure 2-17   Layered End-to-End Communication

Standards and Protocols

In this chapter we have been looking at standards that govern interactions between hardware and software processes at the same layer but on different hosts. Such standards are called **protocols,** a term we have been using for such standards throughout this chapter.

> Protocols are standards that govern interactions between hardware and software processes at the same layer but on different hosts.

This terminology is shown in the names of the standards we have seen in this chapter, including the Internet *Protocol,* the Transmission Control *Protocol,* the Hypertext Transfer *Protocol,* and the Simple Mail Transfer *Protocol.* With a very few exceptions, all of the standards we will see in this book are protocols, whether they have "protocol" in their names or not.

Figure 2-18 shows the relationship between protocols and vertical communication on various devices. The protocols are shown as dotted lines to indicate that protocols provide *logical communication* between different processes at the same layer but on different hosts.

As noted earlier, application layer protocols and transport layer protocols are end-to-end protocols that only involve processes on the source and destination host. Lower-layer protocols are hop-by-hop protocols that involve processes on intermediate routers (internet layer processes) and on intermediate switches (data link layer processes).

TEST YOUR UNDERSTANDING

21.   a) What is a protocol? b) Are all standards protocols? Explain. c) Are almost all of the standards described in this book protocols? d) Why are protocols said to govern logical communication?

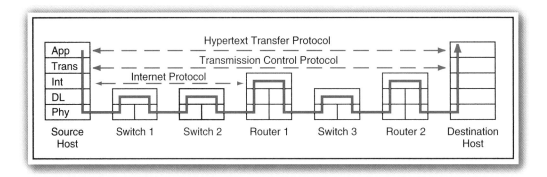

Figure 2-18 Protocols

STANDARDS ARCHITECTURES

It might be nice if there were only one set of standards that governed all network equipment. The reality, however, is that there are several competing **standards architectures,** which are families of related standards that collectively allow an application program on one machine on an internet to communicate with another application program on another machine on that same internet.

Standards architectures are families of related standards that collectively allow an application program on one machine on an internet to communicate with another application program on another machine on that same internet.

Unfortunately, hardware and software processes that use standards from one architecture cannot talk to processes that use standards from another architecture at the same layer.

To give a non-network example of competing standards architectures and incompatibility, different countries have different standards for their electrical systems (voltages, cycles per second, plug design, etc.). If you take your computer from one country to another, it may operate at the wrong voltage and with the wrong number of cycles per second. Plugging your PC into a wall socket might damage the computer beyond repair. Fortunately, you probably would not even be able to plug your computer into wall sockets because your power cord would not fit the wall jack.

TCP/IP and OSI Architectures

Although there are several major network standards architectures, two of them dominate actual corporate use: OSI and TCP/IP. Both are illustrated in Figure 2-20. Although they sometimes are viewed as competitors, we will see that they actually work together in most networks.

	OSI	TCP/IP
Standards Agency(ies)	ISO (International Organization for Standardization) ITU-T (International Telecommunications Union Telecommunications Standards Sector	IETF (Internet Engineering Task Force)
Dominance	Nearly 100% at physical and data link layers	70% to 80% at the internet, transport, and application layers
Documents are Called	Various	Mostly RFCs (requests for comments)

Notes:
Do not confuse OSI (the architecture) with ISO (the organization).
The acronyms for ISO and ITU-T do not match their names, but these are the official names and acronyms.

Figure 2-19   OSI and TCP/IP (Study Figure)

TCP/IP	OSI	Hybrid TCP/IP–OSI	Broad Purpose
Application	Application	Application (Layer 5)	Applications
	Presentation		
	Session		
Transport	Transport	Transport (Layer 4)	Internetworking
Internet	Network	Internet (Layer 3)	
Use OSI Standards Here	Data Link	Data Link (Layer 2)	Communication within a single LAN or WAN
	Physical	Physical (Layer 1)	

Notes:
The Hybrid TCP/IP-OSI Architecture is used on the Internet and dominates internal corporate networks.
OSI standards are used almost universally at the physical and data link layers (which govern communication within individual networks-LANs and WANs).
TCP/IP is used for 70% to 80% of all corporate traffic above the data link layer.

Figure 2-20   The Hybrid TCP/IP–OSI Architecture

TEST YOUR UNDERSTANDING

22. a) What are the two dominant network standards architectures? b) Are they competitors?

OSI

OSI is the "Reference Model of Open Systems Interconnection." "Reference model" is another name for "architecture." An open system is one that is open to communicating with the rest of the world. In any case, *OSI* is rarely written out.

Standards Agencies: ISO and ITU-T

Standards architectures are created and maintained by organizations called **standards agencies.** Figure 2-19 shows that OSI is governed by two cooperating standards agencies.

➤ One is the **International Organization for Standardization (ISO),** which generally is a strong standards organization for manufacturing, including computer manufacturing.[2]

➤ The other is the **International Telecommunications Union–Telecommunications Standards Sector (ITU-T).**[3] Part of the United Nations, the ITU oversees international telecommunications.

Although ISO or the ITU-T must ratify all OSI standards, other organizations frequently create standards for inclusion in OSI. For instance, we will see in Chapter 4 that the 802.3 Working Group of the IEEE 802 LAN/WAN Standards Committee creates Ethernet standards. These standards are then submitted for ratification to the ITU-T and ISO.

OSI's Dominance at Lower Layers (Physical and Data Link)

Although OSI is a seven-layer standards architecture (see Figure 2-20), standards from its five upper layers are rarely used.

However, at the two lowest layers, the physical and data link layers, OSI standards are used almost universally in networks. These two layers govern transmission within a single network. Almost all single networks—both LANs and WANs—follow OSI standards at the physical and data link layers, regardless of what upper-layer standards they use.

> Almost all single networks—both LANs and WANs—follow OSI standards at the physical and data link layers, regardless of what upper-layer standards they use.

Other standards agencies, recognizing the dominance of OSI at the physical and data link layers, simply specify the use of OSI standards at these layers. They then create standards only for internetworking and applications.

OSI Network and Transport Layers

The network and transport layers of OSI correspond closely to the internet and transport layers of TCP/IP that we saw earlier in this chapter. However, actual OSI standards at these layers are incompatible with actual TCP/IP standards and are rarely used anyway.

---

[2] *By the way, do not confuse OSI and ISO. OSI is an architecture.* ISO is a *standards agency.*
[3] No, the names and acronyms do not match for ISO and ITU-T, but these are the official names and acronyms for these two organizations.

OSI Session Layer

OSI differs most markedly from TCP/IP by having *three* layers for application standards instead of TCP/IP's single application layer.

The **OSI session layer (OSI Layer 5)** initiates and maintains a connection between application programs on different computers. For instance, if a single transaction requires a number of messages, if there is a network connection break, the transmission can begin at the last session layer checkpoint instead of having to restart the transaction entirely. Figure 2-21 illustrates this layer.

The OSI session layer is especially good for database applications, which involve complex series of transactions. TCP/IP does not have a session layer, requiring application programs to create, maintain, and terminate application layer transactions if they have the need to do so. (For instance, HTTP uses cookies to keep track of the state of a transaction series.) However, few application programs have this need, so the OSI session layer tends to add complexity without comparable benefits.

OSI Presentation Layer

The **OSI presentation layer (OSI Layer 6)** is designed to handle data formatting differences between the two computers. For example, most computers format character data (letters, digits, and punctuation signs) using the ASCII code. In contrast, IBM mainframes format them using the EBCDIC code. Computers differ even more widely in the ways they represent numerical data.

Transfer Syntax   As Figure 2-22 shows, the two presentation layer processes communicate using a common **transfer syntax,** which may (or may not) be quite different than either of their internal methods of formatting information. It is as if native English and German speakers, who both also speak French, agree to communicate in French.

Of course, if both parties are native English speakers, they will communicate in English. In OSI, in turn, if two computers use the same internal data formatting, the presentation syntax will be the same as their internal syntaxes.

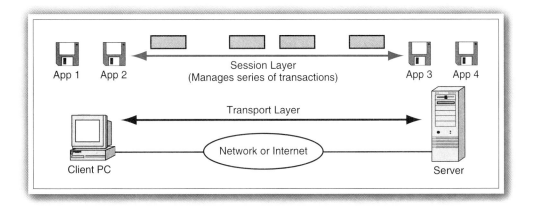

App 1   App 2

Session Layer
(Manages series of transactions)

App 3   App 4

Transport Layer

Network or Internet

Client PC

Server

Figure 2-21   OSI Session Layer

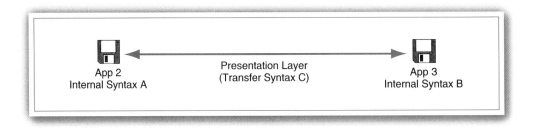

Figure 2-22   OSI Presentation Layer

Compression and Data Formatting   The OSI presentation layer also can handle data compression and data encryption for application data. In TCP/IP standards, application standards have to include their own compression and encryption processes if these services are desired.

File Format Standards   In general, the presentation layer is where many OSI application file format standards are placed, such as those for facsimile.

MIME in TCP/IP   TCP/IP has nothing like the OSI presentation layer's intermediate presentation syntax capabilities. Different TCP/IP applications have different data format standards, causing frequent problems for users. Still, the way the OSI presentation layer works adds a good deal of overhead to each application, and it would rarely be desirable in practice.

In turn, the TCP/IP MIME standard we will see in Chapter 12 at least notifies the receiver of a message's file format.

OSI Application Layer

Remaining application-specific matters are governed by the **OSI application layer (OSI Layer 7).** The OSI application layer, freed from session and presentation matters, focuses on concerns specific to the application in use.

TEST YOUR UNDERSTANDING

23.   a) What standards agencies are responsible for the OSI standards architecture? b) In which layers are OSI standards dominant? c) Name and describe the functions of OSI Layer 5. d) Name and describe the functions of OSI Layer 6.

TCP/IP

The TCP/IP architecture is mandatory on the Internet at the internet, transport, and application layers. TCP/IP is also widely used at these layers by companies for their internal corporate networks.

TCP/IP is named after two of its standards, TCP and IP, which we have looked at briefly in this chapter. However, TCP/IP also has many other standards (as we will see in Chapter 8), making its name rather misleading. Note that TCP/IP is the standards architecture while TCP and IP are individual standards within the architecture.

> Note that TCP/IP is the standards architecture while TCP and IP are individual standards within the architecture.

The Internet Engineering Task Force (IETF)
TCP/IP's standards agency is the **Internet Engineering Task Force (IETF).**[4]

Requests for Comments (RFCs)
Most documents produced by the IETF are given the rather misleading name **requests for comments (RFCs).** Every few years, the IETF publishes a list of which RFCs are **Official Internet Protocol Standards.** RFCs are added to and dropped from this list over time.

The IETF traditionally has been rather informal.[5] Its committees traditionally focus on rough consensus rather than on voting, and technical expertise is the source of most power. There even is a tradition of publishing joke RFCs on April 1 of each year (April Fools' Day). Although corporate participation has somewhat "tamed" the IETF, it remains a fascinating organization, although its members might argue that "organization" is too strong a word.

Dominance at Upper Layers (Internet, Transport, and Application)
As noted earlier, physical and data link layer standards govern the transmission of data within a *single network*. We saw earlier that OSI standards are completely dominant at these layers.

TCP/IP internet and transport layer standards, in turn, govern transmission across an entire internet, ensuring that any two host computers can communicate. TCP/IP application standards ensure that the two application programs on the two hosts can communicate as well.

TCP/IP is dominant in corporate networking above the physical and data link layers. It is not nearly universally used, as OSI standards are at the bottom two layers. TCP/IP is by far the most widely used architecture at upper layers, and its dominance is growing. In most organizations, TCP/IP standards govern 70 percent to 80 percent of internet and transport layer traffic. In a few years, the use of other architectures at upper layers in new products will be increasingly rare. TCP/IP standards are widely used at the application layer as well.

> TCP/IP is dominant in corporate networking at the internet and transport layers, although less dominant than OSI is at the physical and data link layers. In most organizations, TCP/IP standards govern 70 percent to 80 percent of internet and transport layer traffic.

---

[4] Traditionally, the IETF has been viewed as being in competition with ISO and ITU-T for standards development. However, in recent years the IETF and these other organizations have begun to cooperate in standards development. For instance, the IETF is working closely with ITU-T for voice over IP transmission standards, as discussed in Chapter 6.

[5] In 1992, IETF member Dave Clark summarized the situation this way: "We reject kings, presidents, and voting. We believe in rough consensus and running code."

TEST YOUR UNDERSTANDING

24. a) What is the standards agency for TCP/IP? b) What are most of this agency's documents called? c) In which layers is TCP/IP dominant? d) How dominant is TCP/IP today at these layers compared to OSI's dominance at the physical and data link layers? e) Which of the following is an architecture: TCP/IP, TCP, or IP? f) Which of the following are standards: TCP/IP, TCP, or IP?

TCP/IP and OSI

Although TCP/IP and OSI sometimes are viewed as competing standards architectures, most organizations use them together. The most common standards pattern in organizations is to use OSI standards at the physical and data link layers and TCP/IP standards at the internet, transport, and application layers. This is very important for you to keep in mind because this **hybrid TCP/IP–OSI standards architecture** shown in Figure 2-20 will form the basis for most of this book.

The hybrid TCP/IP–OSI standards architecture—with OSI standards at the physical and data link layers and TCP/IP standards at the internet, transport, and application layers—forms the basis for most of this book.

TEST YOUR UNDERSTANDING

25. a) What layers of the hybrid TCP/IP–OSI standards architecture use OSI standards? b) What layers use TCP/IP standards? c) Do LAN standards come from OSI or TCP/IP? (The answer is not explicitly in this section.) d) Do WAN standards come from OSI or TCP/IP? (Again, the answer is not explicitly in this section.)

A Multiprotocol World at Higher Layers

At the same time, quite a few networking products (especially legacy products) in organizations follow other architectures shown in Figure 2-23. Real corporations live and will continue to live for some time in a **multiprotocol** world in which network

IPX/SPX

Used by older Novell NetWare file servers for file and print service

Sometimes used in newer Novell NetWare file servers for consistency with older NetWare servers

SNA (Systems Network Architecture)

Used by IBM mainframe computers

AppleTalk

Used by Apple Macintoshes

Figure 2-23   Other Major Standards Architectures

administrators have to deal with a complex mix of products following different archi-tectures above the data link layer. In this book we must focus on OSI and TCP/IP because they are by far the most important and are becoming ever more so. However, in a typical organization, 20 percent to 30 percent of all upper-layer traffic still uses protocols from other standards architectures.

IPX/SPX
The most widely used non-TCP/IP standards architecture found at upper layers in LANs is the **IPX/SPX architecture.** This architecture is required by all older Novell NetWare file servers and is a widely used option on new NetWare file servers to main-tain consistency with older servers. NetWare is still used in many firms for file and print service, although its use is waning, and a growing number of NetWare users are switching to TCP/IP.

Systems Network Architecture (SNA)
IBM Mainframe computers traditionally use the **Systems Network Architecture (SNA)** standards architecture, which actually predates OSI and TCP/IP. Most firms, however, are transitioning their mainframe communications from SNA to TCP/IP.

AppleTalk
Macintosh computers are designed to use Apple's proprietary **AppleTalk** architecture. Macintoshes are rare in corporations today, so AppleTalk is not widely seen. However, some industries, such as publishing, use Macintoshes and AppleTalk extensively. They do this to frustrate textbook writers.

TEST YOUR UNDERSTANDING

26.   a) Under what circumstances would you encounter IPX/SPX standards? b) SNA stan-dards? c) AppleTalk standards?

Standards at the First Bank of Paradise
Physical and Data Link Layer Standards
At Layers 1 and 2, the First Bank of Paradise uses OSI standards almost exclusively—although the bank still has a few legacy systems built on data link layer standards that predate OSI.

However, this does not mean that selecting specific standards at these layers is simple. In LANs, Ethernet dominates, but the bank still has a few older legacy LANs as well. As we will see in Chapter 7, the bank uses several different OSI standards for its WAN traffic.

Protocols at Higher Layers
At higher layers, there is nothing like the dominance of OSI. As with most other firms, TCP/IP standards govern more than 70 percent of the bank's traffic above the data link layer. However, the bank still has a good deal of SNA traffic because of its IBM mainframe traffic. The bank also has some upper-layer traffic that is governed by stan-dards from other architectures.

At last count, upper-layer transmission for an average branch office used nine different internet and transport layer standards from three different architectures. A tenth upper-level standard used by the average branch was created before standards architectures emerged in the 1960s!

SYNOPSIS

Orientation

In this chapter we looked broadly at standards. Most of this book (and the networking profession in general) focuses on standards.

Standards govern message exchanges. More specifically, they place constraints, on message semantics (meaning), message syntax (format), and message timing (when each party may transmit).

Standards are connection-oriented or connectionless, as Figure 2-24 illustrates. In connection-oriented protocols, there are distinct openings before content messages are sent and distinct closings afterward. There also are sequence numbers. In connectionless protocols, messages are sent without such openings and closings. Connectionless protocols are simpler but lose the advantages of sequencing.

In turn, reliable protocols do error correction, while unreliable protocols do not (although unreliable protocols may do error detection without error correction). In general, standards below the transport layer are unreliable to reduce costs. The transport standard usually is reliable; this allows error correction processes on just the two hosts to correct errors at the transport layer and at lower layers.

Standards are created within broad plans called standards architectures. Most networks use a five-layer architecture in which the physical layer governs transmission between adjacent devices, the data link layer governs transmission between two devices on a single network, the internet layer governs transmission between the source and destination hosts on an internet, the transport layer governs communication between two hosts (and usually corrects errors at lower layers), and the application layer governs interactions between application programs.

Layer	Protocol	Connection Oriented or Connectionless?	Reliable or Unreliable?	Strong or Weak Timing Constraints?
5 (Application)	HTTP	Connectionless	Unreliable	Weak
4 (Transport)	TCP	Connection-oriented	Reliable	Strong
3 (Internet)	IP	Connectionless	Unreliable	Weak
2 (Data Link)	Ethernet	Connectionless	Unreliable	Weak

Figure 2-24   Characteristics of Protocols Discussed in the Chapter

Transmission in single networks—both LANs and WANs—is governed by physical and data link layer standards. Ethernet is a fairly simple standard for physical and data link layer transmission in LANs. Ethernet is connectionless, unreliable, and has no timing constraints when switches are used. This simplicity leads to low costs, and low costs have brought Ethernet to dominance in LANs. The Ethernet frame's syntax usually is represented by showing each field's length in octets. Ethernet addresses are 48 bits long and are expressed for human reading in hexadecimal notation.

Internetworking is governed by internet and transport layer standards. At the internet layer, the Internet Protocol (IP) also is a fairly simple connectionless and unreliable protocol with no timing constraints. IP depends on TCP for error correction and even for the correct ordering of received packets. The IP packet's syntax usually is represented as a series of lines with 32 bits on each line. IP addresses are 32 bits long and are expressed for human reading in dotted decimal notation.

Note that messages at the data link layer are called frames, while messages at the internet layer are called packets. If two hosts are separated by ten networks, there will be ten frames along the way but only one packet.

The Transmission Control Protocol (TCP) is a complex transport layer protocol that is connection-oriented, reliable, and has many timing constraints. The receiving transport process acknowledges every correct TCP segment. If a segment is not acknowledged promptly, the sender retransmits it.

HTTP is a simple connectionless and unreliable protocol, which depends on TCP at the transport layer for reliable data transmission. HTTP has a simple text-based syntax for its headers.

Protocols are standards that govern interactions between hardware and software processes at the same layer but on different hosts. Most standards we will see in this book are protocols.

Standards processes at each layer also communicate vertically with the process in the layer directly above them and the process in the layer directly below them on the same computer. On the sending device, a process usually creates a message and then passes the message down to the next-lower-layer process, which encapsulates the message in the data field of its own message. On the receiving process, the reverse process (decapsulation) is used. Hosts are five-layer devices, while switches are Layer 2 devices and routers are Layer 3 devices.

Standards are created by standards agencies, which first create standards architectures that guide the creation of individual standards. Standards from different architectures are incompatible. The dominant standards architecture used in organizations today is the TCP/IP–OSI architecture with OSI standards being used at the bottom two (physical and data link) layers and TCP/IP standards being used at the upper (internet, transport, and application) layers.

Almost no non-OSI standards are used at the physical and data link layers. However, most firms are multiprotocol environments in which 20 percent to 30 percent of all traffic at upper layers comes from standards architectures other than TCP/IP—usually IPX/SPX (for some Novell NetWare file servers), SNA (for IBM mainframe communications), and AppleTalk (for Macintosh computers).

SYNOPSIS QUESTIONS

1. What three aspects of message exchange do standards govern?
2. a) What is the benefit of reliable protocols? b) What are the benefits of unreliable protocols? (If you refer to costs, be specific.)
3. a) Of the four major protocols discussed in this chapter—HTTP, Ethernet, TCP, and IP—which are connectionless? b) Which are unreliable? c) Which have weak timing constraints? d) What protocol is different than the others on these three dimensions?
4. a) Messages are just long strings of bits. How can a receiver know where fields end in HTTP and in Ethernet? b) Compare and contrast how addresses are represented for humans in Ethernet and IP.
5. a) What layers govern transmission in LANs and WANs? b) What layers govern internetworking? c) At what layer are messages called frames? d) At what layer are they called packets? e) If a source host and a destination host are separated by 23 networks, how many packets will there be? f) How many frames will there be along the way?
6. Which device encapsulates—the sending or the receiving device?
7. a) How are OSI and TCP/IP complementary? b) Compare the dominance of OSI at the physical and data link layers with the dominance of TCP/IP at higher layers.

THOUGHT QUESTIONS

1. How do you think TCP would handle the problem if an acknowledgement were lost, so that the sender retransmitted the unacknowledged TCP segment and therefore the receiving transport process received the same segment twice? By the way, TCP segments have sequence numbers.
2. You can place both TCP/IP clients and servers and IPX clients and servers on the same Ethernet network, and each client will talk to its server. How do you think this is possible? (Hint: Consider the Ethernet frame in Figure 2-12.)

HANDS ON: THE TCP/IP PLAY

Learning Objectives:

By the end of this chapter, you should be able to:

- Have a better understanding of how vertical layering and horizontal communication work together.
- Evaluate whether you are destined to be an actor rather than a technologist.

INTRODUCTION

In Chapter 2 we looked at two types of communication: horizontal and vertical. Horizontal communication occurs between processes *at the same layer* but *on different computers*. These horizontal communications are governed by protocols. We also saw vertical communication between processes *on the same computer at different (but adjacent) layers*. Of course, in any real networked communication, there is both horizontal and vertical communication.

It is extremely difficult to envision how vertical and horizontal communication work together just by reading about them. To help you, this chapter presents a TCP/IP play you can perform with five of your classmates to give you an experiential understanding of how vertical and horizontal exchanges interact.

THE CAST OF CHARACTERS

There are six roles in the play. They are shown in Figure 2a-1. The play should be acted six times, with students rotating among all the roles.

Application Layer

The first two roles are the application processes on the two machines. One computer is a client PC. Its application program is a browser. The browser communicates via HTTP with the webserver application on the other computer, the server.

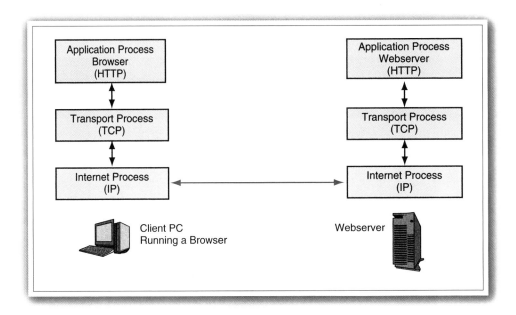

Figure 2a-1 Roles in the TCP/IP Play

Transport Layer

The third and fourth roles are the transport processes on the two machines. HTTP requires TCP at the transport layer. In Chapter 2, Figure 2-6 shows a series of interactions between the TCP processes on a client PC and a server. The job of the group will be to go through a connection with two HTTP request–response cycles.

Internet Layer

The fifth and sixth roles are the internet layer processes on the two machines. Internet layer processes are governed by the simple IP protocol.

Data Link Layer and Physical Layer

Of course, there also are data link layer processes and physical layer processes, but we are going to cheat in order to keep the cast size down. The internet layer processes on the client PC and the server will be allowed to pass IP packets *directly back and forth between them.*

MESSAGES

Application Messages

The application layer actors will use 3-by-5 index cards to represent HTTP messages.

HTTP Request Messages

The browser sends two HTTP response messages. Call the files that you will request Red.htm and Blue.htm. You will need two cards—one for each request. Both cards should have two lines. The first should say GET *filename,* where *filename* is the name of the file being requested (Red.htm or Blue.htm). The second line on each card should say Host: Pukanui.com. Pukanui.com is the name of the webserver.

HTTP Response Messages

The server sends two HTTP response messages. You will need a card for each response message. The card should have two lines. The first should say OK *filename,* where *filename* is the name of the file being requested. The second line should say Enclosed: *filename.* This will represent the file, which is enclosed in the HTTP response message. You will need two index cards—one for Red.htm and one for Blue.htm.

TCP Segments

Letter or Legal Envelope

Transport layer messages—TCP segments—will be represented by envelopes large enough to contain the 3-by-5 index cards representing HTTP application messages.

Port Numbers

As discussed in Chapter 8, each transport process has a port number for each connection. For the client PC, the port number is 55555. For the webserver transport process, the port number is 80.

Client PC TCP Segments

On the envelope representing the TCP segment, the client transport process should begin with three lines—Host: server, Source Port: 55555, and Destination Port: 80.

Server TCP Segments

On the envelope representing the TCP segment, the server transport process should write Host: Client PC, Source Port: 80, and Destination Port: 55555 on the first three lines.

Flag

After the Host, Source Port, and Destination Port lines, envelopes should also have flag lines, which describe the type of TCP segment being sent.

> ➤ **Flag: Data** means that an application message is contained in the envelope. Two of these envelopes should be sufficient for each of the two transport processes.

> ➤ **Flag: ACK** indicates that the TCP segment is a pure acknowledgement. Four of these envelopes should be sufficient for each of the two transport processes.

> ➤ **Flag: Data/ACK** is for TCP segments that both carry data and are acknowledgements. Four of these envelopes will be sufficient for each of the two transport processes.

> ➤ **Flag: SYN, Flag: SYN/ACK,** and **Flag: FIN** are self-explanatory. The client PC transport process will need a SYN and FIN envelope. The server transport process will need a SYN/ACK envelope.

IP Packets

Larger Envelopes

IP packets will be represented by envelopes large enough to hold a TCP segment envelope.

Host IP Addresses

IP packets use host IP addresses. The client IP's address is 1.2.3.4. The server's IP address is 123.4.5.6.

Client IP Packets

In envelopes simulating client-transmitted IP packets, the first three lines should be To: 123.4.5.6, From: 1.2.3.4, and Protocol: TCP. The last tells the internet process on the webserver that the IP packet (large envelope) contains a TCP segment. The client IP process should have six of these envelopes.

Server IP Packets

In envelopes simulating client-transmitted IP packets, the first three lines should be To: 123.4.5.6, From: 1.2.3.4, and Protocol: TCP. The last tells the internet layer process on the client PC that the IP packet (large envelope) contains a TCP segment. The server IP process should have six of these envelopes.

ENACTING THE PLAY

Basic Action

In the play the action will pass from one actor to another. Each actor in control of the action will announce what he or she will be doing. (An emotional reading is not required.)

Other cast members will either concur or object. If the others concur, the actor in control will then take the action, which typically consists of putting something into an envelope and passing the envelope down to the next-lower-level process or taking something out of an envelope and passing the decapsulated message to the next-higher-layer process.

The one exception is that actors playing the internet processes on the two hosts can pass IP packet envelopes directly to each another. This omits lower-layer processes and processes on intermediary devices such as clients and routers. It is done to minimize the number of actors in the cast.

To Begin

For example, the action is initially controlled by the actor playing the client HTTP application process.

➤ The actor playing the role of the HTTP process on the client picks up the HTTP request card for Red.htm and announces that he or she will pass it down to the client transport process.

➤ If other actors concur, the HTTP process will pass the HTTP request card down to the transport layer process.

➤ However, if even one other actor objects, the entire group will discuss whether the actor in control's action is appropriate technically. (However, the actor's dramatic reading style cannot be challenged no matter how bad.)

The actor playing the transport layer process on the client PC will now be in control of the action. He or she will then announce his or her next action. Other members will concur or object.

Deliverables

One cast member in each round—the actor playing the HTTP application process on the webserver—should record the first twenty steps of each round. Each step should include the step number, the process and machine, and a brief but accurate description of what the actor did. This recording is a difficult job, so the recorder has the absolute right to ask all cast members to pause while he or she finishes writing. The teacher may award extra points for classifying whether the round was a drama, comedy, or tragedy. The six write-ups should be handed in. The following table's format should be used.

Step	Actor	Action
1	Browser on Client PC	Takes card that is HTTP Request Message for Red.htm. Passes message down to transport process on client PC.

PHYSICAL LAYER PROPAGATION

Learning Objectives:

By the end of this chapter, you should be able to discuss:

- Binary data representations for important types of data.
- Binary signaling (on/off signaling or using two possible voltage states) versus digital signaling, including the difference between bit rates and baud rates.
- Unshielded twisted pair (UTP) wiring, including relevant propagation effects that must be controlled by limiting cord length and by limiting the untwisting of pairs during connectorization.
- The differences between serial and parallel transmission, including the speed advantage of parallel transmission.
- Optical fiber cabling, including wavelength division multiplexing, relevant propagation effects, and different types of optical fiber cabling and signaling.
- Network topologies, including point-to-point connections, stars, extended stars (hierarchies), rings, meshes, and busses.
- Building wiring for telephone service and data transmission.

INTRODUCTION

The Physical Layer

Chapter 2 presented an overview of layered standards. Most of that chapter focused on the data link, internet, transport, and application layers. In this chapter we will focus more closely on Layer 1, the physical layer, which differs from upper layers in two ways:

- ➤ It is the only layer that does not use messages. It merely transmits bits in isolation.
- ➤ It alone deals with propagation effects that change signals when they travel over transmission lines.

This chapter will cover physical layer signaling and two major transmission media: UTP and optical fiber. We will look at radio propagation in Chapter 5.

SIGNALING

Disturbances and Propagation

Jiggling a Rope

Suppose that you and someone else are standing a few meters apart and are holding a rope tautly between you. Now, close your eyes. The other person jiggles the rope for a fraction of a second. That disturbance will **propagate** (travel) down the rope to your hands. When it arrives, you will feel it.

Signal Creation and Propagation

Something similar happens in network transmission. As Figure 3-1 shows, the transmitter creates a disturbance in a transmission medium—wire, optical fiber, or radio. This disturbance is the signal. The **signal,** then, is a disturbance that propagates down the transmission medium to the other side, where it is received.

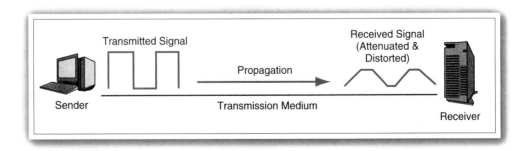

Figure 3-1   Signal and Propagation

Propagation Effects

Note that the received signal is different from the transmitted signal because of **propagation effects,** that is, changes in the signal during propagation. In the figure, the signal has **attenuated** (weakened) and is **distorted** (its shape has been changed). If propagation effects are too large, the receiver will not be able to read the signal correctly.

TEST YOUR UNDERSTANDING

1. a) What is a signal? b) What is propagation? c) What are propagation effects? d) Why are propagation effects bad?

Binary Data

The messages that we saw in the previous chapter are long series of bits. This type of data only has two possible values (ones and zeros), so it is called **binary data** (from the Greek word for two).

Inherently Binary Data

Certain types of data are inherently binary. For instance, Ethernet addresses are 48-bit binary strings.

Encoding Binary Numbers

Some message fields contain numbers. Whole numbers (integers) can be represented as simple Base 2 **binary numbers.** Figure 3-2 illustrates binary counting and arithmetic.

With binary numbers, counting begins with zero. This is a source of frequent confusion.

A 1 is added to each binary number to give the next binary number. There are four simple rules for addition in binary, as Figure 3-2 illustrates.

➤ If you add 0 and 0, you get 0.
➤ If you add 0 and 1, you get 1.
➤ If you add 1 and 1, you get 10 (carry the one).
➤ If you add 1, 1, and 1, you get 11 (carry the one).

These rules are simple to use, as the figure illustrates.

➤ For example, eight is 1000.
➤ Nine adds a one to the final zero, giving a one, so nine is 1001.
➤ Adding one to the final one gives us 10, so ten is 1010.
➤ Eleven adds a one to the final zero, giving 1011.
➤ Twelve adds a one to the final one. With carries, this gives 1100.

Encoding Alternatives Fields

Sometimes a field represents one of several alternatives, such as site names in a corporation or product numbers. How many possible alternatives can a field represent? The answer depends on the field's length. If a field is N bits long, it can represent 2^N possible alternatives.

If a field is N bits long, it can represent 2^N possible alternatives.

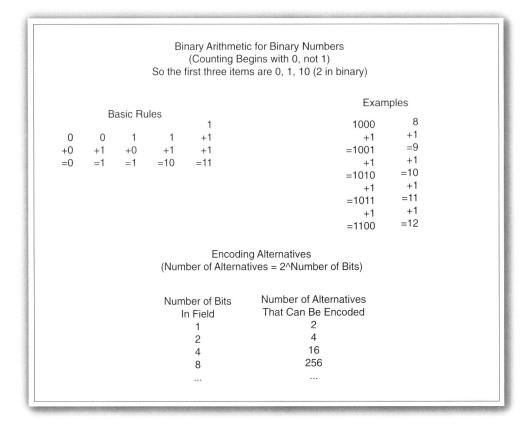

Figure 3-2   Binary Data

> ➤ If the field is only one bit long, it can only represent 2^1 (2) alternatives. For instance, a one-bit gender field might represent female by 1 and male by 0.
> ➤ If the field is two bits long, it can represent 2^2 (4) possibilities, representing each by 00, 01, 10, or 11. For instance, a two-bit season field might represent Spring by 00, Summer by 01, Autumn by 10, and Winter by 11.
> ➤ A one-octet field can represent 2^8 (256) alternatives.[1]
> ➤ A two-octet field can represent 2^{16} (65,536) alternatives.

Numbering Alternatives

Remember that binary counting begins with zero. So if you have 8 bits, there are 256 possibilities. These begin with 0 and end with 255.

---

[1]   Text usually is represented by the ASCII code, whose individual symbols are seven bits long but are usually stored as whole bytes. Seven bits gives 128 possibilities. This is enough for capital letters (A is 1000001), lowercase letters (a is 1100001), digits (3 is 0110011), other characters (@ is 1000000), and printing control (carriage return is 0001101 and line feed is 0001010).

TEST YOUR UNDERSTANDING

2. a) Give the binary representations for 13, 14, 15, 16, and 17 by adding one to successive numbers (12 is 1100). b) Rounding off, about how many possible addresses can you represent with 32-bit IP addresses? (You probably will need a spreadsheet program to answer this question.) c) If you have four bits, how many possibilities can you represent? d) With four bits, what is the smallest binary number? e) What is its decimal equivalent? f) What is the largest binary number if you have four bits? g) What is its decimal equivalent?

Signaling

To be transmitted, the data stream's bits must be converted into signals that will convey the meaning of the data. Figures 3-3 through 3-5 show three basic ways to do signaling for binary data.

On/Off Signaling

The simplest way to signal is to divide time into brief **clock cycles** and to have each clock cycle represent one bit. A signal can be turned on for a one during the clock cycle and off for a zero. Think of turning a flash light on and off to send a signal. Optical fiber generally uses **on/off signaling.**

Binary Signaling

On/off signaling is an example of **binary signaling,** which is characterized by using two possible states to represent information (ones and zeros). The two possible states are *on* and *off.*

Binary signaling is characterized by using two possible states to represent information (ones and zeros).

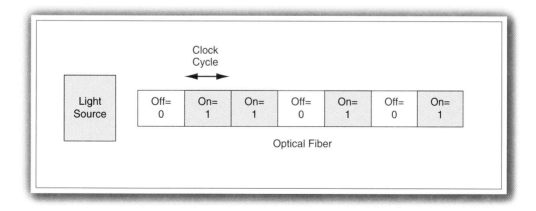

Figure 3-3   On/Off Signaling

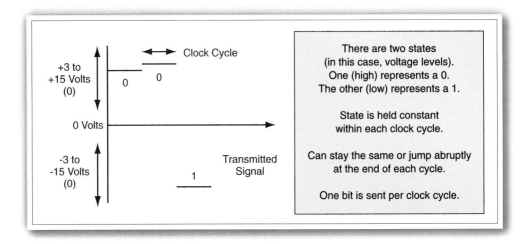

Figure 3-4   Binary Signaling

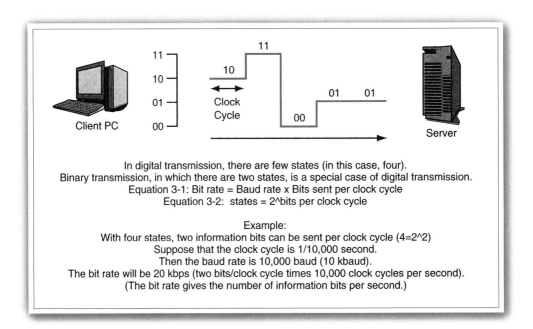

Figure 3-5   Digital Signaling

Binary Voltage Signaling

When signals are sent over wires, it is common to use two different voltages to represent ones and zeros, as Figure 3-4 illustrates. This is also binary signaling.

232 Serial Ports   In this figure a high voltage is anything between 3 and 15 volts. A low voltage, in turn, is anything between negative 3 and negative 15 volts. A high voltage represents a zero, and a low voltage represents a one. No, this is not a misprint. This is the way a **232 serial port** works on your PC.[2]

Constancy During Each Clock Cycle   As in on/off signaling, time is divided into brief clock cycles. The signal is held constant within each clock cycle. This constancy allows the receiver to read the voltage at any time during the clock cycle. Even if the receiver is slightly off on the timing of its read for a bit, the receiver will still read the value correctly.

At the end of each clock cycle, the signal either stays the same or changes to the other voltage level (state).

Relative Immunity to Errors   Binary signal transmission is relatively immune to attenuation errors (loss in signal intensity). If a signal is transmitted at +12 volts and attenuates to +6 volts (a 50 percent loss), it will still be read correctly as a high voltage and therefore a zero.

Digital Signaling

Figure 3-5 illustrates a slightly more complex form of signaling, digital signaling. In binary signaling, there are only *two* possible states (in the case of 232 serial ports, voltage levels) to represent information. In **digital signaling,** in contrast, there are *a few* possible states; the state (voltage level in this example) is held constant during the clock cycle. The figure specifically shows digital signaling with four possible states, but digital signaling can use even more possible states: eight, sixteen, thirty-two, and occasionally (but rarely) more.

Note also that binary signaling is a special case of digital signaling. If "few" is "two," then digital signaling is binary.

In digital signaling, there are *a few* possible states.

Binary signaling is a special case of digital signaling.

Why use more than two possible states? The answer is that you can send multiple bits per clock cycle if you have more than two possible states. Binary transmission only sends one bit per clock cycle (a one or a zero). With four possible states, however, there can be four possibilities. If we let the four voltage levels represent 00, 01, 10, and 11, then we can send two bits per clock cycle—doubling the transmission rate.

Adding more possible states allows even more bits to be sent per clock cycle. However, there are diminishing returns. Each doubling in the number of possible states allows only one more bit to be transmitted per clock cycle. For instance, having

---

[2]   Sometimes called RS-232-C serial ports, they actually follow the newer ANSI/TIA/EIA-232-F standard. In Europe equivalent ports are specified by three standards: V.24, V.28, and ISO-2110.

eight possible states allows only three bits to be transmitted per clock cycle, and having sixteen possible states allows only four bits to be transmitted per clock cycle.

If too many possible states are used, furthermore, the possible states will be very close together. If the signal changes even slightly during transmission, the receiver will record the wrong state. This is why digital signaling only uses a *few* possible states.[3] In essence, sending more bits increases transmission rates but results in more errors. Beyond some number of possible states, the net transmission rate actually falls because of the need to retransmit incorrect transmissions.

Bit Rates and Baud Rates

Bit Rates   In digital data transmission, the rate at which we transmit data is called the **bit rate.** It is measured in bits per second. This is what users care about.[4]

Baud Rate   In turn, the **baud rate** is the number of clock cycles the transmission system uses per second. If there are 1,000 clock cycles per second, the baud rate is 1,000 baud (not 1,000 bauds per second).

The bit rate is the rate at which we transmit data.

The baud rate is the number of clock cycles the transmission system uses per second.

An Example   To compute the bit rate in digital signaling, you must know two equations. The first is simple: the bit rate is the number of bits sent per clock cycle times the baud rate. For instance, if a line changes its state one hundred times per second, the baud rate is 100 baud. If it can send three bits per clock cycle, then the bit rate is 300 bits per second.

Equation 3-1:   Bit rate = Baud rate $\times$ Bits sent per clock cycle

Second, the number of bits sent per clock cycle, in turn, is given by the equation states = $2^{\text{bits per clock cycle}}$, where states is the number of possible states and bits per clock cycle is the number of bits that are sent per clock cycle. If you wish to send three bits per clock cycle, you will need 2^3 possible states, that is, eight. Alternatively, if you have four possible states, 2^2 is four, so you can send 2 bits per second.

Equation 3-2:   states = $2^{\text{bits per clock cycle}}$

For example, if there are 10,000 clock cycles per second, the baud rate is 10,000 baud. If there are four possible states, then Equation 3-2 says that two bits can be sent

---

[3]   Originally, digital signaling used ten states; it was called digital because our ten fingers are called digits.

[4]   Purists note that "speed" is not an appropriate term, because data transmission speed in bits per second does not represent velocity. Signals carrying data at 10 Mbps and 100 Mbps both propagate down a wire or optical fiber at the speed of light in that material (about 60 percent to 70 percent of the speed of light in vacuum). However, the 10 Mbps signal, if binary, can change its stated 10 million times per second, while the 100 Mbps signal, if binary, can change its state 100 million times per second. Higher transmission "speed" is like talking faster, not running faster. However, describing the bit rate as transmission speed is well-established in practice.

per clock cycle. By Equation 3-1, the bit rate will be two bits per clock cycle times 10,000 baud (10 kbaud) or 20,000 bits per second (20 kbps).

TEST YOUR UNDERSTANDING

3. a) In binary 232 serial port transmission for transmission over a binary transmission line, how are ones and zeros represented? b) How does this give resistance to transmission errors? c) A signal is sent at 10 volts in a 232 serial port. What percentage of its strength can it lose before it becomes unreadable?

4. a) Distinguish between binary and digital transmission. b) Is binary transmission digital? c) What is desirable about having multiple possible states instead of just two? d) What is undesirable about having multiple possible states? e) How many more bits can you send every time you double the number of possible states?

5. a) Distinguish between the bit rate and the baud rate. b) When are the two equal? (You'll have to think about this one.) c) If you have 10,000 clock cycles per second and transmit in binary, what is the baud rate? d) What is the bit rate? e) If instead you use sixteen possible voltage levels for digital signaling, what is the baud rate? f) What is the bit rate? g) To transmit 30 kbps over a 10 kbaud line, how many possible states will you need?

UNSHIELDED TWISTED PAIR (UTP) WIRING

Having looked at propagation effects and signaling in general, we will now see how these concepts apply to UTP transmission. Later, we will see how they apply to optical fiber transmission.

UTP Transmission Standards

LANs need transmission links to connect NICs to switches and switches to one another. Transmission media in the United States are governed by the **TIA/EIA-568** standard.[5] We will look first at UTP wiring, which is the most widely used LAN wiring. Later, we will look at optical fiber transmission.

TEST YOUR UNDERSTANDING

6. What is the main standard for transmission media (both UTP and optical fiber)?

4-Pair UTP and RJ-45

The 4-Pair UTP Cable

Ethernet networks typically use **4-pair unshielded**[6] **twisted pair (UTP)** wiring. This name sounds complicated, but the medium is very simple, as Figure 3-6 illustrates.

> ➤ A length of UTP wiring is called a **cord.**
> ➤ Each cord has eight copper wires, organized as four pairs.

---

[5] TIA is the Telecommunications Industry Association. EIA is the Electronic Industrial Alliance. The standard is a cooperative effort.

[6] Some types of wiring use shielding, in which a metal mesh is placed around each pair of wires and around the jacket. This reduces interference, which is discussed later in this chapter. However, shielding is expensive, so almost all UTP wiring used in organizations today is unshielded, although USB cables usually are shielded.

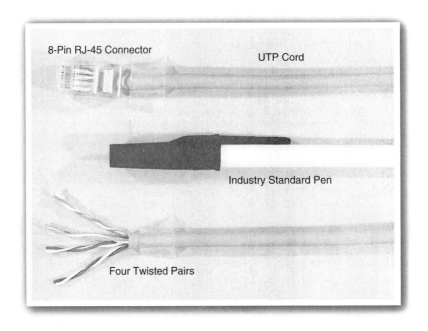

Figure 3-6   Unshielded Twisted-Pair (UTP) Cord with RJ-45
Connector, Pen, and UTP Cord with 4 Pairs Displayed

> ➤ Each wire is covered with **dielectric** (nonconducting) **insulation.** This prevents
> short circuits between the electrical signals traveling on different wires.
> ➤ Each pair's two wires are twisted around each other several times per inch.
> ➤ There is an outer plastic **jacket** that encloses the four pairs.

RJ-45 Connectors

At the two ends of a UTP cord, the wires must be separated and placed within an 8-pin
RJ-45 connector, which is shown in Figure 3-7. The RJ-45 connector at each end of a
4-pair UTP cord snaps into an **RJ-45 jack** (port) in the NIC or the switch.[7]

Easy, Inexpensive, and Rugged

UTP is inexpensive to purchase, easy to **connectorize**[8] (add connectors to), and relatively easy to install. It is also rugged, so that if a chair runs over it accidentally, it probably will not be damaged. Four-pair UTP dominates corporate usage in access links
from the NIC to the first switch because of UTP's low cost and durability (access links
are frequently exposed to harsh treatment in the office).

---

[7] Home telephone connections use a thinner RJ-11 connector and jack. They were designed to terminate
six wires but usually terminate only a single pair.

[8] Yes, "connectorize" is a really ugly term. Believe me, I didn't make it up.

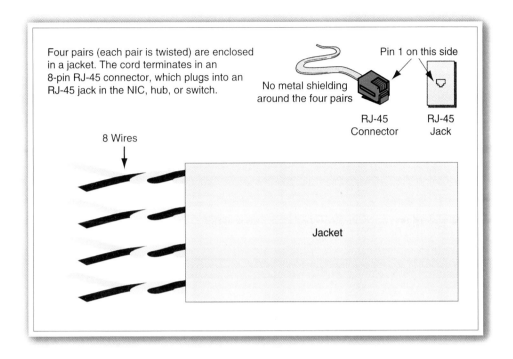

Four pairs (each pair is twisted) are enclosed in a jacket. The cord terminates in an 8-pin RJ-45 connector, which plugs into an RJ-45 jack in the NIC, hub, or switch.

No metal shielding around the four pairs

Pin 1 on this side

RJ-45 Connector

RJ-45 Jack

8 Wires

Jacket

Figure 3-7   4-Pair Unshielded Twisted-Pair Cable with RJ-45 Connector

TEST YOUR UNDERSTANDING

7. a) What is a length of UTP wiring called? b) In 4-pair UTP, how many wires are there in a cord? c) How many pairs? d) What surrounds each wire? e) How are the two wires of each pair arranged? f) What is the outer covering called?

8. Why is 4-pair UTP dominant in LANs for the access line between a NIC and the switch that serves the NIC?

Attenuation and Noise Problems

UTP signals change as they travel down the wires. If they change too much, they will not be readable. As noted earlier in this chapter, there are several of these propagation effects. We will look at the most important propagation effects for UTP transmission, beginning with attenuation and noise.

Attenuation

As Figure 3-8 illustrates, when signals travel, they **attenuate** (grow weaker). To give an analogy, as you walk away from someone who is speaking, his or her voice will grow fainter and fainter. As noted earlier in this chapter, if a signal attenuates too much, the receiver will not be able to recognize it.

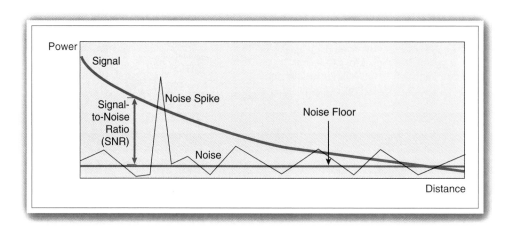

Figure 3-8   Noise and Attenuation

Decibels

Attenuation is measured in **decibels (dB).** The equation for decibels is dB = $10 \log_{10}(P_2/P_1)$, where the Ps are the initial (P_1) and final (P_2) powers. If P_2 is larger than P_1, then the answer will be negative. For voltages, the equation is dB = $20 \log_{10}(V_2/V_1)$.

Three Decibels Is a Halving in Power   Fortunately, you do not have to use this equation much in practice. You can work primarily with two facts. First, a decrease of half in power is a loss of 3 dB. Each additional halving is another loss of 3 dB. For example, a decrease to 1/4 of the original power would be two 3 dB losses or 6 dB. A 9 dB loss would be a decline to 1/8 the original power.

Ten Decibels Is a Reduction to 1/10 the Original Power   Second, falling to 1/10 the initial power is a loss of 10 dB. So falling to 1/100 the original power would be a loss of 20 dB. Incredibly, 20 dB losses can occur in UTP transmission without making the signal unintelligible.

Noise

Electrons within a wire are constantly moving, and moving electrons generate random electromagnetic energy. This random energy is called **noise.** Noise energy adds to the signal energy, so the receiver actually sees the total of the signal plus the noise.

The mean of the noise energy is called the **noise floor**—despite the fact that it is an average and not a minimum, as the name "floor" would suggest.

As a consequence of noise being a random process, there are occasional **noise spikes** that are much higher or lower than the noise floor. If a noise spike is about as large as the signal, the combined signal and noise may be unrecognizable by the receiver.

Attenuation is measured in decibels (dB)
Equations

> The equation for decibels is $dB = 10 \log_{10}(P_2/P_1)$
>> Where the Ps are the two levels of power being compared
>> If P_2 is larger than P_1, then the answer will be negative
> For voltages, the equation is $dB = 20 \log_{10} (V_2/V_1)$

Two Simple Facts

> **3 dB loss is a reduction to 1/12 the original power**
>> 6 dB loss is a decrease to 1/4 the original power
>> 9 dB loss is a decrease to 1/8 the original power
> **10 dB loss is a reduction to 1/10 the original power**
>> 20 dB loss is a decrease to 1/100 the original power
>> 30 dB loss is a decrease to 1/1,000 the original power

Figure 3-9   Decibels (Study Figure)

Noise, Attenuation, and Propagation Distance

If a signal is far larger than the noise floor, then we have a high **signal-to-noise ratio (SNR).** With a high SNR, few random noise spikes will be large enough to cause errors. However, as a signal attenuates during propagation, it falls ever closer to the noise floor. Noise spikes will equal the signal's strength more frequently, so errors will become more frequent. In other words, even if the noise *level* is constant, longer propagation distances create attenuation that results in a lower SNR and therefore more noise *errors*.

> Even if the noise *level* is constant, longer propagation distances result in a lower SNR and therefore more noise *errors*.

Limiting UTP Cord Distance to Limit Attenuation and Noise Problems

Fortunately, attenuation and noise can be controlled fairly simply by limiting the length of UTP cords. The Ethernet standard limits UTP propagation distances to 100 meters. If UTP cords are restricted to 100 meters, the signal still will be comfortably larger than the noise floor, so there will be few noise errors.

TEST YOUR UNDERSTANDING

9.   a) Describe the attenuation problem and why it is important. b) A signal is 1/16 its initial power when it arrives at the receiver. How many decibels has it lost? c) A signal is 1/10,000 of its initial power when it reaches the receiver. How many decibels has it lost? d) Describe the noise problem. e) As a signal propagates down a UTP cord, the noise level is constant. Will greater propagation distance result in fewer noise errors, the same number of noise errors, or more noise errors? Explain.

10.   a) What propagation problems does limiting a UTP cord's length limit to an acceptable level? b) What is the limit on UTP cord length in Ethernet standards?

Electromagnetic Interference (EMI) in UTP Wiring

General EMI

Noise is unwanted electrical energy *within* the propagation medium. In turn, **electromagnetic interference (EMI)**—or more simply, **"interference"**—is unwanted electrical energy coming from *external* devices, such as electrical motors, fluorescent lights, and even nearby UTP cords (which always radiate some of their signal). Like noise, interference adds to the signal and can make the received signal unreadable.

Using Twisted Pair Wiring to Reduce Interference

Fortunately, there is a simple way to reduce EMI to an acceptable level. This is to twist each pair's wires around each other several times per inch, as Figure 3-10 illustrates. Consider what happens over a full twist. Over the first half of the twist, the interference might add to the signal. Over the other half, however, this same interference would subtract from the signal. The interference on the two halves would cancel out, and the net interference would be zero. Does twisting really work this perfectly? No, of course not. However, **twisted pair wiring** dramatically reduces interference, limiting it to an acceptable level. As a historical note, Alexander Graham Bell himself patented twisted-pair wiring as a way to reduce interference in telephone transmission.

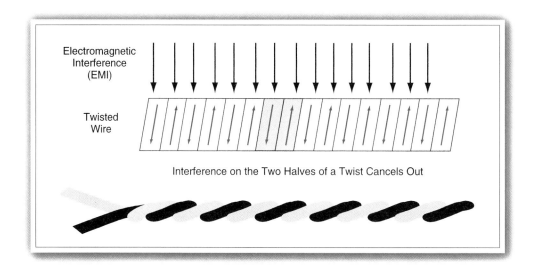

Figure 3-10   Electromagnetic Interference (EMI) and Twisting

Crosstalk Interference

As Figure 3-11 shows, individual pairs in a cord will radiate some of their energy, producing electromagnetic interference in other pairs within the cord. This mutual EMI among wire pairs in a UTP cord is called **crosstalk interference.** It is always present in wire bundles and must be controlled. Fortunately, the twisting of each pair normally keeps crosstalk interference to a reasonable level.

Terminal Crosstalk Interference

Unfortunately, when a UTP cord is connectorized, its wires must be untwisted to fit into the RJ-45 connector, as shown in Figure 3-11. The eight wires are now parallel, so there is no protection from crosstalk interference. Crosstalk interference at the ends of the UTP cord, which is called **terminal crosstalk interference,** usually is much larger than crosstalk interference over the entire rest of the cord.

   Installers must be careful not to untwist UTP wires more than 1.25 cm (half an inch) when adding connectors. This limit will not completely eliminate terminal crosstalk interference, but it limits it to an acceptable level.

Installers must be careful not to untwist UTP wires more than 1.25 cm (half an inch) when adding connectors. This limit will not completely eliminate terminal crosstalk interference, but it limits it to an acceptable level.

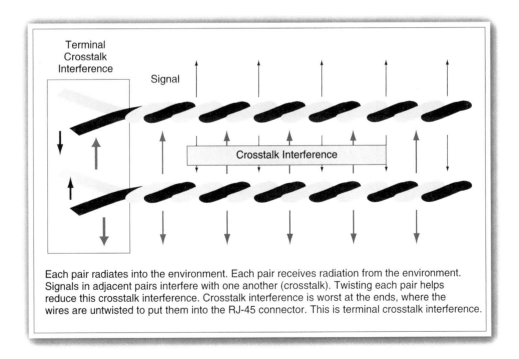

Each pair radiates into the environment. Each pair receives radiation from the environment. Signals in adjacent pairs interfere with one another (crosstalk). Twisting each pair helps reduce this crosstalk interference. Crosstalk interference is worst at the ends, where the wires are untwisted to put them into the RJ-45 connector. This is terminal crosstalk interference.

Figure 3-11   Crosstalk Electromagnetic Interference (EMI) and Terminal Crosstalk Interference

TEST YOUR UNDERSTANDING

11. a) Distinguish between electromagnetic interference (EMI), crosstalk interference, and terminal crosstalk interference. b) How is EMI controlled? c) How is terminal crosstalk interference controlled in general? Explain. d) Does this completely eliminate crosstalk interference?

Serial and Parallel Transmission

Figure 3-12 shows an important distinction in wire communication: serial versus parallel transmission.

Serial Transmission

In the next chapter we will look at Ethernet standards. In slower versions of Ethernet that run at 10 Mbps and 100 Mbps, a single pair in a UTP cord is used to transmit in each direction. Two-way transmission, then, uses two of the four pairs—one in each direction. The other two pairs are not used, although the standard calls for them to be present.

If one pair of wires is used to send a transmission, this is **serial transmission,** because the bits that are transmitted must follow one another in series.

Parallel Transmission

For gigabit Ethernet, however, when the NIC or switch transmits, it transmits on all four pairs in each direction. This means that it can transmit four bits at a time instead of just one. This is **parallel transmission.**

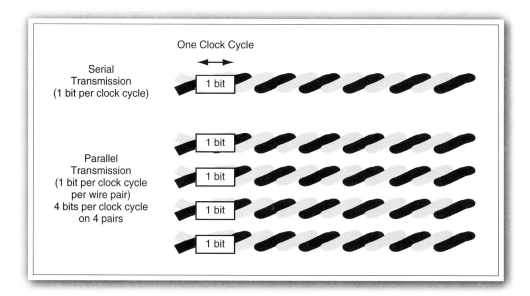

Figure 3-12   Serial Versus Parallel Transmission

Obviously, parallel transmission is faster than serial transmission for the same clock cycle duration. Speed is the key benefit of parallel transmission.

In general, parallel transmission does not mean the use of four transmitting pairs (or grounded wires). It means using more than a single pair of wires or one wire and a ground wire. For example, your computer's internal bus, which carries signals among components in your systems unit, has about 100 wires for high-density parallel transmission. To give another example, parallel cables used to connect PCs to printers use eight wires to carry data in each direction.

TEST YOUR UNDERSTANDING

12. a) Distinguish between serial and parallel transmission. b) What is the main benefit of parallel transmission? c) In parallel transmission, how many pairs (or single wires and ground wires) are used in each direction?

Ethernet Standards Using UTP

We saw in Chapter 1 that small Ethernet networks use 4-pair UTP.

100-Meter Distances
Standards for Ethernet operating at 10 Mbps, 100 Mbps, and one gigabit per second are 10Base-T, 100Base-TX, and 1000Base-T (better known as gigabit Ethernet). All three can work with UTP cords up to 100 meters long.

Category 5e and Category 6 UTP Quality Standards
EIA/TIA-568 governs quality standards for UTP wiring. Higher-quality UTP has higher "category" numbers. Almost all UTP now sold is **Category 5e** (enhanced) or **Category 6. "Cat 5e"** is fine for Ethernet all the way through gigabit Ethernet. In fact, even plain Cat 5 cabling is fine through gigabit Ethernet.

Category 3 and Category 4 UTP Quality Standards
Some older buildings only have Category 3 or Category 4 wiring. These lower-quality wires can only be used for Ethernet 10Base-T, bringing only 10 Mbps to stations.

Category 6 UTP Quality Standard
In 2003, Category 6 UTP was standardized, and many vendors are beginning to sell Cat 6 UTP wiring. Although Category 6 wiring is not needed up to gigabit Ethernet, there may be a Cat 6 standard for 10 Gbps Ethernet. This would be used primarily for trunk lines, not station access lines. However, if vendors begin to offer only Cat 6 wiring, companies may have to purchase this wiring for their networks.

TEST YOUR UNDERSTANDING

13. a) What are the Ethernet standards for transmitting over UTP at 10 Mbps, 100 Mbps, and one gigabit per second? b) How long can UTP cords be in Ethernet networks? (Yes, this is a repeat of an earlier question.) c) What UTP quality standards are commonly sold today? d) What is the lowest UTP quality level acceptable for gigabit Ethernet? e) What happens if a building has only older Cat 3 or Cat 4 wiring? f) Under what circumstance may Cat 6 wiring be needed?

OPTICAL FIBER TRANSMISSION LINKS

Optical fiber is the champagne of transmission media. For runs longer than 100 meters, companies must turn to **optical fiber** cabling, which is illustrated in Figure 3-13.

Why not use optical fiber anywhere? The simple answer is that although fiber can carry signals at high speeds over long distances, optical fiber is more expensive to purchase and install than UTP. In addition, fiber is fragile because it is made of glass.

In LANs, rugged and relatively inexpensive UTP is used normally in access links between the station and the closest switch, as Figure 3-14 illustrates. These distances are short, and ruggedness is important because cords may be treated roughly in office areas. In turn, optical fiber in LANs is used primarily in trunk links between switches and routers, where its fragility is not as much of a problem as it would be in an access link from the switch to the user's desktop. However, when a trunk line spans less than 100 meters, UTP still may be used because of its lower cost.

Typical LAN distances for fiber are 200 m to 2 km. In WANs, special optical fiber cable can run tens of kilometers, but such long fiber runs only occur in carrier networks, not corporate networks. In corporate networks, the distances spanned between switches seldom run more than 300 meters. This allows less expensive types of optical fiber to be used, as we will see later in this chapter.

Optical Fiber Construction

Core

In optical fiber, a transmitter injects light into a very thin glass tube called the **core.** The core normally is 8.3 microns, 50 microns, or 62.5 microns in diameter. (A micron is a millionth of a meter; a human hair is 40 to 120 microns in diameter.)

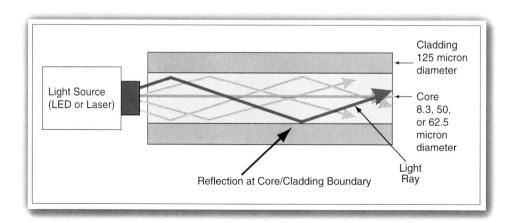

Figure 3-13   Optical Fiber Cord

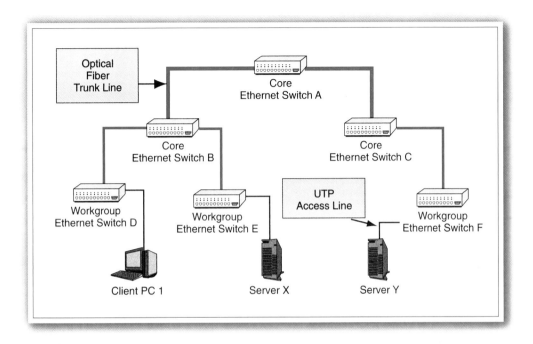

Figure 3-14   UTP in Access Lines and Optical Fiber in Trunk Lines

Signaling

Optical fiber typically uses simple on/off signaling, turning on the light source during a clock cycle to signal a one and turning off the light source during a clock cycle to signal a zero.

Cladding and Perfect Internal Reflection

Surrounding the core is a thicker glass cylinder called the **cladding.** As Figure 3-13 shows, when light begins to spread, it hits the cladding and is reflected back into the core with **perfect internal reflection** so that no light escapes.[9] In contrast, UTP tends to radiate energy out of the wire bundle, causing rapid attenuation. The cladding normally is 125 microns in diameter, regardless of the core diameter.

Strand Thickness

The cladding is surrounded by a **coating** to keep out light and to strengthen the fiber. This coating includes strands of yellow Aramid (Kevlar) yarn to strengthen the fiber. The coating and outer jacket brings the outer diameter of a fiber strand to about 250 microns.

---

[9]   This is based on Snell's Law. The cladding has a slightly lower index of refraction than the core. This difference in index of refraction creates perfect internal reflection.

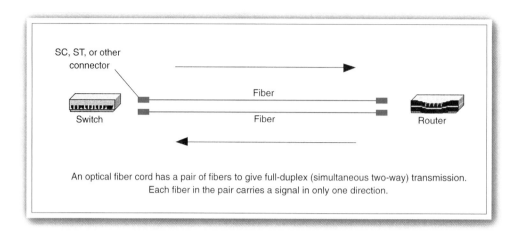

Figure 3-15   Full-Duplex Optical Fiber Cord

Two Strands for Full-Duplex Operation

Note in Figure 3-15 that an **optical fiber cord** has *two* strands of fiber—one strand for transmission in each direction. This gives **full-duplex** communication (simultaneous two-way communication).

Connectors

In UTP, there is only one type of connector—the RJ-45 connector. In optical fiber, things are more complex because there are two popular types of connectors. These two connector types are illustrated in Figure 3-16.

SC Connectors   The **SC connector** is square and snaps into an SC port. The SC connector is very popular and is recommended in the EIA/TIA-568 standard for use in new LANs, although many companies are continuing to install the ST connector, which is also popular.

ST Connectors   The **ST connector** is cylindrical. It pushes into an ST port and then twists to be locked in place. ST is called a *bayonet connector* because bayonets use this type of push-twist-and-lock connection.

Other Connectors   There also are several **small form factor** connectors that are smaller than SC and ST connectors. This reduces the sizes of switches and routers that have many fiber ports. However, small form factor connectors are not standardized, and several types of small form factor connectors are available.

Switches and Routers

Fortunately, almost all switches and routers can be purchased with either SC or ST ports for optical fiber connections. In practice, then, fiber's lack of a single connector standard is not a problem except for small-form-factor connectors.

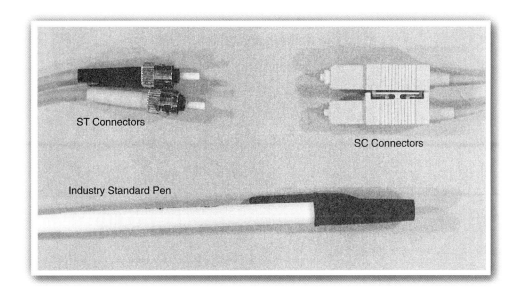

Figure 3-16   Pen and Full-Duplex Optical Fiber Cords with SC and ST Connectors

TEST YOUR UNDERSTANDING

14.   a) In optical fiber transmission, how are ones and zeros usually signaled? b) In optical fiber, what are the roles of the core and the cladding? c) What is the ability to transmit in both directions simultaneously called? d) Why does a fiber cord need two fiber strands for full-duplex transmission? e) What fiber connector is now recommended in new LANs? f) What other fiber connector is popular? g) Does the presence of two major connector standards for fiber cause problems for switch and router purchasing? Explain.

Wavelengths, Wavelength Division Multiplexing (WDM), and Attenuation

Wavelengths

Optical fiber transmits data as light signals. Figure 3-17 illustrates a simple light wave. The physical distance between comparable points (say, from peak to peak) in successive cycles is called the signal's **wavelength.** For light signals, wavelengths are a few **nanometers (nm)** long. A nanometer is one billionth of a meter.

Frequency

Another characteristic of waves shown in Figure 3-17 is frequency. A wave's **frequency** is the number of times the wave goes through a complete cycle per second. Frequencies are measured in hertz. A **hertz (Hz)** is one cycle per second.

Signals propagate at a fixed speed in optical fiber. This means that shorter wavelengths will produce more cycles per second. This can most easily be seen with

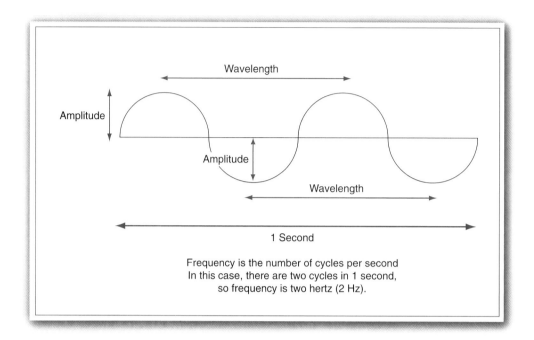

Figure 3-17   Wave Characteristics

stringed instruments, where the wavelength of the sounds made by strings generally is lower for longer strings. On a harp, for instance, the shortest strings make the highest-pitch sounds. In contrast, bass guitars have longer strings than regular guitars.

Wavelength Division Multiplexing

The labor to install optical fiber is extremely expensive. One way to leverage your installed base of optical fiber is to replace signaling equipment at the two ends with signaling equipment that can transmit several light sources at slightly different wavelengths, as Figure 3-18 shows. This allows you to add signal capacity by using slightly more expensive signaling equipment but without incurring the high labor cost of laying new fiber.

This simple idea is given the complex name **wavelength division multiplexing (WDM).** Multiplexing is the mixing of multiple signals in a transmission line. WDM does this by sending each signal at a different wavelength.

Not content with that, marketers created another term, **dense wavelength division multiplexing (DWDM),** for fiber that uses more than forty light sources at different frequencies.[10]

---

[10]   The minimum number of wavelengths to distinguish between WDM and DWDM will increase in the future. Some analysts call systems with fewer than forty wavelengths coarse wavelength division multiplexing.

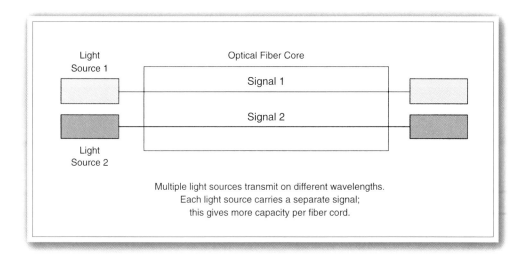

Multiple light sources transmit on different wavelengths.
Each light source carries a separate signal;
this gives more capacity per fiber cord.

Figure 3-18   Wavelength Division Multiplexing (WDM) in Optical Fiber

Historically, wavelength is represented by the Greek character lambda (λ). Consequently, a signal being sent at a particular frequency is called a **lambda,** and a WDM system with sixteen light sources is said to have 16 lambdas.

TEST YOUR UNDERSTANDING

15. a) What is a wave's wavelength? b) What is a wave's frequency? c) Distinguish between traditional optical fiber transmission, WDM, and DWDM. d) How do WDM and DWDM reduce costs? e) In WDM, what is a lambda?

Wavelengths and Attenuation

Although optical fiber can carry signals over long distances, even optical fiber has distance limits, and the least expensive types of optical fiber can carry signals only about twice as far as UTP. It is important to understand the things that affect the distance over which you can send signals with different types of optical fiber.

Attenuation

One concern in optical fiber is attenuation—the loss of signal strength over distance. Although attenuation is far lower in fiber transmission than in UTP transmission, it still exists.

Lower Attenuation at Higher Wavelengths

Fiber attenuation is sensitive to the wavelength of the light transmitted. As wavelength increases, optical fiber attenuation decreases.

As wavelength increases, optical fiber attenuation decreases.

Attenuation

    Decreases with wavelength

        850 nm: better than 0.35 dB/km

        1,300 nm: better than 0.15 dB/km

        1,550 nm: better than 0.05 dB/km

        In comparison, UTP attenuation is only better than 20 dB in 100 meters

    Light source prices increase with wavelength

        850 nm uses inexpensive LEDs

        1,300 nm and 1,550 nm use expensive lasers

    Must balance distance and cost

Modal Bandwidth

    Modal Dispersion

        Light rays only enter at a few angles

        These rays are called modes

        Different modes travel different distances

        The modes from sequential clock cycles tend to overlap, causing problems

        Also called temporal dispersion

    Single-mode fiber

        If core diameter is only 8.3 microns, only one mode will propagate

        Only attenuation is important in single-mode fiber

    If core is thicker, there will be multiple modes

        Fewer modes with 50-micron core than with 62.5-micron core

    Bandwidth

        Signals have a range of bandwidths

        Higher speeds require a broader range of bandwidth

    Modal bandwidth

        Better-quality multimode fiber has more modal bandwidth

        Measured as MHz-km

            If 200 MHz-km, 200 MHz bandwidth allows 1 km cord length

            If 200 MHz-km, 100 MHz bandwidth allows 2 km cord length

            If 500 MHz-km, 250 MHz bandwidth allows 2 km cord length

        For 850 nm, 160 MHz-km to 500 MHz-km modal bandwidth is typical

        For 1,300 nm, 400 MHz-km to 1,000 MHz-km modal bandwidth is typical

        Fiber with greater modal bandwidth costs more

    Key points

        For single-mode fiber, attenuation is the dominant distance limitation

        For multimode fiber, modal dispersion is the dominant distance limitation

Figure 3-19 Optical Fiber Transmission (Study Figure)

In practice, there are three "windows"—wavelength ranges—in which light waves propagate especially well through a glass fiber. These window ranges are centered around 850 nm, 1,300 nm, and 1,550 nm. In comparison, visible light ranges from 400 nm to 700 nm.[11]

Transmission using 850 nm light attenuates the most rapidly, while transmission at 1,550 nm attenuates the least rapidly with distance. At 850 nm, 1,300 nm, and 1,550 nm, attenuation usually is better than 0.35 dB/km, 0.15 dB/km, and 0.05 dB/km, respectively. For UTP, in contrast, attenuation is only better than about 20 dB in 100 meters.

Higher Transceiver Costs at Longer Wavelengths

Falling attenuation with increasing wavelength would seem to suggest that using longer wavelengths is automatically the best choice. However, there is also the matter of cost. **Transceivers** (transmitter/receivers) operating at 850 nm are the least expensive because they can use inexpensive light emitting diode (LED) light sources like those in your wristwatch. Transceivers operating at 1,300 nm or 1,500 nm, in contrast, must use more expensive laser light sources. In the end, the best choice is to use the least expensive wavelength that will meet distance requirements for a fiber run.

TEST YOUR UNDERSTANDING

16. a) What are the three wavelengths for optical fiber transmission? b) Which can span longer distances—fiber systems with longer wavelengths or fiber systems with shorter wavelengths? c) Which are less expensive—fiber systems with longer wavelengths or fiber systems with shorter wavelengths? d) What determines what wavelength you should use for a particular fiber trunk line?

Modes and Optical Fiber

Modes and Modal Dispersion

The main propagation problem for the least expensive optical fiber technology is **modal dispersion.** Modal dispersion in optical fiber also is called **"temporal dispersion"** for reasons that soon will be obvious. As Figure 3-20 shows, in multimode fiber, which is the most common type of fiber in LANs, light rays in a pulse can enter at different angles.

For technical reasons, light really can only enter at a certain number of angles. The light rays allowed to enter the fiber are called **modes.** The single mode going along the axis will travel straight through the core without reflection. However, other modes entering at high angles will be reflected many times as they travel down the fiber, consequently traveling farther. If the difference in the arrival times of various modes is too large, the modes of sequential pulses will begin to overlap in their arrival times, and the signal will be unreadable. This modal dispersion will make it impossible for the receiver to read the signal. For inexpensive fiber, modal dispersion limits transmission distance much more than attenuation.

---

[11] Fiber signaling operates at longer wavelengths than visible light. Higher wavelengths mean lower frequencies, so fiber signaling operates at lower frequencies than visible light. Specifically, fiber signaling operates in the infrared frequency region.

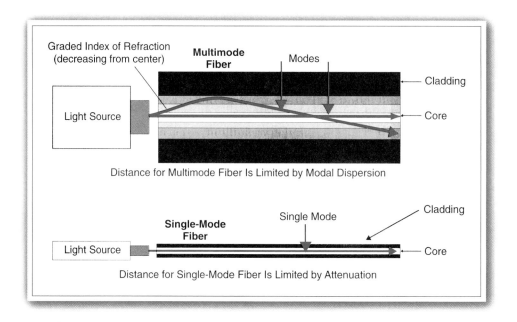

Figure 3-20   Multimode and Single-Mode Optical Fiber

Diameter and Modal Dispersion

The best way to reduce modal dispersion is to reduce the diameter of the fiber. In fact, if the fiber core is sufficiently thin, about 8.3 microns, only the single mode going straight through the fiber will propagate, as shown in Figure 3-20. This is called, understandably, **single-mode fiber.**

Single-mode fiber completely eliminates modal dispersion, but single-mode fiber is very expensive to produce and install (it is very difficult to line up the light source with the core precisely). Although single-mode fiber can span tens of kilometers, such distances are rarely needed in LANs. Single-mode fiber is used almost exclusively by telephone and WAN carriers.

Modes and Multimode Fiber

LAN transmission instead uses **multimode fiber,** which is thicker than single-mode fiber and so allows multiple modes to propagate. In multimode fiber, modal dispersion usually reduces maximum transmission distance to a few hundred meters, but this is sufficient for almost all LAN needs. In addition, it is relatively inexpensive to install multimode fiber because it is easier to align light sources with thick cores than with thin cores (think of threading a needle), and it is easier to splice two strands with thick cores than with thin cores.

Even within multimode fiber, diameter is important. Most multimode fiber has a core diameter of 62.5 microns. However, some multimode fiber has a diameter of 50 microns, allowing somewhat longer transmission distances.

Due to modal dispersion, multimode fiber runs usually are limited to between 200 meters and 2 kilometers. However, these distances are more than sufficient for most LANs, and multimode fiber is less expensive to install than single mode fiber. Consequently, multimode fiber dominates in LANs.

Overall, multimode fiber dominates optical fiber use in LAN transmission.

Bandwidth

Figure 3-17 shows a wave with a single frequency. However, real signals are spread over a range of frequencies called the signal's **bandwidth,** as Figure 3-21 illustrates.

Bandwidth is important because the wider the bandwidth, the higher the possible transmission speed. (We will see the precise relationship between bit rate and bandwidth in Chapter 5.) Consequently, it is important for a fiber not to limit bandwidth, or high-speed signals will not be able to propagate down the fiber. We will look at fiber bandwidth limitations later.

The wider the bandwidth of a transmission system, the faster data can be transmitted.

Graded Index Multimode Fiber and the Modal Bandwidth

As Figure 3-20 shows, all multimode fiber sold today has a **graded index of refraction,** with the index of refraction decreasing from the center of the core to its circumference.

Propagation speed increases as the index of refraction decreases. Consequently, the single mode propagating directly down the middle travels more slowly than modes traveling in the outer areas area of the core. This partially compensates for modal dispersion.

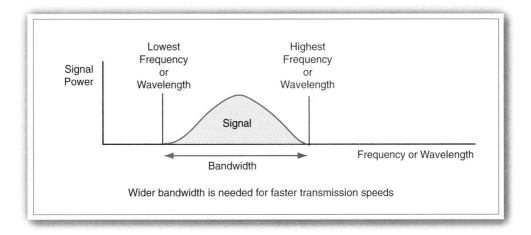

Figure 3-21   Signal Bandwidth

Not all graded index multimode fiber products are equally good. Multimode fiber quality is given by the fiber's **modal bandwidth,** which is the fiber's bandwidth-distance product. For instance, a modal bandwidth of 200 MHz-km means that if your bandwidth is 200 MHz, then you can transmit 1 km. In turn, if your bandwidth is 400 MHz, you can transmit half a kilometer. A greater modal bandwidth means that a multimode fiber can support greater bandwidth over a given distance. Again, recall that a greater bandwidth means greater transmission speed.

Modal bandwidth values depend heavily on wavelength. For instance, at 850 nm, values typically range from 160 MHz-km to 500 MHz-km, while at 1,300 nm, a typical range is 400 MHz-km to 1,000 MHz-km. Higher-quality multimode fiber can carry signals over longer distances, but it also costs more to purchase. In general, modal bandwidth values are increasing over time as manufacturing improves.

For single-mode fiber, modal bandwidth is not applicable because there is only one mode, and so there is no modal dispersion to limit propagation distance. Only attenuation is important.

TEST YOUR UNDERSTANDING

17. a) What are modes? b) What problem do they cause? c) Distinguish between single-mode and multimode fiber in terms of fiber construction. d) Distinguish between single-mode and multimode fiber in terms of transmission distance. e) Distinguish between single-mode and multimode fiber in terms of cost of installation. f) Given your answer to the previous two parts of this question, why does multimode fiber dominate in LANs? g) Where is single-mode fiber most commonly used?

18. a) What is a signal's bandwidth? b) Why is being able to support a wide bandwidth good? c) How does graded index multimode fiber reduce modal dispersion? d) What is modal bandwidth? e) Why is it better to have a greater modal bandwidth? f) If your fiber has a modal bandwidth of 200 MHz-km, how much bandwidth would you have if you transmit a signal 200 meters? g) If you have to transmit 4 km? h) Is modal bandwidth important for single-mode fiber?

Key Points about Fiber

Overall, attenuation is the limiting factor for single-mode fiber. Attenuation is very low, particularly at longer wavelengths, so single-mode fiber can transmit signals over many kilometers. Consequently, expensive single-mode fiber dominates in WAN transmission.

Multimode fiber is less expensive and is sufficient for LAN distances. Consequently, multimode fiber dominates for LAN use. For multimode fiber, modal dispersion is the main limiter of transmission distance. Longer-wavelength light, smaller core diameter (50 micron versus 62.5 micron) and higher-quality fiber (which has greater modal bandwidth) can all increase multimode transmission distance but cost more. Selecting multimode fiber transmission systems involves selecting the least expensive fiber technology for the job.

NETWORK TOPOLOGIES

The term **network topology** refers to the physical arrangement of a network's stations, switches, routers, and transmission lines. Topology, then, is a physical layer concept. Different network (and internet) standards specify different topologies.

Figure 3-22 shows the major topologies found in networking. Some are seen only in older legacy LANs using obsolete technology.

Network topology is the physical arrangement of a network's stations, switches, routers, and transmission lines.

Point-to-Point Topology

The simplest network topology is the **point-to-point topology,** in which two nodes are connected directly. Although some might say that a point-to-point connection is not a

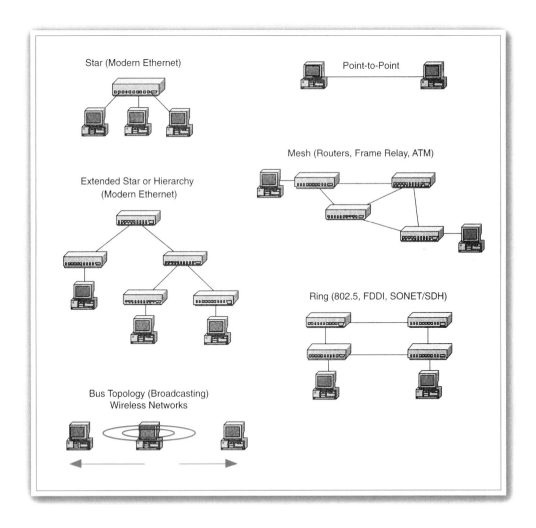

Figure 3-22   Major Topologies

network, WANs often are built of point-to-point private lines provided by the telephone service, and dial-up telephone connections effectively provide point-to-point connections between users and the Internet.

Star Topology and Extended Star (Hierarchy) Topology

These are the topologies used by modern versions of Ethernet, which is the dominant LAN standard. In a simple **star topology,** all wires connect to a single switch. In an **extended star (or hierarchy) topology,** there are multiple layers of switches organized in a hierarchy. We saw an Ethernet star in Chapter 1 and will see Ethernet hierarchies in Chapter 4.

Mesh Topology

In a **mesh topology,** there are many connections among switches, so there are many **alternative routes** to get from one end of the network to the other. We will see mesh topologies with ATM and Frame Relay in Chapter 7 and with routers in Chapter 8.

Ring Topology

In a **ring topology,** stations are connected in a loop. Messages pass in only one direction around the loop. Eventually, all messages pass through all stations. In LANs, the obsolete 802.5 Token-Ring Network and FDDI network technologies used a ring topology. In addition, the worldwide telephone network increasingly uses rings to connect its switches, using the SONET/SDH technology as we will see in Chapter 6.

Bus Topologies

In a **bus topology,** when a station transmits, it broadcasts to all other stations. Wireless LANs, which we will see in Chapter 5, use a bus topology by broadcasting signals. So do Ethernet hubs, which we will see in Chapter 4. The obsolete Ethernet 10Base5 and 10Base2 (see Chapter 4a) technologies also used a bus topology.

TEST YOUR UNDERSTANDING

19.   a) What is a topology? b) At what layer do we find topologies? c) List the major topologies and give the defining characteristics for each. Present each topology in a separate paragraph. d) What technology is associated with each topology?

WIRING THE FIRST BANK OF PARADISE HEADQUARTERS BUILDING

The First Bank of Paradise headquarters building is a typical multistory office building. We will look at wiring in this building. Wiring in other large buildings tends to be similar.

Facilities

Figure 3-23 illustrates the building. Although the building is ten stories tall, only three stories (plus the basement) are shown.

Equipment Room

Wiring begins in the **equipment room,** which is in the building's basement. This room connects the building to the outside world.

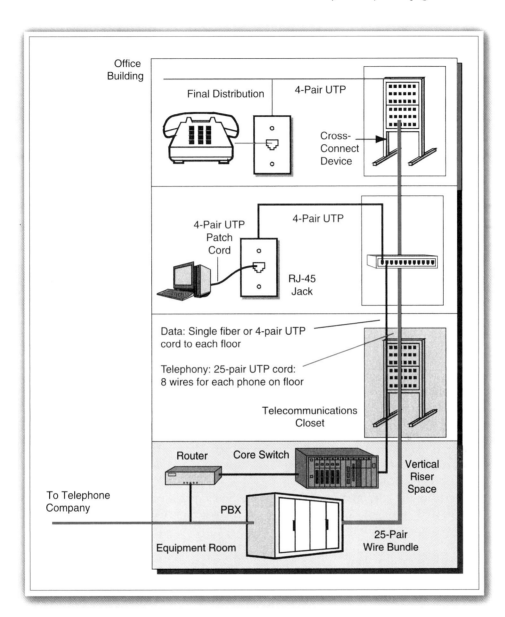

Figure 3-23   First Bank of Paradise Building Wiring

Vertical Risers

From the equipment room, telephone and data cabling has to rise to the building's upper floors. The bank has **vertical riser** spaces between floors. These typically are hard-walled pipes to protect the wiring.

Telecommunications Closets

On each floor there is a **telecommunications closet.** Within the telecommunications closet, cords coming up from the basement are connected to cords that span out horizontally to telephones and computers on that floor.

Telephone Wiring

This building infrastructure was created for telephone wiring. Fortunately, most companies allocated ample space to their equipment rooms, vertical risers, and telecommunications closets. This allows data wiring to use the same spaces.

PBX

Many companies have internal telephone switches called **private branch exchanges (PBXs).** Eight wires must span out from the PBX to each wall outlet in the building. So if the building has ten floors and each floor has 100 RJ-45 telephone wall jacks, 8,000 wires ($10 \times 100 \times 8$) will have to be run from the PBX through the vertical riser space. On each floor, the 800 wires for that floor will be organized as 100 4-pair UTP cords running from the telecommunications closet to telephone wall jacks on that floor. Obviously, careful documentation of where each wire goes is crucial for maintaining sanity.

Vertical Wiring

For vertical wiring runs, telephony typically uses **25-pair UTP cords.** These vertical cords typically terminate in **50-pin octopus connectors.**

Cross-Connect Device

Within the telecommunications closet, the 25-pair vertical cords plug into **cross-connect devices,** which connect the wires from the riser space to 4-pair UTP cords that span out to the wall jacks on each floor.

Horizontal Wiring

The horizontal telephone wiring, as just noted, uses 4-pair UTP. Yes, telephone wiring uses the same 4-pair UTP that data transmission uses. Actually, telephone wiring first introduced 4-pair UTP. Telephony has used 4-pair UTP for several decades.

Data transmission researchers learned how to send data over 4-pair UTP to take advantage of widespread installation expertise for 4-pair UTP. In addition, some companies had excess UTP capacity already installed, so in some cases, it would not even be necessary to install new wiring.

Data Wiring

In data wiring for Ethernet, there normally is a core Ethernet switch in the equipment room, as Figure 3-23 illustrates. This core switch communicates to the outside world via a router.

Vertical Wiring

Vertical data wiring is much simpler than vertical telephone wiring. A single UTP or optical fiber cord runs from a port in the core switch up to a port in the Ethernet workgroup switch on each floor. So if there are ten floors, only ten UTP or optical fiber cords would have to be run through the vertical riser space. The core switch needs a port for the link to each floor. The workgroup switch on each floor needs a single port for the link to the basement and needs a port for each horizontal cable run on each floor.

Horizontal Wiring

For horizontal wiring, however, telephone wiring and data wiring are *identical*. Both use 4-pair UTP. Often, wall outlets have two jacks, one labeled "voice" and the other labeled "data."

Plenum Cabling

Fire regulations require the use of a special type of fire-retardant cabling, called **plenum** cabling, any time cables run through airways (plenums) such as air conditioning ducts. Ordinary sheaths on UTP and optical fiber cords are made of polyvinyl chloride (PVC), which gives off deadly toxins when it burns. If these toxins are released in airways, the toxins will spread rapidly to office areas.

TEST YOUR UNDERSTANDING

20. a) What equipment are you likely to find in a building's equipment room? b) In its telecommunications closets? c) What is the purpose of a PBX? d) What two types of data switches will you find? e) Compare and contrast vertical distribution for telephony and data. f) Compare and contrast horizontal distribution for telephony and data.

21. A 10-story building has 60 voice jacks on each floor and 40 data jacks on each floor. a) For telephony, how many wires will you run through the vertical riser space? b) How many cords will this require? c) For vertical data wiring if you use UTP? d) For vertical data wiring if you use optical fiber? e) On each floor, how many wires will you run horizontally from the telecommunications closet to wall jacks? f) How many cords will this require?

22. a) Where is plenum cabling required? b) Why is plenum cabling needed?

SYNOPSIS

Chapter 2 discussed standards in general. It focused on the data link layer through the application layer because these layers all operate in the same general way—by sending messages. The physical layer, which was the focus of this chapter, is very different. It sends individual bits instead of messages, and it alone governs transmission media and propagation effects.

In this chapter we looked at UTP and optical fiber—the two transmission media that traditionally have dominated in LANs and WANs. We also looked at network topology and building wiring. In Chapter 5, we will look at radio transmission.

To be transmitted, data must be represented as strings of bits (ones and zeros). Whole numbers (integers) typically are represented as binary numbers.

Binary counting begins with zero. You should know how to count in binary by adding one to each previous value. You should also know that a field with N bits allows you to represent 2^N alternatives.

These bits must be converted into signals to propagate down the transmission medium. In simple on/off signaling in optical fiber, light is turned on during a clock cycle for a one and off for a zero. On/off signaling is binary signaling, in which there are two possible states or line conditions (on and off). Binary signaling is also done using two voltage ranges on a wire. One voltage range represents a zero; the other, a one. In digital signaling there are a few possible states (2, 4, 8, 16, etc.). Adding more possible states allows more bits to be transmitted per clock cycle but decreases immunity to attenuation propagation errors. Binary transmission is a special case of digital transmission.

In unshielded twisted pair wiring, a cord consists of four wire pairs. There is an RJ-45 connector at each end. UTP is rugged and inexpensive and so dominates the access links that connect stations to workgroup switches.

The two wires of each pair in a UTP cord are twisted around each other to reduce electromagnetic interference (EMI) problems. Two simple installation expedients are used to keep propagation problems to an acceptable level. First, restricting cable runs to 100 meters usually prevents serious attenuation and noise errors. Second, limiting the untwisting of wires to no more than 0.5 inches (1.25 cm) usually keeps terminal crosstalk interference to an acceptable level.

Earlier versions of Ethernet transmit serially—sending on only one wire pair in each direction. Gigabit Ethernet transmits in parallel, sending on all four wire pairs when it transmits. Parallel transmission is faster than serial transmission.

There are different grades of UTP quality. Ethernet, which dominates LAN transmission, needs Category 5 (Cat 5) or Cat 5e (enhanced) wiring to operate at 100 Mbps or one gigabit per second. Higher-quality Category 6 UTP is not necessary today but may be needed for 10 Gbps Ethernet.

In optical fiber, light signals are injected into a thin glass core. Transmission occurs near three wavelengths: 850 nm, 1,300 nm, and 1,550 nm. Light with longer wavelengths experiences lower attenuation and so can travel farther. However, longer-wavelength light is more expensive to produce.

Single-mode fiber has a core that is only 8.3 microns wide. Only one mode can transmit in single-mode fiber, so there is no problem with modal dispersion. Signals can travel tens of kilometers. LAN use, however, is dominated by less expensive multimode fiber that is 50 microns or 62.5 microns wide. Multimode fiber, as its name suggests, allows light to propagate in different modes that may arrive at slightly different times. This limits multimode cord length, but these limits typically are sufficient for LANs.

Another way to increase distance (in multimode fiber) is to use better-quality fiber, as measured by its modal bandwidth (its bandwidth-distance product). A higher modal bandwidth will allow a signal of a particular bandwidth (which is related to transmission throughput) to travel farther.

One way to increase throughput without laying more fiber is to send multiple light signals through existing fiber cords; this is called wavelength division multiplexing.

Network topologies describe the way that nodes are connected by transmission lines. Examples of topologies are the point-to-point, star, hierarchy, mesh, ring, and bus topologies. As you work through this book, you will see these topologies and their implications for performance and reliability.

We looked at cabling for telephony and data in a multistory building. We saw that there are vertical runs from the equipment room up to each floor's telecommunications closet and horizontal runs from the telecommunications closets to the wall jacks on each floor. We saw that vertical distribution for telephony involves eight wires for each telephone on each floor, while vertical distribution for data involves running only a single UTP or optical fiber cord to each floor's workgroup switch from the basement's core switch. Horizontal runs are the same for telephony and data—a single UTP cord is run from the telecommunications closet to each wall outlet.

SYNOPSIS QUESTIONS

1. a) For binary positive integers, what is the smallest number? b) In binary, four is 100. What are five, six and seven? c) If a field has seven bits, how many alternatives can it represent?
2. a) Distinguish between digital and binary signaling. b) What is the advantage of digital signaling with several possible states over binary signaling? c) What is the disadvantage?
3. a) Describe the organization of a UTP cord. b) What two simple installation expedients can be used to limit propagation problems to an acceptable level in most situations? c) What propagation problems does each method limit?
4. What is the advantage of parallel transmission compared to serial transmission?
5. What is the lowest UTP quality grade acceptable for gigabit Ethernet?
6. a) What are the three main wavelengths for optical fiber transmission? b) What is the advantage of transmitting at longer wavelengths? c) What is the disadvantage of transmitting at longer wavelengths?
7. a) What is the advantage of single-mode fiber compared to multimode fiber? b) Why does multimode fiber transmission dominate in LANs? c) In multimode fiber, why is the modal bandwidth important?
8. a) What is wavelength division multiplexing? b) Why is this superior to simply laying more fiber cords?
9. What is a topology?
10. A multistory building is cabled for telephony and data. a) Contrast vertical distribution for telephony and data. b) Compare horizontal distribution.

THOUGHT QUESTIONS

1. In binary, 35 is 100011. Compute 36 through 40 in binary. (Forty is 101000.)
2. a) A field is 8 bits long. How many values can it represent? b) Repeat for 1, 4, 8, 16, 32, 48, and 64 bits. You may use a spreadsheet program for more than eight bits. (Hint: Each bit doubles the number of possible alternatives.)
3. The clock cycle is one millionth of a second. Eight-level digital signaling is used. a) What is the baud rate? b) How many bits are sent per clock cycle? c) What is the bit rate? d) If your clock cycle is 1/10,000 of a second and you want a bit rate of 50 kbps, how many possible states will you need per clock cycle?

4. a) If power falls to 40 percent of its initial value through attenuation, how many decibels is this? b) If voltage drops to 40% of its initial value through attenuation, how many decibels is this? You may wish to use a spreadsheet program to compute the logarithm.

5. What type of interference is most likely to create problems in UTP transmission?

6. When a teacher lectures in class, is the classroom a full-duplex communication system or a half-duplex communication system?

7. a) How many possible paths are there between stations in a point-to-point topology? b) In a hierarchical topology? c) In a mesh topology with four switches (without backtracking)? (Hint: Draw a picture.)

TROUBLESHOOTING QUESTIONS

1. A tester shows that a UTP cord has too much interference. What might be causing the problem? Give at least two alternative hypotheses and then describe how to test them.

2. What kinds of errors are you likely to encounter if you run a length of UTP cord 200 meters? (Recall that the standard calls for a 100-meter maximum distance.)

HANDS-ON

Chapter 3a discusses how to connectorize bulk UTP cabling. To try it out, you will need a box of bulk UTP cabling, a wire cutter, a wire stripper, a crimper, a bag of RJ-45 connectors, and a tester (because only about half of connections done by novices work). All of this will set you back about $300. Knowing this, the price of UTP patch cables, which are cut, connectorized, and tested at the factory, seems more reasonable, doesn't it?

Getting Current
Go to the book website's New Information and Errors pages for this chapter to get new information since this book went to press and to correct any errors in the text.

CHAPTER 3a

HANDS ON: CUTTING AND CONNECTORIZING UTP[1]

INTRODUCTION

Chapter 3 discussed UTP wiring in general. This chapter discusses how to cut and connectorize (add connectors to) solid UTP wiring.

SOLID AND STRANDED WIRING

Solid-Wire UTP Versus Stranded-Wire UTP

The TIA/EIA-568 standard requires that long runs to wall jacks use **solid-wire UTP,** in which each of the eight wires really is a single solid wire.

However, patch cords running from the wall outlet to a NIC usually are **stranded-wire UTP,** in which each of the eight "wires" really is a bundle of thinner wire strands. So stranded-wire UTP has eight bundles of wires, each bundle in its own insulation and acting like a single wire.

Relative Advantages

Solid wire is needed in long cords because it has lower attenuation than stranded wire. In contrast, stranded-wire UTP cords are more flexible than solid-wire cords, making them ideal for patch cords—especially the one running to the desktop—because they can be bent more and still function. They are more durable than solid-wire UTP cords.

---

[1]   This material is based on the author's lab projects and on the lab project of Prof. Harry Reif of James Madison University.

Solid-Wire UTP
>   Each of the eight wires is a solid wire
>   Low attenuation over long distances
>   Easy to connectorize
>   Inflexible and stiff—not good for runs to the desktop

Stranded-Wire UTP
>   Each of the eight "wires" is itself several thin strands of wire within
>   an insulation tube
>   Flexible and durable—good for runs to the desktop
>   Impossible to connectorize in the field (bought as patch cords)
>   Higher attenuation than solid-wire UTP—Used only in short runs
>>   From wall jack to desktop
>>   Within a telecommunications closet (see Chapter 3)

Figure 3a-1   Solid-Wire and Stranded-Wire UTP (Study Figure)

Adding Connectors

It is relatively easy to add RJ-45 connectors to solid-wire UTP cords. However, it is very difficult to add RJ-45 connectors to stranded-wire cords. Stranded-wire patch cords should be purchased from the factory precut to desired lengths and preconnectorized.

In addition, when purchasing equipment to connectorize solid-wire UTP, it is important to purchase crimpers designed for solid wire.

CUTTING THE CORD

Solid-wire UTP normally comes in a box or spool containing 50 meters or more of wire. The first step is to cut a length of UTP cord that matches your need. It is good to be a little generous with the length. This way, bad connectorization can be fixed by cutting off the connector and adding a new connector to the shortened cord. Also, UTP cords should never be subjected to pulls (strain), and adding a little extra length creates some slack.

STRIPPING THE CORD

Now the cord must be stripped at each end using a **stripping tool** such as the one shown in Figure 3a-2. The installer rotates the stripper once around the cord, scoring (cutting into) the cord jacket (but not cutting through it). The installer then pulls off the scored end of the cord, exposing about 5 cm (about two inches) of the wire pairs.

It is critical not to score the cord too deeply, or the insulation around the individual wires may be cut. This creates short circuits. A really deep cut also will nick the wire, perhaps causing it to snap immediately or later.

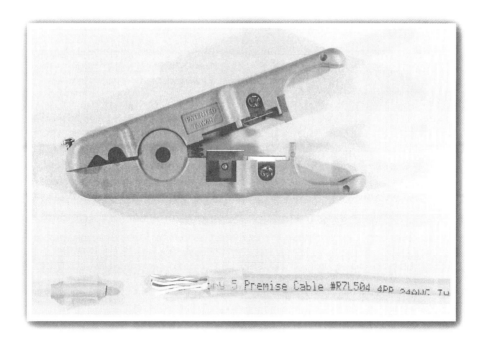

Figure 3a-2   Stripping Tool

WORKING WITH THE EXPOSED PAIRS

Pair Colors

The four pairs each have a color: orange, green, blue, or brown. One wire of the pair usually is a completely solid color. The other usually is white with stripes of the pair's color. For instance, the orange pair has an orange wire and a white wire with orange stripes.

Untwisting the Pairs

The wires of each pair are twisted around each other several times per inch. These must be untwisted after the end of the cord is stripped.

Ordering the Pairs

The wires now must be placed in their correct order, left to right. Figure 3a-3 shows the location of Pin 1 on the RJ-45 connector and on a wall jack or NIC.

Which color wire goes into which connector slot? The two standardized patterns are shown in Figure 3a-4. The T568B pattern is much more common in the United States.

The connectors at both ends of the cord use the same pattern. If the white-orange wire goes into Pin 1 of the connector on one end of the cord, it also goes into Pin 1 of the connector at the other end.

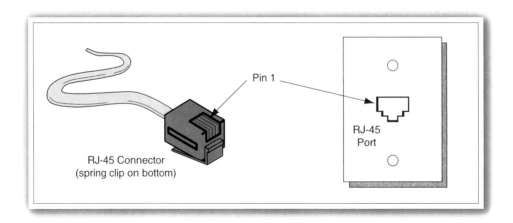

Figure 3a-3   Location of Pin 1 on an RJ-45 Connector and Wall Jack or NIC

Cutting the Wires

The length of the exposed wires must be limited to 1.25 cm (half an inch) or slightly less. After the wires have been arranged in the correct order, a cutter should cut across the wires to make them this length. The cut should be made straight across, so that all wires are of equal length. Otherwise, they will not all reach the end of the connector when they are inserted into it. Wires that do not reach the end will not make electrical contact.

Pin*	T568A	T568B
1	White-Green	White-Orange
2	Green	Orange
3	White-Orange	White-Green
4	Blue	Blue
5	White-Blue	White-Blue
6	Orange	Green
7	White-Brown	White-Brown
8	Brown	Brown

Note: Do not confuse T568A and T568B
pin colors with the TIA/EIA-568 Standard.

Figure 3a-4   T568A and T568B Pin Colors

ADDING THE CONNECTOR

Holding the Connector

The next step is to place the wires in the RJ-45 connector. In one hand, hold the connector, clip side down, with the opening in the back of the connector facing you.

Sliding in the Wires

Now, slide the wires into the connector, making sure that they are in the correct order (white-orange on your left). There are grooves in the connector that will help. Be sure to push the wires all the way to the end or proper electrical contact will not be made with the pins at the end.

Before you crimp the connector, look down at the top of the connector, holding the tip away from you. The first wire on your left should be mostly white. So should every second wire. If they are not, you have inserted your wires incorrectly.[2]

Some Jacket inside the Connector

If you have shortened your wires properly, there will be a little bit of jacket inside the RJ-45 connector.

CRIMPING

Pressing Down

Get a really good **crimping tool** (see Figure 3a-5). Place the connector with the wires in it into the crimp and push down firmly. Good crimping tools have ratchets to reduce the chance of your pushing down too tightly.

Making Electrical Contact

The front of the connector has eight pins running from the top almost to the bottom (spring clip side). When you **crimp** the connector, you force these eight pins through the insulation around each wire and into the wire itself. This seems like a crude electrical connection, and it is. However, it normally works very well. Your wires are now connected to the connector's pins. By the way, this is called an **insulation displacement connection (IDC)** because it cuts through the insulation.

Strain Relief

When you crimp, the crimper also forces a ridge in the back of the RJ-45 connector into the jacket of the cord. This provides **strain relief,** meaning that if someone pulls on the cord (a bad idea), they will be pulling only to the point where the jacket has the ridge forced into it. There will be no strain where the wires connect to the pins.

---

2   Thanks to Jason Okumura, who suggested this way of checking the wires.

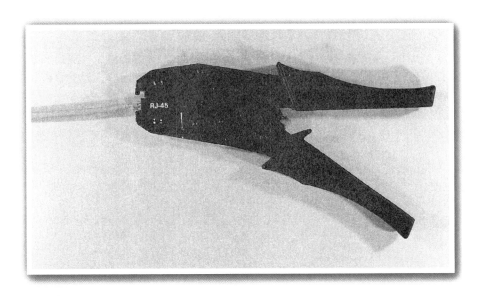

Figure 3a-5   Crimping Tool

TESTING

Purchasing the best UTP cabling means nothing unless you install it properly. Wiring errors are common in the field, so you need to test every cord after you install it. Testing is inexpensive compared to troubleshooting subtle wiring problems later.

Testing with Continuity Testers

The simplest testers are **continuity testers,** which merely test whether the wires are arranged in correct order within the two RJ-45 connectors and are making good electrical contact with the connector. They cost only about $100.

Testing for Signal Quality

Better testers cost $500 to $2,000 but are worth the extra money. In addition to testing for continuity problems, they send **test signals** through the cord to determine whether the cord meets TIA/EIA-568 signal quality requirements. Many include **time domain reflectometry (TDR),** which sends a signal and listens for echoes in order to measure the length of the UTP cord or to find if and where breaks exist in the cord.

TEST YOUR UNDERSTANDING

1. a) Explain the technical difference between solid-wire UTP and stranded-wire UTP. b) In what way is solid-wire UTP better? c) In what way is stranded-wire UTP better? d) Where would you use each? e) Which should only be connectorized at the factory?

2. If you have a wire run of 50 meters, should you cut the cord to 50 meters? Explain.

3. Why do you score the jacket of the cord with the stripping tool instead of cutting all the way through the jacket?

4. a) What are the colors of the four pairs? b) If you are following T568B, which wire goes into Pin 3? c) At the other end of the cord, would the same wire go into Pin 3?

5. After you arrange the wires in their correct order and cut them across, how much of the wires should be exposed from the jacket?

6. a) Describe RJ-45's insulation displacement approach. b) Describe its strain relief approach.

7. a) Should you test every cord in the field after installation? b) For what do inexpensive testers test? c) For what do expensive testers test?

CHAPTER 4

ETHERNET LANs

By the end of this chapter, you should be able to discuss:

- Ethernet physical layer standards and how they affect network design.
- The Ethernet data link layer and the Ethernet MAC layer frame.
- Basic Ethernet data link layer switch operation.
- Advanced aspects of Ethernet switch operation.
- Ethernet switch purchasing criteria.

INTRODUCTION

Dominance in the LAN Market

According to Nortel Networks, 95 percent of LAN ports were Ethernet ports in 2002. That percentage is even higher today. As the dominant LAN technology, Ethernet deserves close attention. In this chapter we will look at Ethernet in detail. In the next chapter, we will look at the other main LAN technology today—wireless LAN technology. Although wireless LAN technology is still small today, it is growing explosively.

Ethernet became the dominant LAN technology because of its simple data link layer operation, which we saw briefly in Chapters 1 and 2 and which we will look at in more detail in this chapter. This simplicity means that Ethernet switches are inexpensive per frame handled. This cost advantage has driven competing technologies out of the corporate LAN market.

TEST YOUR UNDERSTANDING

1. a) What is the dominant LAN technology today? b) Why did it become dominant?

Ethernet Standards Development

Metcalfe and Boggs created Ethernet technology at the Xerox Palo Alto Research Center in the mid-1970s.[1] In the early 1980s, Xerox, Intel, and Digital Equipment Corporation teamed up to produce the first two commercial standards for Ethernet: Ethernet I and Ethernet II.

The 802 Committee
After creating Ethernet II, the three companies passed responsibility for Ethernet standards to the newly created **802 LAN/MAN Standards Committee** of the **Institute for Electrical and Electronics Engineers (IEEE).**

The 802.3 Ethernet Working Group
The **"802 Committee,"** as everybody calls it, is broadly responsible for creating local area network standards and metropolitan area network standards.

The 802 Committee delegates the actual work of developing standards to specific **working groups.** The 802 Committee's **802.3 Working Group,** for example, creates Ethernet-specific standards. We will use the terms "Ethernet" and "802.3" interchangeably in this book.

The 802 Committee has several other working groups. For example, the 802.11 Working Group creates the wireless LAN standards we will see in the next chapter. In turn, the 802.16 Working Group is creating the WiMax standards for wireless subscriber access within a city.

Ethernet Standards are OSI Standards
Ethernet standards are LAN standards, so they are Layer 1 (physical) and Layer 2 (data link) standards. Recall from Chapter 2 that standards at the lowest two layers are

---

[1]   Bob Metcalfe has noted that he conceived of Ethernet after visiting the University of Hawai`i's packet radio Alohanet project. (Bob Metcalfe, "Internet Fogies Reminisce and Argue at Interop Confab," *Infoworld*, September 21, 1992, p. 45.)

always OSI standards. Although the 802.3 Working Group creates Ethernet standards, these standards are not official OSI standards until they are ratified later by ISO. In practice, however, as soon as the 802.3 Working group releases an 802.3 standard, vendors begin building products based on the specification.

TEST YOUR UNDERSTANDING

2. a) What working group creates Ethernet standards? b) To what committee does this working group report? c) In what organization is this committee? d) Are there other working groups? If so, what do they do? (The answer is not explicitly given in the text.) e) Are "Ethernet" and "802.3" used interchangeably in this book? f) Why would you expect Ethernet standards to be OSI standards?

Major Ethernet Physical Layer Standards

As just noted, Ethernet is a LAN technology, so standards must be set at both the physical and data link layers. We will look at 802.3 physical layer standards first.

When the 802.3 committee first created physical layer Ethernet standards, it used technologies that are no longer in use. Figure 4-1 shows the Ethernet physical layer standards currently in use.

Baseband Transmission

As the figure shows, most Ethernet physical layer standards have "Base" in their names. This is short for "baseband." In **baseband** transmission, the signal is simply injected into the transmission medium, as Figure 4-2 illustrates. For UTP, signals consist of voltage changes that propagate down the wires. For fiber, signals are light pulses.

Broadband Transmission

In contrast, in **broadband** transmission, signals are sent in radio channels. However, radio-based broadband transmission is more expensive than baseband transmission. Although the 802.3 Working Group came up with early broadband LAN standards, they did not thrive because of their high cost.

Transmission Speed

The names of Ethernet physical layer standards also indicate transmission speeds— 10 Mbps, 100 Mbps, or more. These are the speeds of NICs and switch ports. For example, a 12-port gigabit Ethernet switch will be able to send and receive at 12 Gbps across all of its ports.

10 Mbps Physical Layer 802.3 Standards

The slowest Ethernet physical layer standards in use today operate at 10 Mbps. The **802.3 10Base-T** standard uses 4-pair UTP wiring. A 10 Mbps fiber standard, **10Base-F** was created but saw little use and is now almost entirely extinct. Even 10Base-T is no longer common today.

100 Mbps Physical Layer 802.3 Standards

Next, the 802.3 Working Group produced two 100 Mbps standards for UTP (T) and optical fiber (F). **100Base-TX** is the dominant technology connecting stations to

Physical Layer Standard	Speed	Maximum Run Length	Medium
UTP			
10Base-T	10 Mbps	100 meters	4-pair Category 3 or higher
100Base-TX	100 Mbps	100 meters	4-pair Category 5 or higher
1000Base-T	1,000 Mbps	100 meters	4-pair Category 5 or higher
Optical Fiber			
100Base-FX	100 Mbps	2 km	62.5/125 multimode, 1,300 nm, switch.
1000Base-SX	1 Gbps	220 m	62.5/125 micron multimode, 850 nm. 160 MHz-km modal bandwidth.
1000Base-SX	1 Gbps	275 m	62.5/125 micron multimode, 850 nm. 200 MHz-km modal bandwidth.
1000Base-SX	1 Gbps	500 m	50/125 micron multimode, 850 nm 400 MHz-km modal bandwidth
1000Base-SX	1 Gbps	550 m	50/125 micron multimode, 850 nm 500 MHz-km modal bandwidth
1000Base-LX	1 Gbps	550 m	62.5/125 micron multimode. 1,300 nm.
1000Base-LX	1 Gbps	5 km	9/125 micron single mode, 1,300 nm.
10GBase-SR/SW	10 Gbps	65 m	62.5/125 micron multimode (850 nm).
10GBase-LX4	10 Gbps	300 m	62.5/125 micron multimode 1310 nm, wave division multiplexing.
10GBase-LR/LW	10 Gbps	10 km	9/125 micron single mode, 1310 nm.
10GBase-ER/EW	10 Gbps	40 km	9/125 micron single mode, 1550 nm.
40 Gbps Ethernet	40 Gbps	Under development	9/125 micron single mode.

Notes:
For 10GBase-x, LAN versions (R) transmit at 10 Gbps. WAN versions (W) transmit at 9.95328 Gbps for carriage over SONET/SDH links (see Chapter 6).
The 40 Gbps Ethernet standards are still under preliminary development.

Figure 4-1   Ethernet Physical Layer Standards

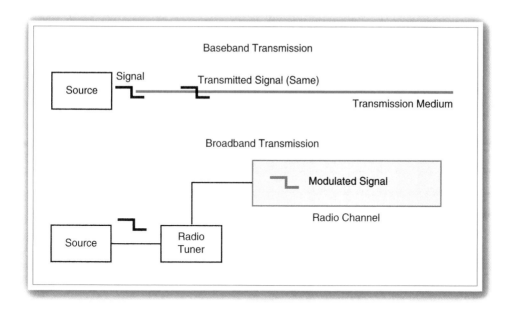

Figure 4-2   Baseband Versus Broadband Transmission

switches today. **100Base-FX,** in contrast, has been almost entirely phased out in favor of faster fiber-based Ethernet standards.[2]

These standards are sometimes called **10/100 Ethernet** because 100Base-TX and 100Base-FX NICs and switches can do **autosensing,** which means that if they detect a 10Base-x NIC or switch at the other end of the connection, they will slow down to 10 Mbps. However, they are full 100 Mbps technologies.

The 100 Mbps Ethernet standards are also referred to as **Fast Ethernet.** Although 100 Mbps is now rather slow, the term Fast Ethernet is still in widespread use for 100 Mbps Ethernet.

Gigabit Ethernet (1000Base-x)

Advancing by another factor of 10, the 802.3 Working Group then produced **gigabit Ethernet (1000Base-x).** This physical layer technology is the dominant trunk line standard for connecting switches to switches, switches to routers, and routers to routers. However, it is increasingly being used to connect servers and some desktops to the switches that serve them.

In addition to a UTP version, **1000Base-T,** there are two fiber versions of gigabit Ethernet. The **1000Base-SX** (short wavelength) version transmits at 850 nm using inexpensive LED light signaling and multimode fiber. This limits transmission

---

[2]   Why TX and FX instead of T and F? The 802.3 Working Group created other 100 Mbps Ethernet standards, but only the TX and FX standards saw market acceptance.

distance to 220 m for older 62.5 micron fiber with a modal bandwidth of 160 MHz-km and to 275 m for newer 62.5 micron fiber with a modal bandwidth of 200 MHz-km.

The **1000Base-LX** (long wavelength) version uses more expensive lasers to transmit at 1,300 nm. It can send data twice as far as SX using multimode fiber. With single-mode fiber, it can send data 5 km. However, gigabit Ethernet LAN usage is dominated by 1000Base-SX, which is less expensive and can span most LAN distances.

10GBase-x

The next step, 10 Gbps technology, uses optical fiber almost exclusively, although the 802.3 Working Group is developing copper wiring standards for short runs within telecommunications closets, where switches to be connected are only a few meters apart.[3] Figure 4-1 shows that the 802.3 Working Group created quite a few 10 Gbps fiber standards.

Several of these standards come in two versions. LAN versions (R) operate at 10 Gbps. WAN versions (W) run over SONET/SDH transmission lines (see Chapter 7). These use a speed of 9.95328, which is the closest SONET/SDH speed to 10 Gbps. The X standards also run at a true 10 Gbps.

In these standards, S indicates 850 nm transmission, while L indicates 1,300 nm transmission, and E (extremely long wavelength) indicates 1,550 nm transmission.

Why focus on SONET/SDH speeds? The answer is that 10 Gbps Ethernet was developed initially for *wide* area networking, or, more correctly, metropolitan area networking within a city. We will see this application in Chapter 7. However, 10 Gbps Ethernet is now used within corporate LANs as well.

40GBase-x

The 802.3 Working Group is now working on the next generation of physical layer Ethernet standards, 40GBase-x. These standards will depart from the traditional practice of increasing speed by a factor of ten in each generation. Like 10GBase-x, they will be developed primarily for WAN carrier networks.

TEST YOUR UNDERSTANDING

3. a) What can you infer from the name 100Base-TX? b) Distinguish between baseband and broadband operation. c) Why does baseband transmission dominate for LANs? d) What is 10/100 Ethernet? e) What is autosensing? f) What is fast Ethernet? g) What is the most widely used 802.3 physical layer standard for connecting stations to switches today? h) What is the most widely used 802.3 standard for connecting switches to other switches today? i) Which versions were created initially for WAN use?

TIA/EIA-568 and IEEE 802.3

TIA/EIA-568: Raw Media

We saw in Chapter 3 that the TIA/EIA-568 standard governs transmission media, including both fiber and UTP. The TIA/EIA-568 standard describes optical fiber core diameters, operating wavelengths, modal bandwidth, and other basic characteristics of fiber and UTP.

---

[3] The 10GBase-CX4 standard is being developed for runs of less than 15 meters. It will use copper wiring in which the pairs are not twisted but in which metal meshes will surround the wires to reduce interference. The 10GBase-T standard probably will use Category 6 UTP and is likely to be able to carry signals more than 30 meters.

IEEE 802.3: Electronics and Signaling

Figure 3-1 shows that the IEEE 802.3 Working Group, in turn, specifies how different types of fiber can be used in various types of Ethernet transmission lines. Ethernet 802.3 specifies the electronics and light signaling at the two ends. In short, it specifies how raw fiber is turned into Ethernet transmission lines.

TEST YOUR UNDERSTANDING

4.   Distinguish between what TIA/EIA-568 and 802.3 standardize at the physical layer.

Link Aggregation (Trunking)

Ethernet transmission capacity usually increases by a factor of ten. What if you only need somewhat more transmission speed than a certain standard? For instance, what if you need 130 Mbps instead of 100 Mbps?

Figure 4-3 illustrates that sometimes a pair of switches is connected by *two* trunk links. This requires the switches to implement the 802.3ad standard. The IEEE calls this **link aggregation.** Many networking professionals call this **trunking** or **bonding.**

Link aggregation allows you to increase trunk speed incrementally by a factor of two or three, instead of by a factor of ten. This incremental growth uses existing ports and is inexpensive compared to upgrading a switch to the next higher Ethernet speed. However, beyond a few trunk links, it is cheaper to move up to the next-higher port speed. After two or three aggregated links, the company should compare the cost of link aggregation with the cost of a ten-fold increase in capacity by moving up to the next Ethernet trunk line speed. Going to a single faster trunk line will also give more room for growth.

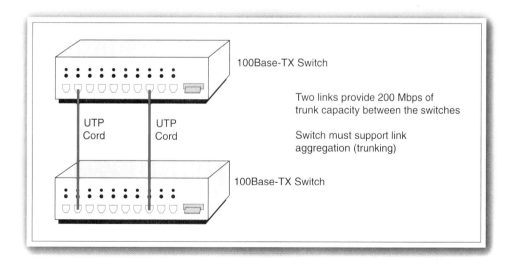

Figure 4-3   Link Aggregation (Trunking)

TEST YOUR UNDERSTANDING

5. a) What is link aggregation (trunking)? b) Why may link aggregation be desirable compared to simply installing a single faster link? c) Why may it not be desirable if you will need several aggregated links to meet capacity requirements?

Ethernet Physical Layer Standards and Network Design

Using Figure 4-1

Note that if you know the speed you need (100 Mbps, 1 Gbps, and so forth), and if you know what distance you need to span, Figure 4-1 will show you what type of transmission link you need. If link aggregation is also available with your switches, you have even more choices.

For instance, if you need a speed of 1 Gbps, and if your two switches are 500 meters apart, you would select 1000Base-LX multimode fiber to minimize cost.

Alternatively, if you are designing a network from scratch, say for a new facility, the options in Figure 4-1 will allow you to consider alternative placements for your switches. If you can place your switches farther apart on average, for instance, you can reduce the total number of switches, and this can save money—although each of the switches in a more dispersed network will cost more because each will have to handle more traffic.

Switches Regenerate Signals to Extend Distance

The 100-meter limit for UTP and the longer distance limits for fiber shown in Figure 4-1 only apply to connections between *a pair of devices*—for example, a station and a switch, two switches, or a switch and a router.

What if longer distances separate the source host and destination host? Figure 4-4 shows a data link with two intermediate switches. In addition to the two 100-meter

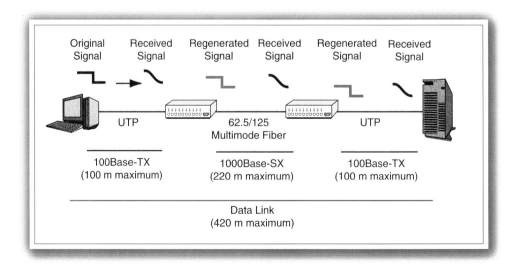

Figure 4-4   Data Link Using Multiple Switches

maximum length UTP access links, there is a 220-meter maximum length 1000Base-SX optical fiber link (using 62.5/125-micron 160-MHz-km modal bandwidth fiber) between the two switches. This setup can span a maximum of 420 meters.

Each switch along the way **regenerates** the signal. If the signal sent by the source host begins as a 1, it is likely to be distorted before it reaches the first switch. The first switch recognizes it as a 1 and generates a clean new 1 signal to send to the second switch. The second switch regenerates the 1 as well.

The key point is that Figure 4-1 shows maximum distances between *pairs of devices, not end-to-end transmission distances.* To deliver frames over long distances, intermediate switches regenerate the signal. There is no maximum end-to-end distance between pairs of stations in an Ethernet network. Although cumulative latency might be a problem with a dozen or more intermediate switches, this is rarely is a problem in real networks.

There is no maximum end-to-end distance between pairs of stations in an Ethernet network.

TEST YOUR UNDERSTANDING

6. a) How might you use Figure 4-1 in network design? b) If more than one type of Ethernet standard in Figure 4-1 can span the distance you need, what would determine which one you choose? c) In Figure 4-1, is the distance shown the maximum for a single physical link or for the end-to-end distance between two stations across multiple switches? d) At what layer or layers is the 802.3 100Base-TX standard defined—physical, data link, or internet? e) How does regeneration allow a firm to create LANs that span very long distances? f) If you need to span 300 meters using 1000Base-SX, what options do you have? (Include the possibility of using an intermediate switch.)

THE ETHERNET FRAME

Layering

The Logical Link Control Layer

When the 802 Committee assumed control over Ethernet standardization, it realized that it would have to standardize non-Ethernet LAN technology as well. Consequently, the 802 Committee divided the data link layer into two layers, as Figure 4-5 illustrates. For the upper part of the data link layer, the committee added a layer of functionality beyond that of individual LAN technologies. This is the **logical link control (LLC)** layer. However, time has proved the added functionality of LLC to be of little value, so it is now largely ignored. As we will see below, it adds an LLC subheader to each 802.3 frame. There is only a single LLC standard, **802.2.**

The 802.3 MAC Layer Standard

The lower part of the data link layer is the MAC layer. **"MAC"** stands for **media access control.** The MAC layer defines functionality specific to a particular LAN technology.

Note in Figure 4-5 that while Ethernet (802.3) has many physical layer standards, it only has a single media access control layer standard, the **802.3 MAC Layer Standard.** This standard primarily defines Ethernet frame organization and NIC and switch operation.

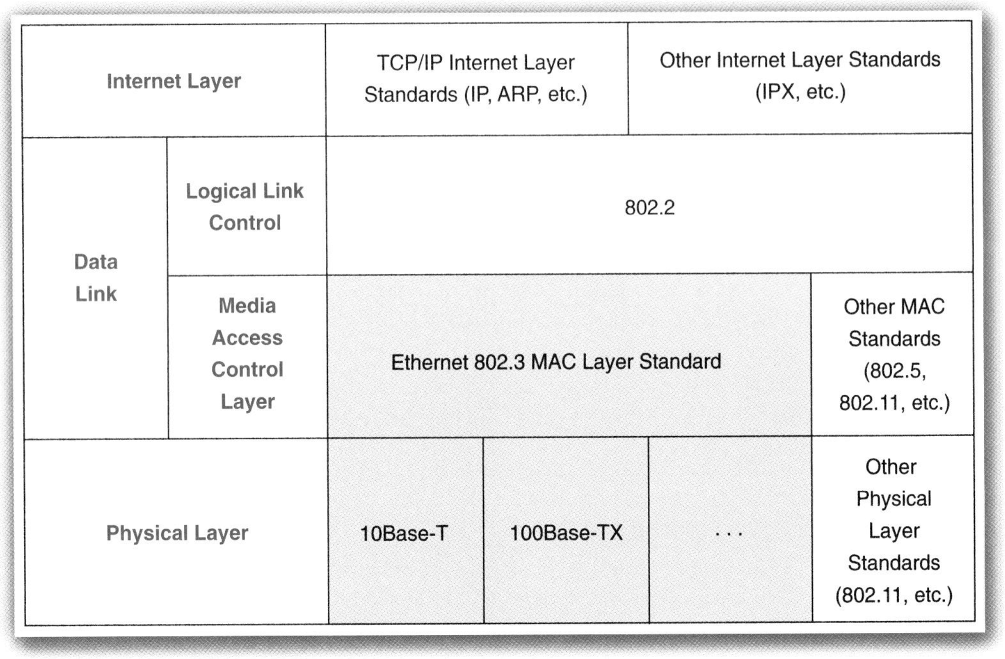

Internet Layer		TCP/IP Internet Layer Standards (IP, ARP, etc.)	Other Internet Layer Standards (IPX, etc.)	
Data Link	Logical Link Control	802.2		
	Media Access Control Layer	Ethernet 802.3 MAC Layer Standard		Other MAC Standards (802.5, 802.11, etc.)
Physical Layer		10Base-T	100Base-TX	. . .	Other Physical Layer Standards (802.11, etc.)

Figure 4-5   Layering in 802 Networks

TEST YOUR UNDERSTANDING

7.   a) Distinguish between the MAC and LLC layers. b) Does Ethernet have multiple physical layer standards? c) Does it have multiple MAC standards? d) What is the name of its single MAC standard?

The Ethernet Frame

Figure 4-6 shows the Ethernet MAC layer frame. We saw it briefly in Chapter 2. We will now look at the Ethernet frame in more depth.

Preamble and Start of Frame Delimiter Fields

Before a play in American football, the quarterback calls out something like "Hut one, hut two, hut three, hike!" This cadence synchronizes all of the offensive players.

In the Ethernet MAC frame, the **preamble field** (7 octets) and **start of frame delimiter field** (1 octet) synchronize the receiving clock to the sender's clock. These fields have a strong rhythm of alternating ones and zeros. The last bit in this sequence is a one instead of the expected zero, to signal that the synchronization is ended.

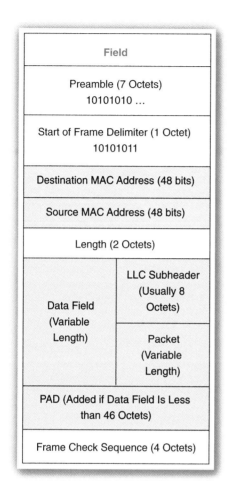

Field
Preamble (7 Octets) 10101010 …
Start of Frame Delimiter (1 Octet) 10101011
Destination MAC Address (48 bits)
Source MAC Address (48 bits)
Length (2 Octets)

Figure 4-6   The Ethernet Frame

Source and Destination Address Fields

Hex Notation   We saw in Chapter 2 that the source and destination address fields are 48 bits long and that while computers work with this raw 48-bit form, humans normally express these addresses in Base 16 **hexadecimal notation,** which is usually called **hex notation.**

➤ First, the 48 bits are divided into twelve 4-bit units, which computer scientists call nibbles.

➤ Second, each nibble is converted into a hexadecimal symbol, using Figure 4-7.

➤ Third, the symbols are written as six pairs with a dash between each pair, for instance B2-CC-67-0D-5E-BA. (Each pair represents one octet.)

4 Bits (Base 2)*	Decimal (Base 10)	Hexadecimal (Base 16)	4 Bits (Base 2)*	Decimal (Base 10)	Hexadecimal (Base 16)
0000	0	0 hex	1000	8	8 hex
0001	1	1 hex	1001	9	9 hex
0010	2	2 hex	1010	10	A hex
0011	3	3 hex	1011	11	B hex
0100	4	4 hex	1100	12	C hex
0101	5	5 hex	1101	13	D hex
0110	6	6 hex	1110	14	E hex
0111	7	7 hex	1111	15	F hex

* 2^4=16 combinations
For example, A1-34-CD-7B-DF hex begins with 10100001 for A1.

Figure 4-7   Hexadecimal Notation

MAC Layer Addresses   Ethernet addresses exist at the MAC layer, so they usually are called **MAC addresses.** They are also called **physical addresses** because Ethernet is implemented by physical devices—network interface cards.

Length Field

The **length field** contains a binary number that gives the length of the data field (not of the entire frame) in octets.

Data Field

The **data field** contains the information that the frame is delivering. The data field usually is far longer than all other fields combined.

LLC Subheader   The data field begins with the **logical link control layer (LLC) subheader.** This normally is 8 octets long. The purpose of the LLC subheader is to describe the type of packet contained in the data field. For instance, if the LLC subheader ends with the code 08-00 hex, the data field contains an IP packet.[4]

---

[4]   The LLC subheader has several fields. In the SNAP version of LLC, which is almost always used, the first six octets are always AA-AA-03 hex. The next three octets are almost always 00-00-00 hex. The final two octets constitute the Ethertype field, which specifies the kind of packet in the data field. Common hexadecimal Ethertype values are 0800 (IP), 8137 (IPX), 809B (AppleTalk), 80D5 (SNA services), and 86DD (IP version 6).

Content   The packet contained in the data field usually is an IP packet. However, it could also be a packet from another standards architecture, say an IPX packet. As long as the source and destination stations understand the packet format, there is no problem.

Minimum and Maximum Lengths   In Ethernet, the maximum data field length is 1,500 octets.[5] There is no minimum length.

PAD Field

Although there is no minimum length for 802.3 MAC layer frame data fields, if the data field is less than 46 octets long, a **PAD field** will be added by the sender so that the total length of the data field and the PAD field is exactly 46 octets long. For instance, if the data field is 26 octets long, the sender will add a 20-octet PAD field. If the data field is 46 octets long or longer, the sender will not add a PAD field.

Frame Check Sequence Field

As noted in Chapter 2, the last field in the Ethernet frame is the frame check sequence field, which is used for error detection. This is a four-octet field. If an error in a transmitted frame is detected, the receiver simply discards the frame.

TEST YOUR UNDERSTANDING

8.   a) What is the purpose of the preamble and start of frame delimiter fields? b) Why are Ethernet addresses called MAC addresses or physical addresses? c) What are the steps in converting 48-bit MAC addresses into hex notation? d) The length field gives the length of what? e) What are the two parts of the data field? f) What is the purpose of the LLC subheader? g) What type of packet is usually carried in the data field? h) What is the maximum length of the data field? i) Who adds the PAD field—the sender or the receiver? j) Is there a minimum length for the data field? k) If the data field is 40 octets long, how long a PAD field must be added? l) If the data field is 400 octets long, how long a PAD field must be added? m) What is the purpose of the frame check sequence field? n) What happens if the receiver detects an error in a frame? o) Convert 11000010 to hex.

BASIC DATA LINK LAYER SWITCH OPERATION

In this section we will discuss the basic data link layer operation of Ethernet switches. This is also governed by the 802.3 MAC layer standard. In the section after this one, we will discuss other aspects of Ethernet switching that a firm may or may not use.

Frame Forwarding with Multiple Ethernet Switches

Figure 4-8 shows an Ethernet LAN with three switches. Larger Ethernet LANs have dozens of switches, but the operation of individual switches is the same when there are only a few switches or many.

---

[5]   Most NICs and switches today can handle nonstandard jumbo frames, which have data fields up to 9,000 octets in length. This allows long messages to be sent with fewer frames. Handling fewer frames per message makes NICs and switches more efficient. However, jumbo frames are bad for voice over IP because their length creates more latency at each switch.

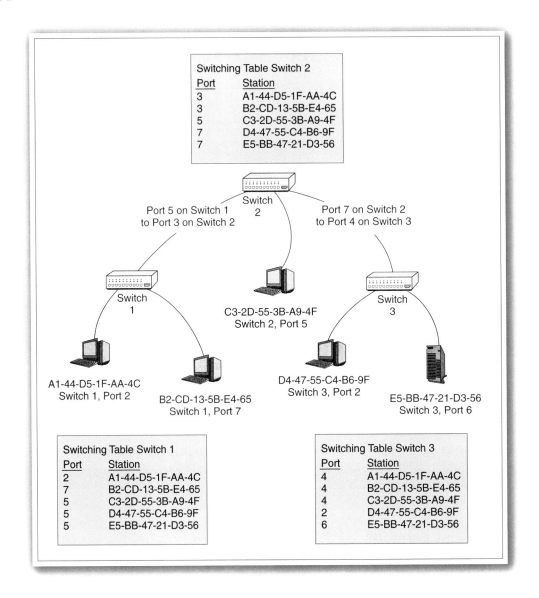

Figure 4-8 Multiswitch Ethernet LAN

On Switch 1

Suppose that Station A1-44-D5-1F-AA-4C on Switch 1 transmits a frame destined for Station E5-BB-47-21-D3-56, which is attached to Switch 3. The frame will have to pass through Switch 2 along the way.

Switch 1 will look at its switching table and note that Station E5-BB-47-21-D3-56 is out Port 5. The switch will send the frame out that port. The link out that port will carry the frame to Switch 2 instead of directly to the destination station.

On Switch 2

Switch 2 will receive the frame in Port 3. It will note that the destination address of the frame is E5-BB-47-21-D3-56. It will look up this address in its own switching table and see that it should send the frame out Port 7. This will take the frame to Switch 3.

On Switch 3

Switch 3 will receive the frame in Port 4. It will note the destination MAC address and look this address up in its own switching table. It will see that Station E5-BB-47-21-D3-56 is out Port 6. It will send the frame out that port, to the destination station. This completes the frame delivery process.

TEST YOUR UNDERSTANDING

9. In a single-switch LAN, the switch reads the address of an incoming frame, looks up an output port in the switching table, and sends the frame out that port. Do individual switches work differently in multiswitch LANs? Explain.

Hubs

Before switches became economical to use, Ethernet LANs generally used simpler devices called hubs. As Figure 4-9 shows, switches send an incoming frame out a single port. In contrast, **hubs** broadcast each arriving bit out all ports except for the port that receives the signal.

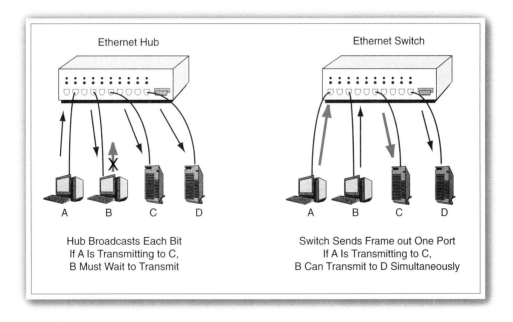

Figure 4-9   Hub Versus Switch Operation

Media Access Control (MAC)

With switches, several stations can transmit at the same time. With a hub, however, if two stations transmit at the same time, their signals will have a **collision** and will be scrambled. Consequently, NICs connected to hubs must be disciplined in when they transmit. Controlling when stations transmit is called **media access control (MAC).**

CSMA/CD

The media access control method used with Ethernet hubs is **carrier sense multiple access with collision detection (CSMA/CD).** Although this sounds complex, it is very simple. Under carrier sense multiple access (CSMA), if a station wants to transmit, it may do so if no station is already transmitting but must wait if another station is already sending. Under collision detection (CD), if there is a collision because two stations sending at the same time, all stations stop transmitting, wait a random period of time, and then try again.

Full-Duplex Operation

NICs that use hubs must work in **half-duplex** mode, meaning that two communicating NICs must take turns transmitting. NICs that use switches instead of hubs use **full-duplex** operation, which means that both parties in a conversation can send and receive simultaneously. In fact, stations that use switches can send any time they choose because there is no danger of collisions with switches. In essence, full-duplex operation means turning off CSMA/CD.[6]

TEST YOUR UNDERSTANDING

10.   a) Why are hubs undesirable? b) What is media access control, and why must NICs that work with hubs use it? c) When NICs work with hubs, what media access control method do they use? d) How does it work?

Hierarchical Switch Topology

Hierarchical Switch Organization

Note that the switches in Figure 4-8 are organized in a **hierarchy,** in which each switch has only one parent switch above it. In fact, Ethernet *requires* a **hierarchical topology** (organization) for its switches. Otherwise, loops would exist, causing frames to circulate endlessly from one switch to another around the loop or causing other problems. Figure 4-10 shows a larger switched Ethernet LAN organized in a hierarchy.

Single Possible Path between End Stations

In a hierarchy, there is only a single possible path between any two end stations. (To see this, select any two stations at the bottom of the hierarchy and trace a path between them. You will see that only one path is possible.)

Workgroup Versus Core Switches

In a hierarchy of Ethernet switches, there are workgroup switches and core switches. Figure 4-10 illustrates these two types of switches.

---

[6]   Oddly, turning off CSMA/CD is referred to as full-duplex CSMA/CD in the 802.3 standard.

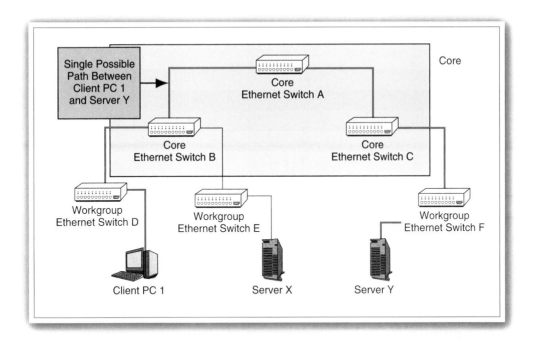

Figure 4-10 Hierarchical Ethernet LAN

➤ **Workgroup switches** are the ones that stations connect to directly via access lines (Switches D, E, and F in Figure 4-10). These switches carry traffic to and from their stations.

➤ **Core switches,** in turn, are switches farther up the hierarchy (Switches A, B, and C in Figure 4-10). They carry traffic via trunk lines between pairs of switches, switches and routers, and pairs of routers. The collection of all core switches plus the trunk lines that connect them is called the network's **core.**

Workgroup switches only handle the traffic of the stations they serve. However, core switches must be able to carry the conversations of dozens, hundreds, or thousands of stations. Consequently, core switches need to have much higher capacity than workgroup switches. Their cost also is much higher.

The dominant port speed for workgroup switches today is 100 Mbps. In contrast, the dominant port speed for core switches today is 1 Gbps, and some core switches already use port speeds of 10 Gbps.

TEST YOUR UNDERSTANDING

11. a) How are switches in an Ethernet LAN organized? b) Because of this organization, how many possible paths can there be between two end stations? c) In Figure 4-10, what is the single possible path between Client PC 1 and Server Y? d) Between Client PC 1 and Server X?

12. a) Distinguish between workgroup switches and core switches in terms of which devices they connect. b) How do they compare in terms of switching capacity? Why? c) How do they compare in terms of port speeds?

Only One Possible Path: Low Switching Cost

We have just seen that a hierarchy only allows one possible path between any two hosts.

If there is only a single possible path between any two stations, it follows that in every switch along the path, the destination address in a frame will appear only once in the switching table—for the specific outgoing port needed to send the frame on its way.

This allows a simple table lookup operation that is very fast and therefore costs little per frame handled. This is what makes Ethernet switches inexpensive. As noted in the introduction, simple switching operation and therefore low cost has led to Ethernet's dominance in LAN technology.

In Chapter 8 we will see that routers have to do much more work when a packet arrives because there are multiple alternative routes between any two hosts. Each of these alternative routes appears as a row in the routing table. Therefore, a router must first identify all possible routes (rows) and then select the best one—instead of simply finding a single match. This additional work per forwarding decision makes routers very expensive for the traffic load they handle. Network professionals say, "Switch where you can; route where you must."

TEST YOUR UNDERSTANDING

13. a) What is the benefit of having a single possible path? Explain in detail. b) Why has Ethernet become the dominant LAN technology?

ADVANCED SWITCH OPERATION

Now that we have discussed basic Ethernet switch operation involved in frame forwarding, we will begin looking at additional aspects of Ethernet switch operation important in larger Ethernet networks.

802.1D: The Spanning Tree Protocol (STP)

Single Points of Failure

Having only a single possible path between any two stations allows rapid frame forwarding and, therefore, low switch cost. Unfortunately, having only a single possible path between any two computers also makes Ethernet vulnerable to **single points of failure,** in which the failure of a single component (switch or trunk line between switches) can cause widespread disruption. To understand this, suppose that Switch 2 in Figure 4-11 fails. Then the stations connected to Switch 1 will not be able to communicate with stations connected to Switch 2 or Switch 3.

Even if the link between Switch 1 and Switch 2 fails, the network will be broken into two parts. Although the two parts might continue to function independently, many firms put most or all of their servers in a centralized server room. Clients on the other side of the broken network would lose much of their ability to continue working. For example, in the figure, Client A1-44-D5-1F-AA-4C, which is connected to Switch 1, cannot reach Server E5-BB-47-21-D3-56, which is connected to Switch 3. External connections also

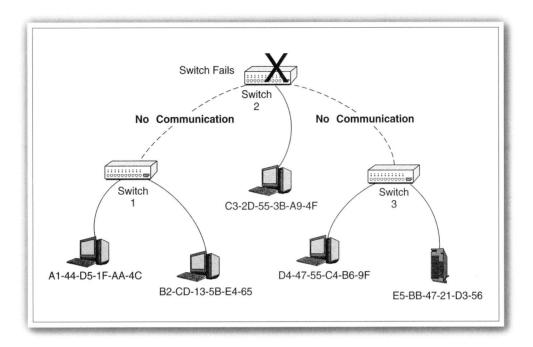

Figure 4-11   Single Point of Failure in a Switch Hierarchy

tend to be confined to a single network point for security reasons. Computers on the wrong side of the divide after a breakdown would lose external access.

Loops
Another problem is that it is very easy to create accidental loops among switches in large networks. Ethernet frame forwarding cannot tolerate loops because loops cause frames to circulate endlessly around the loop.

802.1D Spanning Tree Protocol (STP)
Fortunately, the **802.1D Spanning Tree Protocol (STP),** which is illustrated in Figure 4-12, addresses both single points of failure and loops.

Deactivating Loops
In the **Spanning Tree Protocol (STP),** the switches constantly talk to one another. If they detect a loop, they communicate intensively and find a way to restore a hierarchy by deactivating links that create loops. This is shown in the top half of Figure 4-12.

Redundancy
STP also allows redundancy in a network. The network can be designed for disabled links to be standby links, as shown in the bottom half of Figure 4-12. Here, Switch 2 has failed. The Spanning Tree Protocol allows the switches to detect a break and reconfigure

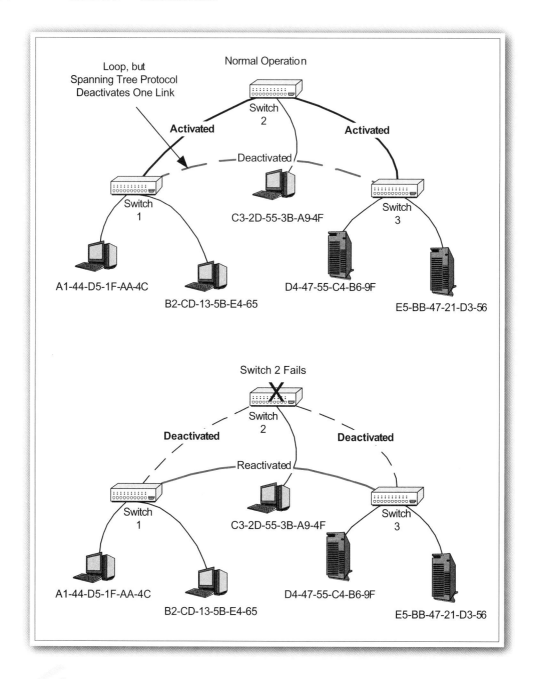

Figure 4-12   802.1D Spanning Tree Protocol

the network. Here, the failed switch is bypassed entirely by the reactivated link, so users of this switch are locked out of the network. However, if a single trunk link leading to the switch had failed, the Spanning Tree Protocol would have recreated a complete hierarchical network among the Ethernet switches.

Rapid Spanning Tree Protocol

The problem with STP is that when a problem occurs, the network converges (reorganizes itself rather slowly)—typically taking 30 to 60 seconds. A newer version of STP, the **Rapid Spanning Tree Protocol (RSTP)** converges in only about a second. Most switches today support RSTP.

TEST YOUR UNDERSTANDING

14. a) Why are loops bad in Ethernet? b) Why is having a single possible path between any two stations in an Ethernet network dangerous? c) What is a single point of failure? d) What can be done to reduce the danger created by the susceptibility of Ethernet to single points of failure? e) How would you use this approach if you have a network that is a strict hierarchy? f) What is the name of the standard that permits this? g) What is convergence? h) How can STP's slow convergence be remedied?

Virtual LANs and Ethernet Switches

Broadcasting

Ethernet NICs normally wish to **unicast,** that is, send a frame to *one* other station. Ethernet switches normally do unicasting, sending the frame out the single port that gets the frame closer to the destination station.

However, sometimes stations wish to **broadcast** messages, that is, send them to *all* other stations, as Figure 4-13 illustrates. Most notably, Novell NetWare servers may advertise their presence every sixty seconds or so by broadcasting a server advertisement message that is designed to go to all other stations. The server that wishes to broadcast will set the frame's destination MAC address to 48 ones (FF-FF-FF-FF-FF-FF). When an Ethernet switch sees this MAC address in the destination address of a frame, it broadcasts the frame out all other ports, like a hub. All NICs, in turn, process frames addressed to FF-FF-FF-FF-FF-FF as if it were their own address.

Congestion

Broadcasting is fine in small networks. However, in a large switched Ethernet LAN with many servers, broadcasting will produce a tremendous amount of traffic and therefore congestion. Quite simply, broadcasting does not scale.

Virtual LANs

As Figure 4-13 illustrates, most Ethernet switches allow stations to be grouped into closed collections of servers and the clients they serve. These collections are called **virtual LANs (VLANs).** When Server E (on VLAN 1) transmits, its frames go only to the clients on its VLAN (Client A and Client C). This reduces congestion.

Standardizing VLANs

Until recently, there was no standard for VLANs, so if you used VLANs, you had to buy all of your Ethernet switches from the same vendor. However, as Figure 4-14

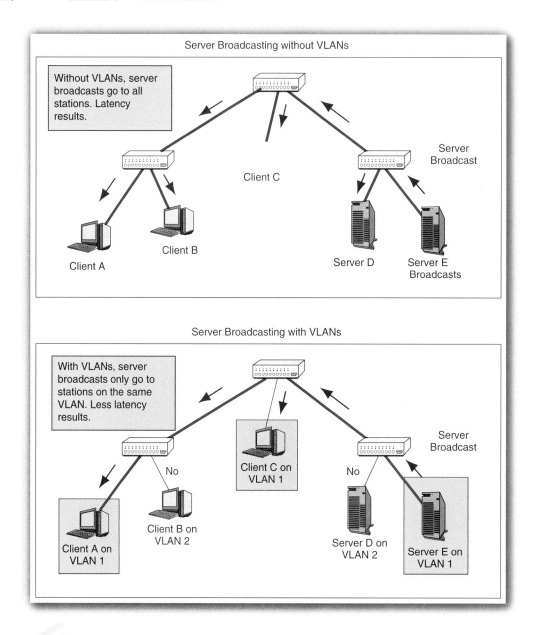

Figure 4-13   Virtual LAN (VLAN) with Ethernet Switches

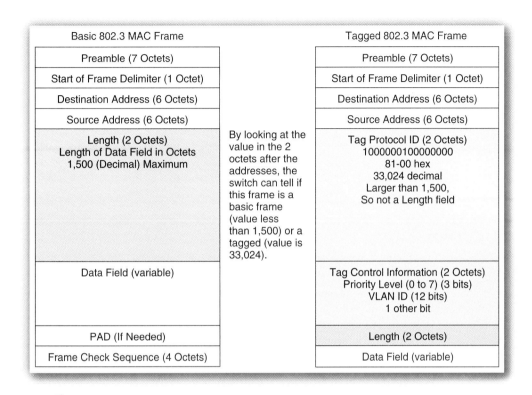

Basic 802.3 MAC Frame		Tagged 802.3 MAC Frame
Preamble (7 Octets)		Preamble (7 Octets)
Start of Frame Delimiter (1 Octet)		Start of Frame Delimiter (1 Octet)
Destination Address (6 Octets)		Destination Address (6 Octets)
Source Address (6 Octets)		Source Address (6 Octets)
Length (2 Octets) Length of Data Field in Octets 1,500 (Decimal) Maximum	By looking at the value in the 2 octets after the addresses, the switch can tell if this frame is a basic frame (value less than 1,500) or a tagged (value is 33,024).	Tag Protocol ID (2 Octets) 1000000100000000 81-00 hex 33,024 decimal Larger than 1,500, So not a Length field
Data Field (variable)		Tag Control Information (2 Octets) Priority Level (0 to 7) (3 bits) VLAN ID (12 bits) 1 other bit
PAD (If Needed)		Length (2 Octets)
Frame Check Sequence (4 Octets)		Data Field (variable)

Figure 4-14   Tagged Ethernet Frame (Governed by 802.1Q)

shows, the **802.1Q** standard is extending the Ethernet MAC layer frame to include two optional **tag fields.**

The first tag field **(Tag Protocol ID)** has the two-octet hexadecimal value 81-00, which simply indicates that the frame is tagged. The second tag field **(Tag Control Information)** contains a 12-bit VLAN ID that it sets to zero if VLANs are not being implemented. If VLANs are being used, each VLAN will be assigned a different VLAN ID. When a station on a VLAN transmits, the station adds the VLAN ID of its VLAN to the Tag Control Information field. The switches will read the VLAN ID to determine how to forward the frame. (The destination Ethernet address is set to 48 ones in broadcasts, so a switch can use *only* the VLAN ID to determine how to forward the frame.) With 12 bits VLAN IDs, there can be 4,095 ($2^{12} - 1$) different VLANs on an Ethernet network.

With VLANs, switches do not use their normal MAC address–port switching tables. Rather, they use a VLAN switching table that associates VLAN ID numbers with one or more ports. Switches from different vendors can all build their VLAN switching tables using standardized VLAN ID numbers. This will allow them to interoperate.

TEST YOUR UNDERSTANDING

15. a) What problem does VLANs address? b) How do they address it? c) When a server on a VLAN broadcasts, what stations receive its message? d) Describe the VLAN tagging standard, 802.1Q.

Handling Momentary Traffic Peaks

Momentary Traffic Peaks

If traffic volume is comfortably below a network's traffic capacity, traffic normally will get through without delay. However, sometimes there are **momentary traffic peaks** that briefly exceed the network's capacity, as Figure 4-15 illustrates. This will create latency, and if traffic peaks last beyond a certain time, some frames will be dropped

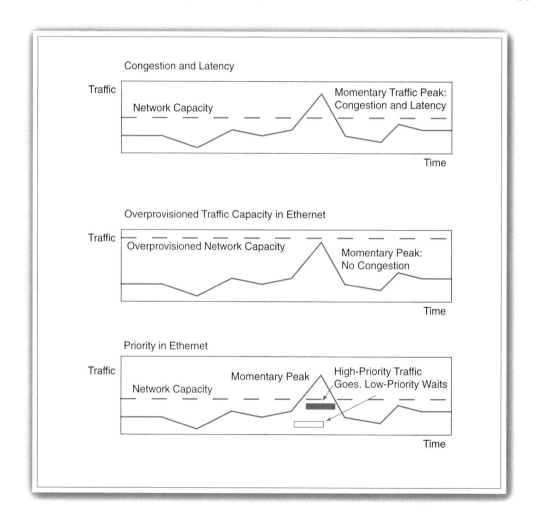

Figure 4-15   Handling Momentary Traffic Peaks with Overprovisioning and Priority

entirely. Although these peaks normally last only a fraction of a second to a few seconds, they can be highly disruptive for some applications, especially voice and video. For many applications, however, such as e-mail, users will not even notice brief delays.

Overprovisioning Ethernet

Most organizations have discovered that the least expensive way to get around peak traffic congestion today is simply to **overprovision** the Ethernet LAN—that is, to install much more capacity in switches and trunk links than will be needed most of the time. If 10Base-T would be sufficient most of the time, for example, they install 100Base-TX. When there are brief traffic bursts, these bursts will very rarely exceed capacity. Although this method wastes capacity most of the time, it works without adding to the cost of switch management. This is why overprovisioning currently is the most economical way of dealing with momentary traffic peaks.

> Overprovisioning currently is the most economical way of dealing with momentary traffic peaks.

Priority in Ethernet

Another way to address momentary traffic peaks is to give **priority** to latency-sensitive traffic, such as voice and video traffic, so that latency-sensitive traffic will go first if there is congestion. Figure 4-14 shows that 802.1Q Ethernet frame tagging can be used to give priority to individual frames. The Tag Control Information field contains not only a 12-bit VLAN ID, but also a 3-bit **priority level** that can be used to give a frame one of eight priority levels from 000 to 111. The definition of these eight priority levels is given in the **802.1p** standard.[7]

When brief traffic peaks occur, higher-priority traffic will go first. If high priority is given to latency-intolerant applications, such as voice communication, and if low priority is given to latency-tolerant traffic, such as e-mail transmission, applications for which latency is a problem will experience no latency, and users of latency-tolerant applications probably will not even notice brief delays during the momentary traffic peak. Priority also guarantees that network control messages get through, which may be crucial during periods of high congestion. Most switches today support priority.

Unfortunately, priority can be difficult to manage. Its management costs have made priority substantially more expensive than overprovisioning Ethernet. In addition, prioritizing traffic can lead to pitched political battles within a firm over whose traffic should have the highest priority.

Momentary Versus Chronic Lack of Capacity

Note that we have been discussing momentary traffic peaks in a network that has sufficient capacity nearly all of the time. This is very different from **chronic lack of capacity,** in which the network lacks adequate capacity much of the time. In such cases, the only good

---

[7] In addition to the 12 VLAN ID bits and the 3 priority bits, the 16-bit Tag Control Information field has a 1-bit canonical format bit. This bit is set to 1 for all networks except 802.5 Token-Ring Networks and FDDI, which are now extremely rare. For those two networks, the canonical format bit is 0. In canonical format, the right-most bit in each byte is sent first. In 802.5 and FDDI, the left-most bit in each byte is sent first. The canonical format bit is rarely used.

solution is to upgrade capacity. Otherwise, either congestion will be high or capacity will have to be rationed—giving it to some applications but denying any capacity to others.

TEST YOUR UNDERSTANDING

16. a) What are momentary traffic peaks? b) What problem do they create? c) In what two ways can Ethernet address momentary traffic peaks? d) What is the advantage of each? e) Overall, what is the most economical way today for most firms to address the problem of momentary traffic peaks in LANs? f) Distinguish between momentary traffic peaks and chronic lack of capacity. g) What must be done if there is a chronic lack of capacity?

PURCHASING SWITCHES

We will end this chapter with a discussion of the issues you will have to deal with when purchasing Ethernet switches. Purchasing an Ethernet switch is a complex task.

Number and Speeds of Ports
      Decide on the number of ports needed and the speed of each
      Usually can buy a switch with your desired configuration
Switching Matrix Throughput (Figure 4-17)
      Aggregate throughput: total speed of switching matrix
      Nonblocking capacity: switching matrix sufficient even if there is maximum input on all ports
      Less than nonblocking capacity is workable
            For core switches, at least 80 percent to 90 percent
            For workgroup switches, at least 20 percent to 30 percent
Store-and-Forward Versus Cut-through Switching (Figure 4-18)
      Store-and-forward Ethernet switches read whole frame before passing it on
      Cut-through Ethernet switches read only some fields before passing it on
      Perspective: Cut-through switches have less latency, but this is rarely important
Jitter (variability in latency from cell to cell)
      Makes voice sound jittery (Figure 4-19)
      Failover speed to backup links if there is a failure
Manageability
      Manager controls many managed switches (Figure 4-20)
      Polling to collect data and problem diagnosis
      Fixing switches remotely by changing their configurations
      Providing network administrator with summary performance data
      Managed switches are substantially more expensive than unmanaged switches
      However, in large networks, the savings in labor costs and rapid response are worth it

Figure 4-16   Switch Purchasing Considerations (Study Figure)

Number and Speeds of Ports

The most basic issue is how many ports you will need to have and what their individual speeds need to be. For instance, you might need a workgroup switch with 12 or 24 100Base-TX ports. To give another example, you might need a core switch with four gigabit Ethernet SC optical fiber ports and two 10GBase-SX SC optical fiber ports. Fortunately, you can buy switches in almost any port and speed configuration you wish. In addition, many switches are modular, meaning that you can buy the basic chassis and insert cards with the ports you require.

TEST YOUR UNDERSTANDING

17. a) What is the first issue to consider in switch purchases? b) Do you have many choices in port numbers and speeds?

Switching Matrix Throughput

As Figure 4-17 illustrates, a switch has a **switching matrix** that connects input ports to output ports.

Aggregate Throughput

A critical consideration in Ethernet switch selection is the switching matrix's **aggregate throughput.** For example, suppose you have a 4-port 100Base-TX switch, as shown in Figure 4-17. In the worst case, all ports may receive data simultaneously at 100 Mbps each for a total of 400 Mbps.

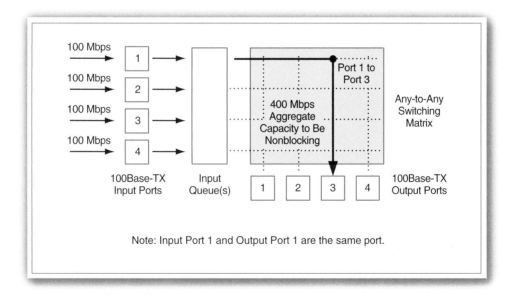

Figure 4-17   Switching Matrix

Nonblocking Capacity

Switches with aggregate throughput large enough to handle even their highest possible input load (maximum input on all ports) are called **nonblocking** switches. For example, a switch that has four 100Base-TX ports must have an aggregate throughput of 400 Mbps to be nonblocking.

Nonblocking Switches

The aggregate capacities of most switches are not fully nonblocking because it is not likely that all ports will be receiving simultaneously. However, if a switching matrix's aggregate throughput is too far below nonblocking capacity, frames will be delayed during peak periods or even lost if the switch's queue holding arriving frames becomes full.[8]

Switching matrix throughput is especially important for switches high in the hierarchy. Each port on such switches is likely to carry traffic to and from many individual stations, so it is common to find most ports sending and receiving at any given moment. Switches high in the hierarchy should be nonblocking or close to nonblocking (80 percent to 90 percent). Even workgroup switches should have aggregate throughput that is at least 20 percent to 30 percent of nonblocking capacity.

TEST YOUR UNDERSTANDING

18. a) What happens if a switch cannot handle the input traffic? b) What is a nonblocking switch? c) A switch has eight gigabit Ethernet ports. What aggregate switch matrix capacity is needed to give nonblocking capacity? Show your calculations. d) To give 80 percent of nonblocking capacity? Show your calculations.

Store-and-Forward Versus Cut-Through Switching

One capability that many vendors tout is the ability to do store-and-forward or cut-through switching. We saw earlier that an Ethernet frame contains multiple fields and that the data field alone can be as large as 1,500 octets long.

Store-and-Forward Ethernet Switches

As Figure 4-18 illustrates, some Ethernet switches wait until they have received the entire frame before sending it out. This is called **store-and-forward** switching. This approach allows them to check each frame for errors and to discard incorrect frames to reduce traffic. Unfortunately, store-and-forward switching adds a slight delay to frame transmission because the entire frame must be processed before any bits are sent out.

Cut-Through Ethernet Switches

In contrast, **cut-through** Ethernet switches examine only some fields in a frame before sending the bits of the frame back out. Obviously, as shown in Figure 4-18, they must at least read the destination address, in order to know what port to use to send the frame back out. This requires reading the preamble, start of frame delimiter, and destination address, for a total of only 14 octets.

---

[8] Obviously, losing frames during periods of high load is undesirable. The 802.3 standard specifies a pause command that the switch can send to the NIC. The pause command asks the NIC to stop transmitting for a brief period of time. This prevents the queues from overflowing, but it still limits throughput. In addition, pause is an optional part of the standard, and it has to be turned on for both the switch and the NIC.

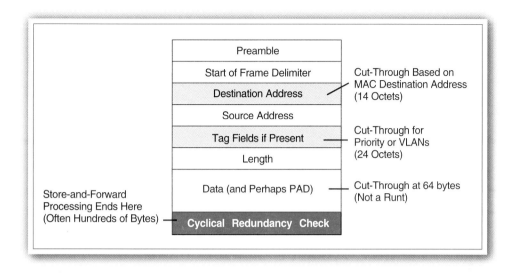

Figure 4-18 Store-and-Forward Versus Cut-Through Switching

Handling VLANs and priority also requires the reading of tag fields if they are used. Finally, some cut-through switches, in the "extreme" case, wait for 64 octets of data because a smaller frame, called a runt, would violate the Ethernet standard.

By examining only a few dozen octets at most, cut-through switching can reduce latency at each switch compared to store-and-forward switching, which typically has to examine hundreds or thousands of octets.

Perspective

Although vendors once touted cut-through operation as a major advantage, the amount of latency added by store-and-forward switching or tends to be negligible today. In addition, most switches today can do both cut-through and store-and-forward operations. Many offer cut-through operation as the default but sample full frames occasionally and change to store-and-forward operation if the error rate becomes too high. Cut-through versus store-and-forward operation is no longer a significant criterion for switch purchasing. However, you need to know these terms to read the vendor literature, and network managers tend to ask about these ideas in job interviews.

TEST YOUR UNDERSTANDING

19. a) Which is likely to have less latency—a cut-through switch or a store-and-forward switch? Why? b) What is the advantage of the other mode of operation? c) Is determining which mode of operation a switch uses a major purchasing criterion today?

Jitter

Jitter is *variability in latency,* as Figure 4-19 shows. If all frames have the same delay, a received voice signal will sound normal. However, if there is substantial variability in

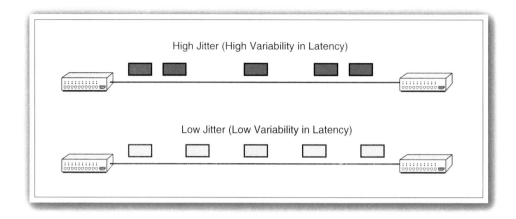

Figure 4-19   Jitter

latency from frame to frame, the voice will appear to speed up and slow down over the space of a few frames. This will make the human voice sound jittery—hence the name "jitter." It is important to consider jitter when reading performance test reports on candidate switches.

TEST YOUR UNDERSTANDING

20.   a) What is jitter? b) Why is jitter bad?

Manageability

The First Bank of Paradise has more than 500 hundred switches. If there is a switch problem, discovering which switch is malfunctioning can be very difficult. Fixing the problem, furthermore, may require traveling to the switch to change its configuration.

Managed Switches and the Manager

As Figure 4-20 shows, the bank mitigates these problems by using only **managed switches.** As the name suggests, these switches have sufficient intelligence that they can be managed from a central computer called the **manager.** In most cases management communication is governed by the Simple Network Management Protocol discussed in Chapter 10.

Polling and Problem Diagnosis

Every few seconds the manager polls each managed switch and asks for a copy of the switch's configuration parameters. If a problem occurs, the manager can discover quickly which switches are not responding and so can narrow down the source of the problem. In many cases the status data collected frequently from the switches can pinpoint the cause of a problem.

Fixing Switches Remotely

In some cases the network administrator can use the manager to fix switch problems remotely by sending commands to the switch. For instance, the manager can command

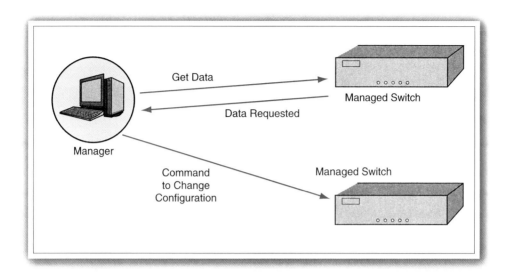

Figure 4-20   Managed Switches

the switch to do a self-test diagnostic. To give another example, the manager can tell the switch to turn off a port suspected of causing problems.

Performance Summary Data
At the broadest level, the manager can present the status data to the network administrator in summarized form, giving the administrator a good feeling for how well the network is functioning and for whether changes will be needed to cope with expected traffic growth.

The Cost of Manageability
Manageability is not cheap. Managed switches are much more expensive than other switches, but in firms with many switches, central management reduces management labor, which is considerable, much more than offsetting its cost. The main benefit of network management is to reduce overall management costs.

TEST YOUR UNDERSTANDING
21.   a) What are managed switches? b) What benefits do they bring?

Physical and Electrical Features

Physical Size
Almost all switches are 19 inches (48 cm) wide. This allows them to fit into standard 19-inch-wide telecommunications racks long used in telephony and later for data switches and routers. In equipment racks, one **U** is 1.75 inches (4.4 cm) in height. Most switches, although not all, are multiples of U. For instance, a 2U switch is 3.50 inches tall.

Form Factor

        Switches fit into standard 19 in (48 cm) wide equipment racks

        Sometimes, racks are built into enclosed equipment cabinets

        Switch heights usually are multiples of 1U (1.75 inches or 4.4 cm)

Port Flexibility

        Fixed-port switches

                No flexibility: the number of ports is fixed

                1 or 2 U tall

                Most workgroup switches are fixed-port switches

        Stackable switches

                Fixed number of ports

                1 or 2 U tall

                High-speed interconnect bus connects stacked switches

                Ports can be added in increments as few as 12

        Modular switches

                1 or 2 U tall

                Contain one or a few slots

                Each slot module contains 1 to 4 ports

        Chassis switches

                Several U tall

                Contain several expansion slots

                Each expansion board contains 6 to 12 ports

                Most core switches are chassis switches

UTP Uplink Ports

        Normal Ethernet RJ-45 switch ports transmit on Pins 3 and 6 and listen on Pins 1 and 2

        If you connect two normal ports on different switches, they will not be able to communicate

        Most switches have at least one uplink port, which transmits on Pins 1 and 2. You can use an ordinary UTP cord to connect a UTP uplink port on one switch to any normal port on a parent switch

Electrical Power

        Switches require electrical power

        Under the 802.3af standard, switches also can provide electrical power over the UTP cord

        Only 12.95 Watts—sufficient for wireless access points (Chapter 5) and IP telephones (Chapter 6) but not for computers

Figure 4-21   Physical and Electrical Features (Study Figure)

Port Flexibility

There are four basic types of switch organization, each giving a different degree of flexibility over how many ports you may have.

➤ **Fixed-port switches,** as their name suggests, give no port flexibility. The ports you buy them with are the ports you will have to live with for the life of the switch. The figure shows a fixed-port switch. They are one or two U tall. Most workgroup switches are fixed-port switches.

➤ **Stackable switches,** like fixed-port switches, also are one or two U tall and have a fixed number of ports. However, as the name suggests, they are designed to be stacked on top of one another. A special **interconnect bus** connects them at speeds that are higher than port-to-port Ethernet connections would permit. With stackable switches, ports can be added in increments as small as 12 ports.

➤ **Modular switches** also are 1 or 2U tall but do not have a fixed number of ports. They have one or more slots into which modules containing one to four ports can be inserted.

➤ **Chassis switches,** at the other extreme, are boxes that are several U tall. The box has **expansion slots** into which modular **expansion boards** can be installed. The expansion boards contain six to twelve ports. The box itself contains a high-speed **backplane bus** that links the ports on all of the expansion cards together. Most core switches are chassis switches.

Uplink Ports

Ethernet NICs transmit on Pins 1 and 2 and listen on Pins 3 and 6. Normal Ethernet RJ-45 switch ports, in turn, transmit on Pins 3 and 6 and listen on Pins 1 and 2. If you connect two normal RJ-45 ports on different switches via a UTP cord, they will not hear each other. To address this problem, most switches have at least one **uplink port,** which transmits on Pins 1 and 2 and listens on Pins 3 and 6. You can use a standard UTP cable to connect a UTP uplink port on one switch to any normal port on its parent switch.

Electricity

Ethernet switches require electrical power. With the new 802.3af standard, they also can provide electrical power to attached devices, over the ordinary UTP cord that stations already use to attach to the switch. This power is very limited—only 12.95 watts at 48 volts. This is sufficient for the wireless access points discussed in Chapter 5 and for the IP telephones discussed in Chapter 6. However, it is not sufficient for computers. Switches providing power can detect an incompatible device automatically and will not attempt to send it electrical power.

TEST YOUR UNDERSTANDING

22.   a) How wide are most Ethernet switches? Why is this so? b) How tall are most Ethernet switches? c) Why are uplink ports needed on Ethernet switches? d) How do they work? e) What does the 802.3af standard permit? f) For what types of devices is this sufficient and not sufficient?

SYNOPSIS

This chapter looked in some depth at Ethernet local area networking. Ethernet is the dominant technology for corporate LANs today. Ethernet's only serious competitor is wireless LANs, which we will see in the next chapter. However, we will see that wireless LANs are not direct competitors to wired Ethernet LANs but rather usually work in conjunction with wired Ethernet LANs.

Ethernet standards are created by the 802.3 Working Group of the IEEE 802 LAN/MAN Standards Committee. Like all networking standards, Ethernet standards exist at both the physical and data link layers. Therefore, they are OSI standards.

The 802.3 Working Group has created many physical layer standards and is still creating better physical layer standards. Speeds range from 10 Mbps to 10 Gbps and are still moving higher. These standards use both 4-pair UTP and optical fiber.

The dominant Ethernet standard for access lines from stations to switches is 100Base-TX, while the dominant standard for switch-to-switch trunk lines is gigabit Ethernet using optical fiber. The newest and fastest standards (10GBase-x and beyond) are being created first for metropolitan area networks (MANs) but are moving into LANs as well. Chapter 7 discusses Ethernet MANs.

Although there are limited distance spans for connecting two switches, switches regenerate signals, so long distances can be spanned by sending signals through multiple switches. Link aggregation permits two or more trunk lines between pairs of switches for additional throughput.

The 802 Committee subdivided the data link layer into two layers. The media access control layer is specific to a particular technology, such as Ethernet or 802.11 wireless LANs. The logical link control layer deals with matters common to all LAN technologies. Ethernet has only a single MAC standard—the 802.3 Media Access Control standard. This standard specifies frame organization and switch operation.

The Ethernet frame has multiple fields. The preamble and start of frame delimiter fields synchronize the receiver's clock with the sender's clock. The destination and source MAC address fields are each 48 bits long, and these are assigned to NICs at the factory. For human memory limitations (and to simplify writing), Ethernet MAC addresses usually are written in hexadecimal format, such as B2-CC-67-0D-5E-BA. The length field specifies the length of the data field (not of the frame as a whole). The data field has two parts: the LLC subheader, which describes the type of packet contained in the data field, and the packet itself. The PAD field is added if the data field is less than 46 octets long to make the data field plus the PAD field 46 octets in length. The frame check sequence field is used by the receiver to check for errors. If the receiver finds an error, it simply discards the frame.

Some older LANs still use Ethernet hubs instead of Ethernet switches. Hubs broadcast incoming bits, so only one station may transmit at a time. Consequently, NICs that use hubs must use CSMA/CD media access control, which only allows NICs to transmit if no other NIC is transmitting and which handles retransmission if there is a collision. NICs that use switches turn off CSMA/CD, giving them full-duplex (simultaneous two-way) transmission.

Ethernet switches are arranged in a hierarchy. This simplifies switching, making Ethernet switches inexpensive. Switches that connect stations to the network are called workgroup switches. Switches higher in the hierarchy are called core switches. Loops among switches (which would break the hierarchy) are forbidden. The Spanning Tree Protocol (802.1D) automatically detects and disables accidental loops. It can also be used to provide backup links in case of link or switch failures. The newer Rapid Spanning Tree Protocol converges faster than the original STP.

Servers often broadcast information to all devices in a network, and other types of broadcasting also occur. This can create a high level of traffic in networks. Virtual local area networks (VLANs) constrain broadcasting to the clients of particular servers, in order to reduce broadcasting traffic. To standardize VLANs (and priority), two tag fields are added to the Ethernet frame, right after the source address. The Tag Control Information field has a 12-bit VLAN number to indicate to which VLAN a particular frame belongs.

Even networks that have sufficient capacity most of the time will experience momentary traffic peaks that exceed their capacity. Overloaded switches may have to drop frames, and congestion will cause latency (delay). The least expensive way to address momentary traffic peaks today is to overprovision the network, that is, to install much larger Ethernet lines and switches than are needed most of the time. A more efficient way to manage resources is to give latency-intolerant applications, such as voice, high priority so that they are transmitted first during periods of congestion, minimizing their latency. Priority management uses the 3-bit priority level in the Tag Control Information field to indicate the priority levels of specific Ethernet frames. Unfortunately, priority is management-intensive.

Purchasing switches is very complex. The most basic issue is the number and speeds of the ports needed. Core switches should have nonblocking or nearly nonblocking capacity, meaning that even if each port is receiving at its maximum speed, the switching matrix will have the capacity needed to switch the input traffic.

Store-and-forward switches forward frames only after the entire frame is received. In contrast, cut-through switches start sending the frame back out after receiving only a few octets. Cut-through switches reduce latency at each switch, but this is rarely important in practice.

Managed switches are more expensive than other switches but can be managed remotely. Using managed switches saves money overall by reducing management labor.

Switches come in various sizes, with varying basic numbers of ports and varying expandability. Some even provide electrical power to the stations they serve.

SYNOPSIS QUESTIONS

1. Which 802 Working Group creates Ethernet standards?
2. a) What is the dominant Ethernet physical layer standard for linking stations to switches? b) For linking switches to other switches? c) Are the newest and fastest Ethernet standards being developed first for LANs or MANs? d) What is link aggregation?

3. a) What are the purposes of the two layers into which the 802 Committee divided the data link layer? b) What does the single Ethernet 802.3 MAC standard standardize?

4. a) List the fields in an Ethernet frame and their purposes. b) How long are MAC addresses? c) Who uses hexadecimal notation for MAC addresses—humans, computers, or both? d) Under what circumstance is a PAD field added? e) What happens if the receiver detects an error in a frame?

5. a) If a NIC attaches to a hub, what must it do when it wishes to transmit? b) What if there is a collision? c) What do NICs do when they are attached to switches?

6. a) Why is it good that Ethernet switches are arranged in a hierarchy? b) Why is it undesirable? c) How does STP help solve reliability problems created by Ethernet's hierarchical organization? d) Why is RSTP desirable?

7. a) What problem do VLANs address? b) How do they address this problem? c) How does tagging help implement VLANs?

8. a) Compare overprovisioning and priority as ways to deal with momentary traffic peaks. b) Which usually costs less?

9. a) What should be considered in purchasing switches? b) What are nonblocking switches, and why are they good? c) What is the advantage of cut-through switches? d) Are managed switches worth their high costs?

THOUGHT QUESTIONS

1. NICs transmit on Pins 1 and 2 and listen on Pins 3 and 6. Switch ports transmit on Pins 3 and 6 and listen on Pins 1 and 2. Uplink ports allow you to connect two switches. Can you think of a way to connect normal RJ-45 ports on two switches?

2. NICs can tell whether an arriving frame is tagged or not simply by looking at it. How can they do so? (Hint: They look at the value in the two octets following the address fields.)

3. If a PAD field is added to an Ethernet frame, the combined data field and PAD will be 46 octets long. How can the receiving NIC tell which part is the data field?

DESIGN QUESTION

1. You will create alternative designs for a network connecting four buildings in an industrial park. Hand in a picture showing your network. There will be a core switch in each building. Building A is the headquarters building. Building B is 85 meters south and 90 meters east of the headquarters building. From the headquarters building, Building C is 150 meters south. Building D is 60 meters west of Building C. Computers in Building A need to communicate with computers in Building B at 60 Mbps. Computers in Building A need to be able to communicate with computers in Building C at 300 Mbps. Computers in Building A must communicate with computers in Building D at 50 Mbps. Computers in Building C must communicate with computers in Building D at 75 Mbps. Building A will be connected directly to Buildings B and C. Building C

will be connected directly to Building D. a) How will you connect Building A to Building B? b) How will you connect Building A to Building C? c) How will you connect Building C to Building D?

HANDS ON

Binary and Hexadecimal Conversions

If you have Microsoft Windows, the Calculator accessory shown in Chapter 1 can convert between binary and hexadecimal notation. Go to the Start button, then to Programs or All Programs, then to Accessories, and then click on Calculator. The Windows Calculator will then pop up.

Binary to Hexadecimal
To convert eight binary bits to hexadecimal (hex), first choose View and click on Scientific to make the Calculator a more advanced scientific calculator. Click on the Bin (binary) radio button and type in the 8-bit binary sequence you wish to convert. Then click on the Hex (hexadecimal) radio button. The hex value for that segment will appear.

Hexadecimal to Binary
To convert hex to binary, go to View and choose Scientific if you have not already done so. Click on Hex to indicate that you are entering a hexadecimal number. Type the number. Now click on Bin to convert this number to binary.

One additional subtlety is that Calculator drops initial zeros. So if you convert 1 hex, you get 1. You must add three initial zeros to make this a 4-bit segment: 0001.

1. a) Convert 1100 to hexadecimal. b) Express the following MAC address in binary: B2-CC-67-0D-5E-BA, leaving a space after every eight bits. c) Express the following MAC address in hex: 11000010 11001100 01100111 00001101 01011110 10111010.

CASE STUDY: REWIRING A COLLEGE BUILDING

Learning Objectives:

By the end of this chapter, you should be able to discuss:

- Legacy Ethernet technologies—10Base5 and 10Base2.
- Design options for upgrading old legacy Ethernet networks.
- How to evaluate design options in terms of technology, cost, and organizational context.

INTRODUCTION

Many firms find themselves upgrading their existing legacy networks rather than building new networks from scratch. Usually, there are several design choices for upgrading old networks. We will look at one major legacy upgrading project and have you consider three design options. We will then ask you to pick one option and justify your choice.

Specifically, we will look at the College of Business Administration (CBA) building at the University of Hawai`i. This building has the shape of a *C*. The farthest parts of the building are 230 meters apart. The building actually consists of several adjacent towers, which are labeled A, B, C, D, E, and G. The F tower had to be demolished a few years after the building was constructed[1] in the mid-1970s, changing the building from a circle to its current *C* shape. The towers range from three to six stories in height.

UTP Wiring

In the 1980s the building was rewired for telecommunications using UTP. In every room there are two RJ-45 wall jacks. One is for telephone use, the other for data use. Unfortunately, the quality of the wiring is well below Cat5e. In fact, it has been characterized as "Cat3 on a good day." In addition, until recently, the university's central telephone services group managed all of this wiring and gave the college no control over data usage. This central control issue has disappeared in recent years, but the wiring quality problem remains.

Ethernet 10Base5 and 10Base2

To circumvent central control, the college created its own data network. Built before 10Base-T was created, the network used earlier Ethernet technologies.

Coaxial Cabling   Older forms of Ethernet used coaxial cabling—the type of cabling you use at home to connect your VCR to your television. Coaxial cabling (coax) has a central wire and a cylindrical outer conductor. The two are separated by insulation.

Ethernet 10Base5   The core of the network was built around a backbone of 10Base5 cabling, as Figure 4a-1 shows. This type of cabling uses a thick backbone cable that can run up to 500 meters and usually runs along the walls of the building. Stations are connected to this backbone cable by a drop cable.

When a station transmits, its signals go down the drop cable to the backbone cable. From a transceiver on the backbone cable, the signal spreads out in all directions, to all stations connected to the backbone. When a signal passes a drop cable, the transceiver passes the signal up to the station. In short, Ethernet 10Base5 is a bus (broadcast) network.

Ethernet 10Base2   Although the network used a 10Base5 core, most users are served by 10Base2 technology. As Figure 4a-2 shows, stations are connected in a daisy chain,

---

[1]   Don't ask.

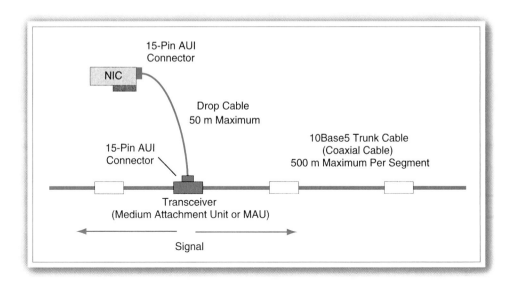

Figure 4a-1   Ethernet 802.3 10Base5 Physical Layer Standard

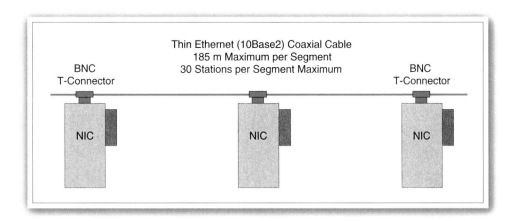

Figure 4a-2   Ethernet 802.3 10Base2 Physical Layer Standard

Figure 4a-3   Ethernet 10Base2 (Close-up)

with each station connected directly to the next. Figure 4a-3 shows a close-up of the T-connector that makes 10Base2 possible.

This also is a bus (broadcast) network. When a station transmits, its signal goes to both of its neighbors, which pass it on to other stations, and so forth, until the signal reaches the end of the daisy chain.

A Mixed Network   How do the 10Base5 core and the 10Base2 segments work together? Figure 4a-4 shows that the CBA used a central 10Base5 backbone whose thick and bright yellow coaxial trunk cable snaked through the corridors of the building. At various points in the building, a repeater connected the 10Base5 trunk cable to 10Base2 daisy chains that run to administrative offices, broom closets (which are used as faculty offices), and classrooms. In addition, the main computer laboratory for students used a 10Base-T hub after the 10Base-T standard was created. A router connects the CBA network to the campus network via optical fiber.

Congestion

The network's performance (shared 10 Mbps) was adequate when it was constructed in the mid-1980s and for several years after that. By the late 1990s, however, the number of computers attached to the network had passed 200, and performance slowed to a crawl. LED collision lights on the repeaters glowed red all day long. Sometimes, the network would freeze up for several seconds at a time. It was time for a new network.

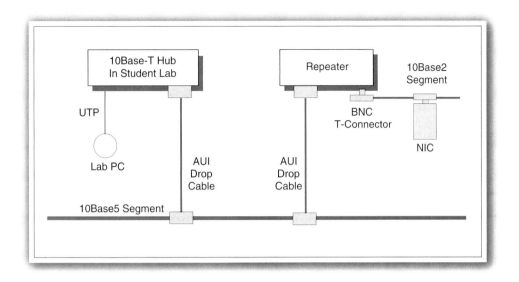

Figure 4a-4   Mixing 10 Mbps Ethernet 802.3 Physical Layer Standards

OPTION 1: STATE-OF-THE-ART REWIRING

The cleanest option is to rewire the entire building with modern technology. From the building's equipment room, optical fiber would run out to telecommunications closets in each tower. From there, Cat5e UTP would run to individual wall jacks. Each user would receive dedicated (unshared) 100 Mbps service to the desktop. This wiring system would be sufficient if switches were upgraded later, even to provide gigabit speed to each desktop.

Optical Fiber Runs

This approach would be very desirable because it would offer high speeds today and minimize rewiring in the future. However, it faces a number of obstacles. The most important is cost. Installing the optical fiber backbone alone would cost $200,000. This cost is somewhat unrepresentative because the building was built without false ceilings, duct spaces for wiring, or wiring closets, which would therefore have to be added. However, this cost would not be far lower if the building's designers had designed the building with telecommunications in mind. The cost of the fiber itself would be a large part of the total in any case; the largest cost of installing fiber would be the cost of labor to install and connect the fiber cords.

UTP Runs

From switches in each tower, Cat5e would run to individual offices and classrooms. The cost to install the UTP wiring after the optical fiber backbone is built was estimated at $250 per wall jack. About 300 wall jacks will have to be installed.

Switches
The switches themselves would cost about $20,000.

Telecommunications Closets
To house the switches and fiber/UTP connections, telecommunications closets would have to be built. They would have to be built as cabinets attached to corridor walls because room was not created in the building design for them. The expensive part would be providing them with the electrical power they would need for the switches and other equipment they would contain. Creating the telecommunications closets would cost about $30,000.

Considerations
One consideration is that the CBA is seeking funds for a new building. Therefore, the CBA is reluctant to invest heavily in a built-from-scratch LAN.

OPTION 2: USING THE EXISTING UTP DATA WIRING

10 Mbps
Testing has shown that most of the building's UTP data wiring is capable of carrying Ethernet at 10 Mbps without excessive error rates. One option is to install Ethernet switches to give unshared 10 Mbps service to each wall jack via the existing 300 data wall jacks.

Assessment
This would end the current congestion problem, which exists because 10Base5 and 10Base2 are shared 10 Mbps technologies. Runs to some offices, however, will violate the 100-meter distance limitation of Ethernet using UTP. In addition, this option would only bring 10 Mbps to each desktop.

Cost
The total cost would be around $75,000. A single large Ethernet switch with at least 300 ports would have to be purchased, and considerable work would be needed to connect existing UTP wire runs to the switch.

OPTION 3: RESEGMENTATION

A Single Collision Domain
The network's main problem is that it is a single large network with more than 200 stations. In Ethernet terminology, it is a single **collision domain.** If any of the 200 stations is transmitting, all other stations must wait. In addition, because of the long distances involved, when one station starts to transmit, signals will not get to other stations for a while. These stations may begin transmitting in the meantime, causing a collision.

Resegmenting the Network
One option is to divide the entire network into four parts, called segments, each of which would have only 60 to 80 stations. A switch would connect the four segments. This is called resegmentation.

In a pure bus network, every station hears every other station, and only one station may transmit at a time. However, the switch isolates the four segments. When a station transmits to a station on the same segment, the Ethernet switch does not forward the frame to other segments. Only the small amount of traffic sent between stations on different segments is passed by the switch. Consequently, stations only hear the traffic of other stations on their own segment, plus a small amount of cross-segment traffic. With only 60 to 80 stations per segment, congestion would not be a serious problem.

Again, however, users would only get 10 Mbps to the desktop. In fact, all stations in a segment would *share* this speed, so the throughput for individual stations would be substantially lower.

Resegmentation can be done for about $30,000 using optical fiber connections from a switch at a central point to a repeater in each of the four areas. The repeater would connect the segment to the switch. The cabling cost would be small in part because three of the optical runs from the central switch to a repeater would be short runs. Also, the network staff would buy fiber precut to the desired length and run the optical fiber through the building without bothering to place ducts or other enclosures for aesthetics or protection from the elements or vandalism.

TEST YOUR UNDERSTANDING

1. Why is the college's data network not adequate for its needs?

2. a) What would be the cost of a completely new network with Category 5e UTP and an optical fiber backbone? b) What would be the components of that cost? c) What would be the total cost per wall jack? d) In what ways is complete rewiring the best option?

3. a) What would be the cost of using existing data UTP lines? b) What would be the components of that cost? c) What would be the cost per wall jack? d) What problems would this option create?

4. a) What would be the cost of a resegmentation? b) What would be the cost per wall jack? c) Would it reduce congestion to an acceptable level? d) What problems would this option create?

5. Put on your consultant's hat. Which option would you recommend and why? You can also offer another option.

WIRELESS LANs

Learning Objectives:

By the end of this chapter, you should be able to discuss:

- Radio signal propagation, including spread spectrum transmission.
- 802.11 Wireless LAN (WLAN) Standards.
- 802.11 Security.
- Bluetooth Personal Area Networks (PANs).
- Emerging wireless technologies.

INTRODUCTION

Today Ethernet dominates corporate LANs. However, Ethernet is a *wired* technology (a general term for networks that use UTP, optical fiber, or both). An Ethernet station must plug into a wall jack or directly into an Ethernet switch. This is a problem for the growing number of notebook computers, personal digital assistants, and other mobile devices used in organizations. To serve mobile users, a new type of local area network is emerging—the **wireless LAN (WLAN),** which uses radio or (rarely) infrared light for physical layer transmission.

Why Wireless LANs?

Rather than being a competitor for wired Ethernet LANs, WLANs today primarily supplement wired LANs. Figure 5-1 shows that mobile users typically connect to devices called access points. These access points have two functions: to control the operation of wireless stations (transmission power, etc.) and to link the mobile user to the firm's wired Ethernet LAN. This link is needed because the servers that mobile client devices need usually are on the wired LAN. It is also needed because the firm's Internet access router usually is on the wired LAN. In other words, although clients need to be mobile, the resources they need to work with typically are on the firm's main wired LAN.

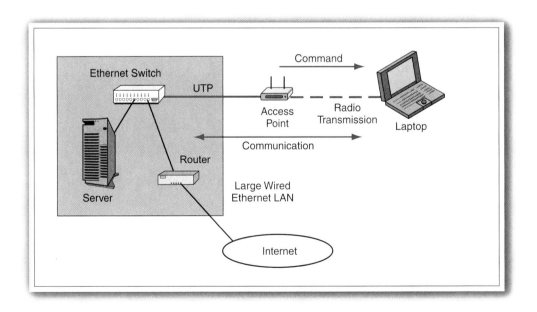

Figure 5-1   Wireless LAN (WLAN) Access Point

Courtesy: D-Link

Figure 5-2 Access Point

802.11 Wireless LANs

The dominant WLAN standards today are the 802.11 standards, which provide rated speeds of up to 54 Mbps (although actual throughput for such networks typically is less than half the rated speed). By placing access points throughout a building, a company can build a large WLAN that can serve mobile users anywhere in the building. Alternatively, a company can begin by creating a few 802.11 **"hot spots,"** which are access points placed where users typically gather, such as conference rooms and cafeterias, before rolling out full wireless LANs.

Bluetooth

Another wireless LAN technology, Bluetooth, has a different purpose. It is designed for personal area networks (PANs) that serve only a few devices carried by a person or around a desk. Bluetooth is primarily a tool for replacing cable connections. For instance, using Bluetooth, you can walk near a printer and print without physically connecting your mobile device and the printer with a printer cable.

TEST YOUR UNDERSTANDING

1. a) Why are WLANs desirable? b) What is the main family of WLAN standards today? c) Are WLANs competitors for wired Ethernet LANs today? d) Why are access points needed in WLANs? e) How can a firm create a large WLAN? f) What are hot spots? g) Contrast 802.11 and Bluetooth technologies.

RADIO SIGNAL PROPAGATION

Before we can talk about wireless LANs in detail, we have to look at radio propagation in general. While wired LAN signals are injected into wires and optical fibers, which have known and controllable propagation characteristics, radio signals are broadcast and encounter many unpredictable conditions that interfere with reception, as we will soon see.

Radio Signals and Antennas

Radio Signals

If you cause an electron to oscillate (move up and down rhythmically), it will generate a weak **electromagnetic signal.** If you can force billions of electrons in an antenna to oscillate in unison, you will create an electromagnetic signal strong enough to be received several meters away or even many kilometers away.

Wave Characteristics

Figure 5-3 shows a simple electromagnetic **radio wave.**

➤ Its **amplitude** is the maximum (or minimum) intensity of the wave. In sound, this corresponds to volume (loudness).

➤ Its **frequency** is the number of complete cycles the wave goes through per second. In sound, frequency corresponds to pitch. One cycle per second is one **hertz (Hz).**

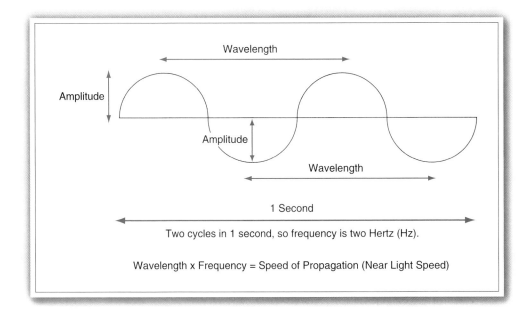

Figure 5-3   Radio Wave Characteristics

Useful radio frequencies for data communications come in the high **megahertz (MHz)** to low **gigahertz (GHz)** range.

➤ As we saw in Chapter 3, a wave's **wavelength** is the physical distance between the corresponding points in adjacent cycles.

Wavelength and frequency are not independent, as Equation 5-1 indicates. This equation shows the relationship between the frequency, which is represented by f, and the wavelength, which is represented by the Greek character lambda (λ). The equation shows that the frequency times the wavelength equals the speed of propagation, c, which in radio is close to the speed of light in a vacuum, 300,000,000 meters per second.

$$\text{Equation 5-1:} \quad f \times \lambda = c$$

Note that frequency and wavelength are inversely related. In other words, a lower frequency (a lower number of hertz) means a longer wavelength, and a higher frequency means a shorter wavelength. Microwave ovens, which operate at an extremely high frequency of about 2.5 GHz, have very short wavelengths, as do wireless LAN signals.

By tradition, frequency is used to describe radio waves, whereas wavelength is used to describe light waves (as we saw in Chapter 3) and services that operate near the wavelengths of light. This is why wavelength is used in optical fiber transmission, but frequency is used in radio LANs and cellular telephone systems. This is also why radio transmission is called radio frequency (RF) transmission.[1]

TEST YOUR UNDERSTANDING

2. a) What are the three characteristics of radio waves? b) What is a hertz? c) How are two radio characteristics related? d) In musical instruments, string length is roughly proportional to the wavelength made by the sound when the string is struck or picked. Are the high-pitch strings on a harp the short strings or the long strings? Explain. e) When would you use frequency to describe propagation? f) When would you use wavelength to describe propagation?

Omnidirectional and Dish Antennas

Radio transmission requires an antenna. Figure 5-4 shows that there are two types of radio antennas: dish antennas and omnidirectional antennas. **Omnidirectional antennas** transmit signals in all directions and receive incoming signals equally well from all directions. **Dish antennas,** in contrast, point in a particular direction, allowing them to send stronger outgoing signals in that direction for the same power and to receive weaker incoming signals from that direction.

Dish antennas are good for long distances because of their focusing ability, but omnidirectional antennas are easier to use. (Imagine if you had to carry a dish with you whenever you carried your cellular phone. You would not even know where to point

---

[1] Theoretically, wireless LANs can also use infrared light, which is electromagnetic radiation with frequencies lower than those of visible light. Your television remote control uses infrared light to communicate with your television. However, infrared technologies to date have been much too slow for LAN use. In addition, infrared is very susceptible to being blocked by objects, even when the infrared light is "diffused" around the room to reduce blockage.

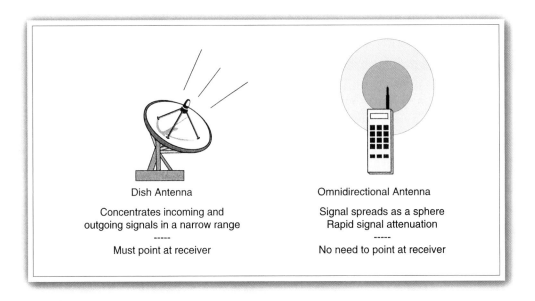

Figure 5-4   Omnidirectional and Dish Antennas

the dish!) WLANs use omnidirectional antennas almost exclusively because distances are short and because users do not always know the location of the nearest access point.

Inverse Square Law Attenuation

When the signal spreads out from an antenna, its strength is spread over the area of a sphere. The area of a sphere is proportional to the square of its radius, so signal strength weakens by an inverse square law ($1/r^2$), as Equation 5-2 illustrates.[2] To give an example, if you triple the distance (r), signal strength (S_2) falls to one-ninth ($1/3^2$) of its original strength (S_1). An inverse square law is very rapid attenuation compared to wire and optical fiber attenuation. You have to be relatively close to your communication partner in radio transmission unless the signal strength is very high.

$$\text{Equation 5-2:}\quad S_2 = S_1/r^2$$

TEST YOUR UNDERSTANDING

3. a) Distinguish between omnidirectional and dish antennas in terms of operation. b) Under what circumstances would you use an omnidirectional antenna? c) Under what circumstances would you use a dish antenna? d) What type of antenna usually is used in WLANs? e) If the signal strength from an omnidirectional radio source is 8 milliwatts at 30 meters, how strong will it be at 60 meters? f) At 120 meters?

---

[2] In fact, research has shown that drop-offs in real environments are even faster in urban areas, being proportional to between $1/r^{3.5}$ and $1/r^5$. Arthur H. M. Ross, "About CDMA Technology: The CDMA Revolution," CDG.org, 1999. http://www.cdg.org/technology/cdma_technology/a_ross/CDMARevolution.asp.

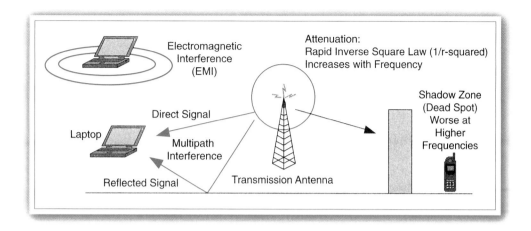

Figure 5-5   Wireless Propagation Problems

Wireless Propagation Problems

Although wireless communication gives mobility, wireless transmission is not very predictable, and serious propagation problems are very common. (Anyone with a cellular telephone already knows this.) When we inject a signal into a UTP cord or optical fiber cord, we can predict losses along the way fairly accurately because the transmission medium has known characteristics. With radio transmission, however, we have much less control over the transmission path between the sender and the receiver. Figure 5-5 illustrates some common wireless propagation problems.

Rapid Attenuation
We have seen that radio signal strength attenuates according to an inverse square law. This is much faster attenuation than in wire or optical fiber transmission.

Shadow Zones (Dead Spots)
To some extent, radio signals can go through and bend around objects. However, if there is a thick wall or a large, dense object blocking the direct path between the sender and receiver, the receiver may be in a **shadow zone (dead spot)** where it cannot receive the signal. If you have a cellular telephone and have tried to use it within buildings, you probably are familiar with this problem.

Multipath Interference
In addition, radio waves tend to bounce off walls, floors, ceilings, and other objects. As Figure 5-5 shows, this may mean that a receiver will receive two or more signals—a direct signal and one or more reflected signals. The different signals will travel different distances and so may be **out of phase** when they reach the receiver (one may be at

its highest amplitude while the other may be at its lowest, giving an average of zero). This **multipath interference** may cause the signal to range from strong to nonexistent in the space of a few centimeters.[3]

Electromagnetic Interference (EMI)

A final major propagation problem in wireless communication is **electromagnetic interference (EMI).** As we saw in Chapter 3, EMI is created by electrical motors and many other devices. Many devices produce electromagnetic interference at frequencies used in data communications, including fluorescent lamps, cordless telephones, microwaves, and especially other wireless networking devices.

Frequency-Dependent Propagation Problems

Some propagation problems depend on frequency. Higher-frequency waves attenuate more rapidly than lower-frequency waves because they are absorbed more rapidly by moisture in the air, leafy vegetation, and other water-bearing obstacles. Consequently, as we will see later, WLAN signals around 5 GHz attenuate much more rapidly than signals around 2.4 GHz.

In addition, shadow zone problems increase with frequency. As frequency increases, radio waves become less able to flow around objects, and even thin objects can stop them.

TEST YOUR UNDERSTANDING

4. a) Which offers more reliable transmission characteristics—UTP and optical fiber transmission or radio transmission? b) Why is attenuation especially bad in radio transmission compared to wire transmission? c) How are shadow zones (dead spots) created? d) Why is multipath interference very sensitive to location? e) List some sources of EMI. f) What propagation problems become worse as frequency increases?

Bands and Bandwidth

The Frequency Spectrum and Service Bands

Before going on, we need to discuss some basic radio transmission terminology. Most basically, the **frequency spectrum** consists of all possible frequencies from zero hertz to infinity, as Figure 5-6 shows.

The frequency spectrum is divided into **service bands** that are dedicated to specific services. For instance, in the United States, the AM radio service band lies between 535 kHz and 1,705 kHz. The FM radio service band, in turn, lies between 88 MHz and 108 MHz. The 2.4 GHz band that we will see below for wireless LANs extends from 2.4000 GHz to 2.4835 GHz.

Channels

Service bands are divided into smaller frequency ranges called **channels.** A different signal can be sent in each channel because signals in different channels do not interfere with one another.

---

[3] Most wireless devices use two or more antennas to average out multipath interference effects. Others have square or fan-shaped antennas. The latter actually are rake antennas, which have many small antennas arranged like the teeth on a garden rake. Both approaches give space diversity.

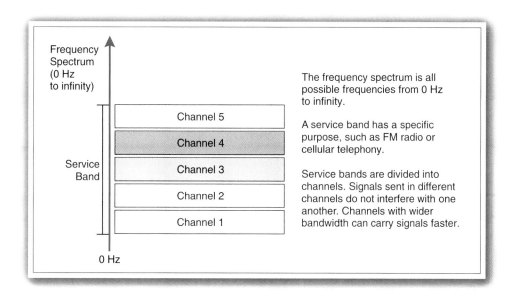

Figure 5-6   The Frequency Spectrum, Service Bands, and Channels

Signal and Channel Bandwidth

Signal Bandwidth   Figure 5-3 shows a signal operating at a single frequency. However, as we saw in Chapter 3, most signals spread over a range of frequencies. The range of frequencies used by a signal is called the **signal bandwidth.** The signal bandwidth is measured by subtracting the lowest frequency from the highest frequency. Signal bandwidth is related to signal speed. The higher the transmission speed in bits per second, the wider the signal's bandwidth will be.

Channel Bandwidth   The channel also has a bandwidth. For instance, if the lowest frequency of a channel is 89.0 MHz and the highest frequency is 89.2 MHz, then the **channel bandwidth** is 0.2 MHz (200 kHz). AM radio channels are 10 kHz wide, whereas FM channels have bandwidths of 200 kHz, and television channels are 6 MHz wide.

The Shannon Equation   The relationship between possible transmission speed and channel bandwidth was quantified by Shannon, who found that the maximum possible transmission speed (*C*) when sending data through a channel is directly proportional to the channel's bandwidth *(B)*, as shown in the **Shannon Equation** (Equation 5-3).[4] The signal-to-noise ratio is also important but is very difficult to modify in practice.

$$\text{Equation 5-3:} \quad C = B \, Log_2 \, (1 + S/N)$$

---

[4]   Claude Shannon, "A Mathematical Theory of Communication," *Bell System Technical Journal,* July 1938, pp. 379–423 and October 28, 1938, pp. 623–656.

Signal Bandwidth

    Figure 5-3 shows a wave operating at a single frequency

    However, most signals are spread over a range of frequencies

    The range between the highest and lowest frequencies is the signal's bandwidth

    Higher transmission speeds need wider signal bandwidth

Channel Bandwidth

    Highest frequency in a channel minus the lowest frequency

    An 88.0 MHz to 88.2 MHz channel has a bandwidth of 0.2 MHz (200 kHz)

    Higher-speed signals need wider bandwidths

Shannon Equation

    $C = B \, \text{Log}_2 \, (1+S/N)$

        C = Maximum possible transmission speed in the channel (bps)

        B = Bandwidth (Hz)

        S/N = Signal-to-Noise Ratio

    Note that doubling the bandwidth doubles the maximum possible transmission speed

    More generally, increasing bandwidth by a factor of X increases the maximum possible speed by X

    Wide bandwidth is the key to fast transmission

    Increasing S/N helps slightly but usually cannot be done to any significant extent

Broadband and Narrowband Channels

    Broadband means wide channel bandwidth and therefore high speed

    Narrowband means narrow channel bandwidth and therefore low speed

    Narrowband is below 100 kbps

    Broadband is above 100 kbps

Channel Bandwidth and Spectrum Scarcity

    Why not make all channels broadband?

    There is a limited amount of spectrum in desirable frequencies

    Making each channel broader than needed would mean having fewer channels or widening the service band

    Service band design requires trade-offs between speed requirements, channel bandwidth, and service band size

The Golden Zone

    Most organizational radio technologies operate in the golden zone in the high megahertz to low gigahertz range

    At higher frequencies, there is more available bandwidth

    At lower frequencies, signals propagate better

    Golden zone frequencies are high enough for there to be large total bandwidth

    Golden zone frequencies are low enough to allow fairly good propagation characteristics

Figure 5-7   Channel Bandwidth and Transmission Speed (Study Figure)

The maximum possible speed is directly proportional to bandwidth, so if you double the bandwidth, for example, you can transmit up to twice as fast.

To transmit at a given speed, you need a channel wide enough to handle that speed. For example, video signals produce many more bits per second than audio signals, so television uses 6 MHz channels, while AM radio only has 10 kHz channels.

Broadband Channels   Channels with large bandwidths are called **broadband** channels. They can carry data very quickly. In contrast, channels with small bandwidths, which are called **narrowband** channels, can only carry data slowly. Speeds below 100 kbps usually are considered to be narrowband speeds, while speeds above 100 kbps are considered to be broadband speeds.

Broadband Channels Versus Spectrum Scarcity

Obviously, we would like each channel to be as wide as possible to provide maximum speed. However, there is a limited amount of frequency spectrum available at useful frequency ranges, so when service bands are created, there are tradeoffs between channel bandwidth, the number of channels, and the width of the service band.

The Golden Zone

Commercial mobile services operate in the high megahertz range to the low gigahertz range. This is called the **golden zone.** At lower frequencies, the spectrum is limited and has been almost entirely assigned. At higher frequencies, radio waves get absorbed rapidly by the air and vegetation. In addition, at higher frequencies, radio waves cannot flow through or around objects as they do at lower frequencies. Consequently, the sender and receiver must have an unobstructed **line of sight** (direct path) between them. Even at the high end of the golden zone, absorption and shadow zone propagation problems are substantial.

The golden zone for commercial mobile services is the high megahertz range to the low gigahertz range.

TEST YOUR UNDERSTANDING

5. Distinguish between a) the frequency spectrum, b) service bands, and c) channels. d) In radio, how can you send different signals without the signals interfering with one another?

6. a) Does a signal travel at a single frequency or over a range of frequencies? b) What is channel bandwidth? c) Why is large channel bandwidth desirable? d) What do we call a system whose channels have large bandwidth? e) What is the Shannon equation? f) What happens to the maximum possible propagation speed in a channel if the bandwidth is tripled? g) Given their relative bandwidths, about how many times as much data is sent per second in television than in AM radio? h) If wide bandwidth is desirable, why do we not use broadband channels for all radio services? i) What is the dividing line between narrowband and broadband speeds?

7. a) What is the golden zone in commercial mobile radio transmission? b) Why are lower frequencies not used? c) For what two reasons is operating at higher frequencies unattractive? d) What is a line-of-sight limitation?

Normal and Spread Spectrum Transmission

Both 802.11 wireless LANs and Bluetooth personal area networks, which we will see later in this chapter, operate in **unlicensed radio bands** that do not require each station to have a license. (Licensed bands, which require each station to have a license and to have its license changed every time it is moved, would not be useful for mobile services.) In the unlicensed bands used for WLANs, regulators mandate the use of a form of transmission called spread spectrum transmission.

Normal Transmission: Just Enough Bandwidth

As noted earlier in our discussion of the Shannon equation, if you want to transmit at a given speed, you will need a channel whose bandwidth is sufficiently wide. To allow as many channels as possible, channel bandwidths in normal radio transmission are limited to the speed requirements of the user's signal, as Figure 5-8 illustrates.

Spread Spectrum Transmission

In contrast to normal radio transmission, which uses channels just wide enough for transmission speed requirements, **spread spectrum transmission** takes the original signal, called a **baseband signal,** and spreads the signal energy over a much broader channel. All wireless LAN systems use spread spectrum transmission.

Note: Height of Box Indicates Bandwidth of Channel

Channel Bandwidth
Required for Signal Speed

Normal Radio: Bandwith is
No Wider than Required
for the Signal's Speed

Spread Spectrum
Transmission:
Channel Bandwidth is
Much Wider than Needed
for the Signal's Speed

Commercial Spread Spectrum Transmission Reduces Certain
Propagation Effects (Multipath Interference and Narrowband EMI);
Does Not Provide Security as in Military Spread Spectrum Systems

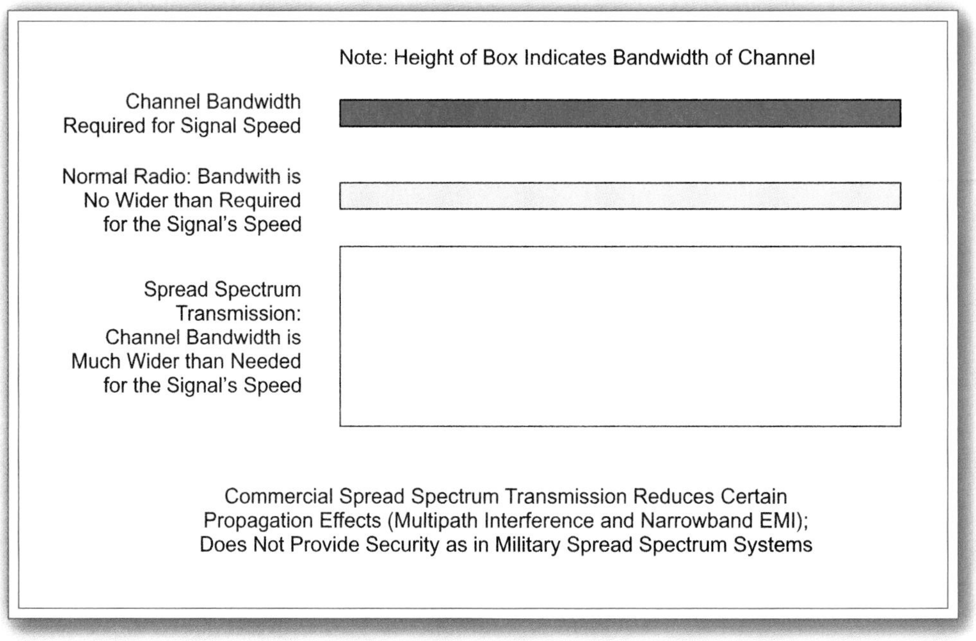

Figure 5-8   Normal Radio Transmission and Spread Spectrum Transmission

Why Spread Spectrum Transmission?

To Reduce Propagation Problems!   Why do we use spread spectrum transmission? The answer is that spread spectrum transmission reduces frequency-specific propagation problems. Many propagation problems occur over a fairly narrow range of frequencies. For instance, electromagnetic interference (EMI) normally occurs over a small range of frequencies, making transmission in those frequencies very bad. At a particular location, in turn, multipath interference also occurs at specific frequencies but not at others. By spreading the transmission over a wide range of frequencies, most of the signal will get through successfully, and the multipath interference or EMI will act like a small amount of noise.

Not for Security   In commercial transmission, security is *not* a reason for doing spread spectrum transmission. The military uses spread spectrum transmission for security, but it does so by revealing certain parameters of spread spectrum transmission only to authorized parties. However, commercial spread spectrum transmission methods must make these parameters publicly known in order to make it easy for two parties to communicate.

Frequency Hopping Spread Spectrum (FHSS)

The simplest form of spread spectrum transmission is **frequency hopping spread spectrum (FHSS).** As Figure 5-9 illustrates, the signal uses only the bandwidth required by the signal but hops frequently within the spread spectrum channel. If the signal runs

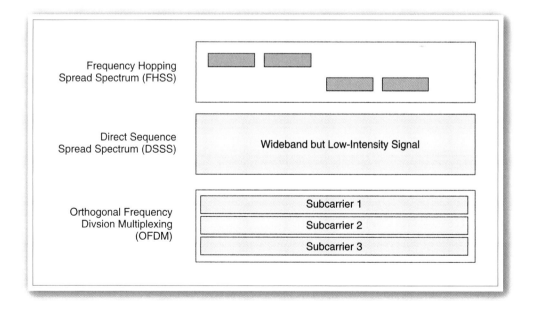

Figure 5-9   Spread Spectrum Transmission Methods

into strong EMI or multipath interference in one part of the broad channel, that part of the message will be lost and must be retransmitted, but parts of the message in other parts of the channel will get through.

FHSS is only useful for relatively low speeds because the transmitter must be retuned with each frequency hop. This takes a small but distinct amount of time, slowing the transmission rate. Consequently, as we will see later, the 802.11 Working Group for wireless LANs only specified frequency hopping spread spectrum transmission for speeds of 1 Mbps and 2 Mbps. Few 802.11 LANs operating at these low speeds were purchased. Bluetooth, also discussed later, currently uses FHSS, which is sufficient for Bluetooth's low speed of 722 kbps.

Direct Sequence Spread Spectrum (DSSS)

Another spread spectrum technique shown in Figure 5-9 is **direct sequence spread spectrum (DSSS)** transmission. In DSSS a signal is spread over the entire bandwidth of a channel. Interference and multipath interference will affect only small parts of the signal, allowing most of the signal to get through for correct reception. Although DSSS is more complex than FHSS, DSSS can support speeds up to 15 Mbps and is used in the 11 Mbps 802.11b wireless LANs we will see later in this chapter.

Orthogonal Frequency Division Multiplexing (OFDM)

There are several spread spectrum transmission methods. However, as transmission speeds move above about 15 Mbps, one form of spread spectrum transmission is beginning to dominate. This is **orthogonal frequency division multiplexing (OFDM),** which Figure 5-9 also illustrates.

In OFDM each broadband channel is divided into many smaller subchannels called **subcarriers.** Parts of each frame are transmitted in each subcarrier.[5] Although this may seem complex, sending data over a single very large channel proves to be much more difficult in practice.

OFDM sends data redundantly across the subcarriers, so if there is impairment in one or even a few subcarriers, all of the data usually will still get through.

OFDM can be used at very high speeds because it is easier to send many signals in many small subcarriers than it is to send one signal very rapidly over a very wide bandwidth. In wireless LANs, both of the 54 Mbps 802.11 wireless LAN standards (802.11b and 802.11g) discussed below use OFDM. The DSL services discussed in Chapter 7 generally also use OFDM, although in DSL service OFDM is called discrete multitone service (DMT).

TEST YOUR UNDERSTANDING

8. a) In normal radio operation, how does channel bandwidth usually relate to the bandwidth required to transmit a data stream of a given speed? b) How does this change in spread spectrum transmission? c) What is the benefit of spread spectrum transmission for business communication? d) Is spread spectrum transmission done for security reasons in commercial WLANs?

---

[5] In the 802.11a and 802.11g wireless LAN standards discussed later, each 20 MHz channel is divided into 52 subcarriers, each 312.5 kHz wide. (This leaves about 3.8 MHz of the channel unused to provide guard bands between subcarriers.) Of these 52 subcarriers, 48 are used to send data and 4 are used to control the transmission.

9. a) Describe FHSS. b) What is its limitation? c) Describe DSSS. d) For what WLAN standard is it used? e) What spread spectrum transmission method is used for 54 Mbps WLANs? f) Describe it.

802.11 WLAN STANDARDS

Wireless LANs replace wires with radio waves. They allow mobile workers to stay connected to the network as they move through an organization. In some cases, such as historical buildings that cannot be changed physically, wireless LANs may be the only way to install networking.

802.11

The IEEE, which creates Ethernet standards through the 802.3 Working Group of the 802 LAN/MAN Standards Committee, also creates wireless LAN standards through a different working group, the **802.11 Working Group.**

TEST YOUR UNDERSTANDING

10. Which IEEE 802 working group creates WLAN standards?

Typical Operation

As Figure 5-10 shows, an 802.11 wireless LAN typically is used to connect a small number of mobile devices to a large wired LAN—typically, an Ethernet LAN—because the servers and Internet access routers that mobile client stations need to use usually are on the wired LAN.

Stations   Each mobile station must have a **wireless NIC.** Mobile stations tend to use PC Card NICs, which simply snap into the station. Fixed stations normally use internal NICs. Both types of stations can use USB wireless NICs, which sit outside the PC and plug into USB ports.

Access Points   When a wireless station wishes to send a frame to a server, it transmits the frame to an **access point.**

The access point is a bridge between wireless stations and the wired LAN. **Bridges** connect two different types of LANs—in the case of 802.11 access points, an 802.11 LAN and an 802.3 LAN.

When a wireless NIC transmits, it places the packet for the server into an 802.11 frame.[6] The access point removes the packet from the 802.11 frame and places it in an 802.3 frame. The access point sends this 802.3 frame to the server. When the server replies, the access point removes the packet from the 802.3 frame coming from the server and forwards the packet to the wireless station in an 802.11 frame.

The access point also controls stations. It assigns transmission power levels to stations within its range and does a number of other supervisory chores.

---

[6] Frames for 802.11 are much more complex than 802.3 Ethernet frames. Much of this complexity is needed because of wireless propagation problems.

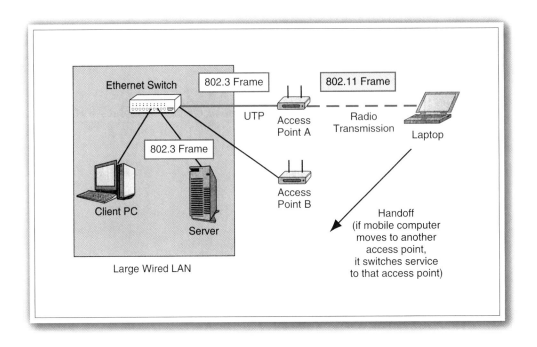

Figure 5-10   Typical 802.11 Wireless LAN Operation with Access Points

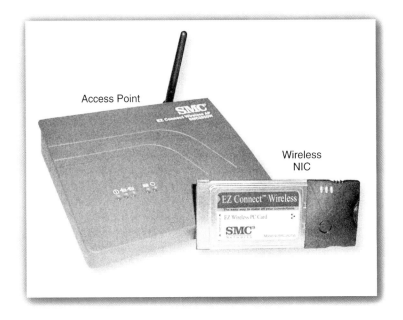

Figure 5-11   802.11 Wireless Access Point and Wireless PC
Card NIC. Courtesy SMC Communications

"Wireless Ethernet"?   Perhaps because it is typically used in conjunction with Ethernet wired LANs, 802.11 sometimes is called **wireless Ethernet.** This may seem to be stretching the definition of Ethernet (802.11 does not even use the 802.3 MAC layer frame). However, Bob Metcalfe, who was one of the creators of Ethernet, supports calling 802.11 "wireless Ethernet" because 802.11 is a direct descendent of Ethernet's core transmission principles.

Handoff   When a mobile station travels too far from an access point, the signal will be too weak to reach the access point. However, if there is a closer access point, the station will be **handed off** to that access point for service. This aspect of 802.11 LANs was standardized as 802.11F in 2003, but vendor interoperability is still limited.

TEST YOUR UNDERSTANDING

11. a) Describe the elements in a typical 802.11 WLAN today. b) Why is a wired LAN still needed if you have a wireless LAN? c) What two things do access points do? d) Why must the access point remove an arriving packet from the frame in which the packet arrives and place it in a different frame when it sends the packet back out? e) If someone says, "wireless Ethernet," what do they probably mean? f) What is a handoff in 802.11? g) Are handoffs standardized? What are the implications of your answer?

CONTROLLING 802.11 TRANSMISSION

If two 802.11 devices (stations or access points) transmit at the same time, their signals will be jumbled together and will be unreadable. The 802.11 standard has two mechanisms to reduce problems with multiple simultaneous transmissions. The first, CSMA/CA+ACK, is mandatory and is always used. The second, RTS/CTS, is optional.

CSMA/CA+ACK Media Access Control

In 802.11 wireless LANs, only one station can transmit at a time in each channel or their signals will mutually interfere and be unreadable. WLANs need a media access control discipline.

Problems in Hearing Collisions

The access point assigns each station a power setting so that the access point can hear all stations equally loudly. Unfortunately, this means that stations cannot necessarily hear one another. They may also be in dead spots relative to one other. Consequently, collision detection mechanisms that depend on stations hearing interference from other stations transmitting at the same time—for instance, the CSMA/CD mechanism discussed in the previous chapter—are impossible.

CSMA/CA

Instead, 802.11 LANs use CSMA/CA, where CA stands for **collision avoidance.** Figure 5-12 describes this process. With CSMA/CA, if the wireless NIC hears a transmission, it must not transmit.

CSMA/CA (Carrier Sense Multiple Access with Collision Avoidance)
    Sender listens for traffic
        If there is traffic, waits
        If there is no traffic,
            If there has been no traffic for less than the critical
                value for time, waits a random amount of time,
                then sends if still no traffic
            If there has been no traffic for more than the critical
                value for time, sends without waiting
ACK (Acknowledgement)
    Receiver immediately sends back an acknowledgement
    If sender does not receive the acknowledgement, retransmits using
        CSMA/CA

Figure 5-12   CSMA/CA+ACK in 802.11 Wireless LANs

If the station does not hear traffic, it considers the last time it heard traffic. If the time is less than some critical value, the station sets a random timer and waits. If there still is no traffic after the random wait, the station may send.

However, if the time since the last transmission exceeds the critical value, the station may transmit immediately.

ACK
Actually, 802.11 uses **CSMA/CA+ACK.** Collisions and other types of signal loss still are possible with CSMA/CA. When an access point receives a frame from a station, or when a station receives a frame from an access point, the receiver immediately sends an acknowledgement frame, an **ACK.** A frame that is not acknowledged is retransmitted.

There is no wait when transmitting an ACK. This ensures that ACKs get through while other stations are waiting.

Inefficient Operation
CSMA/CA+ACK works well, but it is inefficient. Waiting before transmission wastes valuable time. Sending ACKs also is time consuming. Overall, an 802.11 LAN can only deliver throughput (actual speed) of about half the rated speed of its standard, that is, the speed published in the standard.

This throughput, furthermore, is shared by all stations sharing the channel. Individual station throughput will be substantially lower.

In addition, if some stations are far from the access point, their transmission rates will fall far below those of stations near the access point. Even one or two slow stations sharing the access point's throughput will slow throughput for other stations considerably.

Request to Send/Clear to Send (RTS/CTS)

One option in CSMA/CA operation is **request to send/clear to send (RTS/CTS),** which Figure 5-13 illustrates.

Request to Send (RTS)

When a station wishes to send and is able to send because of CSMA/CA, the station may send a **request-to-send (RTS)** message to the access point.

Clear to Send (CTS)

If the access point broadcasts a **clear-to-send (CTS)** message, then other stations must wait, and the station sending the RTS may then transmit, ignoring CSMA/CA.

Only an Option

Although RTS/CTS is widely used, keep in mind that it is only an option, while CSMA/CA is mandatory, at least for initial communication. Also, tests have shown that RTS/CTS can reduce throughput when it is used.

There is one situation in which RTS/CTS is mandatory rather than optional. This is when 802.11b stations operating at 11 Mbps and 802.11g stations operating at 54 Mbps share an access point. In this case the 802.11g station must use request to send/clear to send.

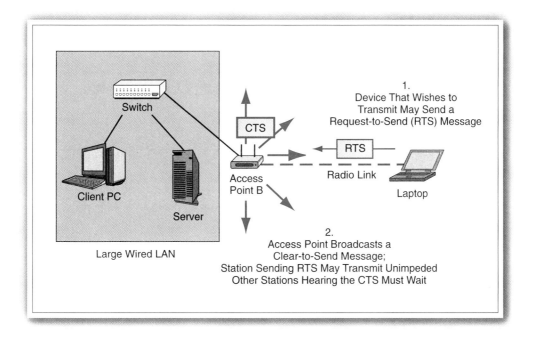

Figure 5-13   Request to Send/Clear to Send (RTS/CTS)

TEST YOUR UNDERSTANDING

12.   a) Why can't wireless networking use CSMA/CD? b) Describe CSMA/CA+ACK. c) Why is CSMA/CA+ACK inefficient? d) Do you think CSMA/CA+ACK governs transmission by NICs, access points, or both? Explain your reasoning. e) Describe RTS/CTS. f) Is CSMA/CA+ACK required or optional? g) Is RTS/CTS required or optional?

802.11 Wireless LAN Standards

As Figure 5-14 shows, the 802.11 Working Group has created several WLAN standards. The figure lists them in order of completion (802.11b was completed before 802.11a). The figure does not show the original 2 Mbps 802.11 standard because it saw little acceptance.

	802.11b	802.11a	802.11g	802.11g if 802.11g access point serves an 802.11b station
Unlicensed Band	2.4 GHz	5 GHz	2.4 GHz	2.4 GHz
Number of Non-Overlapping Channels	3	8 to 14 In future, 19 to 24	3	3
Rated Speed	11 Mbps	54 Mbps	54 Mbps	Not Specified
Actual Throughput, 3 m	6 Mbps	25 Mbps	25 Mbps	12 Mbps
Actual Throughput, 30 m	6 Mbps	12 Mbps	20 Mbps	11 Mbps
Is Throughput Shared by All Stations Using an Access Point?	Yes	Yes	Yes	Yes

Notes: Source for throughput data: Broadcom.com

The number of non-overlapping channels is important because nearby access points should operate on different channels.

For 802.11b and g, the non-overlapping channels are 1, 6, and 11.

For 802.11a in the United States,

There are four non-overlapping channels in the "lower" range in which power is limited to 40 mW and which is reserved for indoor use. These are Channels 36, 40, 44, and 48.

There are four non-overlapping channels in the "middle" range, in which power can be up to 200 mW. These are Channels 52, 56, 60, and 64.

There are four more channels in the "upper" range, which can operate with 800 mW or power. These are Channels 149, 153, 157, and 161. This band allows transmission over distances much longer than 100 meters.

More non-overlapping channels will be added in the future.

Figure 5-14   802.11 Wireless LAN Standards (Table)

How Fast are 802.11 Networks?

Rated Speeds   Figure 5-14 shows that the rated speeds of 802.11 WLANs are fairly high—between 11 Mbps and 54 Mbps. However, these rated speeds are misleading.

Throughput   First, as Figure 5-14 shows, actual throughput usually is considerably lower than stated speeds and tends to fall off rapidly with distance.

Shared Throughput   In addition, 802.11 throughput is *shared* by all stations that wish to transmit at the same time. For instance, if shared throughput is 5 Mbps, stations using an access point shared by ten to twenty stations might see individual throughput of only one or two megabits per second, despite the fact that only a few stations may be transmitting simultaneously.

802.11b

The most widely used 802.11 standard today is 802.11b because it was the first 802.11 standard to become popular. 802.11b has a stated speed of 11 Mbps, although actual throughput shared by all devices is only about half that speed, as the figure shows.

Access points and NICs that follow the 802.11b standard operate in the relatively low-frequency 2.4 GHz radio band. Consequently, throughput does not fall off sharply with distance as it does with 802.11a, which operates at a higher frequency.

However, there is only room for three **nonoverlapping channels** in the 2.4 GHz band.[7] Access points that are near each other should operate on different nonoverlapping channels, as Figure 5-15 shows. Otherwise, they will interfere with each other

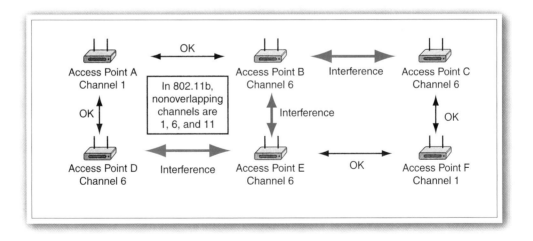

Figure 5-15   Using Different Channels in Nearby Access Points

---

[7]   There actually are 11 "channels" in the United States, but transmissions are spread across multiple channels. In practice, 802.11b stations and access points usually transmit on Channels 1, 6, and 11 in order for their transmissions not to overlap.

somewhat, driving down the throughput each provides. In a large three-dimensional building and with only three nonoverlapping channels, it is impossible to avoid access point channel conflicts with 802.11b.

802.11a and 802.11g

The two newest standards in Figure 5-14 are 802.11a and 802.11g.[8] Both have rated speeds of 54 Mbps.

802.11g: Compatibility and Distance   802.11g is attractive because it operates in the lower 2.4 GHz unlicensed radio band, while 802.11a operates in the higher 5 GHz unlicensed radio band. Therefore, 802.11g throughput falls off more slowly with distance than 802.11a throughput, as Figure 5-14 shows.

In addition, 802.11b and 802.11g NICs can operate together with an 802.11g access point. This makes 802.11g a smooth upgrade path in existing 802.11b networks because old NICs can still be used (although new access points will be needed). Also, 802.11g NICs will drop down to 802.11b operation when they associate with an 802.11b access point.

However, the throughput of 802.11g stations falls dramatically when even one 802.11b station associates with an 802.11g access point. (In Figure 5-14, compare the 802.11g column with the last column.)

802.11a: Many Nonoverlapping Channels   In the 802.11a standard's favor, the 5 GHz band provides many more nonoverlapping channels than 802.11g. Depending on regulations in various countries, the 5 GHz band currently offers at least 8 nonoverlapping channels and may offer 14. In addition, the 5 GHz unlicensed radio band is being expanded, and most countries will soon offer 19 to 24 nonoverlapping channels. With many nonoverlapping channels available, setting access point channels so that no two nearby access points use the same nonoverlapping channel should be straightforward. In addition, future standards could allow individual access points to provide service simultaneously on multiple nonoverlapping channels, offering greater throughput to the stations sharing these access points.

Future Developments

Multistandard Access Points   In the future it probably will not be necessary to choose between 802.11a and 802.11g because many access points will offer both so that they will be able to handle any station following the 802.11a, 802.11b, or 802.11g standard.

Quality of Service (QoS)   Another development to watch for is the **802.11e,** which will standardize quality of service (QoS) for 802.11 WLANs. Primarily, this will allow good-quality voice and streaming audio transmission.

TEST YOUR UNDERSTANDING

13.   a) Distinguish between rated speed, shared throughput, and individual throughput. b) What is the most widely used 802.11 LAN standard today? c) How many nonoverlapping channels does it support? d) Why is the number of nonoverlapping channels

---

[8]   In Europe the HiperLAN2 standard is a competitor for 802.11a in the 5 GHz band. HiperLAN2, unlike 802.11a, offers quality-of-service guarantees.

that can be used important? e) What advantages does upgrading from 802.11b to 802.11g instead of 802.11a bring? f) What advantage does upgrading from 802.11b to 802.11a instead of 802.11g bring? g) Why may it not be necessary to choose between 802.11g and 802.11a in the future? h) Why would 802.11e QoS standards be welcome?

802.11 SECURITY

When companies first began to implement 802.11 WLANs, they discovered to their horror that **drive-by hackers** could park just outside their premises and eavesdrop on their data transmissions. These drive-by hackers also could mount denial-of-service attacks; insert viruses, worms, and spam into the network; and do other mischief. Many companies delayed the deployment of 802.11 WLANs, and many pulled out their early access points and NICs because these devices did not offer good security.

No Security by Default

The worst problem with 802.11 security is simply that security traditionally was *turned off by default* when you installed an older access point or wireless NIC. Consequently, many access points and NICs implement no security at all.[9] Although many new access points and NICs now come with security turned on by default, there still is a large base of completely unprotected access points and NICs.

> The worst problem with wireless LAN security is simply that security traditionally has been turned off by default.

Wired Equivalent Privacy (WEP): Shared Static Keys

When 802.11 was first introduced in 1997, it was introduced with a security mechanism, **wired equivalent privacy (WEP).** We saw in Chapter 4 that when several stations are attached to an Ethernet hub, they can all hear one another, but nobody else can read their traffic. This rather poor security, surprisingly, was the goal of WEP. All WEP transmissions are encrypted, but all stations using the access point will use the same encryption key and so can read each other's messages. So can any hacker who learns the shared key.

Shared Static Keys

Unfortunately, having everyone use the same key is a code-breaker's dream because any key is breakable if a cryptanalyst has enough traffic to analyze. Furthermore, WEP has no standard way to change keys, so most shared keys are never changed. Having all stations use the same **shared static key** ensures a lot of traffic to analyze. Key sharing (aided by other weaknesses in WEP) allows WEP keys to be broken by collecting only 100 megabytes to one gigabyte of data, which can be done in about thirty hours with a typical business access point.[10] Automated drive-by hacking software for WEP key cracking is readily available on the Internet.

---

[9] In 2002 and 2003, people who drove around and collected signals from access points found that only about a third of the access points encountered had WEP enabled. http://www.worldwidewardrive.org

[10] George Ou, "At Last, Real Wireless LAN Security," *TechRepublic,* September 3, 2002. http://techrepublic.com.

Automated Drive-By Hacking

> Can read traffic from outside the corporate walls
>
> Can also send malicious traffic into the network

No Security by Default

> In older products the installation default was to have no security at all

Wired Equivalent Privacy (WEP)

> Initial flawed security developed for 802.11 WLANs
>
> All stations share the same encryption key with the access point
>
> This key is rarely changed
>
> Shared static keys mean that a large volume of traffic is encrypted with the same key
>
> This makes it possible for cryptanalysts to crack the key in hours or days
>
> Once the key is cracked, the attacker can read all messages and can also send attack messages to internal servers without having to go through a firewall
>
> Special software that automates the hacking process is widely available
>
> Adding virtual private networks to protect traffic works but adds to costs

Wireless Protected Access (WPA)

> Interim security mechanism introduced in 2002
>
> Introduced some parts of 802.11i (discussed next) for products built in 2002 and 2003
>
> Often possible to upgrade existing WEP products to WPA

802.11i Security

> Full security mechanism introduced in 2003
>
> Very good security
>
> Temporal Key Integrity Protocol (TKIP)
>
>> Each station gets a separate key for confidentiality
>>
>> This key is changed frequently
>
> Extended Authentication Protocol (EAP)
>
>> Authentication is one party (access point or wireless station) proving its identity to the other party
>>
>> Authenticate with an authentication server (Figure 5-17)
>>
>> Several EAP methods exist and may be used (Figure 5-18)
>
> Compliant products became available in late 2003

The Transition to Strong Security

> We will soon have a mix of no security, WEP, 802.11i, WPA, and other security protocols
>
> Security is only as strong as the weakest link
>
> Legacy equipment that cannot be upgraded to 802.11i will have to be discarded

Rogue Access Points

> Unauthorized access points set up by department or individual
>
> Often have very poor security, leaving a big opening

Figure 5-16   802.11 Security (Study Figure)

Worse yet, access points authenticate stations by means of the shared static key. If a station knows the shared static key, the access point automatically assumes that the station is legitimate. This means that once the attacker cracks the key, he has access to the firm's internal Ethernet network—without going through a firewall. This opens the company to a broad range of attacks.

Adding a VPN

In Chapter 1, we saw that virtual private networks (VPNs) provide security for transmissions traveling over the (nonsecure) Internet. However, VPNs can be used over any nonsecure network, including wireless LANs. Some firms also operate VPNs over their WLANs to add security to this nonsecure environment. Although VPNs are effective in providing security, they are somewhat unwieldy, and corporations want security built into 802.11 standards directly.[11]

802.11i Security

Stung by its initial gaffs, the 802.11 working group has developed a more advanced form of wireless LAN security, **802.11i.** Products using 802.11i security began to ship in late 2003, although the standard was not ratified until 2004.

Temporal Key Integrity Protocol (TKIP)

Most important, 802.11i uses the **Temporal Key Integrity Protocol (TKIP),** which gives each station its own unshared key after authentication and which changes this key frequently. With different keys for each station and with frequent key changes, cryptanalysts cannot collect the volume of traffic data needed to crack an 802.11i key.

Extensible Authentication Protocol (EAP)

In addition, 802.11i strongly authenticates the client station and the access point—requiring each to prove its identity to the other. More specifically, 802.11i uses the **Extensible Authentication Protocol (EAP).** As Figure 5-17 shows, when a station requests service, the access point requires the station to provide authentication data. In the reply (Step 1), the station transmits the authentication data such as a password or proof that it has a digital certificate (discussed in Chapter 9).

Authentication Server   The access point passes this authentication data on to an **authentication server** (Steps 2 and 3). The authentication server checks the authentication data and sends back an affirmation or a rejection message (Steps 4 and 5). The access point can then accept or reject the station.

EAP Allows Multiple Authentication Methods   EAP standardizes communication between the station, access point, and authentication. An organization also needs to select a specific authentication method (EAP offers several). Different authentication methods allow tradeoffs between ease of implementation and degree of authentication security. The easiest and least secure method is EAP-MD5. At the other end of the

---

[11]   In addition, it is possible for the access point to provide service only to NICs whose MAC addresses are on the access point's access control list. This approach is highly labor-intensive and does not scale well to large numbers of users.

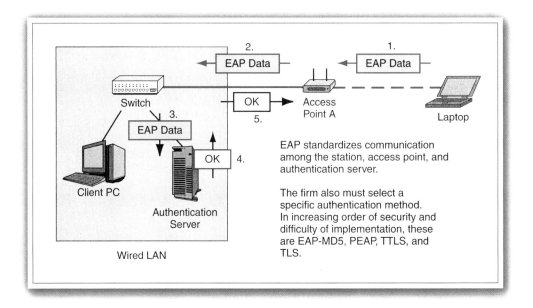

Figure 5-17 Extensible Authentication Protocol (EAP)

spectrum, TLS is both difficult to implement and highly secure. In between are PEAP, which is supported by Microsoft and Cisco, and EAP-TTLS, which has more limited vendor support.

WPA (Wireless Protected Access)

To confuse matters even more, in 2002 and 2003, most vendors shipped 802.11 wireless products that used an interim form of security, **wireless protected access (WPA).** WPA used parts of 802.11i that could be implemented immediately, usually with a firmware upgrade to existing wireless devices.

Transition to Strong Security

Companies are beginning to require 802.11i security in new products they buy. Unfortunately, if a firm has a mix of wireless products having 802.11i, WEP, no security, WPA, and other forms of security,[12] then security will only be as strong as the security of the firm's weakest products. In security, legacy technology cannot be tolerated because it leaves gaping holes open for attackers. Legacy equipment that cannot be upgraded must be discarded.

---

[12] For example, before 802.11i was released, Cisco Systems developed its own proprietary security technology, the Lightweight Extensible Authentication Protocol (LEAP).

Rogue Access Points

Wireless access points are inexpensive and easy to set up. Consequently, departments and individual employees sometimes set up their own access points. These are called **rogue access points** because they are set up outside company control. Rogue access points often have poor security, giving drive-by hackers an easy way into the corporation.

TEST YOUR UNDERSTANDING

14. a) What is the biggest problem today with wireless security? b) What is the advantage of using VPNs with WLANs? c) What are the disadvantages of using VPNs with WLANs? d) Compare WEP and 802.11i keys in terms of key sharing and how often keys are changed. e) Why is it important to avoid key sharing and to change keys frequently? f) Compare WEP and 802.11i authentication. g) How does EAP work? h) Distinguish between EAP and authentication methods. i) What tradeoffs do different authentication methods offer? j) What must be done if a firm has a mix of WEP, WPA, and 802.11i access points? k) What are rogue access points? l) Why are they dangerous?

BLUETOOTH PERSONAL AREA NETWORKS (PANs)

Personal Area Network for Cable Replacement

Although 802.11 is good for fairly large wireless LANs, another wireless networking standard, **Bluetooth,** was created for wireless **personal area networks (PANs),** which

	802.11	Bluetooth
Focus	Local Area Network (LAN)	Personal Area Network (PAN)
Rated Speed (Actual Throughput Will Be Lower)	11 Mbps to 54 Mbps in both directions	722 kbps with back channel of 56 kbps. May increase.
Distance	30 to 100 meters	10 meters
Number of Devices	Limited in practice only by bandwidth and traffic	10 piconets, each with up to 8 devices
Scalability	Good because allows multiple access points	Poor
Cost	Higher	Lower
Battery Drain	Higher	Lower
Application Profiles	No	Yes

Figure 5-18 802.11 Versus Bluetooth

are intended to be used by a single person.[13] Bluetooth basically offers **cable replacement**—a way to get rid of cables between devices.

For example, using Bluetooth, a notebook computer can print wirelessly to a printer and synchronize its files wirelessly with those on a desktop computer. To give another example, a cellphone can print to the same wireless printer and place a call through the firm's telephone system instead of paying to make a cellular call.

Application Profiles

Bluetooth offers one very important thing that 802.11 does not—**application profiles,** which are application-layer standards designed to allow devices to work together automatically, with little or no user intervention. For instance, you may be able to take a Bluetooth-enabled notebook computer to a Bluetooth-enabled printer and print as soon as the two devices recognize each other. The 802.11 standard has nothing like this currently. Unfortunately, the Bluetooth application profiles introduced to date have been rudimentary. In addition, most devices only implement a few of these application profiles, so there is no guarantee that two Bluetooth devices that you wish to connect will be able to work together.

Long Battery Life

Although Bluetooth offers only low speeds and short distances, these limitations mean that radio transmission power is low so that battery life is quite long. This is very important for cellphones and other small devices.

Low Speed and Limited Distance

Bluetooth is very slow. It currently offers a speed of only 722 kbps (with a back channel of 56 kbps) and a maximum distance of 10 meters.[14] Bluetooth's low speed and short distance span mean that it is not a full WLAN technology.

Interference with 802.11 Networks

Another problem is that Bluetooth and 802.11b both operate in the 2.4 GHz unlicensed radio band, so they may interfere with one another. Newer Bluetooth standards have modified Bluetooth's transmission method to reduce this interference, but some interference still remains.

TEST YOUR UNDERSTANDING

15. a) Contrast how 802.11 and Bluetooth are likely to be used in organizations. b) What is a PAN? c) Why are Bluetooth application profiles attractive? d) Why do they not always fulfill their promise? e) What are the speeds of Bluetooth transmission? f) What is the normal maximum propagation distance for Bluetooth? g) What benefit do low speeds and short distances bring? h) What problem may occur if both 802.11b and Bluetooth are used in the same office? i) Will this problem still exist if 802.11a is used instead of 802.11b? (The answer is not explicitly given in the text.)

---

[13] Bluetooth is named after King Harald Bluetooth, a Scandinavian king in the tenth century. As you might guess, Bluetooth was developed in Sweden, but it is now under the control of an international consortium.

[14] This is with standard 1 milliwatt power. This can be raised to 50 meters using 100 milliwatt power, although this drives down battery life and is not necessary for Bluetooth's main purpose, the elimination of cords between devices.

EMERGING WIRELESS TECHNOLOGIES

In this final section, we will look at some emerging wireless LAN developments that are likely to be important in coming years. We will look at them in order of their likely appearance.

Wireless LAN Management

Large organizations soon will have hundreds or even thousands of access points. To keep labor costs under control, organizations need to be able to manage access points remotely, from a central management console. Figure 5-20 illustrates two approaches to centralized access point management.

Smart Access Points
The simplest approach architecturally is to add intelligence to each access point. The central management console can communicate directly with each of these **smart access points** via the firm's Ethernet wired LAN. However, adding management

---

Wireless LAN Management
        Large firms must manage many access points
        Would like to be able to do this centrally
        Smart access points or WLAN switches (Figure 5–20)
        Functions
                Notification of failures
                Support remote access point adjustment
                Constant QoS monitoring
                Send software updates to all access points
                All of this should be as automatic as possible
Radio Frequency IDs (RFIDs)
        Like UPC codes on products but RFID chips can be read from a small distance
        Reduced cost in checkout
        Constant inventory updating
        Real-time data for business
Ultrawideband(UWB)
        Normal spread spectrum bandwidths are a few megahertz
        Ultrawideband (UWB) uses channels more than 100 times larger
        480 Mbps with a distance of about 10 meters
        Wireless TV transmission in homes
        Wireless communication within a telecommunications closet or server room
Fourth-Generation (4G) Stations
        Station that can support multiple radio methods

Figure 5-19   Emerging WLAN Technologies (Study Figure)

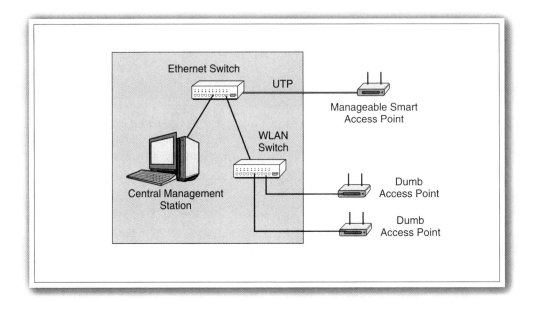

Figure 5-20   Access Point Management Alternatives

processing raises the prices of access points considerably. Using smart access points is an expensive strategy.

Wireless LAN Switches

A second approach illustrated in Figure 5-20 is to use **WLAN switches.** As the figure shows, multiple access points connect to each wireless LAN switch. The intelligence is placed in the WLAN switch rather than in the access points themselves. Vendors who sell WLAN switches claim that this approach reduces total cost because only inexpensive **dumb access points** are needed. However, smart access point vendors dispute WLAN switch cost comparisons.

Wireless LAN Management Functionality

Although approaches to centralized WLAN management vary, most vendors agree on the types of functionality these systems should provide.

➤ They should notify the WLAN administrators of failures immediately, so that malfunctioning access points can be fixed or replaced rapidly.

➤ They should allow remote adjustment, for instance allowing nearby access points to increase their power to compensate for an access point failure. Such adjustments are also needed over time if furniture is moved or if the number of users in an area changes.

➤ They should provide constant transmission quality monitoring to allow WLAN administrators to adjust access point operating parameters constantly.

➤ They should allow software updates to be pushed out to all access points or WLAN switches, bypassing the need to install updates manually.

➤ As far as possible, the management software should be able to work automatically, taking as many actions as possible without user or IT staff intervention.

Radio Frequency IDs (RFIDs)

Most products in grocery stores today have Universal Product Code (UPC) bar codes. The cashier merely runs items by a reader and their sales are recorded. However, scanning UPC codes is somewhat slow and labor-intensive.

In contrast, **radio frequency ID (RFID)** chips can be read by radio over short distances.[15] With RFID chips, checkout requires much less labor. The items have only to pass near an RFID scanner. RFIDs even allow warehouses to have "smart shelves" that provide constant inventory information to a central computer. In general, RFIDs can provide real-time data for many information systems.

Ultrawideband (UWB)

In traditional spread spectrum transmission, channels typically are several megahertz wide. A proposed new spread spectrum transmission approach uses channels that are more than 100 times wider. Understandably, such systems are called **ultrawideband (UWB)** transmission systems.

UWB is attractive because it can provide enormous transmission speed (480 Mbps) over distances of about 10 meters. UWB is capable of providing wireless television transmission within a home, high-speed switch-to-switch transmission within a telecommunications closet, and server-to-switch transmission in a server room. On the negative side, until interference between UWB systems and other wireless technologies is better understood, regulatory approval may be limited.

Fourth-Generation (4G) Stations

In this chapter, we looked at several major radio-based technologies, including 802.11, Bluetooth, RFIDs, and UWB transmission. In the next chapter, we will see second-generation and third-generation cellular radio technologies. Chapter 7 will cover even more wireless transmission alternatives.

If you have a notebook computer, it may have to work with two or more of these technologies in different locations. One solution is to simply buy several wireless NICs conforming to the different environments in which the notebook will have to work. Another solution being researched is the **fourth-generation (4G)** station, which will be able to work with multiple radio standards in multiple bands intrinsically. Some research in this area is focusing on software-defined radio, which can be reconfigured by software instead of requiring multiple hardware implementations for different wireless standards.

---

[15] When the reader scans the RFID chip, the chip comes to life and sends out the information it contains. RFID works in two frequency ranges. In the 30 kHz to 500 kHz range, distances are limited to about a meter. In the unlicensed 900 MHz and 2,400 MHz bands, reading distances are long enough for automated toll booth collection.

TEST YOUR UNDERSTANDING

16. a) Why is centralized access point management desirable? b) What are the two technologies for remote access point management? c) What functions can remote access point management support? d) Why are RFIDs attractive? e) What is UWB transmission? f) For what applications can it be used? g) What are 4G stations?

SYNOPSIS

The mantra of networking has always been "anything, anywhere, any time." With the advent of wireless data transmission, this promise is finally being extended to mobile users.

Wireless networks predominantly use radio transmission. Radio waves are characterized by frequency, wavelength, and amplitude. Wavelength and frequency are inversely related. Real radio signals contain a mix of frequencies. The range of frequencies between the lowest frequency and the highest frequency of a signal is the signal's bandwidth.

Radios send and receive using antennas. Dish antennas concentrate incoming and outgoing signals for long-distance transmission but are bulky and require the receiver to know the direction of the other party. In contrast, omnidirectional antennas send and receive in all directions equally well. If the other party is not too far away, as is the case in WLANs, omnidirectional antennas are attractive.

We saw in Chapter 3 that propagation problems with copper wires and optical fiber are mild and are controllable with good installation discipline. In contrast, we saw in this chapter that radio propagation problems are serious and are difficult to control. First, radio waves attenuate very rapidly, following an inverse square law. Other problems are electromagnetic interference from other devices (including other wireless stations), multipath interference because of reflections off walls and ceilings, and shadow zones (dead spots) where signals cannot reach. Shadow zones are more pronounced at higher frequencies, and attenuation increases with frequency.

The frequency spectrum consists of all frequencies from 0 Hz to infinity. The frequency spectrum is divided into service bands for particular types of services. Service bands are further divided into channels. Different signals can be sent at the same time if they are sent in different channels.

The maximum possible transmission speed within a channel is directly proportional to the channel's bandwidth. Normally, channel bandwidth is set to meet transmission speed requirements. However, spread spectrum transmission uses a much higher bandwidth than the signal requires. Spread spectrum transmission is done to reduce propagation problems, which often occur only at certain frequencies. The fastest wireless LANs use orthogonal frequency division multiplexing (OFDM) spread spectrum transmission, which divides the broadband channel into many smaller subcarriers, each of which carries part of the signal.

The 802.11 Working Group sets most wireless LAN (WLAN) standards. 802.11 LANs normally serve users through access points, which connect wireless stations to resources on the company's wired LAN. The most widespread WLAN technology today is 802.11b, which has a rated speed of 11 Mbps. The 802.11a and 802.11g standards, in turn, have rated speeds of 54 Mbps. Actual throughput (speed delivered to users) is about half of the rated speed near the access point and falls off with distance; in addition, all stations using an access point share this throughput, so individual throughput is even lower.

Signals in 802.11g travel farther than 802.11a signals because 802.11g operates in the lower-frequency 2.4 GHz band while 802.11a operates in the higher-frequency 5 GHz band. On the other hand, 802.11a offers many more nonoverlapping channels, which is important in terms of placing access points so that nearby access points do not interfere with each other by operating on the same channel. 802.11g access points can serve 802.11b stations, but serving even one 802.11b station reduces throughput substantially.

The first 802.11 LANs used weak wired equivalent privacy (WEP) security, and many users did not even turn on this anemic security. Even with WEP enabled, drive-by hackers can easily eavesdrop on conversations and send attack packets into the network. The 802.11 Working Group finally created a robust security standard, 802.11i, but older access points and wireless NICs cannot be upgraded to 802.11i. Older equipment must be discarded or drive-by hackers will be able to enter through older equipment even if newer equipment follows 802.11i.

Bluetooth is a low-speed personal area network (PAN) technology designed to replace wired connections between devices within a few meters of each other.

SYNOPSIS QUESTIONS

1. a) What are the three characteristics of radio waves? b) How are wavelength and frequency related? c) What is a signal's bandwidth?
2. Why do WLANs use omnidirectional antennas?
3. a) What are the four main propagation problems that face WLAN designers? b) Which increase(s) with frequency?
4. a) To send data very quickly, what has to be true about channel bandwidth? b) Compare normal and spread spectrum transmission in terms of bandwidth. c) Why is spread spectrum transmission used? d) What form of spread spectrum transmission is used for very high transmission speeds? e) How does this type of spread spectrum transmission work?
5. a) In 802.11 WLANs, what two things does the access point do? b) What is the most widely used 802.11 technology today, and what is its rated speed? c) Why is 802.11g better than 802.11a? d) Why is 802.11a better than 802.11g? e) What happens if 802.11b and 802.11g wireless stations share an 802.11g access point? f) Distinguish between rated speed and throughput. g) If an access point has a throughput of 25 Mbps, why will the individual stations that use it not be able to obtain this throughput?
6. a) Explain the security weaknesses in early 802.11 wireless LANs. b) What is the new security standard? c) What will have to be done with older access points and wireless NICs for a firm to achieve good wireless LAN security?
7. Compare 802.11 and Bluetooth networks.

THOUGHT QUESTIONS

1. From the information in the text, what is a typical wavelength for microwave ovens? Justify your calculation.

2. Telephone channels have a bandwidth of about 3.1 kHz, as we will see in the next chapter. a) If a telephone channel's signal to noise ratio is 30 dB, how fast can a telephone channel carry data? (Hint: Telephone modems operate at about 30 kbps, so your answer should be close to this.) b) How fast could a telephone channel carry data if the SNR were increased to 40 dB? c) With an SNR of 30 dB, how fast could a telephone channel carry data if the bandwidth were increased to 4 kHz? Show your work or no credit.

3. In a home network, would you use 802.11a or 802.11g? Justify your answer.

4. What advice would you give a company about WLAN security?

DESIGN QUESTION

1. Consider a one-story building that is a square. It will have an access point in each corner. All other access points can hear one another. a) Assign access point channels to the five access points if you are using 802.11b. See Figure 5-14 for a list of possible channels. Try not to have any two stations use the same channel. b) Do it again with 802.11a.

TROUBLESHOOTING QUESTION

1. Your wireless notebook computer works on one desk, but when you move it to a nearby desk, you cannot receive a signal. Come up with at least two hypotheses for why the problem is occurring. How might you test the hypotheses?

THE PUBLIC SWITCHED TELEPHONE NETWORK (PSTN)

Learning Objectives:

By the end of this chapter, you should be able to discuss:

- The importance of telephone service.
- The technology of the public switched telephone network (PSTN), including circuit switching, the access system, the transport core, and signaling.
- The digital nature of the PSTN, except for the analog local loop to residential customer premises; PCM conversion of customer signals at the end office.
- Cellular telephony.
- IP telephony (also called voice over IP or VoIP)
- Telephone carriers.

INTRODUCTION

The Importance of Telephony

In Chapters 4 and 5 we focused on the two main LAN technologies in use today—Ethernet and 802.11 wireless LANs. In this chapter we will begin moving outside the firm's boundaries by looking at the worldwide telephone network, which is officially called the **Public Switched Telephone Network (PSTN).**

New Technologies

In the past, telephone service was widely viewed as an old ho-hum service. Even telephone professionals called it POTS, which stands for "plain old telephone service." However, new technologies, such as third-generation cellular services and voice over IP, now make telephony one of the most exciting areas in networking. In addition, voice and data transmission, which previously traveled over separate networks, are merging into "converged networks" that carry both forms of traffic.

More Complex Management

The management of telephony is also changing. A few years ago, even large firms simply contracted with the local monopoly telephone company and paid their standardized monthly bill. However, competition between many alternative carriers is bringing a growing range of products and services. In addition, telephone expenses are now so high that companies need to manage their telephone costs carefully.

WANs and Telephone Technology and Regulation

Finally, most wide area networks typically are built on top of the PSTN's switches, trunk lines, and access lines. In terms of both technology and carrier regulation, you cannot understand WANs unless you understand telephony. This makes a chapter on telephony a necessary precursor to the next chapter, which deals with WANs.

The Four Elements of the PSTN

Figure 6-1 shows the four technical elements of today's worldwide public switched telephone network: customer premises equipment, access, transport, and signaling.

Customer Premises Equipment

Equipment owned by the customer, including PBXs, internal wiring, and telephone handsets, is called **customer premises equipment.** We looked at customer premises equipment in Chapter 3. For some reason, "customer premises" is always written in plural.

Access

The customer needs an **access line** to reach the PSTN's central transport core. Access lines are also referred to as the **local loop.** The **access system** includes both access lines and **termination equipment** in the **end office** at the edge of the transport core.

Although access systems are relatively simple technologically, they represent a huge capital investment. The hundreds of millions of access lines that run to customer premises cost much more collectively than the trunk lines that run between the internal switches in the transport core. Similarly, there are many more end office switches in the

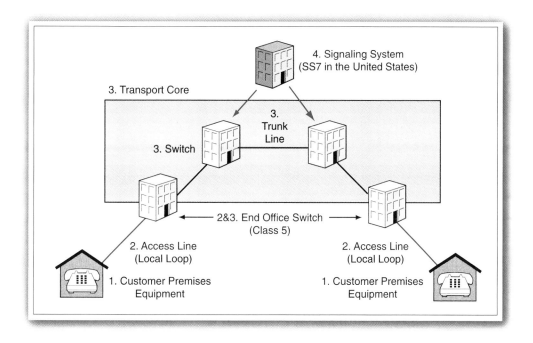

Figure 6-1 Elements of the Public Switched Telephone Network (PSTN)

PSTN than there are internal switches. The huge capital investment in today's access system is an impediment both to change and to the entry of new access competitors.

Transport

Transport means transmission—taking voice signals from one subscriber's access line and delivering them to another customer's access line. Internally, the **transport core** consists of trunk lines and switches. The end office switch is the transition point between the access system and the transport core and is a member of both.

While changes in access line technology have been slow, changes in the transport core have been very rapid (at least by telephony's standards). Changing the transport core represents a smaller investment than changing the access system, and changes in the transport core can save carriers a great deal of money.

Signaling

Finally, **signaling** means the controlling of calling, including setting up a path for a conversation through the transport core, maintaining and terminating the conversation path, collecting billing information, and handling other supervisory functions.

In the PSTN, *transport* is the transmission of voice communication. In contrast, *signaling* is the process of supervising voice communication sessions.

TEST YOUR UNDERSTANDING

1. a) What are the four technical elements in the PSTN? b) What is customer premises equipment? c) What are the parts of the access system? d) What is the local loop? e) What is the transport core? f) What are the two elements of the transport core? g) Which is changing more rapidly—the access system or the transport core? h) Explain why. i) What is signaling? j) In telephony, distinguish between transport and signaling.

Circuit Switching

Circuits

In contrast to the packet-switched networks that we saw in previous chapters, the telephone system has traditionally offered **circuit switching,** in which capacity for a voice conversation is reserved on every switch and trunk line end-to-end between the two subscribers (see Figure 6-2). Although it may be difficult to get a dial tone during natural disasters or on Mother's Day, once a **circuit** (a two-way connection with reserved capacity) is set up, there is no slowing of traffic or latency.

A circuit is a two-way connection with reserved capacity.

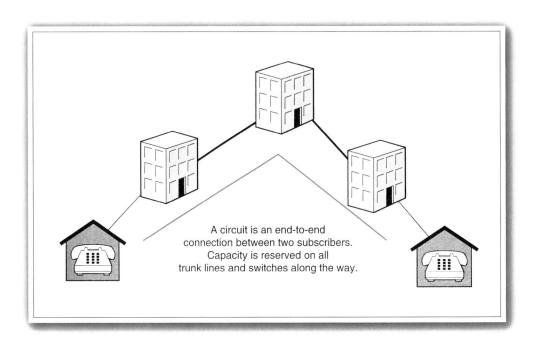

A circuit is an end-to-end
connection between two subscribers.
Capacity is reserved on all
trunk lines and switches along the way.

Figure 6-2   Circuit Switching

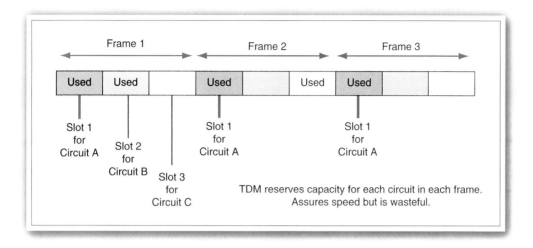

Figure 6-3   Time Division Multiplexing (TDM)

Time Division Multiplexing (TDM)

To provide reserved capacity on trunk lines between switches, telephone carriers typically use **time division multiplexing (TDM)**, which is illustrated in Figure 6-3. First, time is divided into brief periods called **frames.** Second, each frame is divided into even briefer periods, called **slots.** In TDM, a circuit is given the same slot in each frame.

For instance, in the figure, Circuit A is given Slot 1 in every frame. Note that Circuit A uses its slot capacity in every frame shown in the figure. However, Circuits B and C use only some of their slot capacity in two of the three frames. Although TDM provides the reserved capacity required for circuit switching, it wastes unused capacity. Users must pay for this reserved capacity whether they use it or not.

Voice Versus Data Traffic

Voice Traffic   Circuit switching works well for voice. As Figure 6-4 illustrates, voice traffic is fairly constant. In a conversation, one side or the other is talking most of the time. Perhaps 30 percent of the capacity of each full-duplex (two-way) telephone circuit is actually used.

Bursty Data Traffic   However, as Figure 6-4 also shows, data traffic is **bursty,** with short, high-speed bursts separated by long silences. For instance, when you are using a website, your request message is very brief. The response message takes a bit longer to transmit, but, particularly on a broadband connection, transmitting the response message only takes a few seconds. After receiving a webpage, you are likely to look at it for 30 to 60 seconds on average. During this time, no data is transmitted in either direction.

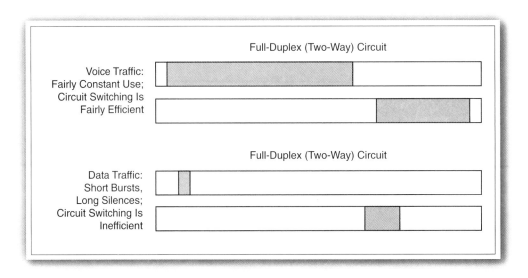

Figure 6-4 Voice and Data Traffic

Other data applications are similarly bursty. Reserved capacity in circuit switching is extremely wasteful for data transmission which typically uses only 5% of capacity.

Dial-Up Circuits

The PSTN provides two types of circuits. From your personal experience, you probably are only familiar with **dial-up circuits.** When you place a call, the PSTN sets up a circuit. When you finish this call, your circuit is ended, and reserved capacity is released for other circuits. You also know that modem-based data transmission over a dial-up telephone circuit is very slow.

Private Line Circuits

Figure 6-5 compares the dial-up circuit with the other type of circuit offered by telephone carriers. This is the **private line circuit,** also called the **leased line circuit.**

Always-on Service In contrast to dial-up circuits, a private line circuit is permanent and **always on.** Once a private line is provisioned (set up) by the telephone company, it is always available for transmission.

High Data Speed Private line circuits also carry data much faster than dial-up circuits. Even the slowest private line circuits carry data at 56 kbps or 64 kbps. The fastest carry data at several gigabits per second.

Multiplexing Multiple Voice Calls Although you probably only use dial-up circuits yourself, businesses primarily use private line circuits. These circuits typically connect a

	Dial-Up Circuits	Private Line Circuits
Operation	Dial-up. Separate circuit for each call.	Permanent circuit, always on
Speed for Carrying Data	Up to 56 kbps	56 kbps to gigabit speeds
Number of Voice Calls	One	Several due to multiplexing

Figure 6-5   Dial-Up Circuits Versus Private Line Circuits

corporate PBX to the nearest end office switch of the public switched telephone network. In doing so, they **multiplex** (mix together on the same line) multiple voice circuits to and from the corporate premises. For instance, the most popular private line, the T1 line, multiplexes 24 voice circuits. In Chapter 7, we will see how private lines carry data as well as multiplexing voice circuits.

TEST YOUR UNDERSTANDING

2. a) What is circuit switching? b) Explain frames and slots in time division multiplexing (TDM). c) How is a circuit allocated capacity on a TDM line? d) What is the advantage of TDM? e) What is the disadvantage?

3. a) Why does circuit switching make sense for voice communication? b) What does it mean that data transmission is bursty? c) Why is burstiness bad for circuit switching?

4. a) What are the differences between dial-up and private line circuits? b) What is multiplexing in the context of telephone calls and private lines?

THE ACCESS SYSTEM

The PSTN's access system is the only part of the PSTN that corporations work with directly. Consequently, we will look at it in the most detail.

The Local Loop

As noted earlier, the local loop, although fairly simple, represents an enormous capital investment and so is very difficult to change. Figure 6-6 shows that three main technologies dominate the local loop, although radio-based local loops and fiber to the home (FTTH) for residential customers may also be important in the future.

Single-Pair Voice-Grade Twisted Pair Wiring

Traditionally, the telephone system has brought a single pair of voice-grade UTP to each subscriber home and office. **One-pair voice-grade** copper has much lower transmission quality than the 4-pair UTP used in LANs.

Technology	Use	Status
1-Pair Voice-Grade UTP	Residences	Already installed
2-Pair Data-Grade UTP	Businesses for high-speed access lines	Must be pulled to the customer premises (this is expensive)
Optical Fiber	Businesses for high-speed access lines	Must be pulled to the customer premises (this is expensive)

Figure 6-6   Local Loop Technologies

2-Pair Data-Grade Twisted Pair Wiring

Even for the slowest private lines, telephone carriers have to run higher-quality access lines, namely **2-pair data-grade** access lines. Note first that two pairs are used—one for communication in each direction. In addition, the wiring is of higher quality than voice-grade UTP. This allows it to carry signals much faster.

Although 2-pair data-grade wiring is very good, the telephone carrier has to pull two new pairs of data-grade wiring UTP to each customer who needs it. The labor to do this is very expensive, and this labor cost translates into high monthly prices. In fact, customers must sign leases for certain periods of time to allow the telephone carrier to recoup its investment. It may also take several weeks to **provision** (install and set up) a 2-pair data-grade UTP access line.

Optical Fiber

For private lines running faster than about 2 Mbps, the telephone carrier has to pull an even more expensive two-strand optical fiber cord to the customer premises. The cost of running fiber to the customer premises is even higher than the cost of running 2-pair data-grade wiring.

TEST YOUR UNDERSTANDING

5. a) Distinguish between the 4-pair UTP wiring used in Ethernet and the UTP wiring in the residential local loop. b) Distinguish between the UTP wiring in the ordinary local loop and the UTP wiring used for lower-speed private lines. c) What technology do higher-speed private lines use? d) What is provisioning?

The End Office Switch

The access line runs from the customer premises to the nearest switch of the telephone company. As Figure 6-1 shows, this is called an **end office switch** or **Class 5 switch** (the telephone network traditionally has used a five-class hierarchy, and end office switches are the lowest switches in the hierarchy).

TEST YOUR UNDERSTANDING

6. a) What is an end office switch? b) Why is it called a Class 5 switch?

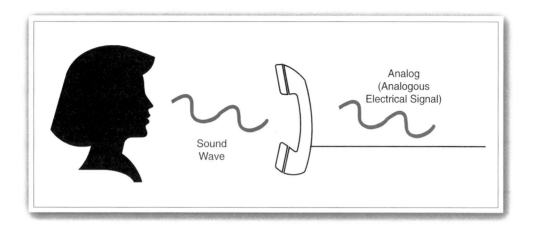

Figure 6-7   Analog Telephone Transmission

Analog-to-Digital Conversion for Analog Local Loops

Analog Voice Signals

In Chapter 3 we saw digital signals. However, traditional telephones produce a different type of signal, an analog signal. As Figure 6-7 shows, when a person speaks into a telephone mouthpiece, their acoustic pressure waves cause a diaphragm to vibrate. This generates an analogous electrical disturbance that propagates down the local loop to the nearest switching office. This **analog signal** rises and falls in intensity smoothly, with no clock cycles and no limited numbers of states as in digital signaling.

An analog signal rises and falls in intensity smoothly, with no clock cycles and no limited numbers of states as in digital signaling.

Mostly Digital

As Figure 6-8 shows, the PSTN transport core, which was originally completely analog, is almost entirely digital today. Almost all of its switches are digital, as are almost all of its trunk lines. Larger businesses even get digital access lines for their local loop communication.

Codecs: Analog-to-Digital and Digital-to-Analog Conversion

On the local loop that connects residential customers to the nearest end office, the customer's telephone sends and receives analog signals, so the end office switch needs equipment to convert between the analog local loop signals and the digital signals of the end office switch.

Figure 6-9 shows that this termination equipment is called a **codec.** Incoming signals from the subscriber go through an **analog-to-digital conversion (ADC)** process, which is called *coding*. In turn, the codec converts digital signals from the switch into

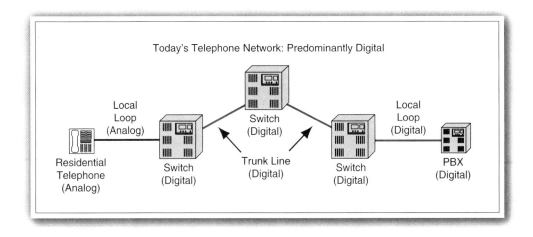

Figure 6-8   The PSTN: Mostly Digital with Analog Local Loops

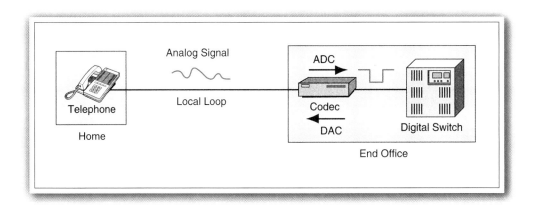

Figure 6-9   Codec at the End Office Switch

analog signals for subscribers. This is the **digital-to-analog conversion (DAC)** process, which is called *decoding*. (Hence the name "codec.")

Analog-to-Digital Conversion

ADC Step 1: Bandpass Filtering   Before the TDM transport core emerged, microwave radio was used heavily for trunk lines. As Figure 6-10 shows, microwave transmission uses **frequency division multiplexing (FDM),** in which the microwave bandwidth is subdivided into channels, each carrying a single circuit.

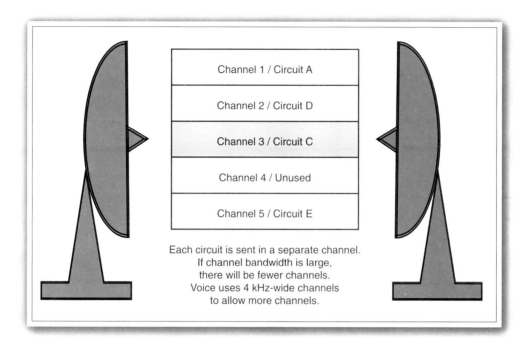

| Channel 1 / Circuit A |
| Channel 2 / Circuit D |
| Channel 3 / Circuit C |
| Channel 4 / Unused |
| Channel 5 / Circuit E |

Each circuit is sent in a separate channel.
If channel bandwidth is large,
there will be fewer channels.
Voice uses 4 kHz-wide channels
to allow more channels.

Figure 6-10   Frequency Division Multiplexing (FDM) in Microwave Transmission

How wide should a microwave channel be? Figure 6-11 shows the frequency spectrum for the human voice. Human hearing can range up to 20 kHz, although for most people, the maximum is substantially lower. Most voice energy, furthermore, comes at frequencies below 4 kHz, so using 20 kHz channels would do fairly little to improve sound quality.

Instead, microwave channel bandwidths were set to 4 kHz. This allowed five times as many voice signals to be carried by a microwave system than a 20 kHz-channel system would have permitted.

To limit voice bandwidth to 4 kHz, termination equipment in the access system passes subscriber incoming signals through a **bandpass filter,** which filters out all signals between 300 Hz and about 3.4 kHz. This gives guard bands around the signal—one below 300 Hz and one between 3.4 kHz and 4 kHz. Although microwave trunk lines are rarely used today, this bandpass filtering continues.

ADC Step 2: Sampling   To digitize voice, the codec ADC **samples** (reads the intensity of) the bandpass-filtered voice signal 8,000 times per second, as Figure 6-11 illustrates. Nyquist showed that if you sample at twice the highest frequency in the signal, you can reproduce the signal with no loss of information.[1] This is why we need 8,000 samples per second (twice 4,000 Hz after bandpass filtering).

[1]   Harry Nyquist, "Certain Topics in Telegraph Transmission Theory," *Trans. AIEE,* Vol. 47, pp. 617–644, Apr. 1928.

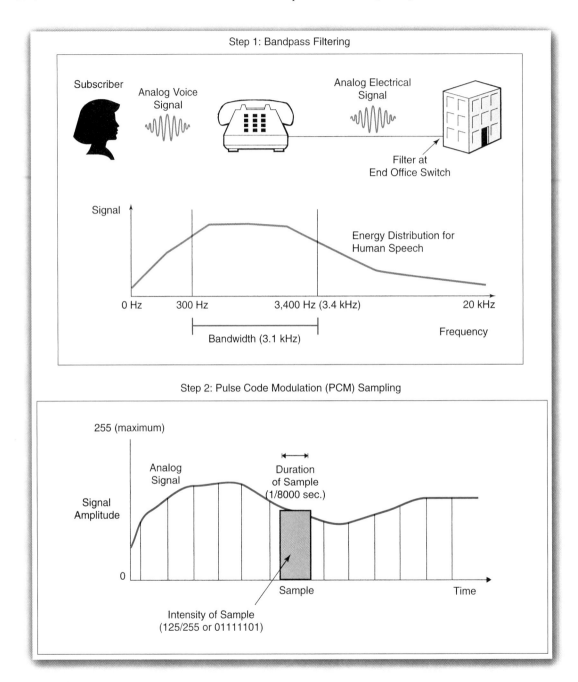

Figure 6-11   Analog-to-Digital Conversion (ADC): Bandpass Filtering and Pulse Code Modulation (PCM)

During each sampling period, the codec measures the intensity of the signal. The ADC represents the intensity of each sample by a number between 0 and 255. For instance, a signal of half the maximum intensity would be represented by 127. With 256 possible values, a single octet of binary data is needed to store each sample's value.

If you multiply 8 bits per sample times 8,000 samples per second, you get 64,000 bits per second. In other words, using this analog-to-digital conversion technique, called **pulse code modulation (PCM),** ADCs produce a data stream of 64 kbps for voice. Consequently, most telephone lines and equipment are built around 64 kbps channels.[2]

64 kbps Versus 56 kbps   In many cases the telephone carrier will "steal" 8 kbps from each channel for supervisory signaling, leaving 56 kbps for transmission. This is why the telephone system is built around units of 56 kbps or 64 kbps.

Digital-to-Analog Conversion (DAC)

ADCs are used for transmissions from the customer premises to the end office switch. In contrast, digital-to-analog converters (DACs) are for converting transmissions from the digital telephone network's core to signals on the analog local loop (see Figure 6-9). Figure 6-12 shows that as the DAC reads each sample, it puts a signal on the local loop that has the intensity indicated for that sample. It keeps the intensity the same for 1/8000 of a second. If the time period per intensity level is very brief, the amplitude changes will sound smooth to the human ear.

Is the resultant signal really analog? Precisely speaking, it still is digital. However, to the human ear, the sampling rate and playback rates are so high that the choppiness of the signal shown in Figure 6-12 is not apparent at all. The final signal *sounds* analog to users, so it is considered to be an analog signal.

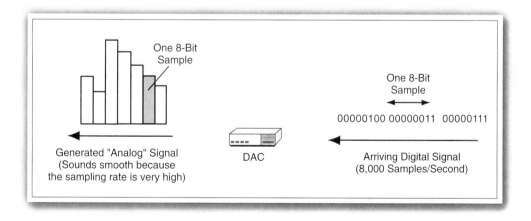

Figure 6-12   Digital-to-Analog Conversion (DAC)

---

[2]   The full PCM process is even more complex, but additional details merely add marginally to sound quality.

TEST YOUR UNDERSTANDING

7. a) Distinguish between analog and digital signals. b) What parts of the telephone system are largely digital today? c) What parts of the telephone system are largely analog today? d) What is the role of the codec in the end office switch? e) Explain why bandpass filtering is done. f) Explain how the ADC generates 64 kbps of data for voice calls when it uses PCM. g) Why do we need DACs? h) How do they work?

Private Lines

Private lines do not need codecs because they carry digital customer signals. Although this eliminates the need for analog–digital conversion, selecting private lines is complex because they come in a wide range of speeds, as we will see in the next chapter.

THE PSTN TRANSPORT CORE AND SIGNALING

Recall that *transport* is the actual transmission of voice in the PSTN, while *signaling* is the control of the PSTN.

The Transport Core

The PSTN transport core consists of switches and trunk line connections to link the switches. As Figure 6-13 shows, the PSTN uses two types of transmission connections.

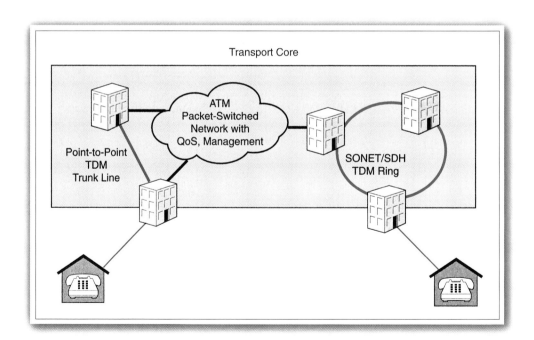

Figure 6-13   TDM and ATM Switch Connections in the PSTN Transport Core

TDM Trunk Lines

Until recently, most trunk lines between switches were TDM trunk lines. TDM trunk lines are well-matched to circuit switching. Each circuit gets reserved capacity (one slot in each frame) on each TDM trunk line between switches along the circuit path.

Note that many trunk lines are point-to-point connections. However, the newest trunk lines, which follow the SONET/SDH standards discussed in the next chapter, operate in a ring or loop. Actually, this technology uses a dual ring, as Figure 6-14 illustrates. Normally, all traffic travels over the main ring. However, if there is a break in the ring between two switches, the ring is wrapped, as shown in the figure. There still is a complete loop, so communication can continue unimpeded. This is important because most telephone outages are due to the accidental cutting of trunk lines by construction vehicles.[3]

Asynchronous Transfer Mode (ATM)

Although TDM has long been synonymous with transmission in the PSTN transport core, many long-distance carriers have already transitioned much of their transmission technology in their transport cores to a *packet-switched* technology called **asynchronous transfer mode (ATM)** technology.[4]

ATM was designed specifically to carry voice. Consequently, ATM provides quality-of-service guarantees for throughput, latency, and jitter. This allows ATM to give a close approximation of reserved capacity without completely reserving capacity.

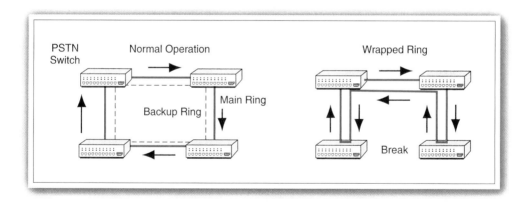

Figure 6-14   SONET/SDH Dual Ring

---

[3]   When Ethernet first emerged, IBM championed another LAN technology, the dual-ring 802.5 Token-Ring Network. This technology proved to be too expensive compared to Ethernet and is rarely found in organizations today. The 100 Mbps Fiber Distributed Data Interface (FDDI) dual-ring standard, in turn, was used in LAN cores during the 1980s, but it was overtaken in this role by the less-expensive 100Base-FX and the faster gigabit Ethernet. FDDI was last produced in 1999.

[4]   Franklin D. Ohrtman, Jr., *Softswitch Architecture for VoIP*, New York: McGraw-Hill, 2003.

These guarantees and the fact that ATM has a management structure that was designed for worldwide service make ATM ideal for long-distance transport core trunking.

TEST YOUR UNDERSTANDING

8. a) List the two main TDM technologies used for connecting switches in the telephone system's transport core. b) How does SONET/SDH's dual ring increase telephone system reliability? c) How is ATM different from previous trunk line technologies? d) Why is ATM good for voice?

Signaling

Again, signaling is the supervision of connections in the PSTN. The ITU-T created **Signaling System 7 (SS7)** as the worldwide standard for supervisory signaling (setting up circuits, maintaining them, tearing them down after a conversation, providing billing information, and providing special services such as three-party calling).[5] The U.S. version of the protocol is ANSI SS7, usually referred to simply as **SS7.** The ETSI version for Europe is called ETSI C7 or **C7.** They are almost the same, so simple gateways can convert between them.

TEST YOUR UNDERSTANDING

9. a) What is the worldwide signaling system for telephony? b) Distinguish between SS7 and C7. c) Does having two versions of the standard cause major problems?

CELLULAR TELEPHONY

Nearly everybody today is familiar with cellular telephony. In most industrialized countries, half or more of all households now have a cellular telephone.[6]

Cells

Cells and Cellsites
Figure 6-15 shows that cellular telephony divides a metropolitan service area into smaller geographical areas called **cells.**

The user has a cellular telephone (also called a **cellphone,** mobile phone, or mobile). Near the middle of each cell is a **cellsite,** which contains a **transceiver** (transmitter/receiver) to receive cellphone signals and to send signals out to the cellphone. The cellsite also supervises each cellphone's operation (setting its power level, initiating calls, terminating calls, and so forth).

---

[5] SS7/C7 actually is a packet-switched technology that operates in parallel with the circuit-switched PSTN but that uses the same transmission lines as the PSTN. SS7/C7 relies on multiple databases of customer information. When a call is set up, the originating telephone carrier queries one of these databases to determine routing information for setting up the service. These databases are also needed to provide advanced services such as toll-free numbers.

[6] Although cellular telephony was first developed in the United States, the U.S. has slightly lower market penetration than most other countries. One reason is that normal telephony is inexpensive in the United States, so moving to cellular service is an expensive choice. Another reason is that when someone calls a cellular phone in the United States, the cellular owner pays; in most countries, the caller pays. These two factors increase the relative price of using a cellular phone compared to using a wireline phone.

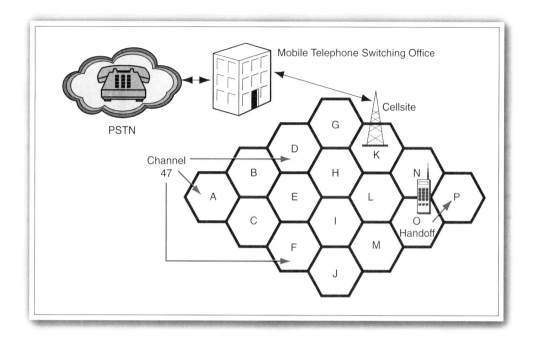

Figure 6-15   Cellular Telephony

Mobile Telephone Switching Office (MTSO)

All of the cellsites in a cellular system connect to a **mobile telephone switching office (MTSO),** which connects cellular customers to one another and to wired telephone users.

The MTSO also controls what happens at each of the cellsites. It determines what to do when people move from one cell to another, including which cellsite should handle a caller when the caller wishes to place a call. (Several cellsites may hear the initial request at different loudness levels; if so, the MTSO selects a service cellsite on the basis of signal loudness—not necessarily on the basis of physical proximity.)

Handoffs Versus Roaming

Handoffs   If a subscriber moves from one cell to another within a system, the MTSO will implement a **handoff** from one cellsite to another. For instance, Figure 6-15 shows a handoff from Cell O to Cell P. The cellphone will change its sending and receiving channels during the handoff, but this occurs too rapidly for users to notice.

Roaming   In contrast, if a subscriber leaves a metropolitan cellular system and goes to another city or country, this is called **roaming.** Roaming requires the destination cellular system to be technologically compatible with the subscriber's cellphone. It also requires administration permission from the destination cellular system. Roaming is as much a business and administrative problem as it is a technical problem.

Handoffs occur when a subscriber moves between cells in a local cellular system. Roaming occurs when a subscriber moves between cellular systems in different cities.

TEST YOUR UNDERSTANDING

10. a) In cellular technology, what is a cell? b) What is a cellsite? c) What are the two functions of the MTSO? d) Trace the path of a call between a cellular subscriber and a wireline (normal telephone) subscriber. e) Distinguish between handoffs and roaming.

Why Cells?

Why not have just one transmitter/receiver in the middle of a metropolitan area instead of dividing the area into cells and dealing with the complexity of cellsites?

Channel Reuse

The answer is **channel reuse.** The number of channels permitted by regulators is limited, and subscriber demand is heavy. Cellular telephony uses each channel multiple times, in different cells in the network. This multiplies the effective channel capacity, allowing more subscribers to be served with the limited number of channels available.

Cellular technology is used because it provides channel reuse—the ability to use the same channel in different cells. This allows cellular systems to support more subscribers.

Traditionally, No Channel Reuse in Adjacent Cells

With traditional cellular technologies, which use frequency division multiplexing to send signals in different channels and sometimes TDM to let several users share a channel, you cannot reuse the same channel in adjacent cells because there will be interference. For instance, in Figure 6-15, suppose you use Channel 47 in Cell A. You cannot use it in Cells B or C. This reduces channel reuse. In general, the number of times you can reuse a channel is only about the number of cells divided by seven. In other words, if you have 20 cells, you can only reuse each channel about 3 (20/7) times.

Channel Reuse in Adjacent Channels with CDMA

Some cellular systems use a new form of cellular technology, **code division multiple access (CDMA).** CDMA is a form of spread spectrum transmission. In contrast to the types of spread spectrum transmission used in 802.11 wireless LANs, which allow only one station to transmit at a time in a channel, CDMA allows multiple stations to transmit at the same time in the same channel. Furthermore, these channels are very wide, so several stations can transmit at the same time.

In addition, CDMA permits stations in adjacent cells to use the same channel without serious interference. In other words, if you have twenty cells, with CDMA, you can reuse each channel twenty times. This allows you to serve far more customers with CDMA than you can with older forms of cellular telephony.

If CDMA is so good, why do only some systems use it? The answer is that it is new. The first CDMA cellular systems were not built until 1993, and even then, they represented a technological and economic risk. However, CDMA has now proven itself, and all future cellular telephone standards will use CDMA.

TEST YOUR UNDERSTANDING

11. a) Why does cellular telephony use cells? b) What is the benefit of channel reuse? c) If I use Channel 3 in a cell, can I reuse that same channel in an adjacent cell with traditional cellular technology? d) Can I reuse Channel 3 in adjacent cells if the cellular system uses CDMA transmission? e) Why does the ability to reuse channels in adjacent cells make CDMA more attractive than traditional cellular technologies?

CELLULAR TELEPHONE GENERATIONS

The first cellular systems did not arrive until the early 1980s, yet in the short time they have been around, they have already gone through two generations of technology and are now entering a third generation. Figure 6-16 illustrates these three generations.

Generation Characteristics

First-Generation Cellular Telephony

First-generation (1G) systems, which were introduced in the early 1980s, had four characteristics.[7]

Generation	First (1G)	Second (2G)	Third (3G)
Year	1980	1990	2002
Technology	Analog	Digital	Digital
Data Transfer Rate	Data transfer is difficult; ~5 kbps	10 kbps	30 kbps to 500 kbps
Channels	~800	~800+2,500	Still Being Defined; Using 2G Channels in the Interim
Cells / Channel Reuse	Large / Medium	Large / Medium and Small / High	Still Being Defined

Figure 6-16   Generations of Cellular Technology

[7] In the United States, some 1G channels were upgraded to Cellular Digital Packet Data (CDPD) service designed specifically for data communications. CDPD is now being phased out in favor of full 2G technology.

➤ They were analog systems. This limited signal quality because analog transmission does not have the resistance to errors that we saw in digital signaling in Chapter 3. In analog systems, if a signal changes by 3 percent, it will be received with a 3 percent error.

➤ They were given only about 50 MHz of spectrum in the 800 or 900 MHz range, limiting the number of subscribers.

➤ They had large cells, so they only had 10 to 50 cells in most cities; this limited channel reuse and the number of possible subscribers.

➤ Although you could buy cellular modems for data transmission, throughput was limited to about 5 kbps—far below today's normal modem speeds of 30 kbps to 50 kbps.

TEST YOUR UNDERSTANDING

12. a) Was the first cellular generation analog or digital? b) About how much spectrum was provided for 1G systems? c) Why was channel reuse limited? d) What data transmission speed did 1G support?

Second-Generation Cellular Telephony

Most cellular systems today are **second-generation (2G)** cellular systems. Although introduced in the early 1990s, only ten years after 1G cellular systems, 2G offered important advances.

Most cellular systems today are second-generation (2G) cellular systems.

➤ Most importantly, 2G systems are digital. This gives good signal quality.

➤ They have been allocated 150 MHz of bandwidth in most countries. This alone tripled the number of potential subscribers compared to 1G systems. In addition, most cellular operators have converted most of their old 1G channels to 2G service.

➤ They operate at higher frequencies (1,800 or 1,900 MHz) than 1G systems, so 2G signals do not travel as far as 1G signals. This requires more cells. Although this is more expensive, it creates more channel reuse.

➤ They provide data transmission at slightly higher throughput (up to 10 kbps) than 1G systems.

Third-Generation Systems

Cellular carriers are beginning to implement a new generation of cellular technology. The main benefit of **third-generation (3G)** cellular systems is that they can carry data at much higher speeds than 2G systems. They will offer rated speeds of 144 kbps to 2 Mbps. In practice, however, actual throughput will be similar to those of telephone modems at the low end to DSL lines and cable modem service at the high end.

To be fully effective, 3G systems will need new spectrum capacity. Some cellular carriers are trying to start with some of their current 2G spectrum. In some cases this allows them to offer full 3G service, but in other cases, it only allows them to offer an intermediate service, called **2.5G** service, which can send and receive data at modem speeds.

Standards Families

Beginning with the second generation, two major standards families have emerged. One began with GSM. The other began with IS-95, which is more widely known as CDMAone.

GSM
> GSM (Global System for Mobile communications)
> > 200 kHz channels, shared by up to eight users via TDM
> > Data transmission speed of approximately 10 kbps
> General Packet Radio Service (GPRS)
> > Uses GSM channels
> > "2.5G": Typical throughput of 20 kbps to 30 kbps
> > Comparable to telephone modem throughput
> EDGE
> > Uses GSM channels
> > "2.5G": Typical throughput of 80 kbps to 125 kbps
> W-CDMA
> > Wideband CDMA
> > 3G service
> > Throughput comparable to DSL and cable modems

Qualcomm CDMA
> CDMAone (IS-95)
> > 2G system used widely in the United States
> > 1.25 MHz channel shared by multiple simultaneous users
> > 10 kbps data transmission
> CDMA2000 (IS-2000)
> > 1x
> > > 30 kbps to 50 kbps throughput in a 1.25 MHz channel
> > > Comparable to telephone modem throughput
> > > Considered to be 3G because rated speed is 144 kbps
> > 1xEV-DO
> > > 100 kbps to 300 kbps throughput
> > > Comparable to cable modem/DSL throughput

Figure 6-17   Cellular Standards Families (Study Figure)

The GSM Family

GSM   For 2G service, nearly the entire world standardized on **GSM (Global System for Mobile communication).** This widespread adoption allows roaming across most of the world with a GSM cellphone.

GSM uses 200 kHz channels—much larger than the 30 kHz channels used in most 1G systems. GSM implements TDM (called time division multiple access or TDMA in radio transmission) to allow up to eight subscribers to share each channel. (The average is fewer.) Although GSM can carry data, it can only do so up to about 10 kbps.

GPRS   Most GSM systems are now upgrading some of their GSM channels to provide **General Packet Radio Service (GPRS).** GPRS can provide data throughput near that of a telephone modem with a typical throughput of 20 kbps to 30 kbps. GPRS is often called a **second-and-a-half-generation (2.5G)** technology because it is a substantial improvement over plain 2G GSM but is not a full third-generation service.

EDGE   Another 2.5G technology for moving beyond GSM is **EDGE** (Enhanced Data Rates for GSM Evolution). EDGE also uses GSM channels but provides a throughput of 80 kbps to 125 kbps. This is higher than telephone modem throughput but lower than DSL/cable modem throughput.

W-CDMA   Both GPRS and EDGE are 2.5G technologies. As user demand grows, GSM operators hope to move to full 3G technology. Most plan to move to the W-CDMA standard, which was developed in Europe and Japan. W-CDMA should provide speeds similar to those of DSL lines and cable modems.

CDMAone

In the United States, the Federal Communications Commission (FCC) decided in the early 1990s to permit open competition in 2G technology.

GSM in the United States   Fortunately, U.S. cellular carriers today now use one of only two technologies. One of these is GSM. Selecting GSM was a low-risk strategy because GSM was a proven technology. Selecting GSM also was an attractive strategy because it permitted global roaming.

CDMA (IS-95)   More aggressive U.S. cellular carriers adopted **CDMA** (specifically, **IS-95**), which did not even become available until 1995. These carriers took economic risks with this new technology with the hope of gaining a larger number of potential subscribers.[8]

CDMA2000 (1x and 2x)   Qualcomm, which created the CDMA technology, has developed a new 3G technology, **CDMA2000** (IS-2000). Qualcomm is taking a staged approach with CDMA2000. This makes it an easy upgrade for IS-95 (2G CDMA) systems.

➤ Its initial 3G step is **CDMA2000 1x,** which offers a rated speed of 144 kbps in a single 1.25 MHz channel. The effective throughput is about 30 kbps to 50 kbps.[9] This 1x technology, then, gives telephone modem throughput.

➤ The next stage is **CDMA2000 1xEV-DO,** which will offer rated speeds of 300 kbps to about one megabit per second, also in a 1.25 MHz channel.[10] In practice, users typically will get throughputs of 100 kbps to 500 kbps—the same as DSL/cable modem systems.

---

[8] What standard does your cellphone carrier use? Among the U.S. cellular vendors using GSM are Cingular, and T-Mobile. CDMA vendors include Qwest Communications, Sprint PCS, US Cellular, and Verizon. For a full list of U.S. carriers, go to www.cellular-news.com and choose "Coverage."
[9] The "1x" reflects the fact that CDMA2000 1x uses a single 1.25 Mbps CDMA channel—the same channel used in 2G CDMA (IS-95). This allows existing IS-95 CDMA systems to be upgraded to CDMA2000 1x or 1xEV-DO. If there is sufficient market demand, CDMA2000 will offer 3x services by consolidating three adjacent 1.25 MHz channels to provide even higher speeds.
[10] EV-DO stands for Evolved Data-Only. Another CDMA2000 standard, 1xEV-DV, can handle both data and voice.

Peak and Off-Peak Speeds

The throughput figures we have been citing are off-peak throughputs. During peak periods, throughput falls considerably, often to only half of the off-peak value.

TEST YOUR UNDERSTANDING

13. a) What generation of cellular technology do most cellular telephones use today? b) List ways in which 2G cellular systems are better than 1G systems. c) What is the main benefit of 3G systems?

14. a) How can GSM systems be upgraded to provide higher-speed data transmission? b) What 3G system will most GSM carriers use? c) Describe the two upgrades for IS-95 technologies. d) Are these 3G technologies? e) Distinguish between peak and off-peak throughputs.

Perspectives on 3G Service

Cellular carriers believe that 3G services will provide much of their revenue growth in the future. However, actual 3G availability today is very limited, and use where it is available is very low.

The Need for Additional Spectrum

Although 2.5G systems and CDMA2000 1x systems can operate in existing frequency bands, if 3G is to be very successful, it will require new service bands to be created.

802.11 Hot Spots

One threat to 3G telephony is the emergence of 802.11 **hot spots,** which are access points offering Internet access in public places such as shopping malls. Hot spots offer throughputs of 5 Mbps to about 25 Mbps—much higher throughputs than 3G systems will offer. If many mobile computer users will spend much of their time near access points at home and work or hot spots in public areas, the market for 3G systems may not emerge.

One possibility is that 802.11 and 3G services will be complementary rather than competitive. If dual-use stations are sold that support both 802.11 and 3G technologies, users will be able to use higher-speed 802.11 when near an access point and slower and more expensive 3G service when too far from access points for service.

TEST YOUR UNDERSTANDING

15. a) Are prospects for 3G service assured? b) Why are new service bands needed? c) How may 802.11 technology hurt 3G acceptance? d) How may it help acceptance?

IP TELEPHONY

Basic Operation

One of the newest areas in telephony is **IP telephony,** which is transmitting telephone signals over IP internets instead of over circuit-switched networks. IP telephony is also called **voice over IP (VoIP).** Both terms are widely used.

Figure 6-18 shows two clients. One is a client PC. The other is an **IP telephone,** which has the electronics to encode voice for digital transmission and to send and

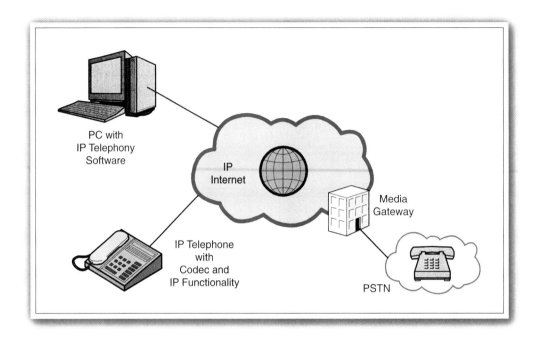

Figure 6-18   IP Telephony

receive packets over an IP internet. With IP telephony, their two users can talk with each other.

In addition, **media gateways** connect IP telephones to ordinary telephones on the public switched telephone network. Media gateways do not simply make connections. They also convert between the signaling (call setup, etc.) formats of the IP telephone system and the PSTN.

TEST YOUR UNDERSTANDING

16.   a) What is IP telephony? b) What is VoIP? c) What is an IP telephone? d) What is a media gateway?

Speech Codecs

As discussed earlier, **speech codecs** convert analog voice signals into digital bit streams. We looked at PCM, which creates 64 kbps data streams with high auditory quality. For VoIP, the default codec standard is G.723.1, which sends voice at 5.3 kbps or 6.3 kbps.

As Figure 6-19 shows, several other speech codecs have been defined by the ITU-T. Some speech codecs compress voice more, reducing traffic transmission requirements and, therefore, costs. However, these codecs usually have lower voice quality.

The fact that there are so many speech codecs is good because it gives organizations more options. However, this flexibility can create problems for interoperability unless planning and vendor selection is done carefully.

Codec	Transmission Rate	Notes
G.711	64 kbps	
G.721	32 kbps	
G.722	48, 56, 64 kbps	
G.722.1	24, 32 kbps	
G.723	5.33, 6.4 kbps	
G.723.1A	5.3, 6.3 kbps	Default for VoIP
G.726	16, 24, 32, 40 kbps	
G.727	32 kbps	Default for voice over Frame Relay
G.728	16 kbps	
G.729AB	8 kbps	Mandatory option for voice over Frame Relay
RPE	13 kbps	Used in GSM

Figure 6-19   Speech Codecs

TEST YOUR UNDERSTANDING

17. a) What do speech codecs do? b) What is the tradeoff to consider when selecting a speech codec standard for use in an IP telephony system? c) What is the drawback of having many speech codecs from which to select?

Transport

Figure 6-20 shows the packet used in VoIP transport. We will look at it piece by piece.

TCP Versus UDP

In Chapters 1 and 2, we saw that TCP is widely used at the transport layer. TCP is a reliable protocol, correcting errors at the transport layer and lower layers as well. However, to do this, TCP has to be a heavyweight protocol requiring openings, closings, and ACKs.

In IP telephony, it is not feasible to correct errors through retransmission. The delay would completely disrupt the voice stream. Instead, if a packet is lost, the receiver either replays the previous packet's sound or interpolates what the lost packet probably contained based on the data contents of earlier and later packets.

Given the impossibility of error correction, VoIP uses a simpler protocol at the transport layer. This is the User Datagram Protocol (UDP), which we will see in more detail in Chapter 8. UDP is connectionless and unreliable. UDP messages are called UDP datagrams.

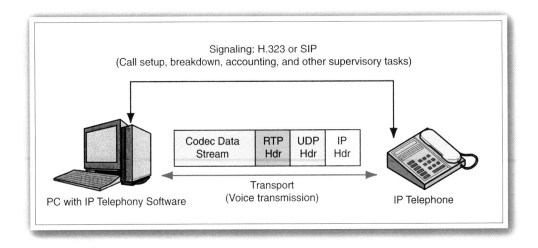

Figure 6-20   IP Telephony Protocols

Real Time Protocol (RTP)

Although UDP is a lightweight protocol, it is not perfect for IP telephony. It does not guarantee that packets will arrive in order, and the timing between UDP datagrams on arrival may be different than it was during transmission. This results in jitter.

Consequently, as Figure 6-20 shows, IP telephony adds a **Real Time Protocol (RTP)** header after the UDP header. RTP headers contain sequence numbers to ensure that the UDP datagrams are placed in proper sequence, and they contain time stamps so that jitter can be eliminated. The receiver collects several packets in a section of RAM called the jitter buffer and plays them out with timing dictated by their RTP time stamps.

Payload

The payload (data field) in the packet is a stream of codec octets.

TEST YOUR UNDERSTANDING

18. a) Draw an IP packet's headers and application message for VoIP transport. b) Why is UDP used instead of TCP at the transport layer in IP telephony? c) Why is an RTP header added? d) What is in the data field of the packet?

Signaling in IP Telephony

RTP is used for the *transport* of voice communication. For *signaling* (supervision), a different protocol is needed. In practice, there are two signaling protocols for IP telephony. The **H.323** signaling protocol, which was created by the ITU-T was popular in earlier systems. Newer systems, while still supporting H.323, also offer the simpler

Session Initiation Protocol (SIP) created by the IETF.[11] SIP is likely to dominate in the future due to its simplicity.

TEST YOUR UNDERSTANDING

19. a) What two signaling protocols are used in IP telephony? b) Why is SIP likely to be dominant in the future?

Concerns for IP Telephony

Cost Savings?

The traditional promise of IP telephony is that it can save money by sending voice over efficient packet-switched networks rather than over inflexible and inefficient circuit-switched networks. However, IP telephony will require upgrades to existing packet-switched networks because voice requires quality-of-service guarantees. This will eliminate at least some potential savings. In addition, the long-distance and international telephone calling rates against which IP telephony costs must be compared have been falling very rapidly. Whether IP telephony will achieve major cost savings is not entirely clear.

Integrated Voice and Data Applications

Given the uncertainty of cost savings, IP telephone advocates are now focusing on the benefits of applications that combine telephone calling with data applications. One obvious example is bringing up customer information automatically on a sales representative's computer when a customer calls.

Maintaining Voice Quality and Availability

Corporate telephone staffs are concerned with maintaining voice quality and availability. They have long been able to deliver voice signals of high quality 99.999 percent of the time. IP telephone networks will provide lower quality and significantly lower availability.

TEST YOUR UNDERSTANDING

20. a) Why may IP telephone systems not reduce costs? b) What benefit may be the most attractive in implementing them? c) What two concerns do corporate telephone staffs have about IP telephony?

Enter the Carriers

One reason why carrier prices have fallen dramatically in recent years is that the carriers themselves have been using VoIP internally to reduce costs. Many carriers are beginning to offer VoIP service directly to users. This gives corporations the choice of internally-built IP telephony systems or carrier services.

TEST YOUR UNDERSTANDING

21. What is the importance of the fact that some carriers now offer VoIP service directly to customers?

---

[11] There is another signaling protocol of note. This does not provide for signaling directly; rather it is a way of controlling the components of signaling systems. This standard was developed jointly by the IETF and ITU-T. IETF calls it MEGACO, while ITU-T calls it H.248, but it is a single standard. H.248/MEGACO permits the centralized management of large IP telephone systems with many signaling elements.

TELEPHONE CARRIERS

Once, almost every nation had a single national telephone carrier. However, the situation has become more complex over time as nations have begun to deregulate telephone service, that is, to permit some competition in order to reduce prices and promote product innovation.

Competition helps corporate customers because telephone prices generally fall as a result of competition. However, to maximize cost savings, companies have to be very smart when they deal with telephone carriers. To do this, a first step is understanding the types of carriers a company will face.

PTTs and AT&T

PTTs and Ministries of Telecommunications

In most countries the single monopoly carrier was historically called the **Public Telephone and Telegraphy authority (PTT).** In the United Kingdom, for example, this was British Telecom, while in Ireland it was Eircom.

To counterbalance the power of the PTT, governments created regulatory bodies generally called **Ministries of Telecommunications.** PTTs provide service, while Ministries of Telecommunications oversee the PTTs.

AT&T, the FCC, and PUCs

The Bell System   In the United States, neither telegraphy nor telephony was made a statutory monopoly. However, telephony quickly became a de facto monopoly when **AT&T,** also known as the **Bell System,** used predatory practices to drive most other competitors out of business. AT&T had a complete long-distance monopoly. For local service, AT&T owned more than 80 percent of all local telephone companies, although when it was developing in the nineteenth century and early twentieth century, it bypassed "unpromising" areas such as Hawai`i and most of Los Angeles.

The RBOCs   In the 1980s AT&T was broken up into a long-distance and manufacturing company that retained the AT&T name and seven **Regional Bell Operating Companies (RBOCs)** that owned most local telephone companies.

Later, mergers among the RBOCs and GTE, which was the largest independent owner of local operating companies, produced today's four dominant owners of local operating companies—Verizon, SBC Communications, BellSouth, and Qwest. These four companies also provide long-distance service in some areas.

The FCC and PUCs   In the United States, the **Federal Communication Commission (FCC)** provides overall regulation for U.S. carriers. However, within individual states, **Public Utilities Commissions (PUCs)** regulate pricing and services.

TEST YOUR UNDERSTANDING

22. a) Do all countries have PTTs? Explain. b) What are the purposes of PTTs and Ministries of Telecommunications? c) Distinguish between the traditional roles of AT&T and the RBOCs. d) Distinguish between the traditional roles of the FCC and PUCs in the United States.

Deregulation

Although telephone carriers had a complete monopoly in the early years, governments began deregulating telephone service in the 1970s. **Deregulation** is the opening of telephone services to competition; it has the potential to reduce costs considerably.

Carriers in the United States

LATAs   Figure 6-21 shows the types of carriers that exist in the United States. Since the breakup of AT&T in 1984, the United States has divided into approximately two hundred service regions called **local access and transport areas (LATAs).**

ILECs and CLECs   Within each LATA, the traditional monopoly telephone company is called the **incumbent local exchange carrier (ILEC).** Competitors are called **competitive local exchange carriers (CLECs).**

> LATAs are geographical regions. ILECs and CLECs are carriers that provide access and transport within LATAs.

IXCs   In contrast, **inter-exchange carriers (IXCs)** carry voice traffic *between* LATAs. Major ILECs are AT&T, MCI, and Sprint.

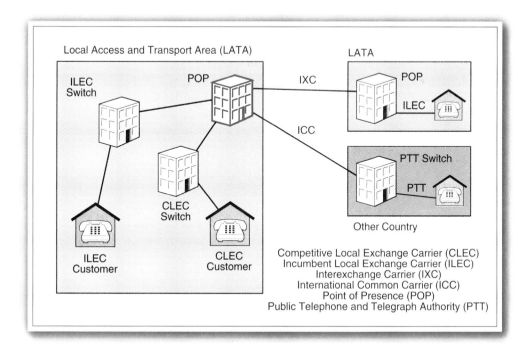

Figure 6-21   Telephone Carriers in the United States

ICCs   ILECs, CLECs, and IXCs are **domestic** carriers that provide service within the United States. Similarly, PTTs provide domestic service within their own countries. In contrast, **international common carriers (ICCs)** provide service *between* countries.

Points of Presence (POPs)

As Figure 6-21 shows, the various carriers that provide service are interconnected at **points of presence (POPs).** Thanks to points of presence, any subscriber to any CLEC or ILEC in one LATA can reach customers of any other CLEC or ILEC in any other LATA. ICCs also link to domestic carriers at POPs.

Deregulation by Service

Customer Premises Equipment   Although it seems odd today, telephone companies used to own all of the wires and telephones in homes and businesses. Today, however, nearly all countries prohibit carriers from owning customer premises equipment. Deregulation for customer premises equipment, in other words, is total in most countries.

Long-Distance and International Calling

In most countries both long-distance and international telephone services have been heavily deregulated. As just noted, long-distance service in the United States is provided by interexchange carriers (IXCs). PTTs in other countries also have domestic competitors for long-distance service.

Local Telephone Service

Local telephone service is the least deregulated aspect of telephony. The need for large investments in access systems and regulatory reluctance to open local telephone service completely (for fear of losing currently subsidized service for the poor and rural customers) have combined to limit local telephone competition.

Some countries now require the traditional monopoly carrier to open its access systems and central offices to competitors for a "reasonable" fee. However, court delays and high "reasonable" fees have limited the effectiveness of this facility-sharing approach.

TEST YOUR UNDERSTANDING

23.   a) Distinguish between LATAs, ILECs, and CLECs. b) What is the role of IXCs relative to LATAs? c) What is the role of ICCs? d) Why are POPs important?

SYNOPSIS

Although telephony is often called "plain old telephone service," it is anything but plain today. Competition is producing lower prices and more service options. It is also generating new technologies that promise to revolutionize data transmission outside the firm and to integrate voice and data networking.

The worldwide telephone network is officially the Public Switched Telephone Network (PSTN). The PSTN has four technological elements:

➤   Customer premises equipment includes the PBX, telephones, and transmission lines in the customer's buildings.

> ➤ The access system is the way customers connect to the PSTN. The access system is the most expensive part of the PSTN, and it is the only part of the PSTN that corporations see directly. The subscriber access line is called the local loop.

> ➤ The transport core carries signals between customer access lines across distances ranging from less than a mile to halfway around the world.

> ➤ Signaling is the way the PSTN sets up connections, breaks down connections, provides billing information to carriers, and handles other supervisory chores.

Almost all data networks use packet switching, but the PSTN uses circuit switching, in which capacity is reserved at the start of a call on all switches and trunk lines along the path (circuit) of the call. Circuit switching is reasonably efficient for voice conversations, in which one side or the other usually is talking. However, it makes little sense for data transmission, which is bursty, with short traffic bursts separated by long silences. Data transmission must pay for reserved capacity even during these long silences.

You personally are most familiar with dial-up circuits, which are set up at the start of a call and broken down afterward. However, most corporations use private lines, which are circuits that are always on, can multiplex many voice calls, and can carry high-speed data. Private line circuits use time division multiplexing, in which time is divided into short periods called frames and in which a time slot in each frame is allocated to each conversation.

The PSTN access system consists of single pairs of voice-grade UTP to residential homes and two pairs of data-grade UTP or an optical fiber cord to businesses for private lines. These carry customer signals to the PSTN carrier's end office switch (Class 5 switch). The PSTN is almost entirely digital internally, but residential customers send and receive analog signals, which rise and fall smoothly in intensity over time.

At the end office switch, a device called a codec converts residential analog subscriber signals into digital signals the transport core can carry. This is analog-to-digital conversion (ADC). In ADC the subscriber's signal is sent through a bandpass filter to filter out all frequencies lower than about 300 Hz and higher than about 3.4 kHz. Pulse code modulation (PCM) samples the bandpass filtered voice signal 8,000 times per second and sends an 8-bit signal for each sample, resulting in a 64 kbps data stream. (This is sometimes 56 kbps because some carriers steal 8 kbps for signaling.) The codec has a digital-to-analog converter (DAC) that reverses the PCM process for telephone systems going to residences.

The transport core consists of switches connected by two types of transmission systems: TDM trunk lines and ATM packet-switched networks. The PSTN's signaling system, Signaling System 7, is SS7 in the United States variant and C7 is the variant in Europe.

Cellular networks divide a region into multiple small areas called cells. This allows the same channel to be used in multiple cells, allowing more subscribers to be served with a limited number of channels. Traditional FDM and TDM cellular systems cannot reuse a channel in adjacent cells, but CDMA systems can, thus allowing more channel reuse. Many current cellular networks are GSM networks that use TDM, but future cellular technologies will be based on CDMA. Today's cellular systems primarily use second-generation (2G) technologies, which are digital but cannot carry data rapidly. New third-generation (3G) systems will allow mobile users to send and receive data at telephone modem and DSL/cable modem speeds.

Today, corporations have separate packet-switched networks for data and circuit-switched networks for voice. However, IP telephony (also called VoIP) promises to integrate all networking via IP transmission over packet-switched networks. Several codecs are available for converting analog voice to digital signals. For transport, these codec data streams are divided into IP packets containing an RTP header, a UDP header, and several bytes of codec data. Signaling technologies include H.323 and the newer and simpler SIP.

To purchase telephone service, a company must deal with telephone carriers. In the United States, local transmission is provided within local access transport areas (LATAs) by ILECs and CLECs, while IXCs provide transmission services between LATAs. Carriers interconnect at points of presence (POPs). In other countries, Public Telephone and Telegraph authorities (PTTs) and new competitors provide service. ICCs provide transmission between countries.

Once, telephony was a service provided only by monopoly carriers. However, telephone service is increasingly deregulated. Customer premises equipment is completely deregulated, and long-distance and international calling are highly deregulated. Local access and transport are the least competitive because of regulatory concerns over open competition and because of the immense investment needed for local access.

SYNOPSIS QUESTIONS

1. Briefly characterize the four major technological elements of the PSTN.
2. a) Why is circuit switching's reserved capacity reasonably efficient for voice but bad for data transmission? b) Distinguish between dial-up and private line circuits. c) What is TDM?
3. a) List the three access technologies in common use today. b) Which now run to residential homes?
4. a) Why are codecs needed at end office switches? b) What are the two steps in ADC? c) How does PCM produce 64 kbps data streams?
5. a) Describe the technologies for connecting switches in the transport core. b) What is SS7?
6. a) What is the purpose of dividing regions into small areas called cells? b) How do FDM, TDM, and CDMA technologies vary in channel reuse? c) What is the promise of 3G cellular systems?
7. a) What is the promise of IP telephony? b) Is there a single codec standard or are there multiple codec standards for IP telephony? c) What are the parts of an IP telephony transport packet? d) What are the two signaling standards for IP telephony?
8. a) Distinguish among LATAs, ILECs, and CLECs. b) What is the role of IXCs? c) What is the role of POPs? d) What are PTTs, and what is their purpose? e) What is the role of ICCs? f) What aspect of telephony is the least deregulated? Why?

THOUGHT QUESTION

1. In this chapter, you saw how PCM generates 64 kbps of data when it digitizes voice. For audio CDs (which store information digitally), a PCM-like algorithm was also used. However, instead of cutting off sounds above 3.4 kHz, music digitization uses a 20 kHz cutoff to capture the higher-pitched sounds of musical instruments. Music digitization also uses 16 bits per sample instead of the 8 bits per sample used by voice to give more precise volume representation. Furthermore, music is presented in stereo, so there are two 20 kHz channels to digitize. Audio CDs were designed to store one hour of digitized music. Compute how big audio CDs need to be. Remember to convert your bit rates into bytes per second, and remember that there are 1,024 bytes in a kilobyte and 1,024 kilobytes in a megabyte. (Hint: The first CD-ROM disks, which were based on audio disk technology, had 550 MB of capacity. You should get a number reasonably close to this.) Present your answer as a spreadsheet. Copy and paste the spreadsheet into your answer page.

CHAPTER 7

WIDE AREA NETWORKS (WANs)

Learning Objectives:

By the end of this chapter, you should be able to discuss:

- Differences between LANs and WANs, including the high cost of WANs per bit transmitted and, consequently, the dominance of low-speed transmission (56 kbps to a few megabits per second) in WAN service.

- The three purposes of WANs: remote individual access, site-to-site corporate networking, and Internet access.

- Individual Internet access with telephone modems, private lines, DSLs, cable modems, and eventually wireless access.

- Site-to-site networks with private line networks and Frame Relay, ATM, and metropolitan area Ethernet Public Switched Data Networks (PSDNs).

- Virtual private networks (VPNs) using SSL/TLS, PPTP, and IPsec.

INTRODUCTION

Chapter 6 began to take us beyond the customer premises, to telephone services provided by carriers. This chapter looks at carrier services for wide area data networking.

WANs and the Telephone Network

Many of these **wide area network (WAN)** services are built on top of the telephone network's technology that we saw in the last chapter. Sometimes end-user companies lease circuits from the telephone company to carry their internal data. In other cases WAN carriers lease telephone circuits, add their own switching, and offer data networking services to end-user corporations.

Reasons to Build a WAN

There are three main purposes for WANs.

The Telephone Network
      WAN technology often is built on top of the PSTN
WAN Purposes
      Provide remote access to individuals who are off-site
      Link sites within the same corporation
      Internet access
WAN Technologies
      Technologies for Individual Internet Access
          Telephone modems
          DSL lines
          Cable modems
          Wireless Internet access
      Site-to-Site Transmission within a Firm
          Private line networks
          Public switched data networks (PSDNs)
          Virtual Private Networks (VPNs)
             Transmission over the Internet with added security
          Low cost per bit transmitted
High Costs and Low Speeds
      High cost per bit transmitted compared to LANs
      Consequently, lower speeds (most commonly 56 kbps to a few megabits per second)
Carriers
      Beyond their physical premises, companies must use the services of regulated carriers for transmission

Figure 7-1   Wide Area Networks (WANs) (Study Figure)

➤ The first is to provide remote access to customers or to individual employees who are working at home or traveling.

➤ The second is to link two or more sites within the same corporation. Given the large amount of site-to-site communication in most firms, this is the dominant WAN application.

➤ The third is to provide access to the Internet.

WAN Technologies

In this chapter we will look at specific WAN technologies.

➤ For Internet access, we will see how individual users can use slow telephone modems, private lines, DSLs, cable modems, and eventually wireless access.

➤ For site-to-site corporate networking, we will look at networks of private lines, public switched data networks, and VPNs.

High Costs and Low Speeds

LAN users are accustomed to 10 Mbps or 100 Mbps unshared speed to the desktop. In contrast, long-distance communication is much more expensive per bit transmitted, so companies usually content themselves with slower transmission speeds in WANs. Most WAN communication links operate at between 56 kbps and a few megabits per second, and this throughput often is shared by multiple simultaneous users.

Most WAN communication links operate at between 56 kbps and a few megabits per second, and this throughput often is shared by multiple simultaneous users.

Carriers

A company can build its own LANs because these LANs run through the company's own buildings and land. However, you cannot lay wires through your neighbor's yard, and neither can corporations. Transmission beyond the customer premises requires the use of regulated **carriers.**

TEST YOUR UNDERSTANDING

1. a) How are telephony and wide area networking related? b) What are the three main purposes for WANs? c) Compare LAN and WAN transmission speeds. d) Why are they different? e) What are carriers, and why must they be used?

INDIVIDUAL INTERNET ACCESS

Before we look at technologies for complete corporate WANs, we will look at wide area transmission technologies for a more limited purpose: individual Internet access.

Telephone Modem Communication

Telephone Modems

The simplest data transmission method is to use a **telephone modem,** which converts digital data into an analog signal that can travel over the local loop. Nearly every home and business has a telephone line, so telephone modem users can connect nearly anywhere.

Modulation and Demodulation

As discussed in the previous chapter, the telephone system expects analog signals from home subscribers. However, computers generate digital signals. A translation device called a modem is needed to translate between digital computer signals and analog local loop signals.[1]

More specifically, telephone modems **modulate** digital customer signals—convert them to analog signals that can travel over the analog local loop. They also **demodulate** (convert from analog to digital) the analog signals coming from the telephone carrier to the customer's digital computer. This is why they are called modems.

Figure 7-2 shows a simple form of modulation called **amplitude modulation.** In this approach the modem transmits one of two analog signals—a high-amplitude (loud) signal or a low-amplitude (soft) signal. Suppose that the high-amplitude signal indicates a one and the low-amplitude signal represents a zero. To send 1011 in four successive clock cycles, the modem would transmit loud-soft-loud-loud. Modern modems use more complex forms of modulation. However, such details are hidden from both users and corporate network professionals. (Module B has these details.)

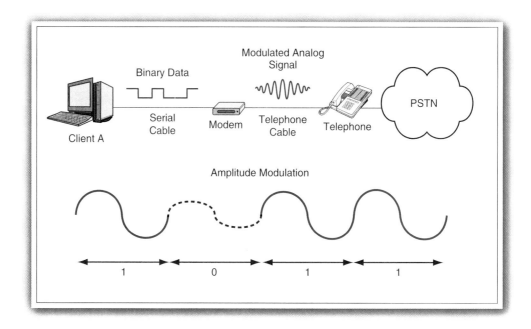

Figure 7-2   Amplitude Modulation

[1]   At the end of the local loop, the end office switch will convert the modem's analog signal back to digital signals to travel through the transport core.

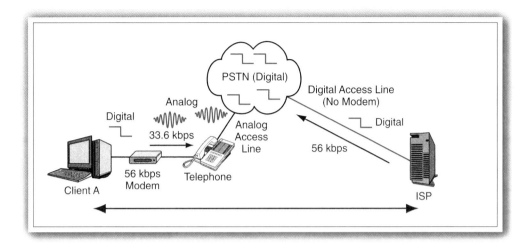

Figure 7-3   Telephone Modem Communication

ISP Equipment

As Figure 7-3 illustrates, only one of the two communicating partners—the user— has an analog access line. The modem allows the user to transmit at 33.6 kbps. This limit exists because of bandpass filtering at the ADC. The other side—the ISP—must have a digital private access line to the telephone network. This allows the ISP to trans- mit at 56 kbps, because there is no bandpass filtering on digital access lines.

Problems

The main problem with telephone modems, of course, is their low speed. A telephone modem is suitable for e-mail, but it is painful for World Wide Web applications and for large file transfers. Older V.34 modems, which are now uncommon, are limited to 33.6 kbps in both directions.

The other problem with modems is that they tie up your telephone line. If you need to call someone, you must first disconnect from the Internet. If someone calls you, in turn, you may get kicked off your Internet connection.

TEST YOUR UNDERSTANDING

    2. a) What is modulation? b) What is demodulation? c) How are ones and zeros repre- sented in analog modulation? d) In telephone modem communication, does the user have a modem at home? e) Does the ISP have a modem at its premises? f) What is the biggest problem with telephone modems? g) What is the second biggest problem?

Digital Subscriber Lines (DSLs)

High-Speed Data over Existing Voice-Grade Copper

Private lines are extremely expensive because new transmission media have to be pulled to the residences (as discussed in the previous chapter). Even for the lowest- speed private lines, pulling the required 2-pair data-grade UTP is extremely expensive.

	ADSL	HDSL	HDSL2	SHDSL
Uses Existing 1-Pair Voice Grade UTP Telephone Access Line to Customer Premises? (1)	Yes	Yes	Yes	Yes
Available Everywhere? (2)	No	No	No	No
Downstream Throughput	256 kbps-1,000 Mbps	768 kbps	1,544 kbps	384 kbps-2.3 Mbps
Upstream Throughput	60 kbps-256 Mbps	768 kbps	1,544 kbps	384 kbps-2.3 Mbps
Symmetrical Throughput?	No	Yes	Yes	Yes
Target Market	Residences	Businesses	Businesses	Businesses
Strong Throughput Guarantees?	No	Yes	Yes	Yes

Figure 7-4   Digital Subscriber Lines (DSLs) (Study Figure)

What about sending data over the single pair of voice-grade UTP that already runs to each residential subscriber's premises? This would eliminate the high cost of pulling new cabling.

Until recently, sending high-speed data over these low-quality existing wires was impossible. However, new technologies now allow this. This is called **digital subscriber line (DSL)** transmission because it involves sending digital signals over the residential customer's existing single-pair UTP voice-grade copper access line.

A digital subscriber line provides digital data signaling over the residential customer's existing single-pair UTP voice-grade copper access line.

ADSL

There are several different types of DSLs. The type designed to go into residential homes is the **asymmetric digital subscriber line (ADSL),** which offers high downstream throughputs (256 kbps to 1 Mbps) but limited upstream speeds (typically 60 kbps to 256 kbps). This asymmetry is good for World Wide Web access and FTP downloading. For most other services, such as e-mail, upstream speeds are sufficient.

As Figure 7-5 shows, the ADSL customer plugs a splitter into each telephone wall jack. The splitter separates data signals and voice signals. This permits simultaneous voice and data transmission.

At the end office of the telephone company, there is a **DSL access multiplexer (DSLAM),** which sends voice signals to the ordinary PSTN and sends data to a data network, such as an ATM network (ATM is discussed later in this chapter).

Although the access line to the customer is not shared, multiple DSL users share the capacity of the DSLAM. Most DSL carriers do not install enough capacity in the

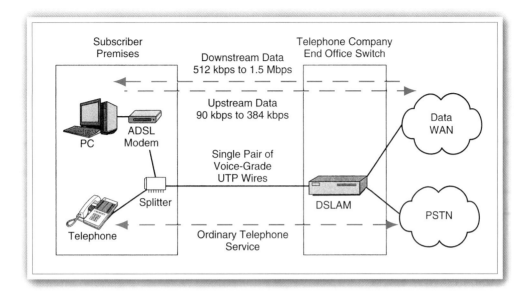

Figure 7-5   Asymmetric Digital Subscriber Line (ADSL)

DSLAM to give everybody the maximum speed if user traffic is heavy. Consequently, ADSL subscribers often find their throughput falling during peak usage periods.

HDSL

ADSL is fine for home Internet access, but heavy business use requires symmetric speeds and higher speeds. The most popular business DSL is the **high-rate digital subscriber line (HDSL).** This standard allows symmetric transmission at 768 kbps in both directions (half of a T1's speed). A newer version, **HDSL2,** transmits at 1.544 Mbps in both directions. Like all DSLs, both use a single voice-grade twisted pair.

SHDSL

The next step in business DSL is likely to be **SHDSL (super-high-rate DSL),** which can operate symmetrically over a single voice-grade twisted pair and over a speed range of 384 kbps to 2.3 Mbps. In addition to offering a wide range of speeds and a higher top speed than HDSL2, SHDSL also can operate over somewhat longer distances.

Quality-of-Service Guarantees

Generally, there are no hard guarantees for ADSL speeds, which are aimed at the tolerant home market. However, throughputs for HDSL, HDSL2, and SHDSL generally come with QoS guarantees because they are sold to businesses, which require predictable service. Of course, meeting these guarantees requires more stringent engineering and management by the carrier and so increases carrier costs. This leads to higher prices for HDSL, HDSL2, and SHDSL.

Limits

Although modern signal processing now allows us to send high-speed data over the ordinary voice-grade lines of the local loop, there are limits on what voice-grade lines can do. Many local loop lines will not support DSL service. The most common reason is that they are too far from the end office of the telephone company, but there are other limitations as well. Overall, DSLs are attractive, but they cannot be obtained everywhere.

Fiber to the Home (FTTH)

Sending data over voice-grade copper is possible, but it is very difficult technologically, and transmission speed over voice-grade copper is necessarily limited. As noted in the previous chapter, carriers are now beginning to experiment with bringing optical fiber to individual homes and businesses. This fiber to the home (FTTH) technology could bring high megabit speeds to individual subscribers—enough for services like television on demand.

However, to provide fiber to even a small fraction of all homes and businesses would require an enormous investment in fiber and installation labor. This would lead to high prices. In addition, the entire transport core would have to be upgraded to support the traffic if there were many FTTH users. When FTTH comes, it is not likely to be cheap.[2]

TEST YOUR UNDERSTANDING

3. a) How do private lines and DSLs differ in terms of transmission media? b) Describe ADSL speeds. c) Describe DSL technology. d) Does ADSL disable your telephone line when you are using the Internet? e) Does sharing affect ADSL speed? f) Describe HDSL and HDSL2 in terms of speed. g) Describe SHDSL in terms of speed. h) Which DSL services usually offer performance guarantees? i) What performance parameters do they guarantee? j) Can you always get a DSL to your premises? k) What is the biggest determining factor for whether you can? l) What is the potential benefit of FTTH? m) What is deterring FTTH's rapid widespread implementation?

Cable Modem Service

Cable Modems

Cable television operators, besides bringing television into homes, also offer high-speed Internet access, as Figure 7-6 indicates. From the cable television company's headquarters building, called the head end, optical fiber runs out to neighborhoods with about five hundred households apiece. Within the neighborhood, a standard coaxial television cable comes into the subscriber's house.

The coaxial cable coming into the home plugs into a device called a **cable modem.** The cable modem has a coaxial cable port plus either an RJ-45 port or a USB port to connect to your PC or access router.

---

[2] Currently, only about 10 percent of businesses have fiber access, and hardly any homes have fiber access lines. In practice, FTTH transmits a number of subscriber signals over a single fiber via wavelength division multiplexing. These WDM fibers go to a neighborhood splitter, which sends each lambda on to an individual home via a lower-cost fiber.

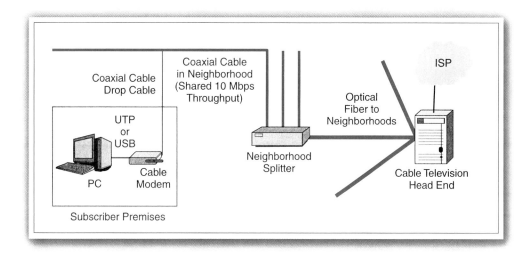

Figure 7-6   Cable Modem Service

High Shared Speed

Cable modem service normally provides downstream speeds of up to 10 Mbps,[3] although this speed is shared by multiple users in a neighborhood. In general, individual cable modem throughput is similar to ADSL throughput, but cable modem throughput depends heavily on how many users in the neighborhood are sharing the aggregate capacity at any moment.

TEST YOUR UNDERSTANDING

4. a) Describe cable modem technology. b) What is typical cable modem throughput? c) Compare speed reductions because of sharing in DSL service and cable modem service.

Wireless Access Systems

If it is too expensive to replace the wires in the local loop, why not eliminate them entirely and connect users to the transport core with radio for Internet access? The answer is that while technology makes wireless access systems possible, such systems tend to suffer from propagation effects, limited available bandwidth, and high costs.

Fixed and Mobile Wireless Access

In **fixed wireless service,** the user is at a fixed location. This allows the customer to use a dish antenna for efficient reception and transmission. In **mobile wireless access,** service is provided to moving customers. Mobile subscribers must have omnidirectional

---

[3]   This is with the widely used DOCIS 1.0 standard. DOCIS 2.0 will raise this to 30 Mbps and should increase throughput considerably.

Fixed Versus Mobile
>   Fixed: Requires a dish antenna for efficiency
>   Mobile: Requires an omnidirectional antenna for mobility

Satellie Versus Terrestrial Wireless Service
>   Satellite
>>   Expensive because of transmission distance
>>   Expensive because satellites are expensive to launch and maintain
>   Terrestrial
>>   Earth-based stations

3G Cellular Service
>   Telephone modem and DSL/cable modem speeds

802.16 WiMAX
>   One of several terrestrial standards now under development
>   Fixed
>>   20 Mbps up to 50 km (30 miles)
>   Mobile
>>   802.16e under development
>>   3 Mbps to 16 Mbps for mobile users

Figure 7-7   Wireless Access Systems (Study Figure)

antennas, because they cannot carry dishes around with them and would not know where to point them if they did.

Satellite Versus Terrestrial Wireless Service

Another important distinction is whether service is provided by **satellite transmission** or **terrestrial** (earth-based systems). Satellite systems are more expensive than terrestrial services because signals must travel farther and because satellites are expensive to launch and maintain.

3G Cellular Service

In the last chapter, we saw that 3G cellular systems will offer telephone modem or DSL/cable modem speeds to mobile customers.

802.16 WiMAX

The IEEE is now working on several terrestrial wireless access standards. The most developed is **802.16,** which is widely known as **WiMAX.** WiMAX was designed to serve fixed customers. It provides 70 Mbps service over a distance of up to 50 km (30 miles). This will make it useful for providing service to large corporate customers, 802.11 hot spot providers, and, eventually, to individual homes.

The 802.16e WiMAX standard now under development will extend WiMax service to mobile users, providing speeds of between 3 Mbps and 16 Mbps.

TEST YOUR UNDERSTANDING

5. a) What are the drawbacks of wireless systems for access service? b) Distinguish between fixed and mobile wireless access. c) What type of antenna do you use for fixed access? d) For mobile access? e) What is the advantage of using satellites for wireless access? f) Distinguish between 3G cellular service and WiMAX.

POINT-TO-POINT PRIVATE LINE NETWORKS

So far, we have been looking at services for individual Internet access. Now we will begin looking at technologies for building complete corporate WANs. We will start with networks built with private lines, which provide point-to-point, permanent, always-on, and fast service.

Private Line Networks for Voice and Data

Private Telephone Networks

Figure 7-8 shows that companies have traditionally used private lines to connect their PBXs at various sites. This allows any telephone at any site to call any other telephone at any other site. Although private lines are expensive, this arrangement is almost always much cheaper than using normal dial-up service to place calls long-distance between sites.

Private Line Data Networks

Figure 7-8 also shows an internal corporate data network using private lines. If the voice and data parts of the figure seem similar, this reflects the fact that data networking using private lines is based on the technology used for private line telephone networks. The main difference is that data networks use routers at each site rather than PBXs.

TEST YOUR UNDERSTANDING

6. Distinguish between private line voice networks and private line data networks.

Private Line Network Topologies

Should many or all pairs of sites be connected to each other, or should there be as few connections as possible? How the organization links its sites to one another is the network's topology.

Full Mesh Topology Private Line Networks

Figure 7-9 shows two topological extremes for building private line networks. The first is a **full mesh topology,** which provides direct connections between every pair of sites. This provides many redundant paths so that if one site or private line fails, communication will continue unimpeded.

Unfortunately, as the number of sites increases, the cost of a full mesh grows exponentially. For example, if there are N sites, a pure mesh will require $N(N-1)/2$ private lines. So a 5-site pure mesh will require 10 private lines ($5\times4/2$), a 10-site pure mesh will require 45 private lines, and a 20-site pure mesh will require 190 private lines. Full meshes, while reliable, are prohibitively expensive if a company has many sites.

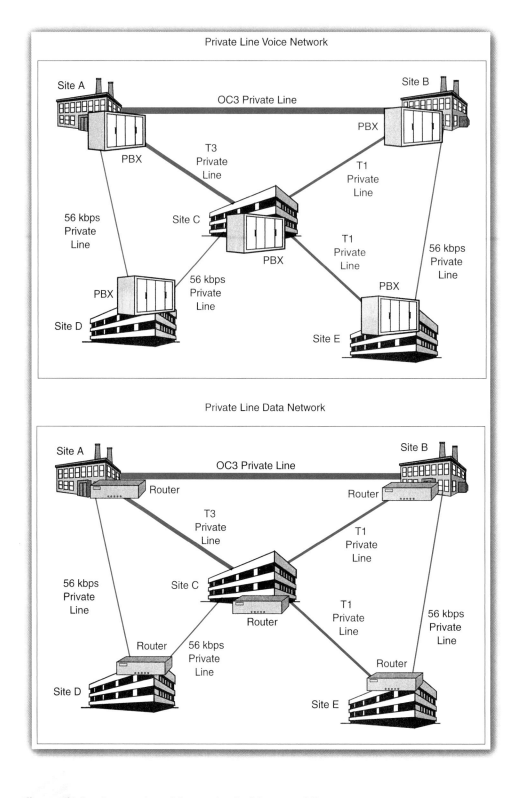

Figure 7-8 Private Line Networks for Voice and Data

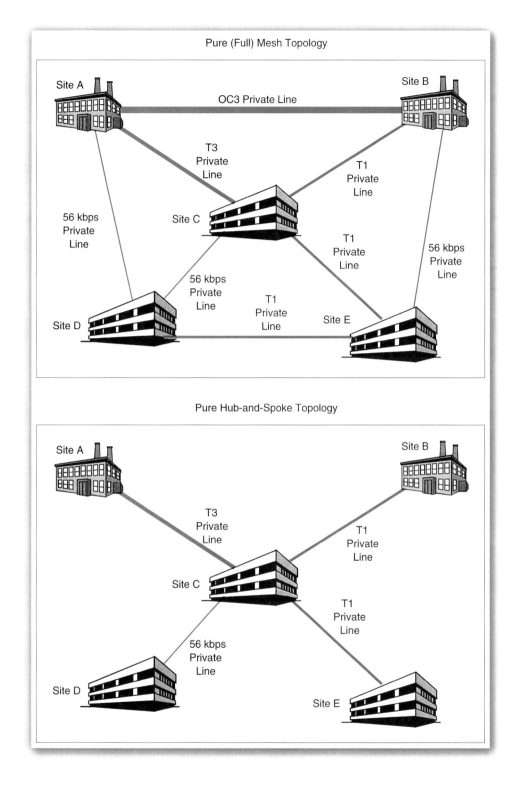

Figure 7-9   Mesh and Hub-and-Spoke Private Line Data Network Topologies

Hub-and-Spoke Private Line Networks

The second extreme topology for building private line networks is the **pure hub-and-spoke topology.** This is illustrated in Figure 7-9. In a pure hub-and-spoke topology, all communication goes through one site. This dramatically reduces the number of private lines required to connect all sites compared to a full mesh and so minimizes cost. However, it also reduces reliability. If a line fails, there are no alternative paths for reaching an affected site. More disastrously, if the hub site fails, the entire network goes down.

Mixed Designs

As you might suspect, full meshes and hub-and-spoke topologies represent the extremes of cost and reliability. Most real networks use a mix of these two pure topologies. Real networks must trade off reliability against cost.

TEST YOUR UNDERSTANDING

7.   a) What is the advantage of a full mesh private line network? b) What is the disadvantage? c) What is the advantage of a pure hub-and-spoke private line network? d) What is the disadvantage? e) Do most private line networks use a full mesh or a pure hub-and-spoke topology?

Private Line Speeds

We have already noted that private lines vary in speed from 56 kbps to a few gigabits per second. We will now be more specific by looking at the types of private lines actually offered by telephone carriers.

Figure 7-10 shows that different parts of the world use different standards for private lines below 50 Mbps. The figure shows lower private line speeds in the United States, Europe, and Japan.

56 kbps and 64 kbps

The lowest-speed lines in these hierarchies operate at 56 kbps or 64 kbps. This is sufficient to carry a single voice call or to send low-speed data.

T1, E1, and J1

At the next level of the hierarchy, the T1 line in the United States operates at 1.544 Mbps and can multiplex 24 voice calls. Japan's J1 line is identical, while the European E1 line operates at 2.048 Mbps, making it capable of multiplexing 30 conversations.[4]

Fractional T1/E1/J1 Private Lines

As noted earlier, most demand for WAN service lies between 56 kbps and a few megabits per second. The gap between 56 kbps and 1.544 Mbps to 2.048 Mbps private line speeds is fairly large, so many U.S. carriers offer **fractional T1** private lines operating at 128 kbps, 256 kbps, 384 kbps, 512 kbps, or 768 kbps. These provide intermediate speeds at intermediate prices. Not shown in the figure, carriers in some other countries also offer fractional E1 and J1 lines. T1/E1/J1 and fractional T1/E1/J1 lines are the most widely used private lines.

---

[4]   Although are T2, E2, and J2 standards, they are not offered commercially.

North American Digital Hierarchy

Line	Speed	Multiplexed Voice Calls	Typical Transmission Medium
56 kbps	56 kbps	1	2-Pair Data-Grade UTP
T1	1.544 Mbps	24	2-Pair Data-Grade UTP
Fractional T1	128 kbps, 256 kbps, 384 kbps, 512 kbps, 768 kbps	Varies	2-Pair Data-Grade UTP
Bonded T1s (multiple T1s acting as a single line)	Small multiples of 1.544 Mbps	Varies	2-Pair Data-Grade UTP
T3	44.736 Mbps	672	Optical Fiber

CEPT Hierarchy

Line	Speed	Multiplexed Voice Calls	Typical Transmission Medium
64 kbps	64 kbps	1	2-Pair Data-Grade UTP
E1	2.048 Mbps	30	2-Pair Data-Grade UTP
E3	34.368	480	Optical Fiber

Japanese Hierarchy

Line	Speed	Multiplexed Voice Calls	Typical Transmission Medium
64 kbps	64 kbps	1	2-Pair Data-Grade UTP
J1	1.544 Mbps	24	2-Pair Data-Grade UTP
J3	32.064 Mbps	480	Optical Fiber

SONET/SDH Speeds

Line	Speed (Mbps)	Multiplexed Voice Calls	Typical Transmission Medium
OC3/STM1	155.52	2,016	Optical Fiber
OC12/STM4	622.08	6,048	Optical Fiber
OC48/STM16	2,488.32	18,144	Optical Fiber
OC192/STM64	9,953.28	54,432	Optical Fiber
OC768/STM256	39,813.12	163,296	Optical Fiber

Figure 7-10 Private Line Speeds

Bonded T1s

Sometimes, a firm needs somewhat more than a single T1 line but does not need the much higher speed of a T3 line. Often, a company can **bond** a few T1s to get a few multiples of 1.544 Mbps. This is like link aggregation in Ethernet, which we saw in Chapter 4. There also is bonding with E1 and J1 lines, although this is not shown in the figure.

T3, E3, and J3

The next level of the hierarchy is the T3 line in the United States. This operates at 44.736 Mbps and can multiplex 672 voice calls. The E3 line operates at 34.368 Mbps and can multiplex 480 voice calls. The J3 line also multiplexes 480 voice calls but operates at a slightly different speed, 32.064 Mbps.

SONET/SDH

Beyond T3/E3/J3 lines, the world is nearly standardized on a single technology or, more correctly, on two compatible technologies. These are **SONET (Synchronous Optical Network)** in North America and **SDH (Synchronous Digital Hierarchy)** in Europe. Other parts of the world select one or the other.

Figure 7-10 shows that SONET/SDH speeds are multiples of 51.84 Mbps, which in these technologies can multiplex 672 voice calls—the same as a T3 line. SONET speeds are given by **OC (optical carrier)** numbers, while SDH speeds are given by **STM (synchronous transfer mode)** numbers.

Note that the lowest official SONET/SDH speed is 155.52 Mbps and that speeds range up to several gigabits per second. Note also that the SONET speed nearest to 10 Gbps is 9,953.28 Mbps. Ethernet uses this speed for WAN usage so that it can transmit data over physical layer SONET lines.

TEST YOUR UNDERSTANDING

8. a) Below about what speed are there different private line standards in different parts of the world? b) At what speeds do the slowest private lines run? c) What is the exact speed of a T1 line? d) How many simultaneous voice calls can it multiplex? e) What are the speeds of comparable private lines in Europe and Japan? f) Why are fractional T1, E1, and J1 speeds desirable? g) List common fractional T1 speeds. h) What are the most widely used private lines? i) How many voice calls can be multiplexed on T3, E3, and J3 private lines? j) What private line standards are used above 50 Mbps? k) All SONET and SDH lines are multiples of what speed?

PUBLIC SWITCHED DATA NETWORKS (PSDNs)

Private Lines in Private Line Data Networks

Earlier, we looked at two topologies for building private line data networks, namely mesh and hub-and-spoke topologies. Both approaches use many private lines, and these private lines must span long distances—all the way between sites. This is very expensive.

Public Switched Data Network (PSDN) Access Lines

In contrast, Figure 7-11 illustrates a **public switched data network (PSDN).** Using a PSDN, the user needs only one private line per site. This private line has to run only from the site to the PSDN's nearest access point, called a **point of presence (POP).**[5]

This means that if you have ten sites, you only need ten private lines. Furthermore, most PSDN carriers have many POPs, so the few private lines that are needed tend to span only short distances.

The PSDN Cloud

The PSDN's transport core usually is represented as a **cloud.** This reflects the fact that although the PSDN has internal switches and trunk lines, the customer does not have to know how things work inside the PSDN cloud. The PSDN carrier handles almost all of the management work that customers have to do when running their own private line networks. Customers merely have to send and receive data to the PSDN cloud in the correct format. Although PSDN carrier prices reflect their management costs, there are strong **economies of scale** in managing very large PSDNs instead of individual corporate private line networks, meaning that it is cheaper to manage the traffic of many firms than of one firm. There also are very large economies of scale in technology. These economies of scale allow low PSDN prices compared to the costs of running private line networks.

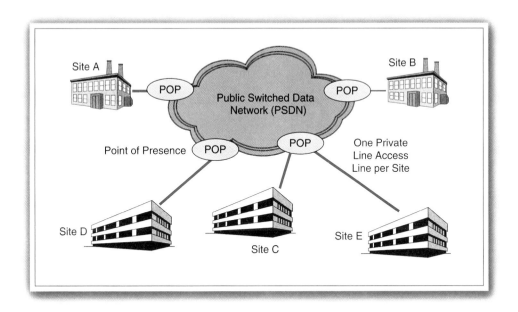

Figure 7-11   Public Switched Data Network (PSDN)

---

[5] In Chapter 6 we saw that the term point of presence (POP) is also used in telephony as a place where various LECs, IXCs, and ICCs interconnect.

Service Level Agreements (SLAs)

Most PSDNs offer **service level agreements (SLAs),** which are quality-of-service guarantees for throughput, availability, latency, error rate, and other matters. For instance, an SLA may guarantee a latency of no more than 100 ms 99.99 percent of the time. Although SLAs are very nice to have, they add considerably to the price of a service because PSDN vendors need to allocate more resources to the customer to ensure that SLA guarantees are met.

The Market Situation

Although private lines and PSDNs will both remain popular in the foreseeable future, PSDN use is growing rapidly, whereas the number of private line installations is growing more slowly and actually is falling in the long-distance market. In new corporate networks, private lines are used primarily to connect very large sites because private line prices may still be competitive with PSDN prices for high-volume connections.

Virtual Circuit Operation

Figure 7-12 shows that PSDN switches are connected in a mesh topology. In any mesh topology, whether partial or full, there are multiple alternative paths for frames to use to go from a source POP to a destination POP.

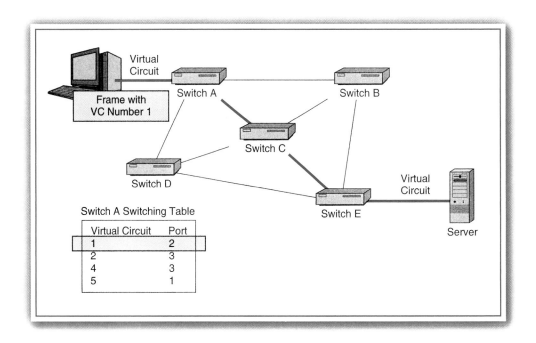

Figure 7-12   Virtual Circuit

Selecting Best Possible Paths through Meshes

Selecting the best possible path for each frame through a PSDN mesh would be complex and, therefore, expensive. In fact, if the best possible path had to be computed for each frame at each switch along its path, PSDN switches would have to do so much work that they would be prohibitively expensive.

Virtual Circuits

Instead, PSDNs select the best possible path between two computers or sites *before transmission begins.* The actual transmission will flow along this path, called the **virtual circuit.** As Figure 7-12 shows, the switch merely makes a switching decision based upon the virtual circuit number in the frame's header. This virtual circuit lookup is very fast compared to the work needed to select the best path for each frame.

PSDN Frame Headers Have Virtual Circuit Numbers
Rather than Destination Addresses

Note that PSDN frames do not have destination addresses in their headers. Rather, each frame has a virtual circuit number in its header.

TEST YOUR UNDERSTANDING

9. a) Describe PSDN technology. b) Do customers need private lines if they use PSDNs? c) Compare private line costs for private line networks and PSDNs. d) Why is the PSDN transport core drawn as a cloud? e) Why do PSDNs tend to cost less than private line networks? f) What things do SLAs guarantee? g) Why would an SLA guarantee maximum latency rather than minimum latency?

10. a) Why are virtual circuits used? b) With virtual circuits, on what does a switch base its forwarding decision when a frame arrives? c) Do PSDN frames have destination addresses or virtual circuit numbers in their headers?

FRAME RELAY

The Most Popular PSDN

The most popular PSDN today is Frame Relay. Frame Relay operates at 56 kbps to about 40 Mbps, with most customers operating well below that top speed. This is consistent with the needs of most corporations—56 kbps to a few megabits per second. Furthermore, when Frame Relay service is compared to networks of private lines, Frame Relay is almost always less expensive.

Components

Figure 7-13 shows the main elements of a Frame Relay network: access devices, access lines, ports at POPs, PVCs, and management.

Access Devices

Each user site needs an access device. This is either a router or a dedicated **Frame Relay Access Device (FRAD).**

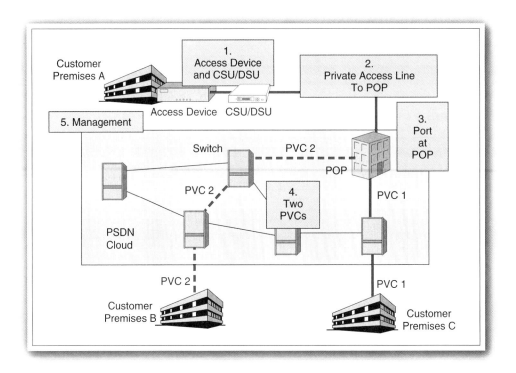

Figure 7-13   Frame Relay Network

CSU/DSU

The port on the router or FRAD that terminates the private access line going to the Frame Relay network must have a physical layer device called a **CSU/DSU.** The **CSU (channel service unit)** is designed to protect the telephone network from improper voltages. In turn, the **DSU (data service unit)** formats the data in the way the private line requires.

Access Lines and Points of Presence (POP)

From the customer premises, the customer needs a private line to use as an **access line** to the nearest **point of presence (POP)** of the Frame Relay network. The POP is the entry point to the Frame Relay network. If a carrier has many POPs, then access lines will be relatively short and therefore relatively inexpensive.

Port Speed

At the POP, user transmission speed is limited to a certain port speed. To transmit faster, the user has to select a port speed suitable for his or her transmission needs. As you would suspect, faster ports cost more. In fact, port speed usually is the most expensive pricing element in Frame Relay service. Selecting a port speed that is sufficient but not extravagant is critical in Frame Relay network design.

> Port speed usually is the most expensive pricing element in Frame Relay service.

Virtual Circuits

Between each pair of sites that wish to communicate, there must be a virtual circuit. Collectively, virtual circuit charges usually are the second largest element in Frame Relay prices.

> Virtual circuit charges usually are collectively the second most expensive pricing element in Frame Relay service.

Management

Most Frame Relay vendors offer **managed Frame Relay** networks. Although PSDN carriers automatically do internal management, managed Frame Relay services also take on most of the remaining management tasks that customers still must do. They provide traffic reports and actively manage day-to-day traffic to look for problems and get them fixed. They also manage corporate access devices.

TEST YOUR UNDERSTANDING

11. a) List the elements in a Frame Relay network. b) Briefly explain the purpose of each. c) Which usually is the most expensive component in Frame Relay pricing? d) Which usually is the second most expensive component?

Virtual Circuits and DLCIs

Recall that PSDNs use virtual circuits, and PSDN frame headers have virtual circuit numbers rather than destination addresses. As Figure 7-14 shows, this virtual circuit number in Frame Relay is the **Data Link Control Identifier** or **DLCI** (pronounced "dull´-see"). A DLCI is 10 bits long.[6] The switch looks up this DLCI in its virtual circuit switching table and sends the frame out the indicated port.[7]

Permanent Virtual Circuits (PVCs) and Switched Virtual Circuits

Normally, the virtual circuit between corporate sites is set up once and kept in place for weeks, months, or years at a time. Such virtual circuits are called **permanent virtual circuits (PVCs).** Many Frame Relay vendors also offer **switched virtual circuits (SVCs)** that are set up just before a call and last only for the duration of the call. SVCs require more carrier work per call and so are more expensive.

TEST YOUR UNDERSTANDING

12. a) What is the name of the Frame Relay virtual circuit number, and how long is it? b) Distinguish between PVCs and SVCs. c) Which is more expensive?

---

[6] There is a mechanism for creating larger DLCIs using "AE" bits. However, according to Ericsson, no switch manufacturer implements this mechanism.

[7] Stations also have true Frame Relay addresses governed by the E.164 standard. These addresses are used to set up virtual circuits. Consequently, if an equipment failure renders a virtual circuit inoperable, a new virtual circuit can be set up using the E.164 addresses.

0	1	2	3	4	5	6	7	Bits
0	1	1	1	1	1	1	0	Start flag field indicating the start of the frame
First 6 bits of DLCI						C/R	AE	First six bits of the 10-bit Data Link Control Identifier (virtual circuit number)
						0/1	0	C/R (command/response); rarely used except when carrying SNA traffic Address Extension bit is 0 to indicate that this is not the last address octet
Last 4 bits of DLCI				FECN	BECN	DE	AE	Final four bits of the 10-bit DLCI
						1	1	FECN and BECN notify the receiver of the frame that there has been congestion so the transmission rate should be reduced; rarely used Discard eligible bit must be set to 1 if the frame is being transmitted beyond the committed information rate (Bc) Address Extension bit is 1 to indicate that this is the last address octet
Information Field (Variable Length)								Typically an IP packet
Frame Check Sequence (First 8 bits)								2-octet error check sequence
Frame Check Sequence (Last 8 bits)								Frames with errors are discarded
0	1	1	1	1	1	1	0	End flag field indicating the end of the frame

Figure 7-14   Frame Relay Frame

Designing a Company's Frame Relay Network

Although the Frame Relay carrier will handle most or all implementation work, companies have to design their Frame Relay networks, including what PVC speeds, port speed, and access line speed they will use for each site.

Traffic Requirements

The first step, of course, is that you begin with your requirements. You need to estimate the traffic volume that will flow between each pair of sites.

Determining One Site's PVC Needs

Next you look at each site individually. For that site you first ask what connection speeds that site needs to other sites. You usually find that you need PVCs from that site to each of several other sites. These PVCs must be fast enough to serve the traffic flowing between that pair of sites.

For instance, suppose that you determine, based on traffic patterns, that you need to provide 56 kbps service from the site you are considering to five other sites. If so, you will need five 56 kbps PVCs from the site, one to each of the five other sites.

Determining the Site's Port Speed Requirement

Next, you determine the *port speed* requirement for the site. The port speed must be fast enough to handle your PVCs. If you have five 56 kbps PVCs running from one site to other sites, this would suggest that you would need a port speed of at least 280 kbps (5 times 56 kbps) to serve all the PVCs.

Port Speed Oversubscription

However, not all of these PVCs will be carrying data at their maximum rates simultaneously. A common rule of thumb—that must of course be modified by actual traffic patterns in real firms—is that about 70 percent of aggregate PVC capacity is sufficient for a site's port speed. So instead of needing 280 kbps of capacity in our port, we would need only about 70 percent of this amount, or 196 kbps. A 256 kbps port speed would be sufficient if it were available. If that port speed were not available, you would have to move up to the *next-higher* port speed. Having port speeds less than the sum of PVC speeds is called **oversubscription.**

PVC Excess Burst Speeds

Sometimes, Frame Relay vendors offer two-part PVC speeds. The first part is the **committed information rate (CIR)** that is guaranteed by the Frame Relay carrier for the PVC. The other is an **excess burst speed** beyond the CIR.[8] PVC bursts up to this speed can take place only for brief periods of time, and if there is congestion, bits beyond the CIR are dropped. However, if other PVCs are not using their full CIRs, excess bursting allows PVCs that need more capacity momentarily to use the capacity not being used by other PVC CIRs.

With excess burst speeds, companies do not use oversubscription for the sum of the PVCs, but they do take excess burst speeds into account. For instance, if four 128 kbps PVCs are needed, the PVCs may be ordered with CIRs of 128 kbps and excess burst speeds of 256 kbps. A port speed of 512 kbps (4 times 128 kbps) might suffice for the four PVCs. This is certainly fast enough for the CIRs. In addition, one or more of the PVCs often will be able to do at least some excess bursting.

There is no rough rule of thumb for excess burst speeds like the 70 percent oversubscription rate for single-speed services. However, the port speed usually is chosen

---

[8] In practice, rates are based on a very brief period of time, Tc, which varies among Frame Relay vendors. During the period Tc, a firm may send a burst of Bc bits with roughly guaranteed throughput. (Bc/Tc is the committed information rate.) In addition, the firm can send an extra burst of Be bits as long as Be/Tc does not exceed the excess information rate.

to be at least as fast as the sum of the PVCs' CIRs, and an even faster port speed might be selected. Prices and service conditions for bursting tend to be complex and require a good knowledge of data traffic patterns between the sites.

Determining a Site's Private Line Speed Requirement

Your last design task for the site is to determine your private line speed requirement. Port charges are higher than private line charges, so you need a private line at least as fast as or faster than your port speed. For example, if your port speed is 1 Mbps, you would need a T1 line operating at 1.544 Mbps.

Repeat for Other Sites

So far, we have looked at Frame Relay requirements for a single site. The steps we just went through—determining requirements for speeds to other sites, specifying PVCs, selecting a port speed, and selecting the speed of the private access line to the POP—must be repeated for each side.

TEST YOUR UNDERSTANDING

13.   a) From a site, you need one PVC of 256 kbps and one of 56 kbps. Without using over-subscription, what port speed would you select if your choices were 56 kbps, 128 kbps, 256 kbps, 384 kbps, and 1 Mbps? Explain. b) Using oversubscription, what port speed would you select if you had the same choices? Explain. c) If you have a port speed of 150 kbps, what private line would you select if your choices were 56 kbps, 128 kbps, 256 kbps, and 1 Mbps? Explain. d) Why may it be possible to have four PVCs with 128 kbps speed requirements and 256 kbps burst speeds with a 512 kbps port?

ASYNCHRONOUS TRANSFER MODE

For PSDN service at speeds greater than Frame Relay can provide, corporations can turn to **asynchronous transfer mode (ATM)** service. ATM services reach gigabits per second. They may extend down to one megabit per second, but most usage will be much faster.

Not a Competitor for Frame Relay

It might seem that Frame Relay and ATM are competitors. In practice, however, almost all carriers offer both Frame Relay and ATM. They recommend Frame Relay for customers with lower-speed needs and ATM for customers with higher-speed needs. In fact, some vendors have interconnected ATM and Frame Relay networks so customers can connect low-speed sites with Frame Relay and high-speed sites with ATM.

Designed for SONET/SDH

ATM was designed to run over SONET/SDH at the physical layer. Although ATM can run over other physical layer technologies, SONET/SDH supports the high speeds that are ATM's forte.

Cell Switching

Most network protocols have variable-length data fields. This gives flexibility, but switches must do a number of calculations when dealing with variable-length frames.

This adds to the work a switch must do and, therefore, its cost. It also creates a bit of latency at each switch.

Fixed-Length Cells to Reduce Switch Costs and Latency

To reduce switch processing costs and latency, ATM uses fixed-length frames. Fixed-length frames are called **cells,** so ATM is referred to as a **cell-switching** technology.

Short Cells to Reduce Latency

Furthermore, as Figure 7-15 shows, ATM cells are very short. They consist of a 5-octet header and a 48-octet data field, which ATM calls the **payload.** This makes ATM cells much shorter than typical frames in Frame Relay, Ethernet, or other network protocols. Having short cells also reduces latency. Often switches must process entire frames before sending them back out. Shorter frames can be sent back out more quickly.

ATM Quality-of-Service Guarantees

ATM supports several different classes of service that receive different guarantees. For voice, ATM can set strict limits on latency and jitter. This makes ATM ideal for voice traffic. Data, however, usually are given no guarantees. In fact, the capacity that has to be reserved to give QoS guarantees to voice means that data traffic gets only leftovers. For pure data transmission, ATM's ability to provide QoS guarantees is not a benefit.

Manageability, Complexity, and Cost

ATM was created to become the transport mechanism for the worldwide PSTN. In fact, as we saw in Chapter 6, most long-distance companies have already moved at least partway to having ATM transport cores.

Bit 1	Bit 2	Bit 3	Bit 4	Bit 5	Bit 6	Bit 7	Bit 8
Generic Flow Control				Virtual Path Identifier			
Virtual Path Identifier				Virtual Channel Identifier			
Virtual Channel Identifier							
Virtual Channel Identifier				Payload Type Indicator		Reserved	Cell Loss Priority
Header Error Check							
Payload (48 Octets)							

This is the cell structure for the user–network interface (UNI) between the switches of customers and carriers. There is a different cell structure for the network–network interface (NNI) between two ATM carriers.

Figure 7-15   ATM Cell

This required the creation of an extremely sophisticated set of ATM management protocols. This sophistication is good because it allows enormous networks to be managed. However, sophistication also means complexity and high cost.

Market Strengths

As just noted, ATM has become very important in the telephone system's transport core. As corporate demands for WAN speeds increase as a result of growing needs and falling prices, many Frame Relay users may migrate to ATM.

TEST YOUR UNDERSTANDING

14. a) Compare Frame Relay and ATM speed ranges. b) Are Frame Relay and ATM competitors? Explain. c) In ATM, what is a cell? d) Why does ATM use short cells? Explain your answer. e) Compare what ATM has to offer to voice and data service. f) Why does ATM have strong management tools? g) Why is ATM's sophistication good? h) Why is it problematic? i) Why is ATM usage likely to grow in the future?

METROPOLITAN AREA ETHERNET

Metropolitan Area Networking

Ethernet now dominates local area networking. However, the newest versions of Ethernet—10 Gbps Ethernet and 40 Gbps Ethernet—are being designed for WAN use. More specifically, they are being used in **metropolitan area networks (MANs),** which span single urban areas. **Metropolitan area Ethernet,** also called **metro Ethernet,** is still very new, but it is already beginning to spread rapidly.

E-Line and E-LAN

Metro Ethernet is offered in two forms. **E-line** services provide point-to-point connections, like private lines. In turn, **e-LAN** services link multiple sites simultaneously. To the switches at each site, e-LAN service simply looks like a set of additional trunk lines linking the sites.

Attractions of Metropolitan Area Ethernet

Low Cost

Although there are several things about metropolitan area Ethernet that are attractive, the most important is its low cost. As it does in LANs, Ethernet's simplicity reduces switching costs in MANs. Overall, metro Ethernet is much cheaper than Frame Relay or ATM for comparable speeds.

High Speeds

In addition, metropolitan area Ethernet offers very high speeds. While Frame Relay offers speeds up to a few tens of megabits per second, metro Ethernet offers speeds up to 10 Gbps at only slightly higher cost and will soon offer 40 Gbps.

Familiar Technology

A third advantage of metropolitan area Ethernet is that firms can use the standard Ethernet interface they already know well instead of having to master Frame Relay, ATM, or other new interfaces.

Carrier Class Service

However, metro Ethernet has not completely developed the quality-of-service and traffic management tools needed to offer true **carrier class service.** Until these tools are finished, corporations will be hesitant to use Ethernet for very large metro networks.

TEST YOUR UNDERSTANDING

15.   a) What is metropolitan area Ethernet? b) Distinguish between e-line and e-LAN service. c) Why is metro Ethernet attractive? d) Why are companies hesitant to create large metro Ethernet LANs?

VIRTUAL PRIVATE NETWORKS (VPNs)

The Attractiveness of Internet Transmission

Although Frame Relay and ATM PSDNs are attractive, most firms already have all of their sites connected to the Internet. The fees paid to ISPs are attractive per bit transmitted compared to those paid to PSDN vendors. Consequently, companies would like to use the Internet for WAN transmission.

However, the Internet is a nonsecure environment, so corporate transmission over the Internet has to be cryptographically protected. As Figure 7-16 shows, **virtual private networks (VPNs)** use the Internet with added security for data transmission.

Virtual private networks (VPNs) use the Internet with added security for data transmission.

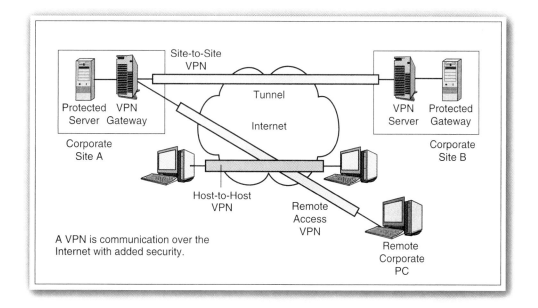

Figure 7-16   Virtual Private Network (VPN)

VPNs, then, should offer much lower costs than private line networks or PSDNs, while offering adequate security.

Companies can build their own VPNs by adding equipment at their sites and then actively managing their VPNs. Alternatively, they can get managed VPNs from carriers. Carriers install and do active management of these VPNs; essentially, managed VPNs are ways for companies to outsource their VPNs.

Types of VPNs

Remote Access VPNs
Some VPNs are used for remote access—connecting an individual user to a corporate site.

Site-to-Site VPNs
Other VPNs are used for site-to-site transmission. They connect LANs at different sites, and they carry the traffic of many users. Consequently, site-to-site VPNs will eventually dominate VPN usage.

Host-to-Host VPNs
Finally, VPNs can be set up directly between two hosts. This allows two employees to communicate securely.

VPN Standards

There are three basic standards for virtual VPN security. The first two, SSL/TLS and PPTP, are useful only for remote access VPNs. The third, IPsec, should eventually dominate VPN transmission.

SSL/TLS
The simplest VPN security standard to implement is **SSL/TLS.** This standard was originally created as **Secure Sockets Layer (SSL)** by Netscape. It was later taken over by the IETF and renamed **Transport Layer Security (TLS).** We will call it SSL/TLS because it is still called by both names.

As Figure 7-17 shows, SSL/TLS provides a secure connection at the transport layer. This protects all applications above it. However, SSL/TLS only protects applications that are **SSL/TLS-aware,** that is, have been modified to work with SSL/TLS. All browsers and webservers are SSL/TLS-aware. Some e-mail systems also are SSL/TLS aware. Few other applications are.

Despite being limited to a few applications, some firms need only remote Web access. These firms are likely to use SSL/TLS, which is easy to implement because every browser and webserver application program has SSL/TLS built in.

In SSL/TLS, one issue is how to authenticate the user, that is, require the user to prove his or her identity. One SSL/TLS option for corporations is to do no authentication for the client; this opens SSL/TLS-based systems to many attacks. Webserver application programs can supplement this SSL/TLS weakness by adding passwords themselves, but password security is not strong security.

The other option is for corporations to use a digital certificate for each client. This provides very strong security, but as we noted in Chapter 5, implementing client digital certificates is very difficult.

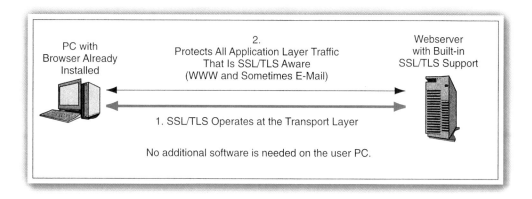

Figure 7-17   SSL/TLS Remote Access VPN

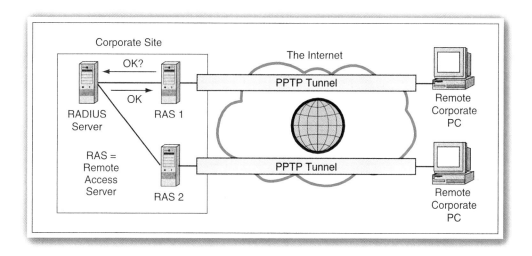

Figure 7-18   PPTP Remote Access VPN

PPTP Remote Access VPN

Figure 7-18 shows another remote access VPN security standard, the **Point-to-Point Tunneling Protocol (PPTP).** PPTP works at the data link layer, and it protects all messages above the data link layer. It provides protection **transparently,** that is, without having to modify applications. PPTP only offers moderate security, including a dependence on password authentication for clients. On the positive side, PPTP has been built into Windows clients since Windows 98. This means that corporations do not have to install any new software on their many clients.

As the figure shows, the remote user connects to a **remote access server (RAS)** which authenticates the user (determines the user's identity). As in the 802.11i security we saw in Chapter 5, remote access servers typically check with an authentication server for authentication information. Typically, this is a RADIUS server.

IPsec

The most sophisticated VPN technology is a set of standards collectively called **IP security (IPsec)**.[9] As its name suggests, IPsec operates at the internet layer. It provides security to all upper layer protocols transparently.

Pros and Cons   IPsec offers very strong security and should eventually dominate remote access VPN transmission, site-to-site VPN transmission, and internal IP transmission as well.

However, IPsec requires clients to have digital certificates. As just noted, giving each client computer a digital certificate is expensive and difficult to manage.

In addition, IPsec has been built into Windows only since Windows 2000. For earlier versions of Windows, IPsec software has to be added, and installing software on many clients is very expensive and difficult.

Transport Mode   Figure 7-19 shows that IPsec has two modes of operation. In **transport mode,** the two computers that are communicating implement IPsec. This gives strong end-to-end security, but it requires IPsec configuration and a digital certificate on all machines. For PCs with versions of Windows older than Windows 2000, it also requires an operating system upgrade or the addition of IPsec software.

Tunnel Mode   In contrast, in **tunnel mode,** the IPsec connection extends only between **IPsec gateways** at the two sites. This provides no protection within sites, but the use of tunnel mode IPsec gateways offers transparent security. The two hosts do not have to implement IPsec security and, in fact, do not even have to know that IPsec is being used between the IPsec gateways.

TEST YOUR UNDERSTANDING

16. a) What is a VPN? b) Why are VPNs attractive? c) Why are managed VPNs attractive? d) What is a remote access VPN? e) What are site-to-site VPNs? f) Why are site-to-site VPNs likely to become the largest corporate use for VPNs? g) What are host-to-host VPNs?

17. a) How is SSL/TLS limited? b) Why is it attractive? c) Under what circumstances is it likely to be used?

18. a) Is PPTP for remote access VPNs or site-to-site VPNs? b) Describe PPTP authentication. c) Why is PPTP attractive? (Give two reasons.) d) At what layer does it operate? e) What communications does it protect?

19. a) At what layer does IPsec operate? b) What layers does it protect? c) Describe IPsec tunnel mode. d) What is the main advantage of tunnel mode? e) What is the main disadvantage of tunnel mode? f) Describe IPsec transport mode. g) What is the main advantage of transport mode? h) What is the main disadvantage of transport mode? i) Describe IPsec authentication. j) Is IPsec for remote access or site-to-site VPNs?

---

[9] Pronounced "EYE-pee-seck."

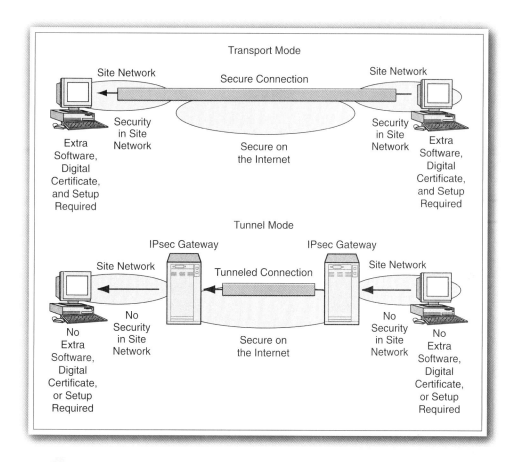

Figure 7-19   IPsec: Transport and Tunnel Modes

20. a) Of the three VPN security technologies in this section, which provides transparent security to higher layers? b) Which tends to require the installation of software on many client PCs? c) Which has the strongest authentication? d) Which would you use for an intranet that gives employees remote access to a highly sensitive Webserver via the Internet? (This is not a trivial question.) Justify your answer.

SYNOPSIS

Corporations build wide area networks (WANs) for individual remote access, site-to-site transmission, and Internet access. Among the technologies they use are telephone modems for low speeds, networks of private lines, public switched data networks (PSDNs), metropolitan area radio transmission (rarely), and virtual private networks (VPNs).

High Prices and Low Speeds

Your personal experience probably has been limited primarily to LAN transmission where cheap transmission leads to high speeds. You must adjust your thinking for wide area networks (WANs), where long distances make the price per transmitted bit very high, which in turns leads to companies limiting themselves primarily to low speeds—most typically between 56 kbps and a few megabits per second. Although faster transmission systems exist, 56 kbps to a few megabits per second is the range of greatest corporate demand.

Internet Access

For Internet access, telephone modems typically transmit at up to 33.6 kbps and can receive at up to 56 kbps. This is very slow, and telephone modems tie up your telephone so that you cannot call out or receive calls.

For faster Internet access, private lines can be used for Internet access, but they are very expensive because they require the local telephone company to pull two pairs of data-grade wire to the subscriber's premises. In contrast, digital subscriber line (DSL) services send and receive broadband data over the existing single pair of voice-grade UTP that already runs to each residential subscriber's premises. For residential users, ADSL provides downstream speeds of 256 kbps to 1 Mbps and upstream speeds of 60 kbps to 256 kbps. Cable television providers offer cable modem service, which is similar in asymmetrical throughput and price. Businesses use DSL services with symmetric speeds; these business-class DSL services also offer guarantees that residentially-focused ADSL does not.

In the future, wireless Internet access may become popular, although its price is likely to be high compared to land-line access.

Site-to-Site Networking with Private Lines

For site-to-site networking, companies have traditionally turned to networks of private lines. They did this first for telephone services, using private lines to connect PBXs at different sites. For data networking, they replaced the PBXs with routers. Most private line networks mix the characteristics of full mesh topologies, which are reliable but expensive, and pure hub-and-spoke topologies, which are inexpensive but have many single points of failure.

Companies that use private lines typically use 56 kbps/64 kbps private lines, T1/E1/J1 private lines operating at 1.5 Mbps to 2 Mbps, and fractional private lines below T1/E1/J1 speeds. However, if they have some connections that require much higher speeds, they can use T3/E3/J3 lines operating at roughly 30 Mbps to 50 Mbps, or SONET/SDH lines operating at 156 Mbps to several gigabits per second.

Site-to-Site Networking with PSDNs

With public switched data networks (PSDNs), the PSDN carrier does most of the transmission and management work. Companies merely need access devices at their sites (typically routers) and a single private line from each site to the PSDN carrier's nearest point of presence. Frame Relay provides speeds of 56 kbps to 40 Mbps, with most corporations using the lower end of this range. ATM offers speeds of 1 Mbps to several gigabits per second, with low-megabit speeds being fairly uncommon. For voice traffic, ATM offers stringent latency and jitter control, but if offers no special QoS benefits for data traffic.

Both Frame Relay and ATM use virtual circuits to simplify the operation of switches and, therefore, minimize switching costs. Switches base forwarding decisions on virtual circuit numbers rather than on destination addresses.

Designing a Frame Relay network is a complex task, but it is crucial for effective and inexpensive operation. Based on traffic requirements, the firm specifies PVC speeds between each pair of sites. The sum of the PVCs for a specific site determines its need for the POP port speed and for the speed of the access line. Transmission is bursty, so the POP speed can be somewhat less than the total of PVC speeds for a site. Alternatively, the CIRs of the PVCs can be set to fit the port speed, but PVCs can be allowed to send bursts beyond the CIRs when other PVCs are not using their shares of the port speed.

Metropolitan area Ethernet, which extends Ethernet beyond the corporate borders for transmission within an urban area, is very new. It offers the potential to slash transmission costs compared to ATM. However, it may not thrive until quality-of-service standards and general management standards are created to allow Ethernet to work in the large but highly price-sensitive world of WAN transmission.

Virtual Private Networks (VPNs)

A final option for both remote access and site-to-site networking is the virtual private network (VPN), which uses the Internet to lower transmission costs but adds security to protect sensitive conversations. For remote access, SSL/TLS VPNs can work if the company is primarily using Web-based services. For more general remote access, there is PPTP. The IPsec standard offers much better security than SSL/TLS and PPTP and so should eventually dominate for both remote access and site-to-site transmission. However, IPsec requires that digital certificates be provided to all clients, and this is expensive. In addition, older versions of Microsoft Windows do not support IPsec, but nearly all versions support SSL/TLS and PPTP. This means that SSL/TLS and PPTP do not require software upgrades on many of a firm's clients.

SYNOPSIS QUESTIONS

1. a) In what broad ways is WAN transmission different from LAN transmission? b) What is the speed range of greatest corporate demand today for WAN transmission lines?
2. What are the two problems with telephone modem Internet access?
3. a) How does DSL technology differ from private line technology? b) What are typical ADSL speeds? c) How do ADSL and cable modem service compare in terms of speed and price? d) Which types of DSL receive quality of service guarantees? e) How do ADSL and cable modem throughputs compare?
4. What is likely to be the disadvantage of wireless Internet access services?
5. a) How do private line data networks differ from private line voice networks? b) What are the relative advantages and disadvantages of full mesh and pure hub-and-spoke topologies? c) Are most private line networks pure meshes or pure hub-and-spoke networks?
6. a) Which private line services operate in the range of greatest corporate demand for WAN speeds? b) Which are much faster? c) Why do both use virtual circuits? (Give a detailed answer.)

7. a) How do private line networks and PSDNs differ in terms of the work a firm must do to operate and manage them? b) How do Frame Relay and ATM compare in terms of speed? c) What are SLAs? d) Compare ATM guarantees for voice and data.

8. What are the steps in designing a Frame Relay network?

9. a) Why is metropolitan area Ethernet attractive for site-to-site networking? b) Why is more work needed in standards development before Ethernet can work effectively in the WAN environment?

10. a) What is a VPN? b) Why are VPNs attractive? c) What VPN standards can be used for remote access? d) Why is SSL/TLS more attractive for remote access VPNs than IPsec? e) Why should IPsec eventually dominate for both remote access and site-to-site VPNs? f) What is holding back IPsec?

THOUGHT QUESTION

1. Several Internet access systems are asymmetric, with higher downstream speeds than upstream speeds. a) Is this good for webservice? b) Does it matter for e-mail? c) Is it good for a server? d) Is it good for videoconferencing?

TROUBLESHOOTING QUESTION

1. You purchase 512 kbps ADSL service. Your download speed typically is only about half this rate. What may be happening? Create at least five hypotheses, not all involving the access line. How would you test each of them?

DESIGN QUESTION

1. A company has four sites. Its headquarters is in Harrisburgh, Virginia. It has branch offices in Montreal, Canada, Tel Aviv, Israel, and Dublin, Ireland. Each branch office needs to communicate with each other branch site at 40 kbps. The headquarters needs to communicate with each of its branches at 256 kbps. You will use Frame Relay to connect them. a) What PVCs will you need from the headquarters site? b) From each branch office site? c) What port speeds will you need for the various sites, using overprovisioning? (Choices are 128 kbps, 256 kbps, 384 kbps, 512 kbps, 768 kbps, 1 Mbps, and 4 Mbps.) d) How many private lines will you need? e) How fast do the private lines have to be for various sites? (Choices are 128 kbps, 256 kbps, 384 kbps 512 kbps, 768 kbps, 1 Mbps, and 4 Mbps. Hint: Draw a picture of the sites and their PVCs.)

PROJECTS

1. **Getting Current.** Go to the book website's New Information and Errors pages for this chapter to get new information since this book went to press and to correct any errors in the text.

7a

CASE STUDY: FIRST BANK OF PARADISE'S WIDE AREA NETWORKS

INTRODUCTION

The First Bank of Paradise (FBP) is a mid-sized bank that operates primarily within the state of Hawai`i, although it has one affiliate office on Da Kine Island in the South Pacific.

FBP is a mid-sized bank, but it is not a small company. The bank has annual revenues of $4 billion. It has 50 branches and 350 ATMs. It has more than 500 switches, 400 routers, 2,000 desktop and notebook PCs, 200 Windows servers, 30 Unix servers, and 10 obsolete Novell NetWare file servers. Its information systems staff has 150 employees.

FIRST
BANK
OF
PARADISE

ORGANIZATIONAL UNITS

Major Facilities

Figure 7a-1 shows that FBP has three major facilities, all located on the island of Oah`u.

➤ **Headquarters** is a downtown office building that houses the administrative staff.

➤ **Operations** is a building in an industrial area that houses the bank's mainframe operations and other back-office technical functions. It also has most of the bank's IT staff, including its networking staff.

➤ **North Shore** is a backup facility. If Operations fails, North Shore can take over within minutes. North Shore is located in an otherwise agricultural area.

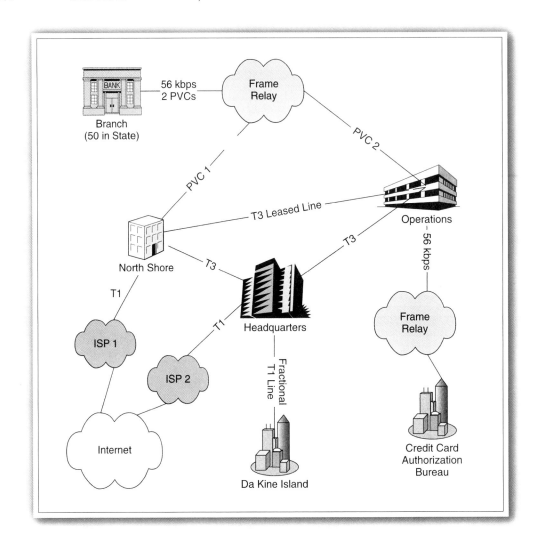

Figure 7a-1   First Bank of Paradise Wide Area Networks

Branches

Although branches are small buildings, they are technologically complex, primarily because the devices there use diverse network protocols. The automated teller machine at a branch uses SNA protocols to talk with the mainframe computer at Operations. The teller terminals use different SNA protocols to talk to the Operations mainframe. File servers require IPX/SPX communication, and branch offices that need Internet access require TCP/IP.

At each branch, there is a Cisco 2600 router to connect the branch to Operations and Backup. This is a multiprotocol router capable of handling the many protocols used at the internet and transport layers in branch office communication.

External Organizations

First Bank of Paradise has to deal with several organizations outside the company. Figure 7a-1 shows only one of these—a connection to a credit card authorization bureau. In fact, FBP deals with more than a dozen outside support vendors, each in a different way. Fortunately, the credit card authorization firm uses TCP/IP, which simplifies matters.

THE FBP WIDE AREA NETWORK (WAN)

Figure 7a-1 shows the complex group of WANs that the bank uses to hold together this geographically dispersed and technologically diverse collection of sites.

T3 Lines

A mesh of T3 lines connects major facilities, as Figure 7a-1 shows. T3 private lines operate at 44.7 Mbps, providing "fat pipes" between these facilities.

Branch Connections

Branches are connected to the major facilities in two ways. Most of the time, they communicate via a Frame Relay network. For each branch, there are two 56 kbps PVCs. One PVC leads to Operations, the other to North Shore.

Da Kine Island Affiliate Branch

For the Da Kine Island affiliate branch, the firm has a 128 kbps fractional T1 digital private line.

Credit Card Service

FBP connects to the credit card processing company using a 56 kbps Frame Relay network connection. This gives adequate speed.

Branch LANs

Branch offices have Ethernet networks. Each branch has a single 48-port 100Base-TX switch connected to the branch's Cisco 2600 border router.

Internet Access

For Internet access, FBP uses two separate ISPs, connecting to each via a T1 private line.

ANTICIPATED CHANGES

Outsourcing

The bank is anticipating two major changes. First, it plans to outsource about 60 percent of its internal operations to a bank processing company in Northridge, California.

Fractional T1 Lines to Branches

Also, in a reversal of past trends, Frame Relay vendors in Hawai`i have been raising their rates in recent years to seek higher profit margins. At the same time, the local telephone company has been dropping its rates on private lines dramatically in response to a strong long-term drop in demand for these circuits. The bank believes that it can bring a 256 kbps fractional T1 connection to each branch economically.

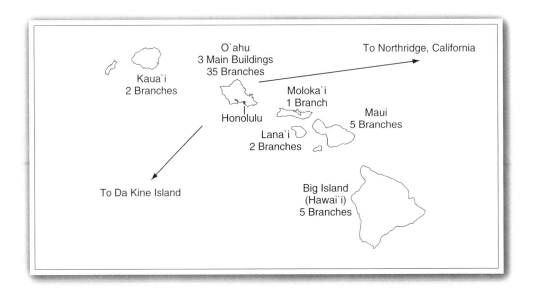

Figure 7a-2   First Bank of Paradise Locations

The bank's main buildings are located on the island of O`ahu. The bank also has 35 branch offices on O`ahu, which is the most populous island. The bank also does business on five "outer islands"—Maui, Kaua`i, Moloka`i, Lana`i, and the Big Island (the Island of Hawai`i). Maui and the Big Island have five branches each. Moloka`i has one branch.

TEST YOUR UNDERSTANDING

1. a) List all examples of redundancy in the FBP network. b) What is the goal of redundancy?
2. a) Why do you think two access points were created instead of one? b) Why are there only two access points to the Internet?
3. Do you think the bank uses the same Frame Relay network to connect its branches as it uses to connect to its credit card processing center?
4. a) Why do you think the bank uses a fractional T1 line to its Da Kine Island branch instead of a full T1 line? b) Instead of a Frame Relay connection?
5. Why do you think the bank uses T3 lines to link its major facilities instead of using ATM?
6. Why do branches need highly capable routers?

DESIGN QUESTIONS

7. What type of connection do you think the bank should have to Northridge, California?
8. Create a rough design for a private line network that would bring a 256 kbps private line to each of the bank's fifty branch offices. Be economical, but ensure that there is redundancy in interisland connections. Assume that connections within an island do not need redundancy because of the high traditional reliability of private lines.

CHAPTER 8

TCP/IP INTERNETWORKING

Learning Objectives:

By the end of this chapter, you should be able to discuss:

- Basic principles of router operation, router standards, and multiprotocol routing.
- How routers make routing decisions for incoming packets using a routing table, including network and subnet masking and the selection of the best route.
- Other important TCP/IP standards, including MPLS, routing protocols, DNS, ICMP, IPv6, TCP, and UDP.
- The differences between IP routers, Layer 3 switches, and Layer 4 switches.

INTRODUCTION

In Chapters 4, 5, and 7, we looked at single LANs and single WANs. However, most corporations have many networks that must be connected into corporate-wide internets. These corporate internets link clients and servers on different networks across the firm.

Corporate internets link clients and servers on different networks across the firm.

Then, of course, there is the global Internet that has revolutionized information exchange around the world. Many of the Internet's thousands of networks are themselves large internets.

TCP/IP RECAP

The TCP/IP Architecture and the IETF

We first looked at TCP/IP in Chapters 1 and 2. Recall from Chapter 2 that TCP/IP standards are set by the Internet Engineering Task Force (IETF). TCP/IP is an architecture for setting individual standards. Figure 8-1 shows a few of the standards the IETF has created within this architecture.

5 Application	User Applications			Supervisory Applications		
	HTTP	SMTP	Many Others	DNS	Routing Protocols	Many Others
4 Transport	TCP			UDP		
3 Internet	IP				ICMP	ARP
2 Data Link	None: Use OSI Standards					
1 Physical	None: Use OSI Standards					

Note: Shaded protocols are discussed in this chapter.

Figure 8-1   Major TCP/IP Standards

IP at the Internet Layer

Recall also from Chapter 2 that internetworking deals with two layers. The internet layer moves packets from the source host to the destination host across a series of routers. Figure 8-1 shows that the primary standard at the internet layer is the Internet Protocol (IP). Figure 8-2 shows that IP is a simple (connectionless and unreliable) standard. This complexity minimizes the work that each router has to do along the way, thereby minimizing routing costs.

Reliable Heavyweight TCP at the Transport Layer

In turn, TCP at the transport layer corrects any errors at the internet layer and lower layers as well. As we saw in Chapter 2, when the destination host transport process receives a TCP supervisory or data segment, it sends back an acknowledgement. If the source host transport process does not receive an acknowledgement for a TCP segment, it resends the segment. TCP is both connection-oriented and reliable, making it a heavyweight protocol. However, the work of implementing TCP is only done on the source and destination hosts, not on the many routers between them.

Unreliable Lightweight UDP at the Transport Layer

In Chapter 6 we saw that TCP/IP offers an alternative to heavyweight TCP at the transport layer. This is the User Datagram Protocol (UDP). Like IP, UDP is a simple (connectionless and unreliable) protocol.

TEST YOUR UNDERSTANDING

1. a) Compare TCP and IP. b) Compare TCP and UDP.

Layer	Protocol	Connection-Oriented or Connectionless?	Reliable or Unreliable?	Lightweight or Heavyweight?
4 (Transport)	TCP	Connection-oriented	Reliable	Heavyweight
4 (Transport)	UDP	Connectionless	Unreliable	Lightweight
3 (Internet)	IP	Connectionless	Unreliable	Lightweight

Figure 8-2   IP, TCP, and UDP

IP ROUTING

In this section we will look at how routers make decisions about **forwarding** packets—in other words, about deciding which port to use to send an arriving packet back out. This is known as **routing.** We will see that router forwarding decisions are much more complex than the Ethernet switching decisions we saw in Chapters 1 and 4. As a consequence of this complexity, routers do more work with arriving packets than switches do. Consequently, routers are more expensive than switches. A widely quoted network adage reflects this cost difference: "Switch where you can; route where you must."

Hierarchical IP Addressing

To understand the routing of IP packets, it is necessary to understand IP addresses. In Chapter 1 we saw that IP addresses are 32 bits long. However, IP addresses are not simple 32-bit strings.

Hierarchical Addressing

As Figure 8-3 shows, IP addresses are **hierarchical.** They usually consist of three parts that locate a host in progressively smaller parts of the Internet. These are the network, subnet, and host parts. As we will see, this hierarchical addressing simplifies routing tables.

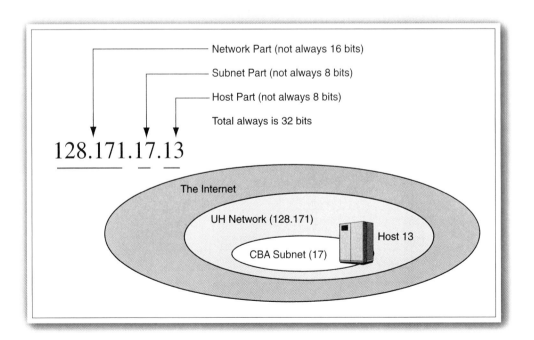

Figure 8-3   Hierarchical IP Address

Network Part

An IP address's **network part** identifies the host's network on the Internet. **Internet networks** are owned by single organizations, such as corporations, universities, and ISPs. In the IP address shown in Figure 8-3, the network part is 128.171. It is 16 bits long. All host IP addresses within this network begin with 128.171. This is the network part for the University of Hawai`i network on the Internet.

Note that "network" in this context does not mean a single network—a single LAN or WAN. The University of Hawai`i network itself consists of many single networks and routers at multiple locations around the state. In IP addressing, **"network"** is an organizational concept—a group of hosts, single networks, and routers owned by a single organization.

Subnet Part

Most large organizations further divide their networks into smaller units called **subnets.** Following the network part bits in an IP address come the **subnet part** bits. The subnet part bits specify a particular subnet within the network.

For instance, Figure 8-3 shows that in the IP address 128.171.17.13, the first 16 bits (128.171) correspond to the network part, and the next eight bits (17) correspond to a subnet on this network. Subnet 17 is the College of Business Administration subnet within the University of Hawai`i network. All host IP addresses within this subnet begin with 128.171.17.

Host Part

The remaining bits in the 32-bit IP address identify a particular host on the subnet. In Figure 8-3, the **host part** is 13. This corresponds to a particular host, 128.171.17.13, on the College of Business Administration subnet.

Variable Part Lengths

In Figure 8-3, the network part was 16 bits long, the subnet part was 8 bits long, and the host part was 8 bits long. However, this is only an example. In general, network parts, subnet parts, and host parts vary in length. For instance, if you see the IP address 60.47.7.23, you may have an 8-bit network part of 60, an 8-bit subnet part of 47, and a 16-bit host part of 7.23. In fact, parts may not even break conveniently at 8-bit boundaries. The only thing you know for certain when looking at an IP address is that it is 32 bits long.

Routers, Networks, and Subnets

Border Routers Connect Different Networks   As Figure 8-4 illustrates, networks and subnets are very important in router operation. Here we see a simple site internet. The figure shows that a **border router** connects different networks. This border router connects the 192.168.x.x network within the firm and the 60.x.x.x network of its Internet service provider.

A border router connects different networks.

Internal Routers Connect Different Subnets   The site network also has an **internal router.** An internal router connects different subnets within a firm—in this case, the

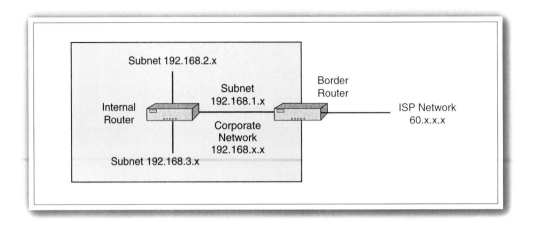

Figure 8-4   Border Router, Internal Router, Networks, and Subnets

192.168.1.x, 192.168.2.x, and 192.168.3.x subnets. Many sites have multiple internal routers to link the site's subnets.

An internal router connects different subnets within a firm.

TEST YOUR UNDERSTANDING

2. a) What is routing? b) What are the three parts of an IP address? c) How long is each part? d) What is the total length of an IP address? e) Distinguish between what border routers and internal routers connect.

A Small Internet

Figure 8-4 does not show much detail. Figure 8-5 looks in more detail at part of a larger site internet. Its focus is internal Router B, which has four ports. Router ports are called **interfaces.**

Routers Connect Different Subnets

Router B is an internal router, so each of its interfaces connects to a different subnet. These are subnets 172.30.19.x, 172.30.20.x, 172.30.21.x, and 172.30.22.x.

In the case of an internal router, each interface connects to a different subnet.

Subnets Can Have Different Technologies

Note that subnets can have different technologies. Two subnets in the figure are Ethernet networks. The other two are point-to-point 802.11 connections using dish antennas.[1]

---
[1] Although we did not discuss 802.11 point-to-point transmission in Chapter 5, it is a legal use of the 802.11 standard.

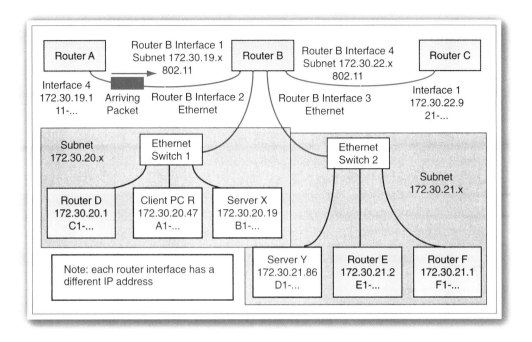

Figure 8-5   Part of an Internet

Different Interfaces on a Router Have Different IP Addresses and Different Data Link Layer Addresses

Each interface on a router has a different data link layer address *and* a different IP address. Each interface connects to a different subnet, and different subnets have different subnet parts. Consequently, although the interfaces on a router may have the same network part, they *must* have different subnet parts (as well as host parts that are assignable on their specific subnets).

> Each interface on a router has a different IP address and a different data link layer address.

Interface 1 on Router B

Interface 1 on Router B connects to Router A with a point-to-point 802.11 connection. This radio link connects to Interface 4 on Router A. This interface on Router A has the MAC address[2] 11-. . . and the IP address 172.30.19.1.

To send a packet to Router A, Router B puts the packet into a frame addressed to Router A's Interface 4 MAC address (11-. . .) and sends this frame out Interface 1.

---

[2] Yes, 802.11 has MAC addresses, just like 802.3 networks.

Interface 4 on Router B

Interface 4 on Router B has a similar connection to Interface 1 on Router C. That interface has MAC address 21-. . . and IP address 172.30.22.9.

To send a packet to Router C, Router B puts the packet into a frame addressed to Router C's MAC address (21-. . .) and sends this frame out Interface 4.

Ethernet Subnets

Finally, the Interfaces 2 and 3 on Router B connect to subnets 172.30.20.x and 172.30.21.x, respectively. Both of these subnets are Ethernet networks.

Interface 2 on Router B

To send a packet to Router D, Router B must place the packet in a frame addressed to Router D's MAC address on that subnet (C1-. . .) and send the frame out Interface 2. The frame will go to Ethernet Switch 1, which will pass the frame on to Router D.

Similarly, to send a packet to Server X, Router B must place the packet in a frame addressed to the server's MAC address. (B1-. . .) and send the frame out Interface 2. The frame will go to Ethernet Switch 1, which will pass the frame on to Server X.

Interface 3 on Router B

Similarly, to send a packet to Router E, Router B must place the packet in a frame addressed to Router E's MAC address on that subnet (E1-. . .) and send the frame out Interface 3. The frame will go to Ethernet Switch 2, which will pass the frame on to Router E.

Packets going to Router F also go in frames sent out Interface 3, but they are addressed to the MAC address of Router F on that subnet (F1-. . .). This frame will go to Ethernet Switch 2, which will pass it on to Router F.

TEST YOUR UNDERSTANDING

3. In Figure 8-5, a) What kind of Router is Router B? b) To how many subnets does it connect? c) Do the subnets have different technologies? d) How many IP addresses must Router B have? e) How can Router B send a packet to Server Y, which is the arriving IP packet's destination host? f) How can Router B send a packet to next-hop router Router C?

Router Forwarding

Figure 8-5 shows the basic function of routers—forwarding each arriving IP packet to another router or to the packet's destination host. The figure shows a packet arriving in Interface 1 of Router B. The router will send the packet back out Interface 2, 3, or 4.

Forwarding to a Destination Host

Suppose that the packet's destination IP address is 172.30.20.47. This is Client PC R. Using its routing table, which is discussed later in the chapter, Router B will realize that the destination host is on subnet 172.30.20.x. It will place the packet in a frame addressed to Client PC R's MAC layer address (A1-. . .).

Forwarding to a Next-Hop Router

In most cases, however, the IP destination address of an arriving packet will not be on one of Router B's subnets. In this case, the router will send the packet on to another router and pass responsibility for the packet to this **next-hop router.**

For example, if Router B receives a packet to the IP destination address 60.217.87.320, the router will realize that the destination IP address is not on one of the subnets attached to the router. It will pass the packet along to either Router C, Router D, Router E, or Router F depending on its routing table (see box below).

TEST YOUR UNDERSTANDING

4.   To what two types of devices do routers forward packets?

Multiprotocol Routing

We have been focusing on IP routing. In the real world, of course, most routers must be **multiprotocol routers** that can handle not only TCP/IP internetworking protocols, but also internetworking protocols from IPX/SPX, SNA, and other standards architectures, as Figure 8-6 shows. In the next section we will look only at IP routing. Multiply the complexity of the next section by a factor of five or ten to understand the complexity of real-world routing (and another reason routers are so much more expensive than switches).

TEST YOUR UNDERSTANDING

5.   a) What are multiprotocol routers? b) Why are they more complex (and, therefore, more expensive) than IP-only routers?

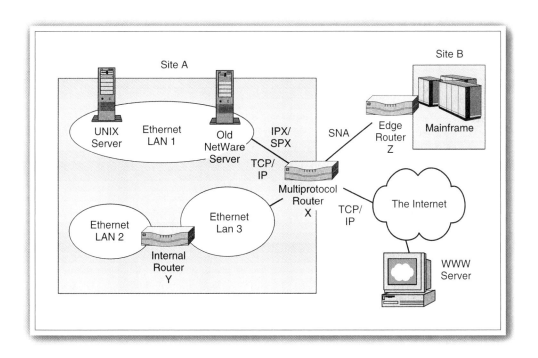

Figure 8-6   Multiprotocol Routing

BOX: IP ROUTING TABLES

We have been talking about router operation in general. In this section we will look in detail at how router forwarding is determined.

SWITCHING VERSUS ROUTING TABLES

Ethernet Hierarchies and Simple Switching Tables

In Chapters 1 and 4, we saw that Ethernet switching is very simple. Ethernet's hierarchical architecture permits only a single possible path between any two endpoints. This makes Ethernet switching tables very simple. As Figure 8-7 shows, the single possible path is represented by a single row in the Ethernet switching table. This can be found quickly, so an Ethernet switch does little work per frame. This makes Ethernet switching inexpensive.

Meshes and IP Routing Tables

In contrast, routers are organized in meshes. This gives more reliability because it allows many possible alternative routes between endpoints. Figure 8-7 shows that in a routing table, each alternative route to a network or subnet is represented by a different row. Consequently, to process a packet, a router must first find all rows representing alternative routes. It must then pick the best alternative route from this list. This requires quite a bit of work per packet, making routing more expensive than switching.

TEST YOUR UNDERSTANDING [BASED ON THE BOX "IP ROUTING TABLES"]

6. Why are routing tables more complex than Ethernet switching tables?

ROUTING TABLES

Routers base packet forwarding decisions on information in their routing tables. Figure 8-8 shows a typical routing table. It is much more complex than the Ethernet switching table we saw in Chapters 1 and 4.

Rows

As just noted, each row in the routing table represents one alternative route. However, this is not an alternative route to an *individual host*. Rather it is a route *to the network or subnet containing the host.*

Having only one row for all packets going to a particular network or subnet simplifies routing tables. Even with these aggregate rows, however, core routers in the Internet have 200,000 to 400,000 rows in their routing tables. If core routers needed even one row for each IP address on the Internet, each router forwarding decision would take days instead of microseconds.

Destination Network or Subnet Column

The **destination network or subnet** column shows the destination network or subnet for the route described in that row. More specifically, it shows the destination network's network part or the destination subnet's network plus subnet parts, followed by zeros. For instance, in the first row, 128.171.0.0 indicates the route to the University of Hawai`i network (128.171.x.x). In turn, Row 7, 128.171.17.0, indicates the route to the College of Business Administration subnet (128.171.17.x).

Mask

Network and subnet masks The next column is the **mask** column. For routes to networks, this is a network mask. It consists of ones in the network part and zeros in the subnet part. For instance, a network mask for the University of Hawai`i network (128.171.x.x) would have 16 ones followed by 16 zeros. In turn, a subnet mask to the College of Business Administration subnet (128.171.17.x) of the University of Hawai`i would have 24 ones followed by 8 zeros.

BOX: IP ROUTING TABLES *(continued)*

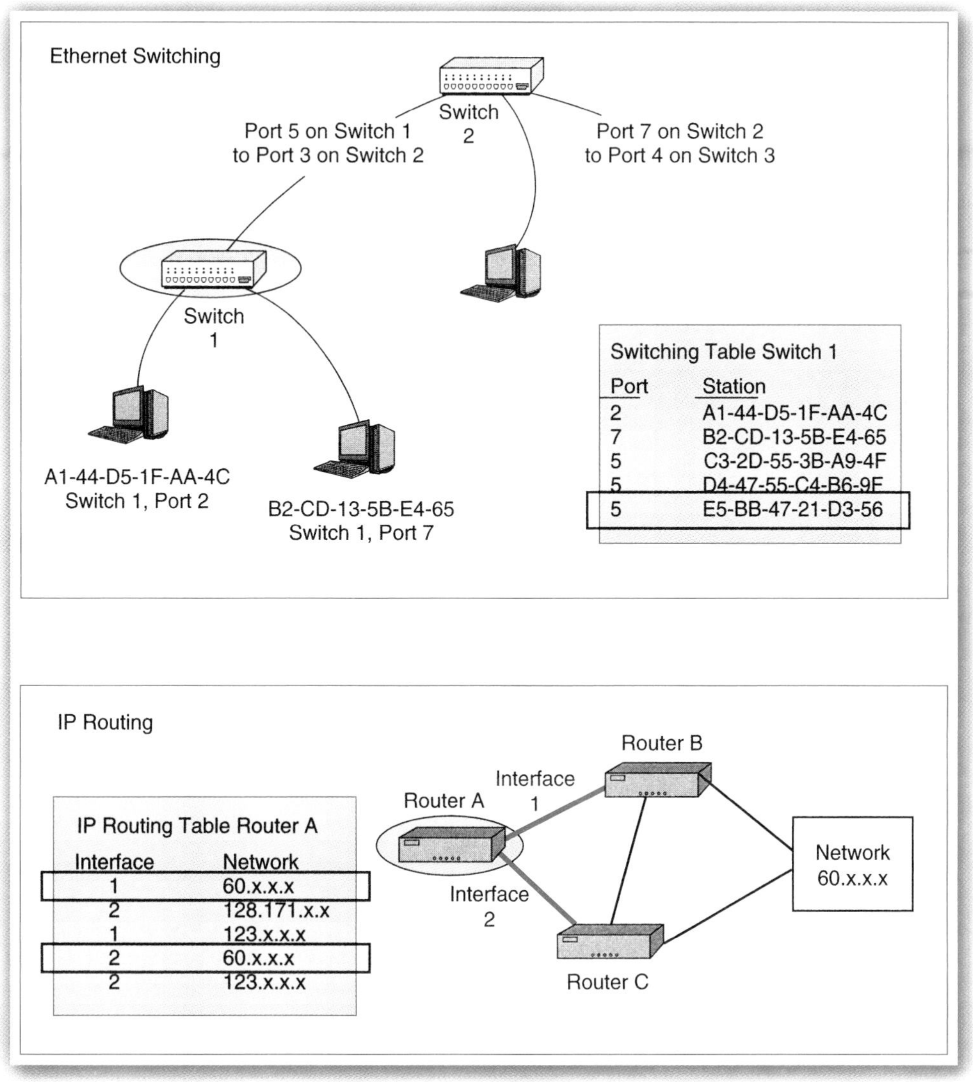

Ethernet Switching

Switch 2

Port 5 on Switch 1
to Port 3 on Switch 2

Port 7 on Switch 2
to Port 4 on Switch 3

Switch 1

A1-44-D5-1F-AA-4C
Switch 1, Port 2

B2-CD-13-5B-E4-65
Switch 1, Port 7

Switching Table Switch 1

Port	Station
2	A1-44-D5-1F-AA-4C
7	B2-CD-13-5B-E4-65
5	C3-2D-55-3B-A9-4F
5	D4-47-55-C4-B6-9F
5	E5-BB-47-21-D3-56

IP Routing

Router B

Interface 1

Router A

IP Routing Table Router A

Interface	Network
1	60.x.x.x
2	128.171.x.x
1	123.x.x.x
2	60.x.x.x
2	123.x.x.x

Interface 2

Router C

Network
60.x.x.x

Figure 8-7   Ethernet Switching Versus IP Routing

BOX: IP ROUTING TABLES (continued)

Row	Destination Network or Subnet	Mask (/Prefix)	Metric (Cost)	Interface	Next-Hop Router
1	128.171.0.0	255.255.0.0 (/16)	47	2	G
2	172.30.33.0	255.255.255.0 (/24)	0	1	Local
3	60.168.6.0	255.255.255.0 (/24)	12	2	G
4	123.0.0.0	255.0.0.0 (/8)	33	2	G
5	172.29.8.0	255.255.255.0 (/24)	34	1	F
6	172.40.6.0	255.255.255.0 (/24)	47	3	H
7	128.171.17.0	255.255.255.0 (/24)	55	3	H
8	172.29.8.0	255.255.255.0 (/24)	20	3	H
9	172.12.6.0	255.255.255.0 (/24)	23	1	F
10	172.30.47.0	255.255.255.0 (/24)	9	2	Local
11	172.30.12.0	255.255.255.0 (/24)	3	3	Local
12	123.241.0.0	255.255.0.0 (/16)	16	2	G
13	0.0.0.0	0.0.0.0 (/0)	5	3	H

Figure 8-8   Routing Table

Basic rule How are masks used? Figure 8-9 shows the basic rule of masking:

➤ When a bit (1 or 0) is masked with a one, the result contains the original bit (1 or 0).

➤ When a bit is masked with a zero, the result is always zero.

For instance, if you have 10101010 and mask it with 11110000, the result is 10100000.

Masks in dotted decimal notation When masks are given in dotted decimal notation, it is important to note that 8 zeros have the decimal value 0, while 8 ones have the decimal value 255. So a mask with 16 ones and 16 zeros is 255.255.0.0 in dotted decimal notation. A mask with 8 ones followed by 24 zeros would be 255.0.0.0 in dotted decimal notation.

Masks in prefix notation Masks are also expressed in prefix notation. The prefix is the number of initial ones in the mask. So the network mask 255.0.0.0 is /8 in prefix notation. For 255.255.255.0, in turn, the prefix is /24.

Applying Masks

As Figure 8-9 shows, applying a mask is very simple. If you mask an IP address, you get back the

BOX: IP ROUTING TABLES *(continued)*

1. Basic Rule					2. Octets	
Information Bit	1	0	1	0	Binary	Decimal
Mask Bit	1	1	0	0	00000000	0
Result	1	0	0	0	11111111	255

3. Example 1					4. Example 2				
IP Address	172.	30.	22.	0	IP Address	172.	30.	22.	0
Mask	255.	0.	0.	0	Mask	255.	255.	0.	0
Result	172.	0.	0.	0	Result	172.	30.	0.	0

Where mask bits are one, the result gives the original IP address bits
Where mask bits are one, the result contains zeros

Figure 8-9   Masking

IP address bits where there are ones in a mask. However, you get zeros where the mask bits are zeros. This is like spray painting through a stencil. You get the color where the stencil is cut out (ones) and nothing where it is solid (zeros).

In the figure the IP address in the first example is 172.30.22.0. If you mask it with 255.0.0.0, you get 172.0.0.0. In the second example, if you mask it with 255.255.0.0, you get 172.30.0.0.

In these examples network parts, subnet parts, and masks have all broken neatly at 8-bit boundaries. However, this is not always the case. For instance, the network part might be 22 bits long, and the subnet part might be 4 bits long, leaving 6 bits for the host part. In this chapter, we will avoid network and subnet parts that do not break at octet boundaries because they add computational complexity without changing basic concepts. Computers find all masking equally easily.

TEST YOUR UNDERSTANDING [BASED ON THE BOX "IP ROUTING TABLES"]

7. a) In a routing table, what does a row represent? b) What does the destination column specify? c) A subnet mask has 24 ones followed by 8 zeros. Write this mask in dotted decimal notation. d) Write the mask in prefix notation. e) An arriving packet has destination IP address 123.87.54.226. Apply the mask 255.255.255.0 to it. What is the result? f) What is the result if you apply the mask /8?

ROUTER OPERATION

Now that we have looked at network/subnet parts and masking, we can begin to look in detail at how routers use routing tables to make routing decisions on individual packets. We will see that they do this by applying three processes to each arriving packet (see Figure 8-10 for more detail).

BOX: IP ROUTING TABLES *(continued)*

For Each Packet
    For every row in the routing table, test for a match
        Take destination IP address in packet
            Mask it with the Mask value in that row
        Take the result
            Compare it with the destination network/subnet value in that row
            If it matches
                Add the row to the list of matching rows for that packet
                Otherwise, ignore the row
        Go on to the next row, all the way to the last row
    After checking for a match in all rows, find the best-match row
        If only one match, select that row as best match
        If several rows match, select the row with the longest match
        If tied longest length of match, select row with best metric
            May be the smallest value (say if metric is cost)
            May be the largest value (say if metric is speed)
    Finally, send the packet out according to directions in the best-match row
        Send it out interface listed in best row to network or subnet out that port
        On that network or subnet, send packet to the
            Next-hop router value in the best row
            Destination host if next-hop router value in best row says "local"

Figure 8-10 Routing Algorithm

➤ First, the destination address in the IP packet is matched against all rows in the routing table. Matching rows are noted.

➤ Second, the best-match row is selected.

➤ Third, the packet is sent out the best-match row's indicated interface to the indicated next-hop router (or destination host) on the subnet connected to that interface.

TEST YOUR UNDERSTANDING [BASED ON THE BOX "IP ROUTING TABLES"]

8. What are the three steps in routing an arriving packet?

FINDING ROW MATCHES

Row Matches

The first step in the algorithm is to find all matching rows. To determine if a row matches, take the arriving IP packet's destination network or subnet address and mask it with the mask value in the row. Compare the result to the destination network or subnet column's value. If the two are the same, the row is a match.

Note that the routing algorithm *does not stop when a match is found*. The router will compare the destination IP address against *every row*

BOX: IP ROUTING TABLES (continued)

in the routing table regardless of how many matches it finds. Many Internet core routers within ISPs have 200,000 to 400,000 rows, so the requirement that every single row must be tested adds enormously to cost.

For example, suppose that a packet arrives for destination host 123.5.6.3. First, the router will compare this IP address with Row 1 in Figure 8-8, where the mask is 255.255.0.0. The result of masking the destination IP address is 123.5.0.0. This does not match the destination value of 128.171.0.0. The row is not a match. Row 2 and Row 3 will not match either when the per-row calculations in Figure 8-8 are executed.

Row	Destination Network or Subnet	Mask	Metric (Cost)	Interface	Next-Hop Router
1	128.171.0.0	255.255.0.0 (/16)	47	2	G
2	172.30.33.0	255.255.255.0 (/24)	0	1	Local
3	60.168.6.0	255.255.255.0 (/24)	12	2	G
4	123.0.0.0	255.0.0.0 (/8)	33	2	G

Now the router comes to Row 4, which has the mask 255.0.0.0. When the router applies this mask to the destination IP address, 123.5.6.3, the result is 123.0.0.0. This matches the row's destination value, 123.0.0.0. This row is a match. The router adds it to its list of matches (only one so far) and goes on to the next row, Row 5.

FINDING THE BEST MATCH

After considering all rows, a router needs to select the **best-match row** among the row matches it found.

Single Match

If there is only a single match, the router's problem is solved. The router selects that match. For instance, Row 13 in Figure 8-8 has a mask of 0.0.0.0. This is guaranteed to match *any* IP address because anything masked by 0.0.0.0 will give the result 0.0.0.0, which is the value in the destination column. This is called the **default row** because it will be selected automatically if nothing else matches. The router named in the next-hop router column is called the **default router.** In this routing table, the default router is H. If a router does not have any other matching row, it will send the packet to the default router.

Row	Destination Network or Subnet	Mask	Metric (Cost)	Interface	Next-Hop Router
13	0.0.0.0	0.0.0.0 (/0)	5	3	H

BOX: IP ROUTING TABLES (continued)

Longest Match

When there are multiple matches, the router will choose the row with the **longest match.** For instance, a packet with IP destination address 128.171.17.13 will match both Row 1 (128.171.0.0/16) and Row 7 (128.171.17.0/24). In this example, the router will choose Row 7 because it has the longest match (24 bits instead of 16 bits).

Row	Destination Network or Subnet	Mask	Metric (Cost)	Interface	Next-Hop Router
1	128.171.0.0	255.255.0.0 (/16)	47	2	G
7	128.171.17.0	255.255.255.0 (/24)	55	3	H

The longest match is chosen because it usually gets a packet closest to its destination. For instance, Row 1 only gets a packet to the 128.171 network. Row 7's route will get the packet farther, to the 128.171.17 subnet.

Metric

What if there are two rows with matches of equal length? Rows 5 and 8 have the same destination part and the same mask, so any packet that matches one will match both.

Row	Destination Network or Subnet	Mask	Metric (Cost)	Interface	Next-Hop Router
5	172.29.8.0	255.255.255.0 (/24)	34	1	F
8	172.29.8.0	255.255.255.0 (/24)	20	3	H

If equal-length matches occur, then the router needs additional information to distinguish between the alternative routes they represent. The router must look at the **metric** column in each row. This column gives a number describing the desirability of the route represented by the row.

Sometimes select the minimum metric value In Figure 8-8, the metric column gives the cost of the route to the network or subnet listed in the destination column. Typically, this would not be a monetary cost, but rather would be the number of router hops between the router making the decision and the destination network or subnet. Fewer hops are better, so cost measured as hops should be minimized.

For instance, the metric in Row 5 is 34, whereas the metric in Row 8 is 20. The router would choose Row 8, which has a lower cost.

Sometimes select the maximum metric value Cost is not the only possible metric, however. For example, suppose the metric in Figure 8-8 represented the transmission speed along a route. In this case,

BOX: IP ROUTING TABLES *(continued)*

more transmission speed would be better. The router would choose Row 5 because of its *larger* speed metric.[3]

AFTER THE BEST-MATCH ROW IS SELECTED

After the best-match row is selected, the router looks at the remaining columns in the best-match row to determine what to do with the packet.

Interface

Each outgoing port is an interface. It represents a particular network or subnet attached to the router. If Row 1 is selected, the router will send the IP packet out Interface 2. If Row 2 is selected, the router will send the packet out Interface 1.

Row	Destination Network or Subnet	Mask	Metric (Cost)	Interface	Next-Hop Router
1	128.171.0.0	255.255.0.0 (/16)	47	2	G
2	172.30.33.0	255.255.255.0 (/24)	0	1	Local

Next-Hop Router

In an Ethernet switching table, once a row is selected, the switch sends the frame out the port number contained in the row. Figure 8-11 shows that the situation in routing is a bit more complex. Although the best-match row will indicate the interface (port) out which the packet should be sent, another step is needed—deciding which host or next-hop router on the subnet out the indicated interface should receive the packet. In the figure, there are three possible destinations for packets on the subnet out the indicated interface. One is a host. The other two are routers. This is why the best-match row must specify both an interface and a host or next-hop router. In Figure 8-8, if Row 1 is selected, the router will send the packet out Interface 2 to next-hop router, Router G.

Row	Destination Network or Subnet	Mask	Metric (Cost)	Interface	Next-Hop Router
1	128.171.0.0	255.255.0.0 (/16)	47	2	G

---

[3] If two rows have the same network/subnet, mask, and metric, the IP standard does not specify what a router should do if there is a tie. Some routers randomly select one row. Others send half of all traffic over one route and the other half over the other route.

BOX: IP ROUTING TABLES (continued)

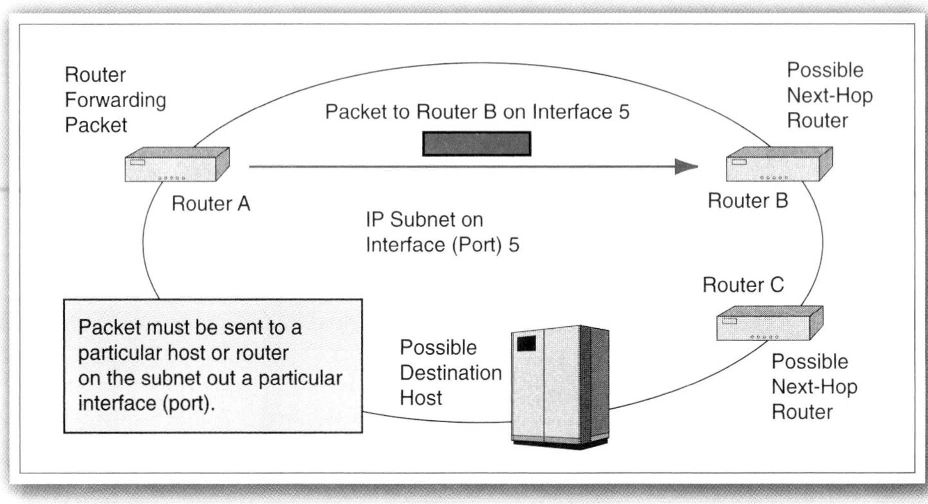

Figure 8-11   Interface and Next-Hop Router

Local Delivery

What if the destination host is on the selected network or subnet? Then instead of putting a next-hop router's IP address in the next-hop router field, the router displays the value **"Local"** (or something else, depending on the router) to indicate that no next-hop router is needed. The router will send the packet to the destination host on the network or subnet out the selected interface. If Row 2 is selected, the router will send the packet out Interface 1 to the destination host on that network or subnet.

Row	Destination Network or Subnet	Mask	Metric (Cost)	Interface	Next-Hop Router
2	172.30.33.0	255.255.255.0 (/24)	0	1	Local

TEST YOUR UNDERSTANDING [BASED ON THE BOX "IP ROUTING TABLES"]

9. For the router whose routing table is shown in Figure 8-8, a packet arrives for the destination address 172.35.6.17. a) Which rows match? b) What will the router do with the packet?

10. For the router whose routing table is shown in Figure 8-8, a packet arrives for the destination address 172.40.6.3. a) Which *two*

BOX: IP ROUTING TABLES (continued)

rows match? b) Which is the best match? Why? c) What will the router do with the packet?

11. For the router whose routing table is shown in Figure 8-8, a packet arrives for the destination address 123.241.3.5. a) Which three rows match? b) Which is the best match? Why? c) What will the router do with the packet?

12. For the router whose routing table is shown in Figure 8-8, a packet arrives for the destination address 172.29.8.34. The metric is cost. a) Which rows match? b) Which is the best match? Why? c) What will the router do with the packet? d) Answer parts a, b, and c again supposing that the metric is speed instead of cost.

13. a) Will the default row automatically be a match for every packet? Explain. b) Why will the default row not be selected as the best match if there is any other match?

14. a) In routing tables, what does "next-hop router" mean? b) What does it mean if the next-hop router column says "local"?

RECAP ON ROUTING DECISIONS

It is easy to get lost in the steps taken by routers when they compare the destination address of an incoming packet to the rows in their routing table. We summarize the key points in Figure 8-10.

For Each Arriving Packet

Note that this process is *repeated for each and every arriving IP packet*. Even if a thousand successive IP packets are going to the same destination, they will be handled independently.

Evaluating Every Row

Also keep in mind that the destination IP address is compared to *each and every row in the routing table*. If there are 300,000 rows, as there are in many backbone Internet routers, there must be 300,000 comparisons. In Ethernet and ATM switching tables, there is only one possible match; it can be found quickly, and when it is found, the process stops. With multiple alternative routes, there may be many matches.

High Cost

The router needs to route each IP packet separately and to evaluate every row in the routing table for each packet. This process places heavy processing burdens on routers, making them much more expensive and slower than switches per message handled.

TEST YOUR UNDERSTANDING [BASED ON THE BOX "IP ROUTING TABLES"]

15. A thousand IP packets addressed to the same destination IP address arrive at a router. The router has 300,000 rows. Only the second row matches the IP address. (Ignore the default row match.) How many rows must the router examine to forward all the IP packets? Show your work.

STANDARDS RELATED TO IP

In this section we will briefly look at several TCP/IP standards that are related to IP.

Routing Protocols

How does a router know if a mask is a network mask or a subnet mask? Actually, it does not care. It simply uses the mask value in the row to decide what bits to look at in the destination addresses of arriving IP packets.

As Figure 8-12 shows, routers transmit routing table information to one another using **routing protocols.** This process provides destination, mask, and metric information.

Note that TCP/IP uses the term **routing** in two different but related ways. First, the process of forwarding arriving packets is called routing. Second, the process of exchanging information for building routing tables is called routing.

In TCP/IP the term routing is used in two ways—for IP packet forwarding and for the exchange of routing table information through routing protocols.

TEST YOUR UNDERSTANDING

16. a) What is the purpose of routing protocols? b) What information do they exchange? c) In what two ways does TCP/IP use the term "routing"?

Multiprotocol Label Switching (MPLS)

Problems with IP

Routers are extremely costly per packet forwarded because they have to compare the packet's destination address to every row in the long routing table and then must pick

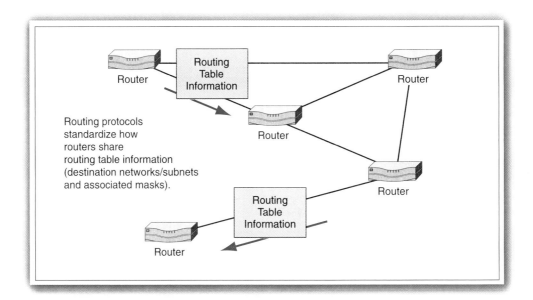

Figure 8-12   Routing Protocols

the best-match row. Another problem with traditional router forwarding is that there is no way to do traffic engineering that controls the details of transmission paths. For example, there is no way to prevent certain connections between routers from becoming overloaded or to give special priority to voice traffic or other time-sensitive traffic.

Label Header
A majority of ISPs are now using a traffic management tool called **multiprotocol label switching (MPLS)** for at least some of the traffic they carry. As Figure 8-13 shows, MPLS places a **label header** before the IP header (and after the frame header).

Label Switching for Cost Reduction
When a labeled packet arrives, a **label-switching router** does not go through the traditional routing calculation processes. Instead, it merely reads the **label number** from the label header. It then looks into the **label-switching table** to find the interface associated with the label number. It sends the packet out that interface. This is extremely fast compared to traditional routing calculations and so dramatically lowers the cost of routers. Lowering router costs is the main attraction of MPLS, but there are other advantages as well.

One Label Number, Many Packets
All packets between two host IP addresses could be assigned the same label number. This would be like creating a virtual circuit at the Internet layer. Obviously, this would

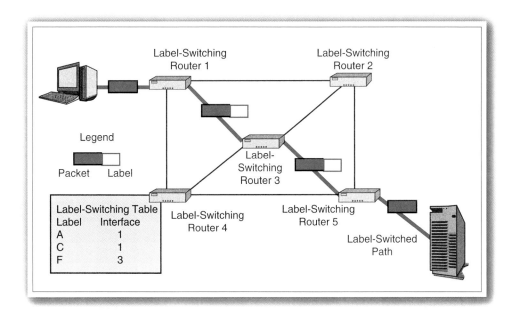

Figure 8-13   Multiprotocol Label Switching (MPLS)

reduce costs dramatically compared to looking at each row in the routing table and then selecting the best match for each packet. In fact, all traffic between two sites might be assigned the same label number because it would all be going to and from the same site.

Quality of Service

Another possibility is that traffic between two sites might be assigned one of two different label numbers. One label number might be for IP telephony, which is latency-sensitive, and the other might be for latency-tolerant traffic between the sites. For latency-sensitive traffic, it is even possible to reserve capacity at routers along the selected path.

Traffic Load Balancing

MPLS can also be used to balance traffic, for instance, to move some traffic from a heavily congested link between two routers to an alternative route that uses different and less-congested links.

Limitations

Although MPLS is very widely used and shows great promise, it is still a work in progress. Protocols for setting up label-switching routes, reserving capacity for specifying quality of service, and carrying out other managerial actions are still evolving.

TEST YOUR UNDERSTANDING

17. a) How does MPLS work? b) What is its main attraction? c) What are its other attractions? d) How are label-switching routes like virtual circuits? e) How is MPLS still limited?

Domain Name System (DNS)

As we saw in Chapter 1, if a user types in a target host's host name, the user's PC will contact its local Domain Name System (DNS) server. The DNS server will return the IP address for the target host or will contact other DNS servers to get this information. The user's PC can then send IP packets to the target host.

What Is a Domain?

Figure 8-14 shows that the **Domain Name System (DNS)** and its servers are not limited to providing information about host names. A **domain** is a group of resources (routers, single networks, and hosts) under the control of an organization. The figure shows that domains are hierarchical, with host names being at the bottom of the hierarchy.

A domain is a group of resources (routers, single networks, and hosts) under the control of an organization.

Top-Level Domains

At the top of the hierarchy is the **root,** which consists of all domain names. Under this are **top-level domains** that categorize the domain by organization type (.com, .net, .edu, .biz, .info, etc.) or by country (.uk, .ca, .ie, .au, .jp, .ch, etc.).

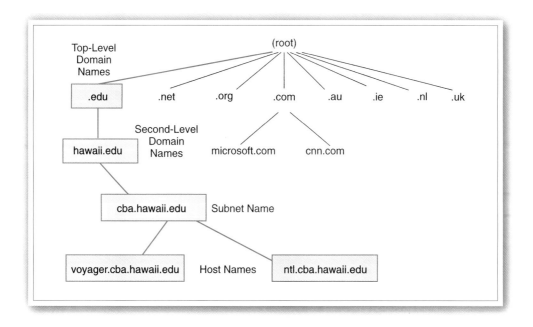

Figure 8-14   Domain Name System (DNS) Hierarchy

Second-Level Domains

Under top-level domains are **second-level domains,** which usually specify a particular organization (microsoft.com, hawaii.edu, cnn.com, etc.). Sometimes, however, specific products, such as movies, get their own second-level domain names. Competition for good second-level domain names is fierce.

To get a second-level domain name, you can go to any domain name registrar. It may pay to shop around because different registrars charge different annual fees for a domain name. The website of the Internet Corporation for Assigned Names and Numbers (ICANN), which oversees the domain name system, is http://www.icann.org. This website lists registrars. You can go to any registrar and see if a domain name you want is available.

Getting a second-level domain name is only the beginning. Each organization that receives a second-level domain name must have a DNS server to host its domain name record. Large organizations have their own internal DNS servers that contain information on all subnet and host names. Individuals and small businesses that use webhosting services depend on the webhosting company to provide this DNS service.

Further Qualifications

Domains can be further qualified. For instance, within hawaii.edu, which is the University of Hawai`i, there is cba.hawaii.edu, which is the College of Business Administration. Within cba.hawaii.edu is voyager.cba.hawaii.edu, which is a specific host within the college.

A Comprehensive Naming System

Overall, DNS gives a comprehensive system for creating unique hierarchical names on the Internet.

TEST YOUR UNDERSTANDING

18. a) Is the Domain Name System only used to send back IP addresses for given host names? b) What is a domain? c) Which level of domain name do corporations most wish to have? d) How can an organization get second-level domain names?

Internet Control Message Protocol (ICMP) for Supervisory Messages

Supervisory Messages at the Internet Layer

IP is only concerned with packet delivery. For supervisory messages at the internet layer, the Internet Engineering Task Force (IETF) created the **Internet Control Message Protocol (ICMP).** IP and ICMP work closely together. As Figure 8-15 shows, IP encapsulates ICMP messages in the IP data field, delivering them to their target host or router. There are no higher-layer headers or messages.

Error Advisement

IP is an unreliable protocol. It offers no error correction. If the router or the destination host finds an error, it discards the packet. Although there is no retransmission, the router or host that finds the error usually sends an **ICMP error message** to the source device to inform it that an error has occurred, as Figure 8-15 illustrates. It is

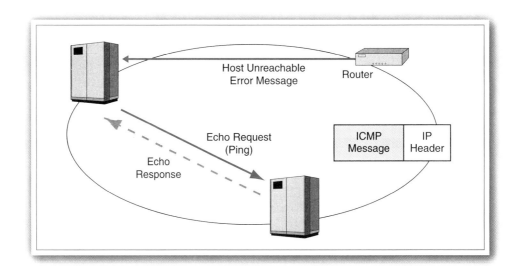

Figure 8-15   Internet Control Message Protocol (ICMP) for Supervisory Messages

then up to the device to decide what to do. This is **error advisement** rather than error correction. There is no mechanism for the retransmission of lost or damaged packets.

Echo (Ping)

Perhaps the most famous ICMP message type is the **ICMP echo** message. One host or router can send an echo request message to another. If the target device's internet process is able to do so, it will send back an echo reply message.

Sending an echo request is often called **pinging** the target host, because it is similar to a submarine pinging a ship to see if it is there. Echo is a good diagnostic tool because if there are network difficulties, a logical early step in diagnosis is to ping many hosts and routers to see if they can be reached.

TEST YOUR UNDERSTANDING

19.   a) For what general class of messages is ICMP used? b) How are ICMP messages encapsulated? c) An Ethernet frame containing an ICMP message arrives at a host. List the frame's headers, messages, and trailers at all layers. List them in the order they will be seen by the receiver. For each header or trailer, specify the standard used to create the header or message (for example, Ethernet 802.3 MAC layer header). d) Explain error advisement in ICMP. e) Explain the purpose of ICMP echo messages. f) Sending an ICMP echo message is called _____ the target host.

IPv4 Fields

Today most routers on the Internet and private internets are governed by the **IP version 4 (IPv4)** standard. (There were no versions 0 through 3.) Figure 8-16 shows the IPv4 packet. Its first four bits contain the value 0100 (binary for 4) to indicate that the packet is formatted according to IPv4. Although we have looked already at some of the fields in IPv4, here we will look at some that we have not seen yet.

Time to Live (TTL)

In the early days of the ARPANET, which was the precursor to the Internet, packets that were misaddressed would circulate endlessly among packet switches in search of their nonexistent destinations. To deal with this problem, IP added a **time to live (TTL)** field that is given a value by the source host. Different operating systems have different TTL defaults. Most insert TTL values between 64 and 128. Each router along the way decrements the TTL field by one. A router decrementing the TTL to zero will discard it, although it will try to send back an ICMP error advisement message to the source host.

Protocol

The **protocol** field tells the contents of the data field. If the protocol field value is 1, the IP packet carries an ICMP message in its data field. TCP and UDP have protocol values 6 and 17, respectively.

Identification, Flags, and Fragment Offset

If a router wishes to forward a packet to a particular network and the network's maximum packet size is too small for the packet, the router can fragment the packet into two or more smaller packets. Each fragmented packet receives the same **identification** field value that the source host put into the original IP packet's header.

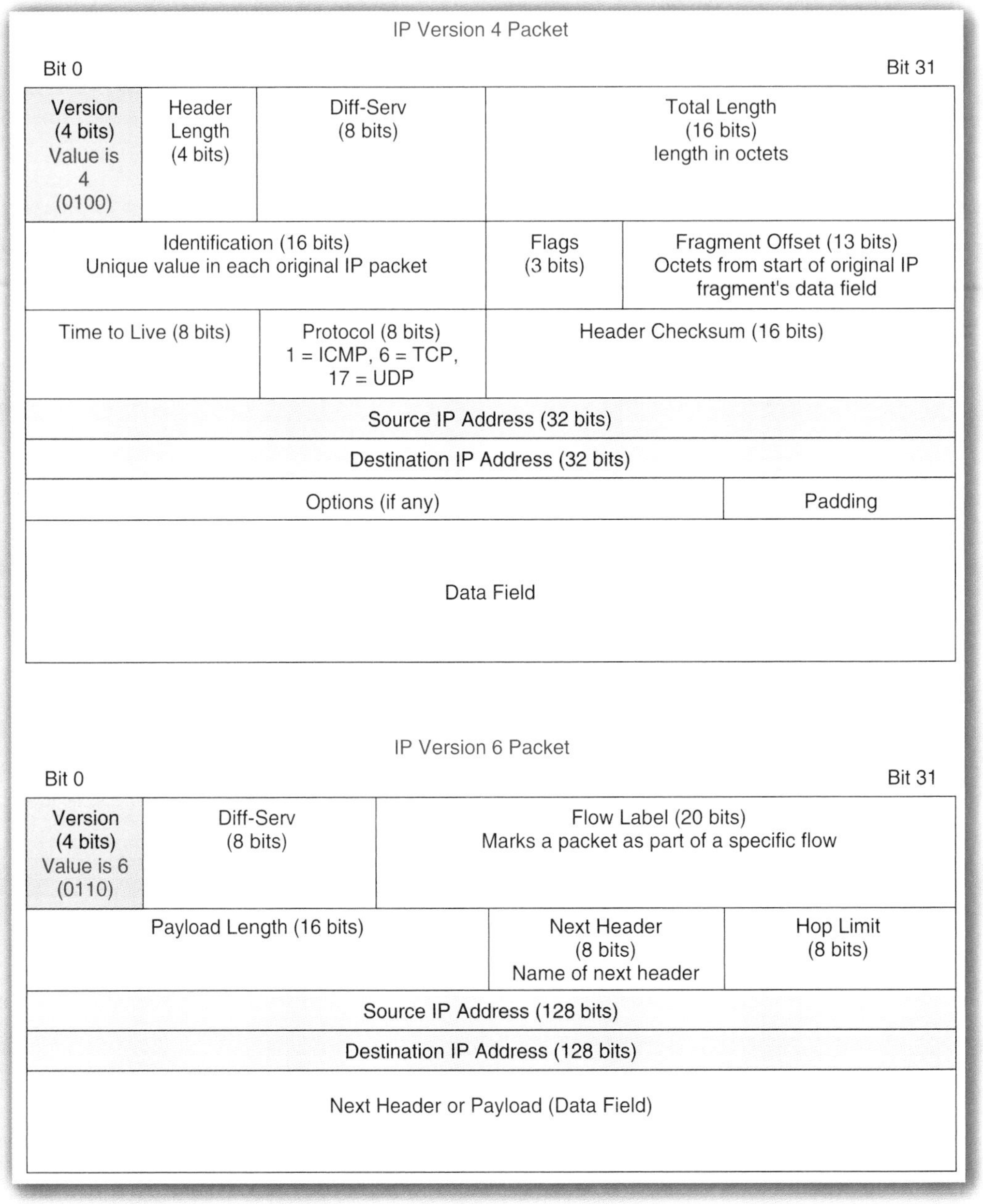

IP Version 4 Packet

Bit 0   Bit 31

Version (4 bits) Value is 4 (0100)	Header Length (4 bits)	Diff-Serv (8 bits)	Total Length (16 bits) length in octets
Identification (16 bits) Unique value in each original IP packet		Flags (3 bits)	Fragment Offset (13 bits) Octets from start of original IP fragment's data field
Time to Live (8 bits)	Protocol (8 bits) 1 = ICMP, 6 = TCP, 17 = UDP	Header Checksum (16 bits)	
Source IP Address (32 bits)			
Destination IP Address (32 bits)			
Options (if any)			Padding
Data Field			

IP Version 6 Packet

Bit 0   Bit 31

Version (4 bits) Value is 6 (0110)	Diff-Serv (8 bits)	Flow Label (20 bits) Marks a packet as part of a specific flow
Payload Length (16 bits)	Next Header (8 bits) Name of next header	Hop Limit (8 bits)
Source IP Address (128 bits)		
Destination IP Address (128 bits)		
Next Header or Payload (Data Field)		

Figure 8-16   IPv4 and IPv6 Packets

The destination host's internet process reassembles the fragmented packet. It places all packets with the same identification field value together. It then places them in order of increasing **fragment offset** size. The more fragments bit is set in all but the last fragment.

Fragmentation is uncommon in IP today and is suspicious when it occurs because it is rarely used legitimately and often is used by attackers.

Options
Similarly, **options** are uncommon in IP today and tend to signal an attack, although IPsec security (discussed in Chapter 7) uses options. If an option does not end at a 32-bit boundary, **padding** is added up to the 32-bit boundary.

Diff-Serv
The **Diff-Serv** field can be used to label IP packets for priority and other service parameters. MPLS is likely to be used instead of this field.

TEST YOUR UNDERSTANDING

20. a) What is the main version of the Internet Protocol in use today? b) What does a router do if it decrements a TTL value to zero? c) What does the protocol value tell the destination host? d) Under what circumstances would the identification, flags, and fragment offset fields be used in IP? e) Why is fragmentation suspicious? f) Why are options suspicious? g) What is the purpose of the Diff-Serv field?

IPv6 Fields

The IETF has standardized a new version of the Internet Protocol, **IP version 6 (IPv6).** As Figure 8-16 shows, IPv6 also begins with a version field. Its value is 0110 (binary for 6). This tells the router that the rest of the packet is formatted according to IPv6.

Address Field
The most important change in IPv6 is an increase in the size of IP address fields from 32 bits to 128 bits. The number of possible IP addresses is 2 raised to a power that is the size of the IP address field. For IPv4, this is 2^{32}. For IPv6, this is 2^{128}. IPv6 will support the enormous increase in demand for IP addresses that we can expect from mobile devices and from the likely evolution of even simple home appliances into addressable IP hosts.

Slow Adoption
IPv6 has been adopted only in a few geographic regions because its main advantage, permitting far more IP addresses, is not too important yet. However, IPv6 is beginning to gather strength, particularly in Asia and Europe, which were short-changed in the original allocation of IPv4 addresses.[4] In addition, the explosion of mobile devices accessing the Internet will soon place heavy stress on the IPv4 IP address space. In 2001 Cisco began to support IPv6 on all of its routers in anticipation of growing

---

[4] North America has 74 percent of all IPv4 addresses. In fact, Stanford University has more IPv4 addresses than China, which now has fewer IP addresses than it has Internet users.

demand. Fortunately, IPv6 packets can be tunneled through IPv4 networks by placing them within IPv4 packets, so the two protocols can (and will) coexist on the Internet for some time to come.

TEST YOUR UNDERSTANDING

21. a) How is IPv6 better than IPv4? b) Why has IPv6 adoption been so slow? c) What forces may drive IPv6's adoption in the future? d) Must IPv6 replace IPv4 all at once? Explain.

THE TRANSMISSION CONTROL PROTOCOL (TCP)

Chapter 2 looked in some depth at the **Transmission Control Protocol (TCP).** In this section we will look at this complex protocol in even more depth. When IP was designed, it was made a very simple "best effort" protocol (although its routing tables are complex). More complex internetworking management tasks were left to TCP. Consequently, network professionals need to understand TCP very well. Figure 8-17 shows the organization of TCP messages, which are called **TCP segments.**

TEST YOUR UNDERSTANDING

22. a) Why is TCP complex? b) Why is it important for networking professionals to understand TCP? c) What are TCP messages called?

Sequence Numbers

Each TCP segment has a unique 32-bit sequence number that increases with each segment. This allows the receiving transport process to put arriving TCP segments in order if IP delivers them out of order.

Flags Fields

The figure shows that TCP has six single-bit flags fields. These allow the receiving transport process to know the kind of segment it is receiving.

Acknowledgements

Recall from Chapter 2 that when a transport process receives a correct TCP segment, it sends an ACK message to its partner. In an ACK message, the **ACK bit** is **set,** that is, given the value 1. In segments that do not contain an acknowledgement, the ACK bit is **not set,** that is, given the value 0. If the ACK bit is set, the acknowledgement number field value is filled in to indicate which TCP segment is being acknowledged.

When a one-bit flag field is set, its value is one. If the bit is not set, its value is zero.

Openings with SYN Segments

Recall that TCP connections open with what is called a three-way handshake. First, one transport process sends a SYN (synchronization) message to open the connection. The other transport process sends back a SYN/ACK. Finally, the side that is opening

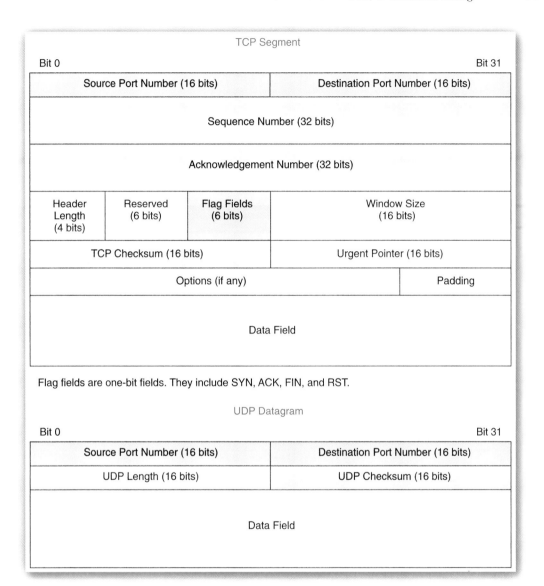

Figure 8-17   TCP Segment and UDP Datagram

the connection sends back an ACK. A synchronization message consists of a header without a body. In a synchronization message, the **SYN bit** is set.

Closings with FIN Segments
Each side ends the connection by sending a FIN message in which the FIN bit is set. This is called a four-way close because four segments must be sent, as Figure 8-18 illustrates.

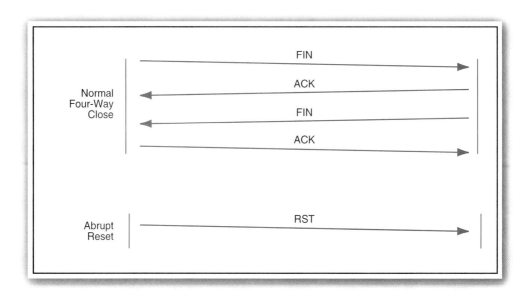

Figure 8-18   Normal Four-Way Closes and Abrupt Resets in TCP

Abrupt Resets

Figure 8-18 shows that TCP also allows a second way to close connections. This is the abrupt reset. It is like hanging up during the middle of a telephone conversation. Either side can send a reset message with the **RST bit** set. This is a one-way close. There is no acknowledgement or response from the other side.

TEST YOUR UNDERSTANDING

23.   a) Why are sequence numbers good? b) If someone says that a one-bit flag is set, what does this mean? c) If the ACK bit is set, what other field must be filled in? d) What is a SYN segment? e) Distinguish between four-way closes and abrupt resets.

Port Numbers

As Figure 8-17 shows, both TCP and UDP have **port number** fields. These fields are used differently by clients and servers. In both cases, however, they tell the transport process what application process sent the data in the data field or should receive the data in the data field. This is necessary because computers can run multiple applications at the same time.

Servers

For servers, the port number field indicates which application program on the server should receive the message. Major applications have **well-known port numbers** that are usually (but not always) used. For instance, the well-known TCP port number for HTTP is 80. For FTP, TCP Port 21 is used for supervisory communication with an

FTP application program, and TCP Port 20 is used to send data segments. Telnet uses TCP Port 23, whereas TCP Port 25 is used for Simple Mail Transfer Protocol (SMTP) messages in e-mail. UDP has its own well-known port numbers. The well-known port numbers run from 0 to 1023.

Figure 8-19 shows that every time the client sends a message to a server, the client places the port number of the application in the destination port number field. In this figure the server is a webserver, and the port number is 80. When the server responds, it places the port number of the application (80) in the source port number field.

Clients

Clients do something different. Whenever a client connects to an application program on a server, it creates an **ephemeral port number.** According to IETF rules, this port number should be between 49153 and 65535. However, many operating systems ignore these rules and use other ephemeral port numbers. In our examples we will use port numbers that follow IETF client port numbering rules.

In the figure the ephemeral port number is 50047 for client communication with the webserver. When the client transmits, it places this in the source port field of the TCP or UDP header. The server, in return, places this ephemeral port number in all destination port number fields of TCP segments or UDP datagrams it sends to the client.

A client may maintain multiple connections to different application programs on different servers. In the figure, for example, the client has a connection to an SMTP mail server as well as to the webserver. It will give each connection a different ephemeral port number to identify each connection. For the SMTP connection, the client has randomly chosen the ephemeral port number to be 60003.

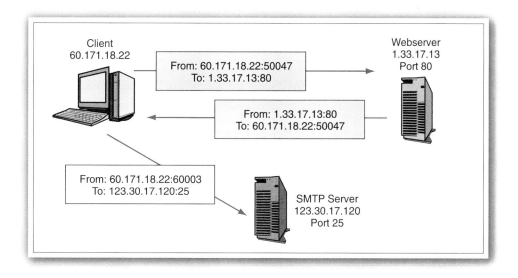

Figure 8-19   Use of TCP (and UDP) Port Numbers

Sockets

The combination of an IP address and a port number designates a specific connection to a specific application on a specific host. This combination is called a **socket.** It is written as an IP address, a colon, and then a port number, for instance, 128.171.17.13:80.

A socket is written as an IP address, a colon, and then a port number. It designates a specific application on a specific host.

TEST YOUR UNDERSTANDING

24. A host sends a TCP segment with source port number 25 and destination port number 64562. a) Is the source host a server or a client? Explain. b) If the host is a server, what kind of server is it? c) Is the destination host a server or a client? Explain.

25. a) What is a socket? b) How is it written? c) When the SMTP server in Figure 8-19 transmits to the client PC, what is the source socket? d) The destination socket?

THE USER DATAGRAM PROTOCOL (UDP)

UDP is a simple (connectionless and unreliable) protocol. We saw in Chapter 6 that IP telephony uses UDP to carry voice packets because there is no time to wait for retransmissions. In Chapter 10 we will see that the Simple Network Management Protocol uses UDP for a different reason—to reduce network traffic. UDP does not have openings, closings, or acknowledgements and so produces substantially less traffic than TCP.

As a consequence of UDP's simple operation, the syntax of the UDP datagram shown in Figure 8-17 is very simple. After two port number fields, which we just saw in the previous section, there are only two more header fields. There is a **length** field so that the receiving transport process can process the datagram properly. There also is a **UDP checksum** field that allows the receiver to check for errors in this UDP datagram. If an error is found, the UDP datagram is discarded.

TEST YOUR UNDERSTANDING

26. a) What are the four fields in a UDP header? b) Describe the third. c) Describe the fourth.

LAYER 3 AND LAYER 4 SWITCHES

In Chapters 4 and 5 we saw why switches are so fast and inexpensive. In this chapter we saw why routers are so slow and expensive. However, just as nature abhors a vacuum, technology abhors a sharp distinction. New devices called Layer 3 switches sit between routers and traditional Layer 2 (data link layer) switches for single networks in terms of speed and price.

Layer 3 Switches

Layer 3 Switches really are routers. They forward IP packets using routing tables, implement routing protocols to exchange routing table information, and do other things that routers do. They are very fast simply because they do all processing in hardware, which is much faster than traditional software-based routing.[5] They are also less expensive to purchase than traditional routers. Overall, then, the term "switch" is highly inaccurate—a "gift" of marketers who used the term "switch" because switches traditionally have been faster than routers.

Layer 3 switches cost only about as much as Ethernet switches and are almost as fast. However, the cost of managing any router—including Layer 3 switches—is much higher than the cost of managing Ethernet switches. Still, Layer 3 switches are much less expensive overall than traditional software-based routers.

Limited-Functionality Routers

However, there are limits today on what can be done in hardware. Consequently, Layer 3 switches today do not have all the functionality of full routers.

Instead of being full multiprotocol routers, for example, they often handle only IP or perhaps IP and IPX. In addition, they often have only Ethernet interfaces.

For many organizations, such as banks, which have multiple internal protocols from different architectures, Layer 3 switches cannot be used at all. However, where the limited functionality of Layer 3 switches is sufficient for an organization, they are ideal.

Often, for instance, Layer 3 switches are used as internal routers in an organization that has standardized on TCP/IP for internal communication. Figure 8-20 shows

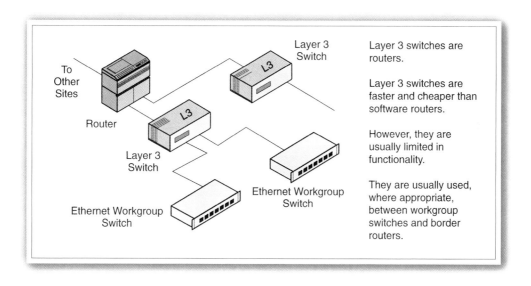

Figure 8-20   Layer 3 Switches and Routers in Site Internets

---

[5] Programs consist of multiple statements. Often, dozens of program statements may need to be loaded and run to accomplish a simple task. If the task can be done in hardware, in contrast, no time is needed to "load" software statements or write out results to memory. In addition, parallel processing usually allows hardware to do functions in fewer steps than software implementations. On the negative side, it is far more expensive to create hardware to implement required functionality than to write a program to do the work.

that in many site networks, Layer 3 switches are pushing routers to the edges of the site because the ability of routers to support multiple WAN protocols is crucial at borders. In contrast, within the site, the low cost of Layer 3 switches usually makes them dominant in the core above the workgroup switch.

Layer 4 Switches

As noted earlier in this chapter, TCP and UDP headers have port number fields that indicate the application that created the encapsulated application layer message and the application layer program that should receive the encapsulated application message.

Layer 4 switches examine the port number field of each arriving packet's encapsulated TCP segment. This allows them to switch packets based on the application they contain. Specifically, this allows Layer 4 switches to give priority to or even to deny forwarding to IP packets from certain applications. For example, TCP segments to and from an SMTP mail server (Port 25) might be given low priority during times of congestion because e-mail is insensitive to moderate latency.

TEST YOUR UNDERSTANDING

27. a) Are Layer 3 switches really routers? b) How are they better than traditional software-based routers? c) How are they not as good? Give a full explanation. d) Where would you use Layer 3 switches? e) Where would you not use Layer 3 switches?

28. a) What are Layer 4 switches? b) What field do Layer 4 switches examine? c) Why are Layer 4 switches good?

SYNOPSIS

TCP/IP is a family of standards created by the Internet Engineering Task Force (IETF). IP is TCP/IP's main standard at the Internet layer. IP is a lightweight (unreliable and connectionless) protocol. At the transport layer, TCP/IP offers two standards: TCP, which is a heavyweight protocol (reliable and connection-oriented), and UDP, which is a lightweight protocol like IP.

Routers forward packets through an internet. Border routers move packets between the outside world and an internal site network. Internal routers work within sites. In the Internet backbone, core routers handle massive traffic flows. Ports in routers are called interfaces. Different interfaces may connect to different types of networks, for instance, Ethernet or Frame Relay networks. Most routers are multiprotocol routers, which can handle not only TCP/IP internetworking protocols, but also internetworking protocols from IPX/SPX, SNA, and other architectures.

Routers are designed to work in a mesh topology. This creates alternative routes through the internet. Alternative routes are good for reliability. However, the router has to consider the best route for each arriving packet, and this is time consuming and therefore expensive.

To make a routing decision (deciding which interface to use to send an incoming packet back out), a router uses a routing table. Each row in the routing table represents a route to a particular network or subnet. All packets to that network or subnet are governed by the one row.

To use the routing table, the router compares the destination IP address of the incoming packet to every row in the routing table. At each row, the router masks the IP address with that row's network or subnet mask, then compares the result to the destination network/subnet column. If there is a match, the row is added to the list of matches. After looking at all rows, the router selects the best-match row. It sends the packet out the interface indicated in that row, to the next-hop router or destination host specified in the best-match row.

Routers build their routing tables by listening to other routers. Routers frequently exchange messages giving information stored in their routing tables. These messages are governed by one of several available routing protocols.

Selecting the best route for each arriving packet is very time consuming and expensive. ISPs and some firms are beginning to implement multiprotocol label switching (MPLS), which essentially creates virtual circuits to reduce costs. MPLS adds a label before the IP packet, and label-switching routers base forwarding decisions on the label's value, which they look up in a simple label-switching table. In addition to reducing router costs, MPLS can be used to provide quality of service for voice and other time-critical traffic flows. MPLS can also be used to do load balancing across different routes for efficiency.

Servers have both host names and IP addresses. Domain name system (DNS) servers provide IP addresses for users who know only a target host's host name. However, DNS is a broader service. It is a hierarchical system for naming domains, which are collections of resources on the Internet. Second-level domains, such as cnn.com, typically are prized by corporations.

IP itself does not have supervisory messages. For internet layer supervisory messages, hosts and routers use the Internet Control Message Protocol (ICMP). ICMP messages are carried in the data fields of IP packets.

IP version 4 has a number of important fields besides the source and destination address fields. The time to live (TTL) field ensures that packets that are misaddressed do not circulate endlessly around the Internet. The protocol field describes the contents of the data field—ICMP message, TCP segment, UDP datagram, and so forth. IP version 6 will offer many more addresses thanks to its 128-bit address fields.

The Transmission Control Protocol (TCP) has sequence numbers that allow the receiving transport process to place arriving TCP segments in order. The TCP header has several flag fields that indicate whether the segment is a SYN, FIN, ACK, or RST segment. Resets close a connection with a single segment instead of the normal four.

Both TCP and UDP have 16-bit source and destination port number fields that tell the transport process which application process sent or should receive the contents in the segment data field. Major applications have well-known port numbers. For instance, the well-known server port number of HTTP is Port 80. Clients, in contrast, have ephemeral port numbers that they select randomly for each connection.

Traditionally, there has been a sharp distinction between fast and inexpensive switches and slow and expensive routers. However, Layer 3 switches bridge this gap. They are true routers but operate in hardware rather than in software as traditional routers do. However, hardware operation can only implement some of the functionality of routers. In particular, Layer 3 routers are not multi-protocol routers. In addition, Layer 3 switches are routers and so are more costly to manage than Layer 2 switches.

SYNOPSIS QUESTIONS

1. a) What are the three main internetworking standards in TCP/IP? b) Compare the two transport layer standards.
2. a) Distinguish between internal, border, and core routers. b) What are interfaces? c) Do the subnets attached to different interfaces have to use the same network protocols (physical and data link layer protocols)? d) Why are most routers multiprotocol routers?
3. a) What topology are routers designed to use? b) Why is this topology good? c) Why is it bad?
4. a) What does each row in a routing table represent? b) List the process the router goes through to forward each arriving IP packet.
5. How do routers get the information for their routing tables?
6. a) Why is MPLS desirable? b) How does it work?
7. a) In what sense is DNS more than a way to look up IP addresses for target hosts? b) Which level domain name is desired by corporations?
8. a) What is the purpose of the IPv4 TTL field? b) Of the IPv4 protocol field? c) What is the advantage of IPv6?
9. a) What is the purpose of TCP sequence numbers? b) What are flag fields? c) Distinguish between normal TCP closes and abrupt resets.
10. a) What is the purpose of port number fields? b) What type of port numbers do major applications use? c) What type of port numbers do clients use?
11. a) What are the advantages of Layer 3 switches compared to traditional routers? b) What are their disadvantages?

THOUGHT QUESTIONS

1. Give a non-network example of hierarchical addressing and discuss how it reduces the amount of work needed in physical delivery.
2. (Based on the box "IP Routing Tables") What would be good metrics for routing tables to use beyond those mentioned in the chapter (cost and speed)?
3. A router forwards a packet to a next-hop router (Router S) for delivery. a) What is in the destination address in the IP packet header's destination address field—the IP address of Router S or the IP address of the destination host? b) In what field of what header will Router S's network address (its data link layer address on its single network) be found, if it is to be found at all? c) In what field of what header will Router S's IP address be found, if it is to be found at all?
4. How do you think TCP would handle the problem if an acknowledgement were lost, so that the sender retransmitted the unacknowledged TCP segment and, therefore, the receiving transport process received the same segment twice?
5. A client PC has two simultaneous connections to the same webserver application program on a webserver. What will be different between the TCP segments that the client sends on the two connections?

6. (Based on the box "IP Routing Tables") a) In Figure 8-8, to what subnets does the router connect? b) Which interface connects to which subnet? c) What is the default router—the router that should be sent the packet if there are no matches except in the default row?

TROUBLESHOOTING QUESTION

1. You suspect that the failure of a router or of a transmission line connecting routers has left some of your important servers unavailable to clients at your site. How could you narrow down the location of the problem using what you learned in this chapter?

PROJECTS

1. **Getting Current.** Go to the book website's New Information and Errors pages for this chapter to get new information since this book went to press and to correct any errors in the text.

HANDS ON: PACKET CAPTURE AND ANALYSIS
WITH WinDUMP AND TCPDUMP

Learning Objectives:

By the end of this chapter, you should be able to discuss:

- The purpose of WinDUMP and TCPDUMP.
- How to obtain and run WinDUMP.
- How to read "simple" WinDUMP output.
- How to read hexadecimal WinDUMP output.

WHAT ARE WinDUMP AND TCPDUMP?

WinDUMP and TCPDUMP are packet capture and analysis programs. As the names suggest, they capture packets entering and leaving your computer and afterward allow you to look at the contents of selected fields in individual packets.

More specifically, after these programs capture packets, they print one or more lines per packet, as Figure 8a-1 illustrates for an HTTP request–response cycle. The printout is dense and looks forbidding, but with a little practice, it becomes easily readable.

This detailed information can help you troubleshoot problems and detect hacker activity. Using WinDUMP/TCPDUMP is also a great way to solidify your understanding of TCP.

Historically, TCPDUMP was created to run on Unix machines, and it has been wildly popular among Unix administrators. TCPDUMP was ported to Windows computers as WinDUMP. Although WinDUMP is not as mature or as full-featured as TCPDUMP, it will run on the Windows clients that most students have.

Command prompt>tcpdump www2.pukanui.com

7:50.10.500020 10.0.5.3.62030 > www2.pukanui.com.http: S 800000050:800000050(0) win 4086 <mss1460>

7:50.10.500030 www2.pukanui.com.http > 10.0.5.3.62030 : S 300000030:300000030(0) ack 800000051 win 8760 <mss1460>

7:50.10.500040 10.0.5.3.62030 > www2.pukanui.com.http: . ack 1 win 4086

7:50.10.500050 10.0.5.3.62030 > www2.pukanui.com.http: P 1:100(100)

7:50.10.500060 www2.pukanui.com.http > 10.0.5.3.62030 : . ack 101 win 9000

7:50.10.500070 www2.pukanui.com.http > 10.0.5.3.62030 : . 1:1000(999)

7:50.10.500080 10.0.5.3.62030 > www2.pukanui.com.http: . ack 1001 win 4086

7:50.10.500090 www2.pukanui.com.http > 10.0.5.3.62030 : P 1001:2000(999)

7:50.10.500100 10.0.5.3.62030 > www2.pukanui.com.http: . ack 2001 win 4086

7:50.10.500110 10.0.5.3.62030 > www2.pukanui.com.http: R

Figure 8a-1   ASCII WinDUMP Printout

WORKING WITH WinDUMP

Installing WinDUMP

To get a copy of WinDUMP, go to http://www.tcpdump.org/. This website has directions for downloading the program to your computer. Websites do not always remain alive, so if this website is not working, you might have to do an Internet search to find WinDUMP.

WinDUMP does not work by itself. It requires a library of packet capture programs collectively known as WinPCAP. Fortunately, websites that help you download WinDUMP also help you download WinPCAP. During installation, install WinPCAP first, then WinDUMP.

Running WinDUMP

As Figure 8a-1 shows, you begin a WinDUMP session by going to the command line. You do this by clicking Start, then Run, and then typing cmd or command. You can then give WinDUMP commands. In this example the command is the following:

Command prompt>tcpdump www2.pukanui.com -c 40

Note that the command is "tcpdump," not windump. TCPDUMP may have been ported over to WinDUMP, but its Unix commands have been kept.

If you simply type the command "tcpdump" without options, all packets going into and out of the computer interface will be captured. However, the tcpdump program has many options to control what packets are captured and how they are displayed. In the example shown in Figure 8a-1, for instance, only packets to and from the specified host, www2.pukanui.com, will be captured. The packet count, c, is set to 40, meaning that only the first 40 packets will be captured.

To get full information about TCPDUMP, do an Internet search for "tcpdump man page." This will take you to a detailed Unix manual page. Most of what you read there will work with WinDUMP.

Getting Data to Capture

Of course, if no packets go to or from www2.pukanui.com, there will be nothing to capture. Without closing your command prompt window, open your browser and go to a website. This will download the home page, giving you at least one HTTP request–response cycle. (If a page has graphics and other elements, each is a separate file, so several request–response cycles will be captured. In this case, capturing only 40 packets will show you only the connection opening and some of the subsequent packet exchanges.

READING WinDUMP OUTPUT

To see how WinDUMP works (and to help you solidify and extend your understanding of IP and especially TCP, we will look at the WinDUMP output in Figure 8a-1.

Opening the TCP Connection

The first three packets in Figure 8a-1 open a TCP connection between a client PC and a server. This is the classic SYN–SYN/ACK–SYN three-way handshake.

SYN

The first packet carries a SYN segment from the client PC running WinDUMP to a webserver.

> 7:50.10.500020 10.0.5.3.62030 > www2.pukanui.com.http: S 800000050:800000050(0) win 4086 <mss1460>

➤ The printout begins with a time stamp, 7:50.10.500020. This gives time to the millionth of a second.

➤ Next comes the source host's IP address and port number. The IP address is 10.0.5.3. The source host's port number, 62030, is an ephemeral port; so the source host must be a client PC. It is the computer running WinDUMP.

➤ Next comes the destination host's host name and port number. Unless you tell it otherwise, WinDUMP looks up the host name for the source and destination IP addresses and inserts them in the printout. The client, 10.0.5.3, does not have a host name, so its IP address is used in the packet printout. The destination host is www2.pukanui.com. The port number is the well-known port for http (80). Unless you tell it otherwise, WinDUMP substitutes protocol names when it sees well-known port numbers.

➤ The "S" indicates that the SYN flag is set. Flags except for ACK are shown in this position. If no flag is set, a period is shown instead of flags.

➤ Next comes the odd-looking 800000050:800000050(0). When a host begins a TCP session, it randomly generates an initial sequence number. The client has generated an initial sequence number of 800000050 for this session. A SYN message is a pure supervisory message containing no data. The 800000050:800000050(0) shows that the data field has a length of zero (0) because this is a supervisory segment containing no data.

➤ The win 4086 part of the printout shows that the client has told the server to use a window of 4,086 bytes. The server can transmit only 4,086 bytes of data before getting a window extension.

➤ Data within angle brackets describe options. Here the client advertises a maximum segment size (MSS) of 1460. This tells the receiver of the packet (the webserver) to place no more than 1,460 bytes of data in TCP data fields.

SYN/ACK

Now the webserver replies with its SYN/ACK message.

> 7:50.10.500030 www2.pukanui.com.http > 10.0.5.3.62030 : S 300000030:300000030(0) ack 800000051 win 8760 <mss1460>

➤ The host designations are reversed to indicate that the webserver is sending to the client.

➤ The S flag is again set.

➤ The webserver's initial sequence number is 300000030, and the packet carries no data.

➤ The "ack" indicates that this TCP segment contains an acknowledgement. The acknowledgement number is 800000051. The acknowledgement number is always one byte larger than the last data byte in the segment being acknowledged (800000050). It specifies the next byte of data that is expected after the segment being acknowledged.

➤ The window again indicates how many more bytes may be transmitted before the window size is increased. In this case, the window size is 8760. The client may transmit through byte 800000051 plus 8,760, or 800008811.

ACK

The client now sends back an ACK.

7:50.10.500040 10.0.5.3.62030 > www2.pukanui.com.http: . ack 1 win 4086

➤ Notice that there is no flag other than ack, so a period (.) is placed in the flags position.

➤ Also notice that the ack is 1, not 300000031, as you would expect (one more than the last data byte received). The packet really does have 300000031 as the acknowledgement. However, to make the printout easier to read, WinDUMP subtracts the initial sequence number, leaving 1. All subsequent TCP data indications will be based on the bytes sent by a transport process since the beginning of the session.

The HTTP Request Message

Now the client sends an HTTP request message. This consists of a single packet.

7:50.10.500050 10.0.5.3.62030 > www2.pukanui.com.http: P 1:100(100)

➤ The flags field shows P, for push. This tells the receiver that a full application message is contained in the message, and that the transport process should pass the application message to the application layer.

➤ The 1:100(100) says that the HTTP request message contains data bytes 1 through 100—100 bytes in total.

The webserver responds with an acknowledgement. The acknowledgement number, as always, gives the next data byte the webserver expects to see (101).

7:50.10.500060 www2.pukanui.com.http > 10.0.5.3.62030 : . ack 101 win 9000

The HTTP Response Message

Now the webserver sends back an HTTP response message. This response message is too large to fit into a single packet. Consequently, the webserver sends the HTTP response message in two packets, each of which is acknowledged separately.

7:50.10.500070 www2.pukanui.com.http > 10.0.5.3.62030 : . 1:1000(999)
7:50.10.500080 10.0.5.3.62030 > www2.pukanui.com.http: . ack 1001 win 4086
7:50.10.500090 www2.pukanui.com.http > 10.0.5.3.62030 : P 1001:2000(999)
7:50.10.500100 10.0.5.3.62030 > www2.pukanui.com.http: . ack 2001 win 4086

➤ The packets containing the HTTP response message do not have any flag fields set, so a period (.) appears where you would see a flag if one had been included in the segment.

➤ Note that the first packet contains bytes 1 through 1000 of the HTTP response message, while the second packet contains remaining bytes, 1001 through 2000.

➤ Note also that the second packet containing the HTTP response message contains a P flag field. This is the push flag, which tells the receiving transport process that all data has been delivered, so that the data should be pushed up to the application program. The first response packet does not have a push flag because it only delivers the first part of the HTTP response message. A push is not needed again until all the data is received.

Ending the Connection

        7:50.10.500110 10.0.5.3.62030 > www2.pukanui.com.http: R

In Chapter 8 we saw how TCP connections should end with a four-way close in which FINs are sent and acknowledged. In Figure 8a-1, however, the client, for no obvious reason, has sent a reset message (flag R) to the webserver. A reset message abruptly terminates the connection. Neither side transmits again.

SOME POPULAR WinDUMP OPTIONS

Major Options

WinDUMP's tcpdump command has a large number of options. We have already seen two of them.

- ➤ Giving a hostname or IP address specifies that packets to or from only that address should be captured.
- ➤ The "-c" option specifies how many packets should be captured.

The following are a few other commonly used options. The tcpdump command has many more.

- ➤ e: Print the data link layer header fields on each line.
- ➤ i: Specify an interface (NIC) if a computer has more than one.
- ➤ n: Do not convert IP addresses into host names (reduces capture processing work).
- ➤ N: Print only the host part of a hostname. This makes output more condensed and perhaps easier to read.
- ➤ q: Quiet. Print less information for each packet in ASCII.
- ➤ s: Snaplen. Specifies how many octets will be shown for each packet. The default is 68.
- ➤ t: Do not print the time stamp on each line.
- ➤ v: Verbose output—more detail than the normal output. There also are vv and vvv options for more verbose and incredibly verbose output.
- ➤ w: Writes the raw packets to a file. A space and then a file name must follow the w option.
- ➤ r: Reads data from a file instead of capturing the data. Often used when the data was captured in a file with the w option.
- ➤ The -x option specifies that output should be in hex, while the -X option specifies that output should be in both ASCII and hex. We will see what this means in the next section.

Example

For example, suppose you give the following command.

        tcpdump www2.pukanui.com -c 1 -tN

This tells tcpdump to collect data only for packets going to and from www2.pukanui.com. It also tells tcpdump to suppress time stamps and to show only the host name. Finally, it tells tcpcump to capture only a single packet. The following shows the output you might see:

10.0.5.3.62030 > www2.http: S 800000050:800000050(0) win 4086 <mss1460>

Expression

In the examples we have been using, we have included the name of a host, www2.pukanui.com. This type of option is called an expression. It specifies what packets will be captured and analyzed.

- ➤ Host expressions are the most common. They should be written as "host hostname", but host is the default, so it does not need to be added.
- ➤ The expression "port 80" tells TCPdump to look only at HTTP traffic. The more finely grained expression "src port 80" captures packets only from HTTP servers.
- ➤ It is even possible to have expressions that limit traffic to a particular network, as in "net 128.171", to capture traffic going to or from a particular network.

HEXADECIMAL PRINTOUT

ASCII versus Hex

The type of printout we have been seeing is called ASCII printout, because it contains alphanumeric characters (keyboard characters) stored in the ASCII format. WinDUMP also offers another way to store and see packet information—hexadecimal format. As discussed in Chapter 4, "hex" represents a group of four bits as a symbol from 0 through F. Hex output typically groups hex symbols in groups of two to represent a byte (octet) or in groups of four, to represent two-byte sequences. WinDUMP does the latter.

Hex Output

Figure 8a-2 shows hexadecimal output for a single packet. To turn on hex output, give the -x (lowercase x) option in a tcpdump command. To get both ASCII and hex output, give the -X (uppercase X) option.

IP Fields

To help you read this very dense output, the IP header is shown in boldface. In the following description, the most widely interpreted sequences are shown in boldface.

- ➤ **4500.** The 4 indicates that this is an IPv4 packet. The 5 indicates that the length of the header is 5 times 32 bits, or 20 octets. This is the header length of an IP header without options. Options are rare and suspicious, so anything other than a 5 for the header length should be seen as a caution sign. The 00 is the Diff-Serv octet, which normally is not used and so normally is set to 00. Almost all packets should start with 4500.
- ➤ 00c7. This is the length of the entire IP packet (199 bytes).
- ➤ ff53 0000. This is the identification field value and other information used to reassemble fragmented packets. Fragmentation is rare.

```
4500 00c7 ff53 0000 8006 3d5e b87a 3270

b87a c3d0 F230 0050 0023 37d6 1d37 1302

5018 07d0 b329 0000 ...
```

Figure 8a-2   Hexadecimal Output from WinDUMP

➤ **8006.** The 80 is the one-byte time to live field value (128 in decimal). The 06 is the one-byte protocol field value. The 06 protocol is TCP. This field is needed to interpret the IP data field, which is not always a TCP segment. Protocol 01 is ICMP, for instance, while Protocol 17 is UDP.

➤ 3d5e. This is the header checksum.

➤ **b87a 3270.** This is the IP source address.

➤ **b87a c3d0.** This is the IP destination address.

TCP Fields

In Figure 8a-2, the TCP fields are underlined. The following are the TCP fields. Again, the most widely interpreted fields are shown in boldface.

➤ **F230.** This is the source port number (62000).

➤ **0050.** This is the destination port number (80 decimal).

➤ **0023 37d6.** This is the sequence number.

➤ **1d37 1302.** This is the acknowledgement number.

➤ **50.** 5 is the header length in 32-bit units; as in IP, 5 is the header length without options. In contrast to IP, options are common in TCP. The 0 is from reserved bits. It should always be 0.

➤ **18.** The 1 indicates that the ack bit is set. The 8 indicates that the push bit is set. These two symbols indicate which bits are set. Experienced WinDUMP users learn the most common combinations.

➤ 07d0. This is the window size (2000 in decimal).

➤ b329. This is the checksum.

➤ 0000. This is the urgent pointer. If the urgent (U) bit is set in the flags field, this pointer tells where urgent data begin in the TCP byte sequence.

REVIEW QUESTIONS

GENERAL QUESTIONS

1. What does WinDUMP do?
2. Distinguish between WinDUMP and TCPDUMP.
3. Distinguish between ASCII and hex output.
4. What steps should you take to capture and display data?

For Each of the Following, Specify a Command to Do the Work

5. Show all packets.
6. Show the first 100 packets going to or from dakine.pukanui.com.
7. Repeat the preceding command, this time not showing a time stamp or the full host name.
8. Show all packets going to or from HTTP servers.

INTERPRETATION

9. Interpret the following ASCII printout.

   7:50.10.500099 db.pukanui.com.54890 > www2.pukanui.com.http: 1:21(21) ack 52 win 4086

10. The following hex printout shows an IP packet. a) What type of message is in its data field? b) What is the IP destination address?

    4500 00c7 ff53 0000 8017 3d5e b87a 3270
    b87a c3d0

11. The following hex printout shows an IP packet containing a TCP segment. a) What is the source port (in decimal)? b) Is the ack bit set? (Tell how you can know.)

    4500 00c7 ff53 0000 8006 3d5e b87a 3270
    b87a c3d0 0060 0050 0023 37d6 0000 0000
    5008 07d0 b329 0000 ....

SECURITY

Learning Objectives:

By the end of this chapter, you should be able to discuss:

- Security threats (worms and viruses, hacking, and denial-of-service attacks) and types of attackers.
- Why security is primarily a management issue, not a technical issue; and the plan-protect-response cycle.
- Authentication mechanisms, including passwords, digital certificate authentication, biometrics, and single sign-on.
- Firewall protection, including packet filter firewalls, stateful firewalls, application firewalls, defense in depth, IDS, and IPSs.
- The protection of dialogues by cryptographic systems. The phases of cryptographic systems.
- Responding to successful compromises.

SECURITY THREATS

Dangers

Compromises

Security is one of the biggest concerns in corporations today. Figure 9-1 shows the percentage of companies that experienced successful attacks in 2002. This report underscores the fact that security **compromises** (successful attacks) are both widespread and varied. Nearly all of the firms surveyed had experienced a successful attack in the previous year. Although not all firms could or were willing to report dollar losses from these attacks, the limited data provided in the table make it clear that many compromises are extremely expensive.

TEST YOUR UNDERSTANDING

1. a) What is a compromise? b) Are compromises common or rare?

Had at Least One Security Incident in This Category (May have had several)	Percent Reporting an Incident in 1997	Percent Reporting an Incident in 2003	Number Reporting Quantified Losses in 2002	Average Reported Annual Loss per Firm (Thousands) in 1997	Average Reported Annual Loss per Firm (Thousands) in 2002
Viruses	82%	82%	254	$76	$200
Insider Abuse of Net Access	Not Asked	80%	180	Not Asked	$136
Laptop Theft	58%	59%	250	$38	$47
Unauthorized Access by Insiders	40%	45%	72	NA	$31
Denial of Service	24%	42%	111	$77	$1,427
System Penetration	20%	36%	88	$132	$56
Sabotage	14%	21%	61	$164	$215
Theft of Proprietary Information	20%	21%	61	$954	$2,700
Financial Fraud	12%	15%	61	$958	$329
Telecom Fraud	27%	10%	34	NA	$50
Telecom Eavesdropping	11%	6%	0	NA	NA
Active Wiretap	3%	1%	0	NA	NA

Survey conducted by the Computer Security Institute (www.gocsi.com).
Based on replies from 530 U.S. Computer Security Professionals.
If fewer than twenty firms reported quantified dollar losses, data for the threat are not shown.

Figure 9-1   CSI/FBI Survey

Viruses and Worms

The most basic principle in security is to begin by understanding the organization's needs. In security, this requires a solid knowledge of the **threat environment** the company faces. Before discussing how to prevent attacks, we will look at the main types of attacks that corporations suffer.

The most widespread compromises identified in the CSI/FBI Survey are virus and worm attacks. In the CSI/FBI survey, almost all of the firms experienced at least one compromise from these sources—despite the fact that about 90 percent had antivirus systems in place.

Viruses

Viruses are pieces of executable code that attach themselves to other programs. Within a computer, the virus spreads whenever an infected program runs by attaching itself to other programs.

Viruses
  Pieces of code that attach to other programs
  Virus code executes when infected programs execute
  Infect other programs on the computer
  Spread to other computers by e-mail attachments, webpage downloads, etc.
  Many viruses spread themselves by sending fake messages with infected attachments
  Antivirus programs are needed to scan arriving files
  Users often fail to keep their computer antivirus programs up to date
  Antivirus filtering on the e-mail server works even if users are negligent
Worms
  Complete programs
  Self-propagating worms identify victim hosts, jump to them, and install themselves
  Can do this because hosts have vulnerabilities
  Vendors develop patches for vulnerabilities but companies often fail to apply them
  Worms take advantage of specific vulnerabilities
  Firewalls can stop many worms by forbidding access to most ports
  E-mail worms can get around antivirus filtering
Blended Threats
  Combine the spreading characteristics of viruses and worms
Payloads
  Programs that can do damage to infected hosts
  Erase hard disks, send users to pornography sites if they mistype a URL
  Trojan horses: exploitation programs disguise themselves as system files

Figure 9-2   Viruses and Worms (Study Figure)

Between computers, the virus spreads when an infected program is transferred to another computer via a floppy disk, an e-mail attachment, a webpage download, an unprotected disk share (see Chapter 1b), a peer-to-peer file-sharing transfer, an instant message, or some other propagation vector. More than 90 percent of viruses today speed this process by e-mailing messages with infected attachments to addresses in the infected computer's e-mail directories. If a receiver opens the attachment, the infected program executes and the receiver's programs become infected.

To stop viruses, a company must protect its computers with **antivirus programs** that scan each arriving e-mail message or floppy disk for signatures (patterns) that identify viruses. These antivirus programs also scan for other types of malware (evil software), including worms and Trojan horses.

Worms

Both viruses and worms can create mass epidemics that infect hundreds or even millions of computers. However, they spread in different ways. We have just seen that viruses are pieces of code that must attach themselves to other programs. In contrast, **worms** are full programs that operate by themselves.

Self-Propagating Worms   Worms can propagate from computer to computer in one of two ways. **Self-propagating worms** propagate on their own by seeking out other computers, jumping to them, and installing themselves on these victims.

Self-propagating worms can spread with incredible speed. In 2003 the Blaster worm infested 90 percent of all computers on the entire Internet that were vulnerable to it within ten minutes.

To self-propagate to a target computer, worms require target computers to have specific **vulnerabilities** (security weaknesses). When operating system and application software vendors discover such vulnerabilities, they issue **patches** (software updates) to correct them. However, companies are often slow to apply patches or fail to apply them at all. Firewalls, which we will see later, may stop worms that are attempting to reach a company's computers, but if the worm uses a TCP or UDP port allowed by the firewall, they will still get through.

E-Mail Worms   Self-propagating worms can spread rapidly, but patching and firewalls can stop them. Consequently, **e-mail worms** can spread like viruses, traveling via e-mail attachments. E-mail messages usually get through firewall defenses, allowing e-mail worms to prey on user gullibility as viruses have long done.

Blended Threats

Many attacks are now **blended threats,** which propagate both as viruses and as worms. This allows them to succeed if one propagation method is thwarted.

Payloads

After they spread, viruses and worms may execute pieces of code called **payloads.** In malicious viruses and worms, these payloads can completely erase hard disks and do other significant damage. In other cases they can take the victim to a pornography site whenever the victim mistypes a URL.

Often the payload installs a **Trojan horse** program onto the user's computer. Once installed, the Trojan horse continues to exploit the user indefinitely. A Trojan horse does not spread by itself but relies on a virus, worm, hacker, or gullible user to install it on a computer. As its name suggests, a Trojan horse looks like an ordinary system file, so it is difficult to detect.

TEST YOUR UNDERSTANDING

2. a) How do viruses propagate within computers? b) How do viruses propagate between computers? c) How can viruses be stopped? d) Distinguish between self-propagating worms and e-mail worms. e) How do self-propagating worms propagate? f) How can self-propagating worms be stopped? g) What are blended threats? h) What are payloads? i) What are Trojan horses? j) How does a Trojan horse get on a computer?

Human Break-Ins (Hacking)

Viruses and worms spread fairly randomly with simple fixed attack methods. However, human attackers often wish to break into a specific company's computers. Human adversaries can attack a company with a variety of approaches until they find one that succeeds. This makes human break-ins very dangerous.

Human Break-Ins
   Viruses and worms rely on one main attack method
   Humans can keep trying different approaches until they succeed

Hacking
   Breaking into a computer
   Hacking is intentionally using a computer resource without authorization or in excess of authorization

Scanning Phase
   Send attack probes to map the network and identify possible victim hosts
   Nmap programming is popular (Figure 9-4)

The Exploit
   The actual break-in
   Exploit is the program used to make the break-in
   Super user accounts (administrator and root) can do anything
   If application running with super user privileges is compromised, the attacker gains super user privileges

After the Break-In
   Become invisible by deleting log files
   Create a backdoor (way to get back into the computer)
          Backdoor account—account with a known password and super user privileges
          Backdoor program—program to allow reentry; usually Trojanized
   Do damage at leisure

Figure 9-3   Human Break-Ins (Hacking) (Study Figure)

Hacking

We will use the term hacking to mean breaking into a computer. More specifically, **hacking** is intentionally using a computer resource without authorization or in excess of authorization. Note that it is still hacking if the attacker is given an account and uses the computer for unauthorized purposes.

Hacking is intentionally using a computer resource without authorization or in excess of authorization.

Scanning Phase

When a hacker begins an attack on a firm, he or she usually begins by **scanning** the network. This involves sending **probe packets** into the firm's network. Responses to these probe packets tend to reveal information about the firm's general network design and about its individual computers—including their operating systems.

Figure 9-4 shows output from one popular scanning program, Nmap. It has identified several open ports on a server—port numbers that will accept connection attempts. If the attacker has an exploit for one of these services, the attacker can now attack the server. The scan has also fingerprinted the operating system of the server, because some attacks are operating system-specific.

The Exploit

Once the attacker has identified a potential victim host, the next step is the break-in itself. There are many tools for breaking into a computer, including tools for cracking passwords and other laborious techniques. However, like worms, most hackers exploit known vulnerabilities. They have break-in software that tries to exploit one or more known vulnerabilities. If the target computer has not been patched, the **exploit** (break-in program) can succeed in less than a second.

Administrator and Root   Chapter 1b noted that operating systems have super accounts that have complete privileges on the computer. In Microsoft Windows, this is the Administrator account. In Unix, it is the root account. If the attacker can take over this account, he or she "owns" the system and can do anything he or she wishes to do.

Application Exploits   The most common way of "gaining root" on a server is to exploit a vulnerability in an application program running with super user privileges. If the hacker can take over such an application, he or she can operate with full super user privileges on the victim computer.

After the Break-In

Becoming Invisible   After the attack, the hacker tries to erase the operating system's log files so that the computer's rightful owner cannot trace how the attacker broke in or gather evidence to find and prosecute the attacker. Unless attackers are stopped very quickly, they become very difficult to stop.

Creating a Backdoor   To be able to exploit the computer later, even if the vulnerability that was exploited is fixed, the attacker must create a way back in. This is called a

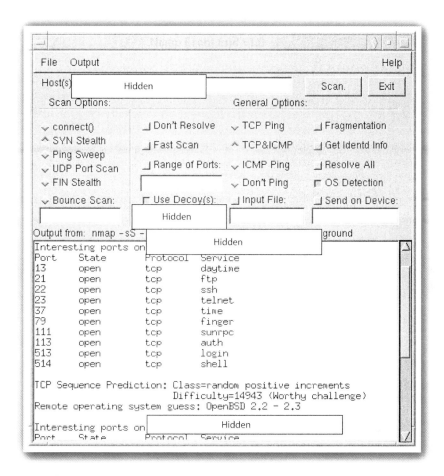

Figure 9-4   Nmap Scanning Output

backdoor. It may simply be a new account, or it may be a Trojan horse program that is difficult to detect.

Doing Damage   The attacker can now read all files on the computer, change them, delete them, or do almost anything else. Hackers often look for trade secrets that they can try to sell to competitors. They also tend to install **host exploitation software** on the computer, for instance, turning the host into a pornography download site or installing software to attack other computers.

TEST YOUR UNDERSTANDING

3. a) List the usual main phases in human break-ins. b) What is hacking? c) Why do hackers send probe packets into networks? d) What is an exploit? e) Why do hackers attempt to take over the Administrator or root account? f) Why are application

exploits often effective in taking over control of a computer? g) What steps does a hacker usually take after a break-in? h) What is a backdoor? i) What is host exploitation software?

Denial-of-Service (DoS) Attacks

Another type of attack, the denial-of-service attack, does not break into a computer, infect it with a virus, or infest it with a worm. Rather, the goal of **denial-of-service (DoS)** attacks is to make a computer unavailable to its users. As Figure 9-5 shows, most DoS attacks involve flooding the victim computer with irrelevant packets. The victim computer becomes so busy processing this flood of attack packets that it cannot process legitimate packets. The overloaded host may even fail. The attack shown in the figure is a **distributed DoS (DDoS) attack.** It allows the hacker to attack the victim using many attack computers on which handler or zombie programs have been installed after break-ins.

TEST YOUR UNDERSTANDING

4. a) What is the purpose of a denial-of-service attack? b) How do distributed DoS attacks work?

Attackers

Curious Hackers

Figure 9-6 shows that there are many types of attackers. In the past, many hackers have been motivated heavily by curiosity.

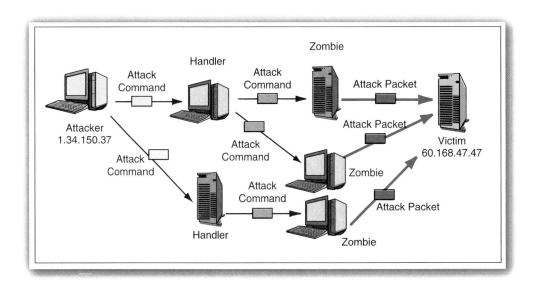

Figure 9-5   Distributed Flooding Denial-of-Service Attack

Traditional attackers:
        Curious hackers
        Disgruntled employees and ex-employees
Growing number of criminal attackers
Potential for far more massive attacks
        Cyberterror attacks by terrorists
        Cyberwar by nations

Figure 9-6   Types of Attackers (Study Figure)

Disgruntled Employees and Ex-Employees
Other attackers have been **disgruntled employees** and **disgruntled ex-employees** who attacked their own firms. Employee-attackers have done major damage but have been few in number.

Criminal Attackers
In addition, there now are many **criminal attackers** who steal credit card numbers to commit credit card fraud, who extort firms, and who steal trade secrets to sell to competitors. The number of criminal attackers is growing rapidly.

Cyberterrorists and National Governments
On the horizon, there is the danger of far more massive **cyberterror** attacks created by terrorists and even worse **cyberwar** attacks by national governments. These could produce damages of hundreds of billions of dollars.

TEST YOUR UNDERSTANDING

5.   a) What are the two traditional types of attackers? b) What new type of attackers is growing rapidly? c) What are cyberterror and cyberwar attacks?

PLANNING

Security Is a Management Issue
People tend to think of security as a technical issue, but security experts are unanimous in noting that security is primarily a management issue. Unless a firm does excellent planning and excellent implementation, the best technology will be wasted.

Security is primarily a management issue, not a technology issue.

TEST YOUR UNDERSTANDING

6.   Why is security primarily a management issue, not a technology issue?

The Plan–Protect–Response Cycle
In general, security management can be thought of in terms of the **plan–protect–respond cycle.** Firms constantly cycle through these three phases when they deal with security issues.

Security Is a Management Issue, Not a Technical Issue
    Without good management, technology cannot be effective
Plan-Protect-Respond Cycle
    Three phases endlessly repeating
    Planning: preparing for defense
    Protecting: implementing planned protections
    Responding: stopping attacks and repairing damage when protections fail
Risk Analysis
    Cost of protections should not exceed probable damage
    Annual probability of damage
        Damage from a successful incident
        Times the annual probability of success
        Gives the probable annual loss
    Cost of protection
        If a protection can reduce the annual probability of damage by a
        certain amount, up to this amount can be spent on the protection
Comprehensive Security
    Attacker only has to find one weakness
    Firm needs comprehensive security to close all avenues of attack
Defense in Depth
    Every protection breaks down sometimes
    Attacker should have to break through several lines of defense to succeed
    Providing this protection is called defense in depth

Figure 9-7   Planning Principles (Study Figure)

Planning

Planning is developing a clear and precise security strategy that will be appropriate for a firm's security threats. Unless planning is done well, there will be many weaknesses that attackers can exploit.

Protecting

Most of the time, the firm is in the **protecting** phase, in which it implements the strategic security plan. This is the time for firewalls, host hardening, and the other protections we will see in the next section.

Responding

Even the best protection will fail sometimes. We will end the chapter with the processes involved in **responding** to (stopping and repairing) compromises.

TEST YOUR UNDERSTANDING

    7.   Briefly describe the three steps in the plan–protect–respond cycle.

Planning Principles

Risk Analysis

In contrast to military security, which often makes massive investments to stop threats, corporate security planners have to ask whether applying a protection against a particular threat is justified economically. For example, if the probable annual loss is $100,000 and security measures to thwart the threat will cost $200,000, firms should not spend the money. Instead, they should accept the probable loss. **Risk analysis** is the process of balancing threats and protection costs.

Two factors must be considered when assessing threats. The first is the damage that the company will suffer if the threat succeeds. The second is the probability of the threat succeeding. If a threat will cost $1,000,000 if it succeeds but only has a 10 percent annual probability of success, then the **probable annual loss** will be $100,000. If protections can cut the probable annual loss in half, up to $50,000 could be justified to implement them.

Comprehensive Security

Corporate security is an example of asymmetrical warfare in which the attacker has a clear advantage. A company must close off all vectors of attack. If it misses even one, and if the attacker finds it, the attacker will succeed. The attacker, in contrast, only has to find one security weakness. Although it is difficult to achieve **comprehensive security,** in which all avenues of attack are closed off, it is essential to come as close as possible.

Defense in Depth

Another critical planning principle is defense in depth. Every protection will break down occasionally. If attackers have to break through only one line of defense, they will succeed during these vulnerable periods. However, if an attacker has to break through two, three, or more lines of defense, a single breakdown of a defense technology will not be enough to allow the attacker to succeed. Having successive lines of defense is called **defense in depth.**

TEST YOUR UNDERSTANDING

8. a) List the three major planning principles. b) What is risk analysis? c) Suppose that an attack would do $100,000 in damage and has a 15 percent annual probability of success. Spending $9,000 on "Measure A" would cut the annual probability of success by 75 percent. Do a risk analysis. Show your work. d) Should the company spend the money? e) Should the company spend the money if Measure A costs $20,000 per year? Again, show your work. f) Why is comprehensive security important? g) What is defense in depth? h) Why is it necessary?

PROTECTION WITH ACCESS CONTROL

A firm has a wide variety of resources on its client PCs and servers. Some of these resources are extremely crucial, others less so. One of the first things a company must do is to enumerate its resources, rank them by sensitivity, and develop an access plan for each resource (or at least for each resource category). This is called an **access control plan.**

Authentication

The first step in access control is **authentication,** which is requiring someone wishing to use a resource to prove his or her identity. As Figure 9-8 illustrates, the user trying to prove his or her identity is the **applicant** (sometimes called the supplicant). The party requiring the applicant to prove his or her identity is the **verifier.** The applicant does this by providing his or her **credentials** (proof of identity) to the verifier. As we saw in Chapter 5, there often is a third party, the **authentication server,** which stores data to help the verifier check the credentials of the applicant. Use of a central authentication server helps provide comprehensive security by ensuring that all verifiers use the same authentication information.

Appropriate Authentication

The type of authentication tools that are used with each resource must be appropriate for its sensitivity. Sensitive personnel information should be protected by very strong authentication methods. For relatively nonsensitive data, less expensive but weak authentication methods may be sufficient.

TEST YOUR UNDERSTANDING

9.   a) What is authentication? b) Distinguish between the applicant and the verifier. c) What are credentials? d) Why are authentication servers used? e) Why must authentication be appropriate for the sensitivity of an asset?

Passwords

The most common authentication method is the **password,** which is a string of characters that a user types to gain access to the resources associated with a certain **username** (account) on a computer.

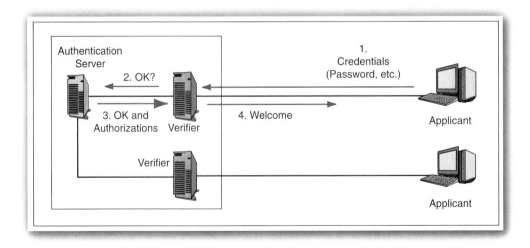

Figure 9-8   Authentication and Authorization

Passwords
    Strings of characters
    Typed to authenticate use of a username (account) on a computer
Benefits
    Ease of use for users (familiar)
    Inexpensive because built into operating systems
Often weak (easy to crack)
    Word and name passwords are common
    Can be cracked quickly with dictionary attack
Passwords should be complex
    Mix case, digits, and other keyboard characters ($, #, etc.)
    Can only be cracked with brute force attacks (trying all possibilities)
Passwords should be long
    Six to eight characters minimum
    Each added character increases the brute force search time by a factor of up to 75

Figure 9-9   Password Authentication (Study Figure)

Ease of Use and Low Cost

People find passwords relatively easy to use. In addition, passwords add no additional cost because operating systems and many applications have built-in password authentication.

Word/Name Passwords and Dictionary Attacks

The problem with passwords is that most users pick very weak passwords. Often, they pick **dictionary words** or the **names** of family members, pets, sports teams, and celebrities. Dictionary-word and name passwords often can be **cracked** (guessed) in a few seconds if the attacker can somehow get a copy of the password file. The attacker uses a **dictionary attack,** trying all words in standard or customized dictionary. There are only a few thousand dictionary words and names in any language, so dictionary attacks can crack dictionary-word and name passwords almost instantly.

Complex Passwords and Brute Force Attacks

Dictionary attacks can be thwarted by making passwords more complex. By using upper- and lowercase, the digits from 0 to 9, and other keyboard symbols such as & and #, the user can make his or her password more difficult to crack (as long as he or she does not simply capitalize the first letter of a common word or name and add a special symbol after the word or name). Complex passwords can be cracked only by **brute force attacks** that try all possible combinations of characters. Brute force cracks take far longer than dictionary attacks.

Password Length

Password length (the number of characters in the password) helps, too. If the password uses a combination of uppercase and lowercase letters, digits, and punctuation symbols, each additional character increases the time needed for a brute force attack by a factor of 75. Passwords should be at least six to eight characters long, and longer passwords are highly desirable.

TEST YOUR UNDERSTANDING

10. a) Distinguish between usernames and passwords. b) Why are passwords widely used? c) What types of passwords are susceptible to dictionary attacks? d) What is a brute force attack? e) What types of passwords can be broken only by brute force attacks? f) Why is password length important?

Digital Certificate Authentication

The gold standard for authentication is digital certificate authentication.

Public Keys, Private Keys, and Digital Certificates

In **digital certificate authentication,** each user is given a **public key,** which, as the name suggests, is not kept secret. This public key is paired with a **private key** that only the user should know. As Figure 9-11 shows, the **digital certificate** gives the name of a true party, the true party's public key, and other information.

Digital Certificate
   User gets secret private key and non-secret public key
   Digital certificates give the name of a true party and his or her public key

Testing a Digital Certificate (Figure 9-13)
   Applicant performs a calculation with his or her private key
   Verifier tests calculation using the public key found in the true party's digital certificate
   If the test succeeds, the applicant must be the true party

Strong Authentication
   The strongest method today

Expensive and Time Consuming to Implement
   Software must be added to clients and servers, and each computer must be configured
   Expensive because there are so many clients in a firm

Client Weaknesses
   Sometimes, only server gets digital certificate
   Client uses passwords or something else

Figure 9-10   Digital Certificate Authentication (Study Figure)

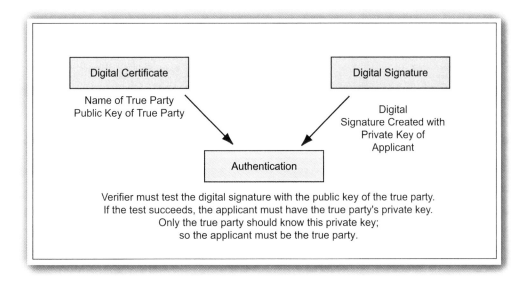

Figure 9-11   Testing a Digital Signature

Operation

For authentication, the applicant uses his or her private key to do a calculation. In Figure 9-11, this is a **digital signature,** which is added to a message to authenticate the sender. The verifier uses the public key contained in the digital certificate of the true party to test the calculation performed by the applicant. If the test is successful, then the applicant must know the true party's private key. This private key is supposed to be kept secret, so the verifier knows that the applicant is the true party named in the digital certificate.

Strong Authentication

Digital certificate authentication is extremely strong because private keys are extremely long and perfectly random. There is no known way to calculate private keys economically if you know the public key in a digital certificate.

Expensive and Time Consuming to Implement

Unfortunately, digital certificate authentication, also known as **public key authentication,** is expensive and time consuming to implement. Each server and client PC must have digital certificate authentication software and a private key installed on it. This requires a great deal of labor. Many companies are reluctant to spend this much money on authentication despite the strength of digital certificate authentication.

TEST YOUR UNDERSTANDING

11.   a) In public key authentication, who should know a user's private key? b) Who should know a user's public key? c) What information does a digital certificate provide? d) Describe how digital certificates are used in authentication. e) In digital certificate authentication, what key does the applicant use to perform a calculation?

f) What key does the verifier use to test the calculation performed by the applicant? g) Is digital certificate authentication strong? h) Why are companies reluctant to implement digital certificate authentication?

Biometrics

A relatively new form of authentication is **biometrics,** which is the use of bodily measurements to identify an applicant. Biometrics holds the promise of simplifying authentication dramatically.

Fingerprint Scanning

The least expensive (and unfortunately least accurate) form of biometric authentication is **fingerprint scanning.** In addition to having substantial **error rates** (normal misidentification rates when the subject is cooperating), many fingerprint scanners can be deceived fairly easily by impostors. Despite its limitations, fingerprint scanning is by far the most widely used biometric authentication method.

Iris Scanners

At the other extreme of the cost and accuracy range, iris scanners use cameras that read the very complex pattern of the applicant's **iris** (colored part of the eye). However, even iris scanners have small error rates and may be deceived.

Biometric Authentication
       Based on bodily measurements
       Promises to dramatically simplify authentication
Fingerprint Scanning
       Simple and inexpensive
       Substantial error rate (misidentification)
       Often can be fooled fairly easily by impostors
Iris Scanners
       Scan the iris (colored part of the eye)
       Irises are complex, so strong authentication
       Expensive
Face Recognition
       Camera allows analysis of facial structure
       Can be done surreptitiously—without the knowledge or consent of person being scanned
       Very high error rate and easy to fool
Error Rates and Deception
       Error and deception rates are higher than vendors claim
       Usefulness of biometrics is uncertain

Figure 9-12   Biometric Authentication (Study Figure)

Face Recognition

Some airports and other public locations now have cameras with **face recognition** systems that scan passersby to identify terrorists or wanted criminals by the characteristics of their faces. This is controversial because it typically is done **surreptitiously,** that is, without the knowledge or explicit permission of passersby. Error rates have been so high that many of these systems are now being removed.

Error Rates and Deception

Error rates refer to the percentage of mistakes made by a biometric system even when users are not practicing deception to fool the system. When an attacker does use deception, his or her probability of being falsely authenticated may be much higher than the error rate. How bad are biometric error rates and vulnerabilities to deception? This currently is an open question, but it is clear that error rates in practice are much higher than those stated by vendors and that many systems are vulnerable to deception. Biometric authentication must be used carefully.

TEST YOUR UNDERSTANDING

12. a) What is biometrics? b) What is its promise? c) Give a pro and a con of fingerprint scanning. d) Give a pro and some cons of iris scanning. e) Give a pro and some cons of face recognition. f) What is surreptitious scanning? g) Distinguish between error rates and deception.

PROTECTION WITH FIREWALLS

Basic Operation

In hostile military environments, travelers must pass through one or more checkpoints. At each checkpoint, his or her credentials will be examined. If the guard finds the credentials insufficient, the guard will stop the arriving person from proceeding and note the violation in a checkpoint log.

Figure 9-13 shows that firewalls operate in similar ways. Whenever a packet arrives, the **firewall** examines the packet; if the firewall identifies the packet as an attack packet, the firewall discards the packet and copies information about the discarded packet into a log file. Firewall managers should read the log file every day to understand the types of attacks coming into the resource that the firewall is protecting.

TEST YOUR UNDERSTANDING

13. a) What do firewalls do when a packet arrives? b) Why is it important to read firewall logs daily?

Packet Filter Firewalls

There are several different types of firewalls. They differ in what they examine when a packet arrives. The simplest is the **packet filter firewall,** which examines fields in the internet and transport headers of individual arriving packets.

Fields Examined

The firewall makes pass/deny decisions based upon the contents of IP, TCP, UDP, and ICMP fields. The most commonly filtered fields are IP addresses and port numbers.

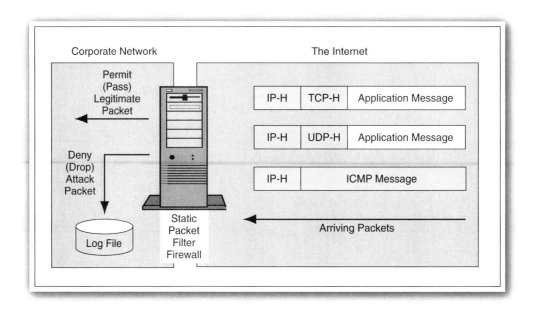

Figure 9-13   Firewall Operation

Access Control Lists (ACLs)

As Figure 9-14 shows, the firewall uses an **access control list (ACL),** which is an ordered list of pass/deny rules. The ACL in our example is very restrictive, only allowing packets to a single well-known port number—Port 80—and then only to a single computer— the company's public webserver. The firewall also allows incoming packets to ephemeral port numbers so that clients can make outgoing connections and receive replies. The last rule, deny all, denies any packet not explicitly passed by an ACL rule.

Single Packets in Isolation

A serious limitation of packet filter firewalls is that they only look at individual packets in isolation. Many attack packets cannot be recognized in isolation, so packet filter firewalls cannot be used as main corporate firewalls. However, they sometimes are placed before the main corporate firewall to "screen out" many simple attacks so that the main corporate firewall will not have to deal with these attacks.

TEST YOUR UNDERSTANDING

14.   a) What fields does a packet filter firewall examine in arriving packets? b) What is an ACL? c) Why is the last rule in an ACL "deny all"? d) Why can't packet filter firewalls be used as main corporate firewalls? e) How can they be used effectively?

Stateful Firewalls

In general, firewalls are liberal about allowing clients to establish outside connections but restrictive in allowing connections from the outside to servers with well-known

1. If destination IP address = 60.47.3.9 AND TCP destination port = 80 OR 443, PASS
   [connection to a public webserver]

2. If ICMP Type = 0, PASS
   [allow incoming echo reply messages]

3. If TCP destination port = 49153 and 65535, PASS
   [allow incoming packets to ephemeral TCP port numbers]

4. If UDP destination port = 49153 and 65535, PASS
   [allow incoming packets to ephemeral UDP port numbers]

5. DENY ALL
   [deny all other packets]

Figure 9-14   Access Control List (ACL) for a Packet Filter Firewall

port numbers. The default behavior of **stateful firewalls** is to allow all connections initiated by internal hosts but to block all connections initiated by external hosts.

> The default behavior of stateful firewalls is to allow all connections initiated by internal hosts but to block all connections initiated by external hosts.

Allowing Internally Generated Connections
Stateful firewalls implement this behavior by asking if each arriving packet is part of an internally initiated connection. Figure 9-15 shows that when a client opens a connection to an external server, the default is to permit this, and the stateful firewall records a connection consisting of the two host IP addresses and port numbers. Returning packets also are passed by the firewall if they are part of an approved connection.

Blocking Externally Generated Connection Attempts
However, if a packet arriving from the outside is not part of an internally initiated connection, the packet is dropped by default. This prevents external attackers from connecting to internal servers.

Stateful ACLs
The default behavior of allowing all internally initiated connections and stopping all externally initiated connections is a good default policy, but exceptions must be made. Fortunately, default behavior can be modified with a stateful inspection access control list. For instance, external connections might be allowed to a particular internal webserver that needs to serve external customers or suppliers.

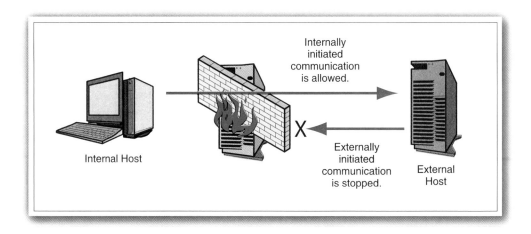

Figure 9-15   Stateful Firewall Default Operation

Simplicity and Low Cost
This is very simple behavior, so stateful firewalls can filter large volumes of traffic inexpensively.

Strong Automatic Security
Stateful firewalls are attractive also because their default behavior provides very strong automatic protection, so users normally do not have to do anything. It is very difficult to attack a stateful firewall from the outside unless ACLs have been set to allow external access on certain ports. The Internet Connection Firewall (ICF) in Windows XP is a stateful firewall (see Chapter 9a).

Corporate Border Firewalls
The strength and efficiency of stateful firewalls make them the primary choice today for corporations' main border firewalls.

> The strength and efficiency of stateful firewalls make them the primary choice today for corporations' main border firewalls.

TEST YOUR UNDERSTANDING

15. a) What is the default behavior for stateful firewalls? b) Describe how a stateful firewall passes packets associated with established connections. c) Why are ACLs needed for stateful firewalls? d) Why are stateful firewalls attractive? e) What type of firewalls do most corporations use for their main border firewalls?

Application Firewalls
Application firewalls, as their name suggests, examine the application layer content of packets.

Application Firewalls
> Examine application layer messages in packets

Application Fidelity
> Requiring the application using a well-known port to be the application that is supposed to use that port
> For instance, if an application uses Port 80, application firewall requires it to be HTTP, not a peer-to-peer file transfer program or something else
> This is called enforcing application fidelity

Limited Content Filtering
> Allow FTP Get commands but stop FTP Put commands
> Do not allow HTTP connections to black listed (banned) websites
> E-mail application server may delete all attachments

Antivirus Scanning
> Few application firewalls do antivirus filtering
> Packets also must be passed through antivirus filtering programs

Figure 9-16   Application Firewalls (Study Figure)

Enforcing Protocol Fidelity

Their main job is to ensure that an application using a particular port is the application it claims to be. This is called ensuring **protocol fidelity.** For instance, if an application attempts to come in over Port 80 (HTTP), the application firewall will ensure that it really is an HTTP application, not a peer-to-peer file-sharing program trying to get around firewall restrictions.

Limited Application Content Filtering

An application firewall also examines some aspects of application message content.

> ➤ FTP Get requests from clients to outside servers might be allowed, while FTP Put messages, which can be used to send files out, might be denied to prevent the transmission of trade secrets outside of the firm.

> ➤ HTTP requests to websites on a corporate **blacklist** of banned websites may be dropped.

> ➤ E-mail application firewalls may delete all attachments in incoming messages.

Not Virus Scanning

You might think that application firewalls do virus scanning. However, this usually is not the case. Separate antivirus programs usually are needed to scan website downloads, e-mail messages and attachments, and other file-transfer mechanisms.

TEST YOUR UNDERSTANDING

16.   a) What is protocol fidelity? b) Why is it needed? c) What type of content filtering do application firewalls do? d) Do application firewalls normally do virus scanning?

Defense in Depth

By contrasting the types of firewalls, we may have made it seem that they are competitors. In fact, they are normally used together, so that attackers will have to break through several different firewalls in a row. This is defense in depth. As Figure 9-17 shows, the border router often acts as a packet filter firewall, screening out many common attacks. The main firewall, in turn, is a stateful firewall. After the stateful firewall, the attacker has to break through an application firewall and then a host firewall on the intended victim host.

TEST YOUR UNDERSTANDING

17.   a) Why are firewalls often implemented in series? b) Describe the firewalls shown in Figure 9-17, listed in the order an attacker would encounter them.

Other Protections

Intrusion Detection Systems (IDSs)

By tradition, firewalls only drop packets that are proven attack packets. They do not drop packets that are merely suspicious because most of these are likely to be innocent packets. To give an analogy, a police officer only arrests someone if the officer has probable cause (a reasonably high level of certainty). However, if someone acts suspiciously, the officer may investigate him or her.

While firewalls drop certain attack packets, **intrusion detection systems (IDSs)** log suspicious packets and actively alert administrators if an attack appears to be underway. The administrator can then check to see if the attack is real.

Unfortunately, IDS technology is fairly immature, and IDSs typically generate large numbers of **false positives** (false alarms) for every real incident they find. Just as the boy who cried wolf too often was ignored when a wolf really did appear, IDSs with

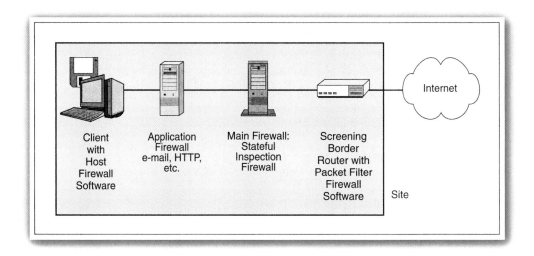

Figure 9-17   Defense in Depth with Firewalls

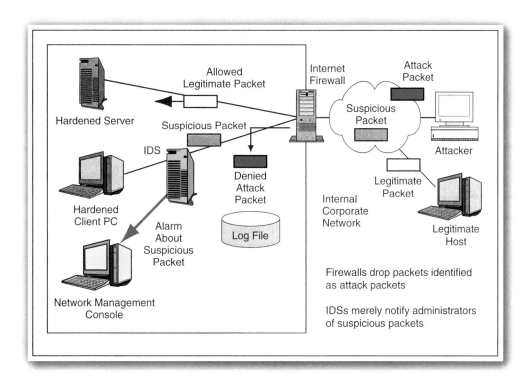

Figure 9-18   Firewalls, Intrusion Detection Systems (IDSs), and Intrusion Prevention Systems (IPSs) (Study Figure)

too many false positives are turned off. Some analysts believe that IDSs built into firewalls will generate fewer false positives.

Intrusion Protection Systems (IPSs)

The newest protection technologies are **intrusion prevention systems (IPSs).** Like firewalls, IPSs drop packets. However, they are more sophisticated than firewalls, examining streams of packets and analyzing both headers and application message contents. Consequently, IPSs are capable of identifying denial-of-service attacks and other complex attacks that firewalls cannot identify or stop. Once an attack is identified, subsequent packets participating in the attack are dropped. Most organizations are moving carefully in considering IPSs because unless an IPS is very precise, it will drop many legitimate packets—in effect, subjecting the firm to a self-inflicted DOS attack.

TEST YOUR UNDERSTANDING

18.   a) Distinguish between what firewalls and IDSs do when they examine packets.
b) What is the main problem with IDSs? c) Distinguish between firewalls and IPSs.
d) Why are companies hesitant to adopt IPSs?

PROTECTION WITH CRYPTOGRAPHIC SYSTEMS

Cryptographic Systems

In Chapter 7 we saw that many companies use virtual private networks (VPNs), which transmit data over the nonsecure Internet with added security. We looked at three VPN standards: SSL/TLS, PPTP, and IPsec. These three standards and many others are called cryptographic systems; as Figure 9-19 shows, **cryptographic systems** provide security to multimessage dialogues with three handshaking phases followed by ongoing communication. The figure specifically illustrates how SSL/TLS works between a browser on a client PC and a webserver program on a webserver.

TEST YOUR UNDERSTANDING

19. a) What do cryptographic systems protect? b) What cryptographic system is illustrated in this section?

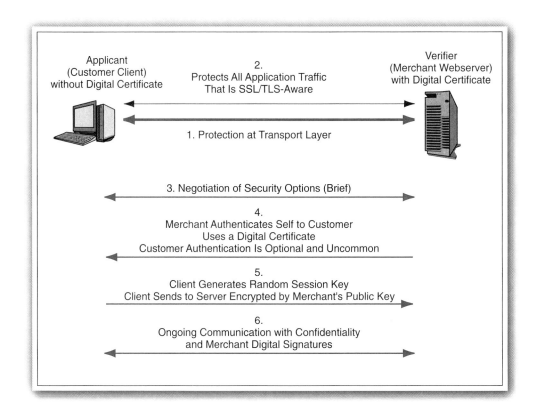

Figure 9-19   Cryptographic System (SSL/TLS)

Negotiation and Authentication

Initial Negotiation

In the first phase, the two parties negotiate how they will do security. In every cryptographic system, the parties must negotiate **optional parameters.** For instance, in SSL/TLS, the client can authenticate itself with a digital certificate or not authenticate itself at all. The two sides must agree on this and other matters.

Authentication

Most cryptographic systems provide **mutual authentication** (authentication by both parties.) In the figure, however, the server authenticates itself to the client, but the client does not authenticate itself to the server. This often is the case in consumer electronic commerce. The server first sends its digital certificate to the client's browser. Later the server adds a digital signature to each message it sends to the client. This **digital signature** contains the results of an authenticating calculation, which the browser checks using the public key contained in the digital certificate (see Figure 9-10).

In addition to providing message-by-message authentication, a digital signature ensures that the message has not been changed en route. This is called **message integrity.**

TEST YOUR UNDERSTANDING

20.   a) Why is a negotiation phase needed in cryptographic systems? b) What is mutual authentication? c) What two protections do digital signatures provide?

Key Exchange

Symmetric Key Encryption for Confidentiality

For ongoing communication, the two sides will **encrypt** each message so that the communication provides **confidentiality,** that is, assurance that interceptors cannot read it. Ongoing communication uses **symmetric key encryption,** in which the two sides use the same key to encrypt messages to each other and to decrypt incoming messages. As Figure 9-20 shows, only a single key is used in exchanges with symmetric key encryption.

Public Key Encryption for Confidentiality

The symmetric key must be kept secret from interceptors. How can the two parties exchange the keys secretly? The answer is that they use public key encryption. As Figure 9-20 shows, in **public key encryption,** each side has a public key and a private key, so there are four keys in total. The sender encrypts messages with the receiver's public key. The receiver, in turn, decrypts incoming messages with the receiver's own private key. Only the receiver knows his or her own private key, so only the receiver can decrypt the message.

Strengths and Weaknesses of Public Key Encryption

Public key encryption does not require the exchange of secret keys—only the exchange of nonsecret digital certificates containing public keys. However, public key encryption can only be done on short messages. Symmetric keys are short enough, so public key encryption can be used to encrypt symmetric keys.

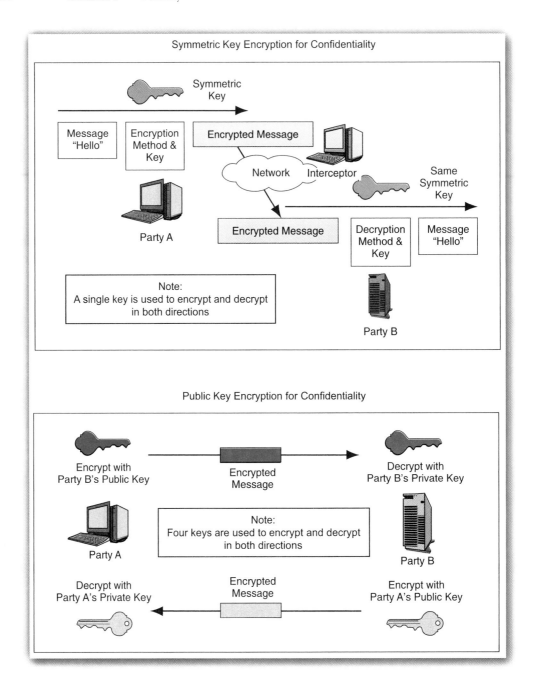

Figure 9-20   Symmetric Key Encryption and Public Key Encryption

Key Exchange in SSL/TLS
In SSL/TLS the client randomly generates a symmetric key. It then encrypts the symmetric key with the server's public key and sends the encrypted key to the server. After the server decrypts the key, both parties have the symmetric key and can do ongoing encryption for confidentiality.

Session Keys
The symmetric key is called a **session key** because it is used for only the current communication session. The next time the browser communicates with the webserver application program, the client will create a new symmetric session key.

TEST YOUR UNDERSTANDING
> 21.   a) Distinguish between symmetric key encryption and public key encryption for confidentiality. b) In two-way dialogues, how many keys are used in symmetric key encryption? c) In public key encryption? d) What kind of encryption can be used for long messages? e) How is key exchange done in SSL/TLS? f) What is a session key?

Ongoing Communication
After the three handshaking phases—negotiation, authentication, and key exchange—both parties have the symmetric session key, and the browser has the server's public key.

In ongoing communication, which is far longer than the initial handshaking phases, the two parties encrypt each message with the symmetric session key, and the browser authenticates each message from the server by applying the server's public key to each server message's digital signature.

TEST YOUR UNDERSTANDING
> 22.   What is the longest phase in cryptographic system security?

OTHER ASPECTS OF PROTECTION

Hardening Hosts

Server Hardening
Even if companies install several layers of firewalls to protect servers, some attack packets will still get through. Consequently, servers have to be **hardened,** that is set up to protect themselves.

Vulnerabilities and Patching   As noted earlier in this chapter, security vulnerabilities are found frequently in both operating systems and application programs. When vulnerabilities are found, vendors develop patches (software updates). If companies install these updates, the companies will be immune to worms and hackers who can exploit these vulnerabilities.

Unfortunately, companies tend to be slow to install patches on servers and sometimes do not install them at all. In contrast, attackers are quick to develop **exploits** (programs that exploit known vulnerabilities). Consequently, hackers often have an easy time breaking into host computers, and worm writers often have massive success.

Hardening Servers and Client PCs
        Setting up computers to protect themselves
        Server Hardening
                Patch vulnerabilities
                Minimize applications running on each server
                Use host firewalls
                Backup so that restoration is possible
        Client PC Hardening
                As with servers, patching vulnerabilities, minimizing applications, having a firewall, and implementing backup
                Also, a good antivirus program that is updated regularly
                Client PC users often make errors or sabotage hardening techniques
Vulnerability Testing
        Protections are difficult to set up correctly
        Vulnerability testing is attacking your system yourself through a consultant
        There must be follow-up to fix vulnerabilities that are discovered

Figure 9-21   Other Aspects of Protection (Study Figure)

Minimizing Applications When servers are installed, certain application programs are set to run each time the computer boots up. In the past, the default has been to install many application programs so that they can be available whenever users need them. However, this gives attackers many targets. Many firms are now implementing the minimum number of applications that the users of a server absolutely need. This makes servers much more difficult to attack.

Host Firewalls Even if a vulnerable program is installed on a server, an attacker cannot attack it unless the attacker can send messages to the server's well-known port number. If the server has a host firewall that permits only messages to the well-known port numbers absolutely required for the server's specific purposes, attackers will not be able to exploit these vulnerabilities.

Backup The most basic computer protection of all is to keep data on servers backed up frequently. Security breaches are inevitable, and recovery may be difficult or impossible unless the data of affected systems have been backed up recently.

Client PC Hardening
The same things that are important in server hardening—minimizing program installation, installing patches, having a firewall, and implementing backup—are important in client PC hardening as well. In addition, clients need to have good antivirus programs.

The problem with client security is that it usually is left to end users, who lack the knowledge and skills needed to make their computers secure. In addition, many end users actively sabotage security, for instance by turning off their antivirus programs to speed the transmission of attachments. Users may even install prohibited software, such as music file-sharing software, which attackers can use to compromise the computers.

TEST YOUR UNDERSTANDING

23.   a) What is host hardening? b) What are the four steps in hardening servers? c) What are the steps in hardening client PCs?

Vulnerability Testing

One problem when creating protections is that it is easy to make mistakes when configuring firewalls and other protections. It is essential to do **vulnerability testing** after configuring protections. In vulnerability testing, the company or a consultant attacks protections in the way a determined attacker would and notes which attacks that should have been stopped succeed.

Vulnerability testing must result in a report that leads to a plan to upgrade protections to remove vulnerabilities. There must be follow-up checking to ensure that the plan actually is carried out.

TEST YOUR UNDERSTANDING

24.   a) What is vulnerability testing? b) Why is it important? c) Describe appropriate steps *after* vulnerability testing is completed.

RESPONSE

The last phase in the plan–protect–respond cycle is response—dealing with the attacks that succeed in getting through the company's protection systems. Although attacks vary widely, there generally are four stages in responding to any attack.

Detecting the Attack

The first stage is detecting the attack. This can be done by the firm's IDS or simply by users reporting apparent problems. Obviously, until an attack is detected, the attacker will be able to do unabated damage.

Stopping the Attack

The second stage is stopping the attack. The longer an attack has to get into the system, the more damage the hacker can do. Reconfiguring corporate firewall ACLs may be able to end the attack. In other cases, attack-specific actions will have to be taken.

Repairing the Damage

The third stage is repairing the damage. In some cases, this is as simple as running a cleanup program or restoring files from backup tapes. In other cases, however, it may involve the reformatting of hard disk drives and the complete reinstallation of software and data.

Response

Dealing with attacks that succeed

Response Phases

Detecting the attack

If not detected, damage will continue unabated

IDS or employee reports are common ways to detect attacks

Stopping the attack

Depends on the attack

Reconfiguring firewalls may work

Repairing the damage

Sometimes as simple as running a cleanup utility

Sometimes, must reformat a server disk and reinstall software

Punishing the attackers

Easier to punish employees than remote attackers

Forensic tools collect data in a manner suitable for legal proceedings

Major Attacks and CSIRTs

Major attacks cannot be handled by the on-duty staff

On-duty staff convenes the computer security incident response team (CSIRT)

CSIRT has people from security, IT, functional departments, and the legal department

Disasters

Natural and attacker-created disasters

Can stop business continuity (operation)

Data backup and recovery are crucial for disaster response

Dedicated backup facilities versus real-time backup between different sites

Business continuity recovery is broader

Protecting employees

Maintaining or reestablishing communication

Providing exact procedures to get the most crucial operations working again in correct order

Figure 9-22   Incident Response (Study Figure)

Punishing the Attacker

The fourth stage is punishing the attacker if possible. This is easiest if the attacker is an employee because remote attackers can be extremely difficult to prosecute even if they can be identified. If legal prosecution is to be pursued, it is critical to use proper **forensic** tools to capture and retain data in ways that fit rules of evidence in court proceedings.

Major Attacks and CSIRTs

Minor attacks can be handled by the on-duty IT staff. However, for major incidents, such as the theft of thousands of credit card numbers from a corporate host, the company must convene the firm's **computer security incident response team (CSIRT).** The team involves the firm's security staff, members of the IT staff, and members of functional departments, including the firm's legal department.

Disasters

When natural disasters, terrorist attacks, or other major attacks occur, **business continuity** (the company's ability to continue operations) may be halted. Companies must have active plans for disaster recovery.

 Disaster recovery is the reestablishment of information technology operations. Many large firms have dedicated backup sites that can be put into operation very quickly, after data and employees have been moved to the backup site. Another option, if a firm has multiple server sites, is to do real-time data backup across sites. If one site fails, the other site can take over immediately or at least very rapidly.

 Business continuity recovery requires a firm to do much more than deal with data and processing recovery. Business continuity teams must deal with protecting employees during disasters, maintaining or reestablishing communication among key employees, and providing the exact procedures needed to get the most critical operations working again in the correct order.

TEST YOUR UNDERSTANDING

25. a) What are the four response phases when attacks occur? b) What is the purpose of forensic tools?

26. a) Why are CSIRTs necessary? b) Should the CSIRT be limited to security staff personnel?

27. a) What is business continuity? b) What is disaster recovery? c) What types of backup sites do firms use in disaster recovery? d) How is business continuity recovery a broader process than disaster recovery?

SYNOPSIS

Attacks

Companies today suffer compromises from many different types of attacks.

➤ Viruses attach themselves to other programs and need human actions to propagate—most commonly by opening e-mail attachments that are infected programs. Worms are full programs; they can spread by e-mail, but self-propagating worms can propagate on their own, taking advantage of unpatched vulnerabilities in victim hosts. Some worms and viruses have damaging payloads. Often, payloads place a Trojan horse program on the victim computer.

➤ Hacking is intentionally using a computer resource without authorization or in excess of authorization. Hacking break-ins require a prolonged series of actions on the part of attacker.

➤ Denial-of-service (DoS) attacks overload victim servers so that they cannot serve users.

Attackers

Traditionally, attackers were curious hackers and disgruntled employees and ex-employees. Now, criminals are attacking in growing numbers. On the horizon, cyberterror attacks by terrorists and and cyberwar attacks by national governments could do unprecedented levels of damage.

Security Management

Security is primarily a management issue, not a technical issue. To implement security, companies go through three phases: planning, protecting (applying protections), and responding (reacting when protection fails).

Planning involves risk analysis (balancing the costs and benefits of protections), comprehensive security (closing all avenues of attack), and defense in depth (providing successive lines of defense in case one line of defense fails).

Authentication

Protection must begin with authentication (proving an applicant's identity to a verifier) and authorization (specifying what the applicant can do to various resources).

➤ Passwords are inexpensive and easy to use, but users typically choose poor passwords that are easy to crack.

➤ Digital certificate authentication at the other extreme is very strong but is complex and expensive to implement.

➤ Biometrics promises to use bodily measurements to authenticate applicants, replacing other forms of authentication. Concerns with biometrics include error rates and deliberate deception by applicants.

Firewalls

Firewalls examine incoming packets and outgoing packets. If they find attack packets, they drop them and record them in a log file.

➤ Packet filter firewalls examine contents in IP headers, TCP headers, UDP headers, and ICMP messages. They pass or drop packets based on rules in their access control lists (ACLs). They examine packets in isolation and consequently cannot catch all attacks.

➤ Stateful firewalls automatically allow internally originated connections and stop externally initiated connections. This provides powerful automatic protection at low cost. ACLs can modify the default behavior of stateful firewalls. Most main firewalls are stateful firewalls.

➤ Application firewalls examine application message content. However, application firewalls normally do not do virus filtering.

At large sites, firms combine multiple types of firewalls to provide defense in depth.

Firewalls sometimes provide intrusion detection and intrusion protection. Intrusion detection notifies security administrators of suspicious packets but does not drop them. Intrusion protection systems do sophisticated analysis on streams of packets to identify and automatically stop attacks.

Cryptographic Systems

Cryptographic systems provide protections to multimessage dialogues.

➤ First, the two parties negotiate how security will be done.

➤ Then they usually mutually authenticate themselves.

➤ Afterward, they exchange symmetric keys. One party randomly generates a symmetric session key and encrypts this key with the public key of the other party. The other party decrypts the symmetric key with its private key.

➤ Finally, the two parties engage in ongoing communication using symmetric key encryption in which both sides encrypt and decrypt messages using a single key.

Symmetric key encryption is almost always used to encrypt messages for confidentiality so that eavesdroppers cannot read the messages. In symmetric key encryption, both sides encrypt and decrypt with a single key. In public key encryption, each side has a private key and a public key. For confidentiality, each side encrypts messages with the public key of the other party. However, public key encryption is rarely used for confidentiality. The one major exception is encryption by a symmetric session key for transmission to the other party in a dialogue.

Host Hardening

Ongoing protection includes server and client hardening to protect them from attacks. Servers can be hardened by patching vulnerabilities, minimizing applications running on the server, installing host firewalls, and backing up servers regularly. Client PC hardening includes the same protections, plus installing an antivirus program.

Vulnerability Testing

Protections are very difficult to set up and configure. Consequently, firms should conduct vulnerability tests in which they or a consultant attempts to attack the firm, in order to identify security weaknesses.

Response

Protections eventually break down. Major incidents require the convening of a computer security incident response team (CSIRT). Disasters can stop business operation (business continuity). Disaster recovery requires getting IT back in operation at a backup site. Business continuity recovery is a broader process involving many parts of the firm, not merely IT. For taking legal actions against attackers, the firm needs to conduct forensic data collection, in which data are collected in a form suitable for court proceedings.

SYNOPSIS QUESTIONS

1. a.) Distinguish between viruses and worms. b.) How can worms propagate in a way that viruses cannot? c.) What are unpatched vulnerabilities? d.) What are payloads? e.) What are Trojan horses?
2. a.) What is hacking? b.) Are break-ins usually done in a single step? c.) What are DoS attacks? d.) Distinguish between traditional types of attackers and newer types of attackers.
3. a.) Is security primarily a technical issue or a management issue? b.) Briefly explain the three phases in the plan–protect–respond cycle. c.) What type of encryption is almost always used for confidentiality? d.) When is public key encryption used for confidentiality? e.) How many keys are used in two-way symmetric key encryption? f.) In two-way public key encryption for confidentiality?
4. a.) What is authentication? b.) What are applicants and verifiers? c.) Give a pro and a con of password authentication. d.) Give a pro and a con of digital certifi-

cate authentication. e.) What is biometric authentication? f.) What are concerns about biometrics? g.) What is single sign-on?

5. a.) What do firewalls do? b.) What do packet filter firewalls examine? c.) How do packet filter firewalls decide what packets to pass or drop? d.) What do stateful firewalls do unless instructed to do differently by an ACL? e.) What do application firewalls examine? f.) Do application firewalls usually do virus filtering? g.) Distinguish between what firewalls, IDSs, and IPSs do.

6. a.) What type of protection is applied to dialogues? b.) What are the four steps in implementing them? c.) How is the exchange of a symmetric session key done? d.) What protection is applied in ongoing communication?

7. a.) What can be done to harden hosts? b.) What must be done beyond these steps to harden client PCs?

8. a.) What does vulnerability testing do? b.) Why is it done?

9. a.) Why is a response phase necessary? b.) What kinds of incidents require the CSIRT to be convened? c.) Distinguish between disaster recovery and business continuity recovery. d.) What is forensic data collection? e.) Why is it needed?

THOUGHT QUESTIONS

1. a.) What form of authentication would you recommend for relatively unimportant resources? Justify your answer. b.) What form of authentication would you recommend for your most sensitive resources?

2. Modify the ACL in Figure 9-14 to permit outside access to an internal FTP server, 60.47.47.119.

3. A file-sharing program tries to get through a firewall by acting like webserver communication. What type of firewall is most likely to be able to stop it?

4. a.) What information would a ping give to hackers in the scanning phase of a break-in? b.) Would the ACL in Figure 9-14 stop a ping attack? Explain.

9a

HANDS-ON: WINDOWS XP HOME SECURITY

Learning Objectives:

By the end of this chapter, you should be able to discuss:

- Windows XP Home versus Windows XP Professional.
- Windows updating.
- Antivirus protection.
- Internet security and privacy options.
- The Windows Internet Connection Firewall (a built-in stateful firewall).
- Packet filter firewall operation with Windows TCP/IP filtering.
- Malware scanning programs.
- Advanced security with Windows XP Professional, including domains, domain controllers, and group policy objects (GPOs).
- Security improvements in Service Pack 2 (SP2).

INTRODUCTION

Windows dominates the market for client PC operating systems, and Windows XP is the dominant version of Windows in both corporations and in homes. Therefore, it is important for you to understand how to deal with Windows XP security.

Windows XP comes in both a Home version and a Professional version. The Home version is designed for residential use, the Professional version for corporate use. Most students have computers running XP Home rather than XP Professional, so we will focus on XP Home security. This will allow students to do hands-on exercises at home. Fortunately, both versions provide basic security in the same way and differ primarily in advanced areas we will cover at the end of this chapter.

The Big Two

We will begin with the "big two" security actions that every desktop user should take:

➤ Updating Windows by downloading security patches.
➤ Adding antivirus software and keeping it current.

Other Security Measures

Windows XP security should not stop with these two steps, however. We will look at several other actions that corporations should consider to protect their desktops:

➤ Setting Internet security and privacy options.
➤ Adding a firewall.
➤ Adding an antispyware program.
➤ Establishing virtual private networks.
➤ Establishing group security policies with Windows XP Professional.

TEST YOUR UNDERSTANDING

1. a) Distinguish between Windows XP Home and Windows XP Professional. b) Why does this chapter deal with Windows XP Home? c) What two security precautions should every client PC user take?

WINDOWS UPDATES

Security Vulnerabilities and Updates (Patches)

As discussed in Chapter 9, when any piece of software ships, it is likely to have unknown security vulnerabilities. When a security vulnerability is discovered, the software vendor releases an update (patch) to remove the vulnerability.[1] Users must download this update or be at risk for attack. This is not a theoretical concern. Attacks on a newly discovered vulnerability typically begin within a few weeks of its discovery. Updating Windows is not a luxury.

---

[1]   Updating Windows has benefits beyond security. Updates also fix bugs in the software and often adds some functionality.

The Need for Windows Updates
  To patch security vulnerabilities
  To fix bugs and add functionality
Options
  Automatic updating turned on by default in Windows XP
  Default is to notify user of updates before downloading and installing
  Option to download but notify user of the need to install
  Option to download and install without user intervention
    Dangerous because problem updates may cause difficulties for users
Other Matters
  Work-arounds (manual) are difficult for end users
  Service packs are cumulative collections of updates
  Service packs must be installed in order of their creation
  Severe updates may be loaded immediately while others wait
Updating Applications
  All applications must be updated as well to eliminate security vulnerabilities
  If an application is taken over, an attacker may be able to take over the computer
  Updating applications is difficult because there are so many of them
  Each will have a different method for users to discover, download, and install updates

Figure 9a-1   Windows Updates (Study Figure)

Turning on and Configuring Automatic Updates

In earlier versions of Windows, users had to do considerable work to download and install updates. In Windows XP, however, users by default are notified whenever an update is ready to be downloaded and installed. This is an enormous improvement and is by itself a good reason to upgrade to Windows XP.

Figure 9a-2 shows how to configure Windows updates. The first step, not shown in the figure, is to right-click on the "My Computer" icon on the desktop. This brings up the "System Properties" dialog box shown in the figure.

When you get to the dialog box, click on the "Automatic Updates" tab. Then click on "Keep my computer up to date." Now you can decide how you should update Windows. You have three choices:

➤ "Notify me before downloading any updates and notify me again before installing them on my computer" ensures that users will not be interrupted as they work, except for receiving brief notifications. This is the default option. Unfortunately, if this option is set, the user may or may not download the update. If this option is widely set, many corporate desktops will go unprotected by updates.

➤ "Download the updates automatically and notify me when they are ready to be installed" is a little better in terms of security. The user will only have to act to do the actual installation. This may be good because most updates require the computer

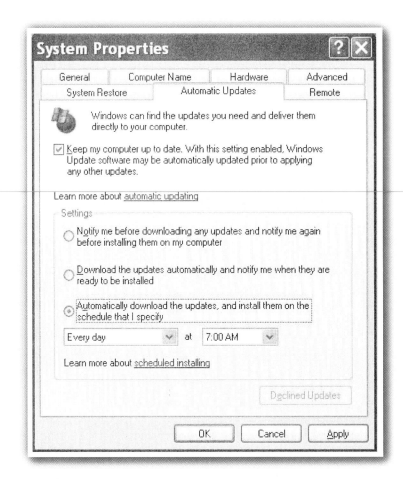

Figure 9a-2   Automatic Windows Updates in System Properties

to be rebooted. Although this rebooting can be delayed, rebooting is disruptive. Consequently, users may wish to defer installation. However, deferring installation can lead to simply not doing the installation at all.

➤ "Automatically download the updates, and install them on the schedule that I specify" is the safest option from a security viewpoint, especially if the "Every day" option is set to ensure prompt updating.

Automatic Updating and Problem Updates

From what we have seen, automatic downloading and updating is the safest option. However, there is one problem. In some cases, updates can be disruptive. For instance, one Windows XP update in 2002 slowed computer operation dramatically. Microsoft did not issue another update to correct this problem for more than a month.

In fact, many companies routinely delay the installation of updates generally, in order to test them before widely installing them. This is understandable, but it leaves a company open to attacks during the testing period. In addition, avoiding automatic installation often means that many computers never have some updates installed. There simply is no good solution to the difficulties raised by **problem updates.**

Work-Arounds

In some cases, installing an update will not fix a security vulnerability. Instead, the company will have to engage in a **work-around,** in which manual changes will have to be made even after an update is installed. In some cases, the work-around involves potentially dangerous actions, such as modifying the system registry. While work-arounds are feasible on servers, which are administered by trained professionals, they make little sense for client PCs.

Service Packs and Severity Ratings

Microsoft issues both individual updates and **service packs,** which combine a number of individual updates into a single cumulative update. On a newly installed computer, it is important to install service packs. There may be several, and they should be installed in order. This takes time but is much faster than installing many individual updates.

Software vendors often have severity ratings for their updates. Updates with high severity ratings correct severe security vulnerabilities. Updates for severe problems should be installed promptly, of course. Many users, however, may wait for service packs to install updates with lower severity ratings.

Updating Applications

Although we have focused on Windows updating, individual application programs also have to be updated regularly. If an attacker can take over an application, he or she often can take over the entire computer. Unfortunately, every application vendor has a different mechanism for updating its programs. Updating applications can be more time consuming and difficult than updating Windows if a user has many applications. If fact, it may be difficult even to discover updates for many applications.

TEST YOUR UNDERSTANDING

2. a) Why are Windows updates crucial for security? b) What updating behavior is the default for Windows XP? c) What are the other two options for updating? d) Which option is the safest from the standpoint of ensuring that updates will be made? e) Why is this option dangerous? f) What are work-arounds, and why are they bad? g) What are service packs? h) In what order should service packs be installed? i) How may severity ratings affect updating choices? j) Why is it essential to update applications as well as the operating system? k) Why is updating all applications on a PC difficult?

ANTIVIRUS SCANNING

Updates primarily provide protection against directly attacking worms and human hackers. They normally do little or nothing to protect against viruses and worms arriving in e-mail attachments. As discussed in Chapter 9, malware attacks are extremely

Importance
     Viruses are widespread
     Every PC needs antivirus software to stop incoming (and outgoing) viruses
Using Antivirus Programs Effectively
     Virus definitions database and program must be updated frequently
     Preferably daily
     Program must be configured to work with user's e-mail, other programs
     Antivirus software must be selected to work with user's applications, including
     peer-to-peer
User Subversion
     Turning off antivirus programs to reduce problems, work faster
     Turning off (or not turning on) automatic dating
     Failing to pay for subscription extensions

Figure 9a-3   Antivirus Scanning (Study Figure)

widespread. According to MessageLabs (www.messagelabs.com), one to two percent of all e-mail messages contain malware, and this percentage continues to rise. Every client PC needs **antivirus software.**

It is difficult to talk specifically about antivirus scanning programs because they differ in their specific operation. However, all have some things in common.

Updates

Most importantly, all programs need to be updated regularly. The antivirus program uses a **virus definitions database** to allow it to identify viruses. As new viruses are found, the virus definitions database must be updated.[2] Otherwise, the antivirus program will provide no protection at all against new viruses. On the contrary, having a nonupdated antivirus program that is no longer effective may give the user a false sense of security.

All antivirus programs have an automatic updating mechanism. However, they vary in how automatic this updating is, and they typically leave it up to the user to decide how often updating is performed. Given the rate of spread of some new viruses, checking for updates every time the computer is turned on is good policy.

Configuration and Breadth of Protection

All antivirus programs will filter e-mail, but the user has to configure the program by telling it what e-mail programs he or she uses. Most antivirus programs also filter website downloads. However, antivirus programs differ considerably in whether they filter peer-to-peer applications such as instant messaging and file downloading. Users need to consider what applications they use when selecting an antivirus program.

---

[2] Updates also tend to extend program functionality.

User Subversion and Failing to Pay for Subscriptions

Even if an antivirus program is installed on a user's PC, users often subvert it. Sometimes they turn it off entirely if they think it is interfering with other programs. In other cases they turn it off to speed up their computer's operation by not waiting for scanning. In yet other cases, users turn off (or do not turn on) automatic updating. More subtly, antivirus program subscriptions have to be paid for annually. Users who fail to pay for subscription extensions lose update protection.

TEST YOUR UNDERSTANDING

3. a) Does updating Windows provide antivirus protection? b) What does? c) What are virus definitions databases? d) Why are updates needed for antivirus programs? e) What is a good schedule for updates to antivirus programs? f) Why must antivirus programs be configured? g) What kind of applications should antivirus programs filter? h) How may users subvert antivirus protection?

OTHER COMMON SECURITY MEASURES

Although handling updates and virus checking is important and time consuming, companies may extend more security protection to Windows XP users.

Internet Explorer Security and Privacy Options

One common step is to develop standards for Internet options for security and privacy (the protection of personal information).

To set Internet options, go to the Control Panel and choose "Network and Internet Connections." This will take you to the "Network and Internet Connections" dialog box shown in Figure 9a-4. At the bottom of the dialog box, select "Internet Options." This will take you to the "Internet Options" dialog box shown in Figure 9a-5.

Security Options
In this figure, the "Security" tab is selected. It lists four content zones for which different security policies can be set:

➤ Internet. This is the default. All websites not in other zones are handled by your security settings for the Internet zone.

➤ Local intranet. This content zone is for webservers inside a company's intranet. These webservers typically are protected by a firewall, so lower security may be appropriate. Users can add local intranet websites to this zone.

➤ Trusted sites. These are highly trusted sites. They usually are given low security barriers.

➤ Restricted sites. These are sites that should not be used or that should be used only with great caution.

Setting Security Options
To change the security settings for a zone, click on the zone and then click on the "Custom Level" button. This will take you to the "Security Settings" dialog box shown in Figure 9a-6. This dialog box has a complex list of settings. Some corporations have check lists for these settings.

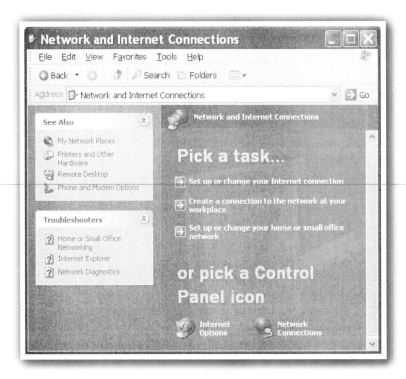

Figure 9a-4   Network and Internet Connections Dialog Box

Privacy Options

Privacy settings are managed through the "Privacy" options tab on the "Internet Options" dialog box. As Figure 9a-7 shows, Windows XP provides a slider control that lets the user select more or less privacy. The default, medium privacy, generally is a good choice. If desired, the Advanced button can allow the user to configure specific privacy options.

TEST YOUR UNDERSTANDING

4.   a) How are security zones used in website security? b) What is the default zone for websites? c) Describe security in other zones. d) How can privacy settings be configured simply?

Internet Connection Firewall (Stateful)

The Need for PC Firewalls

Firewalls are designed to stop attacks. Many of these attacks exploit vulnerabilities, so updating Windows XP regularly helps. However, not all attacks exploit known vulnerabilities. In addition, **zero-day exploits** take advantage of vulnerabilities that have not previously been discovered or for which updates have not been created.

Figure 9a-5   Internet Options Dialog Box Security Tab

Turning on ICF

To protect against such attacks, Windows XP comes with a built-in stateful firewall, the **Internet Connection Firewall (ICF).** To turn ICF on, go through the following steps:

➤ Go to the "Network and Internet Connections" dialog box shown in Figure 9a-4.

➤ Click on "Network Connections" for a list of your network connections. (See Figure 9a-8.)

➤ Right-click on an Internet connection you wish to protect with ICF. (This must be done separately for each connection.) This will bring up the connection's properties dialog box shown in Figure 9a-9.

➤ Click on the "Advanced" tab.

➤ Click the box "Protect my computer and network by limiting or preventing access to this computer from the Internet."

Figure 9a-6   Security Settings Dialog Box

Automatic Operation

The Internet Connection Firewall operates automatically, as do most stateful firewalls. By default, any connection the user initiates is permitted, and any attempt to connect to the computer from the outside will fail. This generally is desired behavior.

> By default, any connection the user initiates is permitted by the Internet Connection Firewall (ICF), and any attempt to connect to the computer from the outside will fail.

Changing Settings

Internally initiated connections are always permitted, and ICF offers no way to change this.

However, to make exceptions to this default behavior for externally initiated connections, the user can click on the Settings button. This takes the user to the "Advanced Settings" dialog box, which has three tabs.

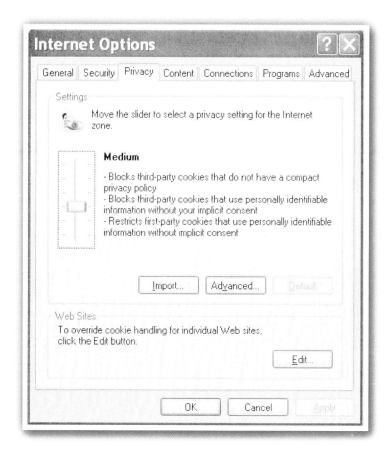

Figure 9a-7   Internet Options Dialog Box Privacy Tab

➤ Services. This permits selected externally initiated connections to server applications running on the user PC, such as an FTP server. Clicking on a service permits outside connections to it.

➤ ICMP. This permits types of incoming ICMP supervisory messages. (The default is to deny all types of incoming ICMP messages.) This default behavior should be changed cautiously because attackers often use ICMP messages to map networks.

➤ Logging. This tab controls the logging of connection information.

Third-Party Firewalls
Although ICF is a good firewall, it is somewhat limited. Third-party firewalls, such as ZoneAlarm, provide much more functionality. In particular, they stop attack packets going out from internally initiated connections. This way, if attack programs find their way onto a user's PC, these programs cannot send their attack packets out through the firewall.

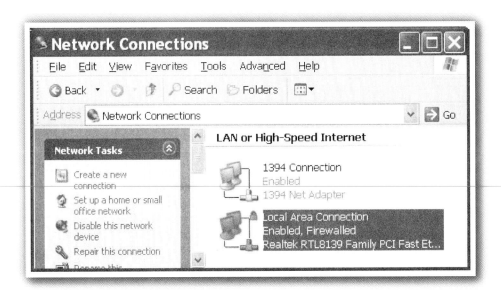

Figure 9a-8   Network Connections Dialog Box

TEST YOUR UNDERSTANDING

5. a) Why are PC firewalls needed? b) What is Microsoft's built-in firewall program? c) What type of firewall is it? d) By default, is it on or off? e) What is its default behavior when turned on? f) Can it filter outgoing communication? g) Can it be set to allow specific types of externally initiated connections? h) Why may it be good to use a third-party firewall?

Packet Filter Firewall (TCP/IP Filtering)

Windows XP also comes with a packet filter firewall. Microsoft calls it "TCP/IP filtering."

Getting to TCP/IP Filtering

To turn on TCP/IP filtering, go to "Network and Internet Connections" (see Figure 9a-4), then to Network Connections. This will show you a list of your network connections. Right-click on a connection, and select properties.

Now you will see a list of protocols used by the network connection. Click on "Internet Protocol (TCP/IP)" to select it. Then click on the "Properties" button.

You are now in the "Internet Protocol (TCP/IP Properties)" dialog box. From here, choose the Advanced button.

In the "Advanced TCP/IP Settings" dialog box that appears, click on the "Options" tab. As Figure 9a-10 shows, TCP/IP filtering appears. Click on it, and select "Properties."

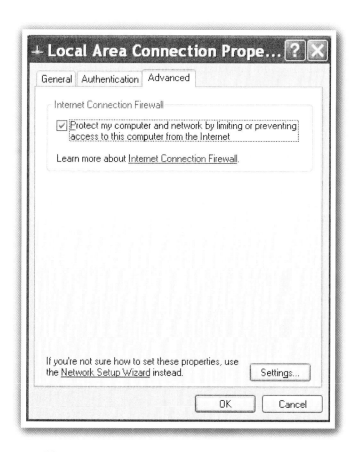

Figure 9a-9   Internet Connection Properties Dialog Box

Configuring TCP/IP Filtering

This takes you to the "TCP/IP Filtering" dialog box shown as Figure 9a-11. Note that you have to click on a box on the top to enable TCP/IP filtering. Clicking the box again turns off TCP/IP filtering. The default is off.

Note also that once you enable TCP/IP filtering, you enable it for all adapters (network and internet connections). In contrast, the Internet Connection Firewall has to be enabled on each connection individually.

By default, all incoming traffic is permitted (TCP/IP filtering ignores outgoing traffic). To specify dropping rules, click on "Permit Only" on the TCP/Ports, UDP Ports, or IP Protocols radio button. Then click on "Add" to specify TCP ports, UDP ports, or IP protocols that should be allowed into the computer. All types of packets not specifically permitted will be dropped automatically. In effect, there is an implicit Deny All rule as the last rule.

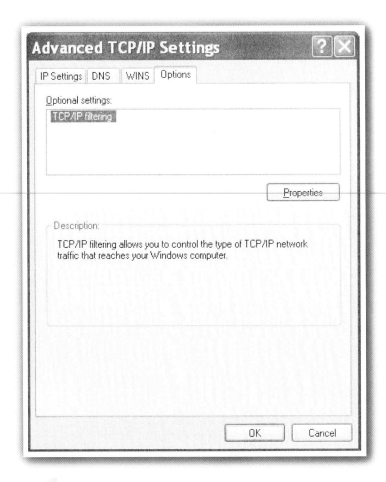

Figure 9a-10   Options in Advanced TCP/IP Settings Dialog Box

Perspective

Windows XP cannot support both TCP/IP filtering and the Internet Connection Firewall simultaneously. In most cases, the Internet Connection Firewall is the better choice. It is automatic and can stop some attacks that TCP/IP filtering cannot. However, TCP/IP filtering is a good hands-on tool for getting practice in setting up a packet filter firewall.

TEST YOUR UNDERSTANDING

6.   a) What is Windows XP's packet filter firewall function called? b) By default, is it on or off? c) When you turn it on, how can you permit specific types of communication? d) What happens to communication you do not specifically permit? e) Is TCP/IP filtering done for each network connection or for all? f) Is ICF the same?

Figure 9a-11   TCP/IP Filtering Configuration

Malware-Scanning Programs

Although antivirus programs and firewalls stop most attacks, some may get through. **Malware-scanning programs** search a user's PC looking for installed **malware (evil software),** including Trojan horse programs and spyware. (Spyware sends information about the user to an outside source.)

People who use these malware-scanning programs find that a large majority of all PCs have some sort of malware installed on them. When the author ran one of these malware-scanning programs, Ad-aware (www.lavasoft.de), on his own system, it removed eighty-two spyware programs and program components. This happened despite the author's use of good security practices. Some security analysts feel that using malware-scanning programs should be added to updating and antivirus filtering as essential security tools for client PCs.

New types of malware are appearing constantly. Consequently, malware scanning programs have to be updated regularly to remain effective.

TEST YOUR UNDERSTANDING

7. a) What is malware? b) What is spyware? c) What is a malware-scanning program? d) Why is a malware-scanning program needed even if precautions to prevent attacks are taken?

Malware
    Evil software
        Viruses and worms
        Trojan horses
        Spyware (reports personal information to outside parties)
    Gets onto client PCs despite security precautions
Malware Scanning Programs Scan for Malware
    Usually find malware
    Must be updated

Figure 9a-12   Malware Scanning Programs (Study Figure)

Establishing Virtual Private Networks

As discussed in Chapter 7, many corporations now use virtual private networks (VPNs), in which the transmission travels over the Internet with added security.

Two Connections

Windows XP permits VPNs in a simple but somewhat confusing way. To use a VPN, *two* connections are needed, as Figure 9a-13 illustrates. First, the user needs a connection to the Internet. Second, once the Internet connection is established, the user establishes a second connection, this time to a security server at a corporate site.

In the "Network and Internet Connections" dialog box (Figure 9a-4), the second choice is "Create a connection to the network at your workplace." This is how you establish a VPN connection to a corporate server.

Setting up the Connection

Selecting this choice starts a New Connection Wizard. During this wizard, you will specify a name for the connection and the host name (or IP address) of the computer to which you wish to connect.

The next screen asks if you wish to have a dial-up connection to the network or if you wish to establish a virtual private network. We will select the VPN alternative, but the dial-up alternative is an important way for many users to get into their networks. The wizard ends with a connection screen, as shown in Figure 9a-14.

Properties

You can set the properties of the connection on the connection screen. Alternatively, you can click on a connection on the "Network Connections" dialog box and right-click on a connection to change its properties. Figure 9a-15 shows the dialog box that appears. The "Security" tab allows the user to establish advanced settings if the user clicks on the "Advanced" radio button and then the Settings button.

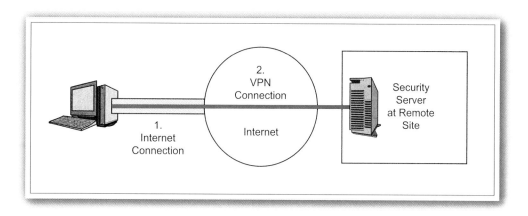

Figure 9a-13   Two Connections for Windows XP VPN

Figure 9a-14   Connection Screen for a VPN

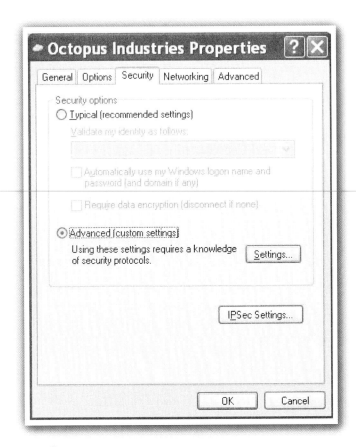

Figure 9a-15   VPN Properties Dialog Box

Figure 9a-16 shows the "Advanced Security Settings" dialog box that appears. This permits the user to specify an authentication protocol for the connection. In addition to basic authentication methods, this dialog box can specify digital certificate authentication and smart cards.

Windows XP Professional
Although Windows XP Home users can establish VPNs to other computers (and allow incoming VPN connections as well), Windows XP Professional, which is designed to work in more sensitive corporate environments, offers more capabilities and options.

TEST YOUR UNDERSTANDING

8. a) What is a VPN? b) What two connections are established to set up a VPN? c) How can you configure VPNs? d) How do Windows XP Professional VPNs differ from Windows XP Home VPNs?

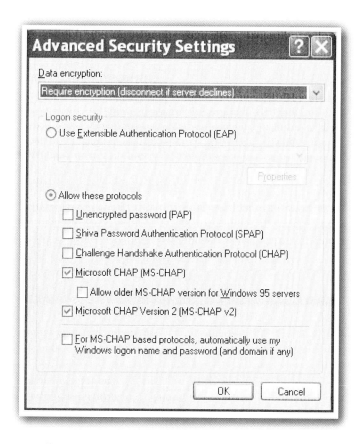

Figure 9a-16   Advanced VPN Security Settings

WINDOWS XP PROFESSIONAL

Windows XP Home, as its name suggests, is designed for home users who connect to the Internet and who may want to share files and printers with other PC users in the same home.

In contrast, Windows XP Professional is designed to be used in the corporate environment. Consistent with the more heightened security environment in corporations than in homes, Windows XP Professional has a number of security advances over XP Home.

In corporate environments, PCs are parts of a larger infrastructure. As Figure 9a-17 shows, organizations arrange their clients and servers into groups of resources called **domains.** The computers in the domain are managed by one or more domain controllers. The domain's clients and other servers (called member servers) are managed by the **domain controller.**

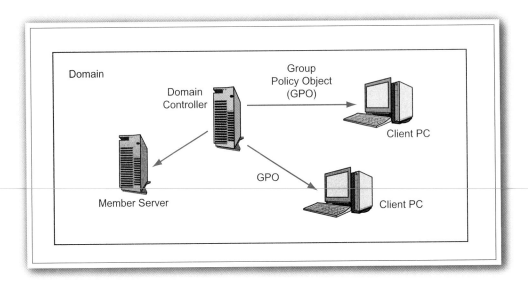

Figure 9a-17   Windows Domain

This domain structure permits a network administrator to exert strong control over individual computers in the domain. In particular, **group policy objects (GPOs)** are policies that govern a specific type of resource, such as client PCs. With GPOs, network administrators can set many security parameters on all PCs within the zone. GPOs can even "lock down" the desktop, controlling what programs can be used and even controlling the layout of the user's desktop.

TEST YOUR UNDERSTANDING

9.   a) What are domains? b) What are domain controllers? c) What are GPOs? d) How can GPOs be used to manage security on Windows XP client computers?

WINDOWS XP SERVICE PACK 2 (SP2)

As this book is going to press, Microsoft is about to release Windows XP Service Pack 2. In contrast to most service packs, which fix bugs, SP2 is primarily a security upgrade.

First, during SP2 installation, users will be forced to decide how they will handle updates. The screen requiring this cannot be bypassed.

Second, the Interconnection Firewall Facility will be substantially upgraded. It will be turned on by default, bringing substantial protection. In addition, the user will have to decide which programs on the computer may communicate over the Internet. Also, the name of the program is likely to be changed to Windows Firewall.

TEST YOUR UNDERSTANDING

10.   What security improvements will SP2 bring?

HANDS-ON EXERCISES

Do the following exercises on your home Windows XP Home computer or on a Windows XP Home computer in your lab.

1. Check the computer's Windows updating. a) What option is selected? b) Comment on the suitability of this option.
2. a) What antivirus program is on the computer? b) How does is it update itself? (You may have to use the help function.) c) What applications is it configured to protect?
3. a) What security option is set for the Internet zone? b) Change this level. c) Then change it back to its original value. d) What level of privacy protection does the computer have? e) Change this level. f) Then change it back to its original value.
4. a) Is the Internet Connection Firewall turned on? b) If not, turn it on. c) Permit incoming FTP connections. d) Reverse the previous two steps.
5. a) Turn on TCP/IP filtering. b) How is it configured? c) Permit incoming HTTP connections but no other incoming connections. d) Reverse these steps.
6. a) Create a VPN connection to a mythical computer at a mythical site. b) Configure the VPN connection to use a digital certificate for authentication.

CHAPTER 10

NETWORK MANAGEMENT

Learning Objectives:

By the end of this chapter, you should be able to discuss:

- Total cost of ownership (TCO) analysis.
- Network simulation.
- Managing IP, including IP subnet planning, private IP addresses, DHCP, DNS, WINS, directory servers, and configuration for switches and routers.
- Network management tools for client PC connectivity, route analysis, and network mapping.
- Remote management with Telnet, SSH, web-based configuration, and TFTP.

NETWORK MANAGEMENT

Up to this point, we have focused primarily on network technology. However, technology is worthless unless the network is well planned and managed. In this chapter we will look at some key issues and skills in network management.

The fact that network management comes late in the book should not be taken as an indication that it is unimportant. Rather, network management comes after discussions of technology because it is impossible to discuss network management in a comprehensive way until the student has a strong understanding of the technologies that network administrators must manage.

COST ANALYSIS

Demand Versus Budget

As Figure 10-1 illustrates, user demand for networking is growing rapidly, but network budgets are either stagnant or growing very slowly. This puts extreme cost pressure on each project. When considering candidate projects, it is critical to develop realistic cost projections.

Labor Costs

In most of this book, we have focused on technology because there are many technical concepts that you must know if you wish to work in networking. However, hardware, software, and carrier services are only a fraction of the total cost of running a

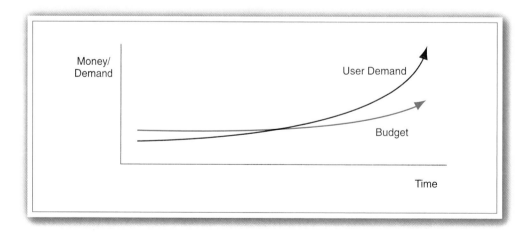

Figure 10-1   Network Demand Versus Budget Trends

The Importance of Costs
  Exploding demand
  Slow budget growth
  Falling hardware costs help, but software costs fall more slowly, and labor
  costs are rising
  Select the least expensive technology that will fully meet user needs
Non-Technology Costs
  Labor costs
        High, and unit labor costs are rising over time
  Carrier fees
Total Cost of Ownership
  Fully configured cost of hardware
  Fully configured cost of software
  Initial installation costs
        Vendor setup costs
        IT and end-user labor
  Ongoing costs
        Upgrades
        Labor costs often exceed all other costs
        Immature products have very high labor costs
  Total cost of ownership (TCO): total of all costs over life span

Figure 10-2   Cost Issues (Study Figure)

network. For instance, capital equipment purchases account for just under half of
the average networking budget.[1]

Labor dominates nontechnology costs. A third of the average corporate networking budget is employee labor. It is expensive to plan and implement a complex
new networking technology, and the long-term labor costs to operate a network system
on a day-to-day basis are more expensive still. In addition, although hardware costs are
falling rapidly, hourly labor costs actually tend to increase over time. Finally, the users
who need the system to do their functional work will also have labor costs to implement and use the system.

Carrier Fees

If a company has wide area networks, it has to pay fees to carriers. These can be very
substantial.

---

1 Network budget percentages are from Sharon Gaudin, "Spending on the Rise," *Network World*, January
   29, 2001. http://www.nwfusion.com/research/2001/0129feat.html.

Total Cost of Ownership (TCO)

Although costs must be managed, this is difficult to do because costs are highly complex and take place over the multiyear time span of most projects. Even so, the costs of projects must be estimated accurately. Companies have lost patience with "guesstimates" that end up in large cost overruns.

Figure 10-3 illustrates a typical cost analysis. Note that it includes several types of costs. Note also that these costs come over the entire life span of a project, not just during its initial planning and implementation stages.

The Fully Configured Cost of Products

We must first look at the **total purchase cost of network products.** Just as PCs are sold as collections of components, most network products, such as switches and routers, come as collections of hardware and software components. A **fully configured** switch or router often costs far more than the **base price** listed in catalogs. When comparing products from different vendors, it is critical to compare fully configured products. Services such as public switched data networks also tend to come with many options that can add considerably to the base price.

Although hardware is visible and, therefore, easy to appreciate, software costs are very important in the total purchase price. In many cases, software costs exceed hardware costs. As in the case of hardware, software often comes as a base price bundle plus many options.

	Year 1	Year 2	Year 3	Year 4	Total
Base Hardware	$200,000	15,000	15,000	15,000	245,000
Hardware Options	85,000	9,000	9,000	9,000	112,000
Base Software	$100,000	10,000	10,000	10,000	130,000
Software Options	50,000	10,000	10,000	10,000	80,000
Technology Subtotal	435,000	44,000	44,000	44,000	567,000
Planning and Development	75,000				75,000
Implementation	50,000				50,000
Ongoing IT Labor	100,000	75,000	75,000	75,000	325,000
Ongoing User Labor	50,000	25,000	25,000	25,000	125,000
Labor Subtotal	275,000	100,000	100,000	100,000	575,000
Total	710,000	144,000	144,000	144,000	1,142,000

Note: The total cost of ownership is $1,142,000.

Figure 10-3 Multiyear Cost Analysis: Total Cost of Ownership (TCO)

Initial Installation Costs

In addition, there typically are substantial costs associated with **initial installation.** Some vendors, especially carriers, charge one-time **setup fees.**

There also are a company's own **initial labor costs** associated with installation. These include the costs of the network staff's labor, and they also include the labor costs of end users.[2] End-user labor costs may rise dramatically during installation periods because of training and disruption. In general, the labor costs to create and implement a new network tend to be very high.

Ongoing Costs

Initial costs, as large as they are, often are far smaller than **ongoing costs.** These may come in the form of hardware or software upgrades over the life of the product selected or in the form of monthly payments to vendors. Also substantial are the ongoing labor costs to operate the system. Ongoing costs, especially ongoing labor costs, may be far larger than initial costs. As in the case of initial costs, the ongoing costs to end users must be taken into account.

Ongoing costs are especially high for new technologies that are not yet **mature.** It may take several years for technologies to become easy to install and use. Immature products often lack utilities that allow easy management; therefore, they require a great deal of expensive labor. It is a good idea to avoid new "bleeding edge" technologies if possible. Perversely, vendors and the networking trade press typically focus on such new and immature technologies, giving a rather distorted picture of the workable options available to network administrators.

Total Cost of Ownership (TCO)

In Figure 10-3, the total cost of the system over its expected four-year lifespan is $1,142,000. This is called the **total cost of ownership (TCO).** This TCO is far higher than the initial base price of the system's hardware ($200,000). As is typical in many systems, labor costs exceed the costs of hardware and software, accounting for about three-quarters of all costs. When comparing projects, it is important to compare TCOs.

TEST YOUR UNDERSTANDING

1. a) Why are base prices misleading? b) Why is multiyear analysis necessary? c) How important are labor costs? d) Why do immature products tend to lead to high labor costs? e) What is TCO? f) Why is it important? g) What are the elements of TCO?

NETWORK SIMULATION

As noted earlier in the chapter, designing a new network or a modified network is very difficult because the designer faces many alternatives and because network components tend to interact in unforeseen ways. Network simulation allows network designers to get a handle on this complexity. In addition, it is far more economical to simulate many alternatives than to build several real systems to study.

---

[2] End users are employees in functional departments, such as marketing or finance, who use networking to perform their functional work more effectively.

Simulation

    What-is versus what-if
    More economical to simulate network alternatives than to build them
    Purposes
            Comparing alternatives to select the best one
            Base case and sensitivity analysis to see what will happen if the values of
            variables were varied over a range
            Anticipating problems, such as bottlenecks
            Planning for growth, to anticipate areas where more capacity is needed
Before the Simulation, Collect Data
    Data must be good
    Otherwise, GIGO (garbage in, garbage out)
    Collect data on the current network
    Forecast growth
The Process
    Based on OPNET IT Guru
    Add nodes to the simulation work area (clients, servers, switches, routers, etc.)
    Specify the topology with transmission lines
    Configure the nodes and transmission lines
    Add applications, which generate traffic data
    Run the simulation for some simulated period of time
    Examine the output to determine implications
    Validate the simulation (compare with reality if possible to see if it is correct)
    What-if analysis
    Application performance analysis (OPNET ACE)

Figure 10-4   Network Simulation (Study Figure)

Network Simulation Purposes

There are several specific reasons to do network simulation.

➤ Comparing alternatives in order to identify the best one.
➤ Sensitivity analysis. This means choosing a most likely situation and seeing how changing ranges of configuration values will affect various performance measures.
➤ Anticipating problems. In the simulation, determining where bottlenecks (points of congestion) will appear.
➤ Planning for growth. Areas where the network will not continue to meet needs as traffic grows must be determined by extrapolating traffic.

TEST YOUR UNDERSTANDING

2. a) Why is network simulation attractive economically? b) For what purposes is network simulation done? c) What is sensitivity analysis?

Before the Simulation: Collecting Data

The best simulation analysis is worthless if the model does not include realistic data. Analysts use the acronym **"GIGO"**—garbage in, garbage out—to emphasize that if input data is bad, the results cannot be accurate. Networking simulation data is never perfect, but simulations are best if the firm collects actual data on its traffic—including how traffic varies by time of day and how much it fluctuates from second to second. Of course, simulations usually deal with the future, so traffic needs to be extrapolated to what it is likely to be in the future.

TEST YOUR UNDERSTANDING

3. a) Explain "GIGO." b) How can a firm get good data for a simulation?

The Process

Once data is collected, the modeler can begin to create the simulation. We will discuss simulation using the popular **OPNET IT Guru** network simulation program.

Adding Nodes

The first step in building a simulation is to place **nodes** (items to be connected by transmission lines) on the simulation work area. These nodes can be clients, servers, switches, routers, and other types of devices. As Figure 10-5 shows, IT Guru has templates with icons of common nodes. You can drag and drop these onto the simulation work area.

Specifying the Topology

The next step is to specify the **topology**—how the nodes are linked together by transmission lines. IT Guru has a template with icons for transmission lines and networks. You select a transmission icon and then decide what nodes it will link. In this case, a Frame Relay network will be used to connect the host computers. You now have something that looks very much like the network you are modeling.

Configuring Elements

However, the nodes and transmission links need to be configured before your simulation of them is complete. For instance, on a router, you will have to specify the speeds of various interfaces. You also will have to configure specific operating parameters, such as the window size field in TCP. Figure 10-6 shows a Frame Relay transmission link being configured so that its outgoing excess burst size (Be) is set to 64 kbps.

Adding Applications

Your nodes and lines are now ready to work, but you need to specify traffic. IT Guru has you do this by specifying applications, the traffic characteristics of these applications, and on which nodes they will run. This is realistic because traffic is created by applications, not by the network.[3] Figure 10-7 shows the final configured simulation model after applications have been added.

---

[3] Somewhat oddly, each application appears as an icon on the simulation working area. It is important not to confuse these software objects with the physical hardware and transmission line objects on the working area.

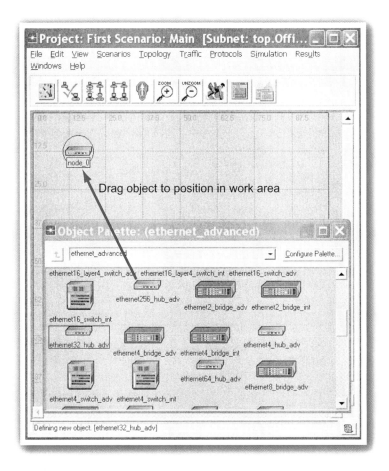

Figure 10-5 OPNET IT Guru Node Template

Running the Simulation

Your simulation is now complete, so the next step is to run it. You can specify a time period over which the simulation is to run—say over an entire simulated day or during a simulated busy period. After you do this, IT Guru will run the simulation using sophisticated statistical and queuing theory methods.

Examining the Output

IT Guru gives you many alternatives for looking at your output. Graphical output is best for searching for trends or anomalies. You often can learn a great deal by looking at the simulation results in detail.

Validating the Simulation

If you are simulating a real network, you can **validate** the model by comparing its performance with that of the real network. If the model gives very different output, you

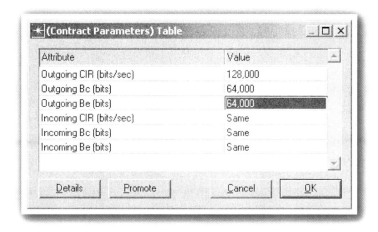

Figure 10-6   Configuring a Frame Relay CIR

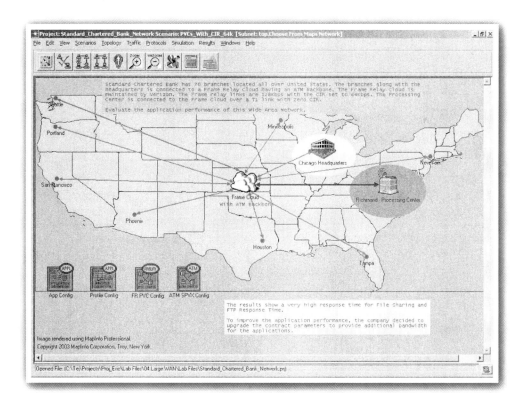

Figure 10-7   Configured Simulation Model

need to revise the model and attempt to validate it again. Of course, for proposed networks, validation generally is impossible.

What-If Analysis

Now comes the real power of simulation—running what-if analyses to see how specific changes would change the results. You might change a single parameter on a single computer (such as a windows size field in TCP). You might change the speed of a transmission line or consider an upgrade for a router. You might see if adding a router in a particular place would get rid of a bottleneck. The possibilities are endless. By trying many alternatives, you can develop a network solution that is optimized for the company's business needs.

　　Figure 10-8 shows a what-if analysis in which two PVC speeds are tried to determine the impact on response times for two applications—FTP and database queries. It shows that PVC committed information rate has a major impact on FTP response time but not on database response time.

Application Analysis

OPNET offers a product related to IT Guru. This is the **Application Characterization Environment (ACE).** While IT Guru focuses primarily on network-level and internet-level performance, ACE allows the modeler to focus on application performance.

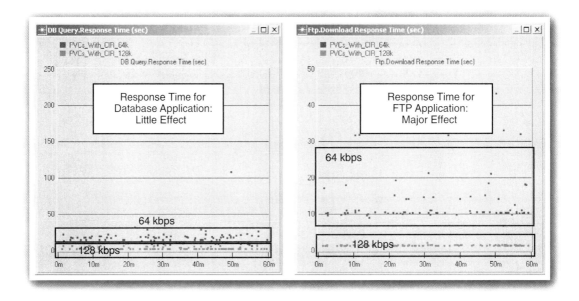

Figure 10-8　What-If Analysis

TEST YOUR UNDERSTANDING

4.   a) What are nodes? b) What is a topology? c) List the steps in building a network simulation. d) What steps should be undertaken after running the simulation? e) What is validation, and why is it important? f) What is what-if analysis? g) Distinguish between IT Guru and ACE.

IP MANAGEMENT

If a firm uses TCP/IP as its internetworking protocol, it must do a considerable amount of work to build and maintain the necessary infrastructure of IP.

IP Subnet Planning

As Chapter 8 discussed, IP addresses are 32 bits long. Each organization is assigned a network part. We saw that the University of Hawai`i's network part (128.171) is 16 bits long. It was up to the university to decide what to do with the remaining 16 bits.

Subnetting at the University of Hawai`i

The university, like most organizations, chose to subnet its IP address space. It divided the 16 bits over which it has discretion into an 8-bit subnet part and an 8-bit host part.

The 2^N-2 Rule

With N bits, you can represent 2^N possibilities. Therefore, with 8 bits, one can represent 2^8 (256) possibilities. This would suggest that the university can have 256 subnets, each with 256 hosts. However, a network, subnet, or host part cannot be all zeros or all ones. Therefore, the university can have only 254 (256−2) subnets, each with only 254 hosts. Figure 10-9 illustrates these calculations.

Step	Description				
1	Total size of IP address (bits)	32			
2	Size of network part assigned to firm (bits)	16		8	
3	Remaining bits for firm to assign	16		24	
4	Selected subnet/host part sizes (bits)	8/8	6/10	12/12	8/16
5	Possible number of subnets (2^N-2)	254 (2^8-2)	62 (2^6-2)	4,094 ($2^{12}-2$)	254 (2^8-2)
6	Possible number of hosts per subnet (2^N-2)	254 (2^8-2)	1,022 ($2^{10}-2$)	4,094 ($2^{12}-2$)	65,534 ($2^{16}-2$)

Figure 10-9   IP Subnetting

In general, if a part is N bits long, it can represent 2^N-2 networks, subnets, or hosts. For example, if a subnet part is 9 bits long, there can be 2^9-2, or 510, subnets. Or if a host part is 5 bits long, there can be 2^5-2, or 30, hosts.

In general, if a part is N bits long, it can represent 2^N-2 networks, subnets, or hosts.

Balancing Subnet and Host Part Sizes

The larger the subnet part, the more subnets there will be. However, the larger the subnet part is made, the smaller the host part must be. This will mean fewer hosts per subnet.

The University of Hawai`i's choice of 8-bit network and subnet parts was useful for many years because no college needed more than 254 hosts. However, many colleges now have more have more than 254 computers, and the limit of 254 hosts required by its subnetting decision has become a serious problem. Several colleges now have two subnets connected by routers. This is expensive and awkward.

The University would have been better served had it selected a smaller subnet part, say 6 bits. As Figure 10-9 shows, this would have allowed 62 college subnets, which probably would have been sufficient. A 6-bit subnet part would give a 10-bit host part, allowing 1,022 hosts per subnet. This would be ample for several years to come.

A Critical Choice

In general, it is critical for corporations to plan their IP subnetting carefully, in order to get the right balance between the sizes of their network and subnet parts.

Using Private IP Addresses

As just noted, companies are assigned network parts for their IP addresses. If this network part is large, then the organization will have comparatively few bits left over to use for subnets and hosts.

However, three IP address ranges have been designated as private IP addresses. These are the three ranges:

➤ 10.x.x.x
➤ 192.168.x.x
➤ 172.16.x.x through 172.31.x.x

Private IP addresses may not be used on the Internet itself. However, they may be used within a firm. Network address translation (NAT), which we saw in Chapter 1, allows internal private IP addresses to be translated into public IP addresses for transmission over the Internet.

The 10.x.x.x range of private IP addresses has only 8 bits in its network part. This leaves 24 bits for the subnet and host parts, allowing the firm to select large subnet parts and large host parts simultaneously.

For instance, as Figure 10-9 shows, if the network and subnet parts are both set at 12 bits, then the firm can have $2^{12}-2$ (4,094) subnets, each with up to 4,094 hosts. Alternatively, if the subnet and host parts had been 8 bits and 16 bits respectively, there could be up to 254 subnets, each with up to 65,534 hosts.

TEST YOUR UNDERSTANDING

5.   a) Why is IP subnet planning important? b) If you have a subnet part of 9 bits, how many subnets can you have? c) Your firm has the 8-bit network part 60. If you need at least 250 subnets, what must your subnet size be? d) How many hosts can you have per subnet? e) Your firm has a 20-bit network part. What subnet part would you select to give at least 10 subnets? f) How many hosts can you have per subnet? g) How are private IP address ranges used? h) What are the three ranges of private IP addresses?

Administrative IP Servers

Managing IP also requires the organization to set up and manage several **administrative IP servers,** which are needed to support IP.

DHCP Servers

Most clients have temporary IP addresses delivered by DHCP servers. These DHCP servers have to be set up and operated by the firm. Some firms have each subnet manager maintain a DHCP server for his or her subnet. At the University of Hawai`i, a strong tradition of decentralization made this a logical choice. At the First Bank of Paradise, however, each site has a single DHCP server—even the headquarters site, which has many subnets. Each DHCP server has a configurable **scope** parameter which determines how many subnets it will serve.

DNS Servers

Large organizations with second-level domain names (such as pukanui.com) are responsible for maintaining a Domain Name System (DNS) server that lists all internal hosts that have host names. Most firms have a primary DNS server and a secondary DNS server for redundancy because losing DNS service will make the network useless for most users. Maintaining the firm's shifting use of host names on the DNS server is a full-time job in large organizations.

WINS

Before Windows 2000 Server, Windows had a different computer naming system associated with Microsoft's NETBIOS protocol. Just as DNS provides IP addresses for host names, **Windows Internet Name Service (WINS)** servers provide IP addresses for NETBIOS computer names.

Suppose an older Windows client or server needs to send a packet to another computer whose NETBIOS name it knows. The computer sends a WINS request message to a WINS server giving the NETBIOS computer name of the target. The WINS server sends back a response message containing the IP address of the target.

Firms that have older clients and servers with older versions of Windows (and this is nearly all firms) need to support WINS servers as well as DNS servers.

Directory Servers and LDAP

Many firms now have directory servers, which centralize information about a firm.

Hierarchical Organization   As Figure 10-10 shows, information in a directory server is arranged hierarchically, much as entries in a DNS server are organized (see Chapter 8).

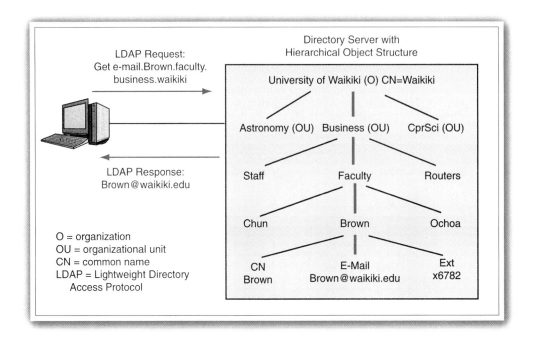

Figure 10-10   Hierarchical Directory Server Name Space

The figure shows the directory structure for the mythical University of Waikiki. The top level is the **organization.** Under the top level, there are schools (**organizational units,** in directory server terminology). In each school, there are faculty, staff, and router categories. Under the faculty category, there are the usernames of faculty members. At the bottom of the hierarchy are the **properties** of individual faculty members, including the faculty member's common name, e-mail address, and telephone extension.

Lightweight Directory Access Protocol (LDAP)   Most directory servers today permit query commands governed by the **Lightweight Directory Access Protocol (LDAP).** The figure shows that LDAP commands specify the path to a property, with individual nodes along the way separated by dots. This is why the request for Brown's e-mail address is specified as the following:

    e-mail.brown.faculty.business.waikiki

Directory Servers and the Networking Staff   Organizations store a great deal of information about themselves in directory servers, including a great deal of networking information. Creating a directory server requires a great deal of planning about what information an organization needs to store and how this information should be arranged hierarchically.

Although creating and managing a directory server goes well beyond networking, the networking staff is often given the task of leading directory server planning projects and managing the directory server on a daily basis.

TEST YOUR UNDERSTANDING

6. a) What major types of servers must be created and maintained to manage IP in a corporate environment? b) What are DHCP scopes? c) Why do most firms have both a primary and a secondary DNS server? d) What is the purpose of WINS? e) For what operating systems is WINS needed? f) How is information in directory servers organized? g) What is the purpose of LDAP? h) If Astronomy has a similar directory organization to Business (in Figure 10-10), give the specification for the telephone extension of Claire Williams (username cwilliams), who is an Astronomy staff member.

Device Configuration

Individual devices on the network—clients, servers, switches, and routers—need to be configured to work with TCP/IP. We saw how to configure Windows XP Home clients for networking in Chapter 1a.

Configuring IOS Devices

Cisco Systems, which dominates the market for switches and routers, has an operating system that it uses on all of its routers and all of its switches. This is the **Internetwork Operating System (IOS).**

Initial Router Configuration

When you turn on a Cisco router for the first time, you are led through a sequence of configuration questions. This allows you to configure much of what needs to be configured about the router. However, although this extended configuration mode gets a router up and running, it does not do all configuration work.

In the initial configuration process, the user specifies many things about the router and its interface, including:

➤ A name for the router.
➤ An enable secret for privileged mode (which is needed to do most configuration tasks).
➤ Whether or not SNMP (discussed later) should be turned on to manage the router.
➤ Whether routing should be set up for IP, IPX, AppleTalk, and other protocols.
➤ For each interface, operating characteristics, an IP address, a subnet mask, and other information.
➤ Finally, whether the configuration should be saved in NVRAM (nonvolatile RAM) for permanent use (until changed).

The Command Line Interface (CLI)

To work with the limited processing power and memory of switches and routers, IOS has to be kept as small as possible. This has necessitated using a **command line interface (CLI),** in which a user has to type highly structured commands, ending each command with Enter. Figure 10-11 shows some steps needed to configure a router in CLI mode rather than through initial configuration.

Command	Comment
Router>enable[Enter]	Router> is the prompt. The ">" shows that the user is in non-privileged mode. Enables privileged mode so that user can take supervisory actions. User must enter the enable secret. All commands end with [Enter]. Enter is not shown in subsequent commands.
Router#hostname julia	Prompt changes to "#" to indicate that user is in privileged mode. User gives the router a name, julia.
julia#config t	Enter configuration mode. The t is an abbreviation for terminal.
julia(config)#int e0	Prompt changes to julia(config) to indicate that the user is in configuration mode. User wishes to configure Ethernet interface 0. (Router has two Ethernet interfaces, 0 and 1.)
julia(config-if)#ip address 10.5.0.6 255.255.0.0	User gives the interface an IP address and a subnet mask. (Every router interface must have a separate IP address.) The subnet is 5.
julia(config-if)#no shutdown	This is an odd one. The command to shut down an interface is "shutdown". Correspondingly, "no shutdown" turns the interface on.
julia(config-if)# *Ctrl-Z*	User types Ctrl-Z (the key combination, not the letters) to end the configuration of e0.
julia(config)#int s1	User wishes to configure serial interface 1. (Router has two serial interfaces, 0 and 1.)
julia(config-if)#ip address 10.6.0.1 255.255.0.0	User gives the interface an IP address and subnet mask. The subnet is 6.
julia(config-if)#no shutdown	Turns on s1.
julia(config-if)# *Ctrl-Z*	Ends the configuration of s1.
julia# router rip	Enables the Router Initiation Protocol (RIP) routing protocol.
julia#disable	Takes user back to non-privileged mode. This prevents anyone getting access to the terminal from making administrative changes to the router.
julia>	

Figure 10-11   Cisco Internetwork Operating System (IOS) Command Line Interface (CLI)

TEST YOUR UNDERSTANDING

7. a) List some configuration tasks for routers. b) What is Cisco's operating system? c) What is a CLI? d) What is the advantage of using a CLI? e) What happens when you first turn on a new Cisco router? f) Does extended configuration when you first turn on a new Cisco router handle all configuration chores? g) Must an IP address and subnet mask be configured for each router interface? h) What is the IOS CLI prompt in nonprivileged mode? i) What is the IOS CLI prompt in privileged mode?

NETWORK MANAGEMENT UTILITIES

As networks grow more complex and geographically dispersed, network administrators need tools to manage their networks. Fortunately, network administrators have a broad spectrum of **network management utilities** that help them manage their networks. We will discuss the broad categories of functionality into which these tools fall. We saw many of these programs in the hands-on exercises in Chapter 1, and we saw more about the ubiquitous ping utility in Chapter 8.

Security Concerns

Usage Policies

Although network management utilities are very useful, the tools that network managers use are the same tools that hackers use to plan their attacks on networks. Companies need **usage policies** for who may use various tools and how they may use them in order to ensure that they are not being used to attack the firm.

Firewalls and Network Management Tools

Another consequence of hacker interest in network management utilities is that border firewalls typically keep network management packets from entering sites. This makes managing multiple sites from a single location difficult.

TEST YOUR UNDERSTANDING

8. a) Why are usage policies for network tools important? b) Why do companies often configure their border firewalls to interfere with network management utilities?

Host Diagnostic Tools

Network Setup Wizard

Figure 10-12 shows several categories of network management utilities. We will begin with host computers, which have to be configured for networking. Client configuration for Windows XP systems was discussed in Chapter 1a. Configuration for other client operating systems is broadly similar; the user or IT installer runs a network setup wizard. Most of the time, this works well and nothing else is necessary. Sometimes, however, more diagnostic help is needed.

Testing the Connection

After configuration, the installer typically tests the connection by double-clicking on the browser and seeing if he or she can go to a known website. It is even better to drop down to the command prompt (see Chapter 1) and ping a distant host. Ping, which we saw in Chapter 1, measures latency very precisely, giving potentially subtle

Security
> Management tools can be used to make attacks
> Policies should limit these tools to certain employees and for certain purposes
> Firewalls block many network management tools to avoid attacks

Host Diagnostic Tools
> Network Setup Wizard works most of the time; need tools if it does not
> Testing the connection
>> Open a connection to a website using a browser
>> Ping a host to see if latency is acceptable
> Loopback testing and ipconfig/winipconfig
>> Go to the command line
>> Ping 127.0.0.1. This is the loopback interface (you ping yourself)
>> For detailed information: ipconfig /all or winipconfig (older versions of Windows)
> Checking the NIC in Windows XP
>> Right click on a connection and select Properties
>> Under the name of the NIC, hit the Configuration button
>> The dialog box that appears will show you the status of the NIC
>> It also offers a Troubleshooting wizard if the NIC is not working
> Packet capture and display programs
>> Capture data on individual packets
>> Allows extremely detailed traffic analysis
>> Look at individual packet data and summaries
>> WinDUMP is a popular packet capture and display program on Windows
> Traffic summarization
>> Shows statistical data on traffic going into and out of the host
>> EtherPeek is a popular commercial traffic summarization program
> Connection analysis
>> At the command line, Netstat shows active connections
>> This can identify problem connections

Route Analysis Tools
> To test the route to another host
> Ping tests if a route to a host exists and what its latency is
> Tracert shows the routers along the way and latencies to them

Network Mapping Tools
> To understand how the network is organized
> Discovering IP addresses with active devices
> Fingerprinting them to determine their operating system (client, server, or router)
> A popular network mapping program is Nmap (Figure 10-6)

Simple Network Management Protocol (SNMP)
> Components: manager, managed devices, agents, objects, RMON probes
> Management information base (MIB)
> Commands, responses, traps
> Set command

Remote Switch and Router Management
> Telnet
> Web interfaces
> SSH
> TFTP

Figure 10-12 Network Management Utilities (Study Figure) indications of problems even if the browser connections succeed.[4] Tracert (Chapter 1) measures latency at various routers along the way to a target host, giving more detailed latency data.

Loopback Testing with Ipconfig or Winipconfig

If testing the connection causes problems or fails to work, it is important to get more diagnostic data. If running the browser, pinging, or tracert cannot get to the target host, the first step is to *ping 127.0.0.1,* which is your computer's **loopback interface.** Essentially, pinging 127.0.0.1 pings your own computer. If it fails, you know that there is a problem with your network setup.

The next logical step in diagnosing a Windows client is to drop down to the command prompt and run *ipconfig /all* or *winipconfig,* depending on your version of windows. This will give you a great deal of information on your Internet connection, including your MAC address, IP address, DNS addresses, subnet mask, and other basic information. This might be enough to identify the problem and suggest how to fix it. For instance, if your computer cannot reach hosts, you may have the wrong IP address for your company's DNS hosts.

Checking the NIC

Another problem may be that the NIC is not functioning properly. To check on the NIC, hit the Start button. Then choose Connect to, and select Show all connections. This will take you to a list of connections.

Right-click on the connection you have established and select Properties. This will take you to a dialog box. Under the name of the NIC, click on the Configure button.

This takes you to a new configuration dialog box for the NIC, as shown in Figure 10-13. The Device Status box will tell you if the NIC is working properly. If the NIC is not working properly, the Troubleshoot button will take you through a wizard to help you diagnose the problem.

Packet Capture and Display Programs

For even more subtle problems, **packet capture and display programs** capture selected packets or all of the packets arriving at a NIC or going out of a NIC. Afterward, you can display key header information for each packet in greater or lesser detail. This packet-by-packet analysis gives you maximum information on traffic going into and out of your computer.

The most popular freeware packet-analysis program is **TCPDUMP** for Unix, which is available as **WinDUMP** for Windows computers.[5] The user runs the program and tells it to collect data for some period of time. TCPDUMP/WinDUMP then presents data on each packet. Chapter 8a discussed WinDUMP/TCPDUMP.

Traffic Summarization

While looking at individual packets can be helpful, it may also be good to go to the other extreme, looking at broad statistical trends over periods of time. Figure 10-14

---

[4] Joseph D. Sloan, *Network Troubleshooting Tools,* Sebastopol, California: O'Reilly & Associates, 2001.
[5] A good link for downloading WinDUMP (and the WinPCAP software you must download first to run WinDUMP) is http://windump.polito.it/. The site also has good documentation on WinDUMP.

Figure 10-13   NIC Configuration Dialog Box

shows some output from **EtherPeek,** a commercial traffic summarization program. EtherPeek works by capturing all packets arriving at and leaving a NIC, then providing a wide spectrum of summarization tools. EtherPeek can show overall trends and also can drill down to specifics.

Connection Analysis

Another popular tool, which is built into the operating system in both Unix and Windows, is **Netstat.** This program shows active connections. In Figure 10-15, there is a single TCP connection, on Port 3290. The connection is closed and in the wait mode. This might be an indication that the computer has a Trojan horse program installed on it. In addition to causing damage in general, Trojan horses can cause serious problems for network connections. Alternatively, this connection might be a legitimate connection.

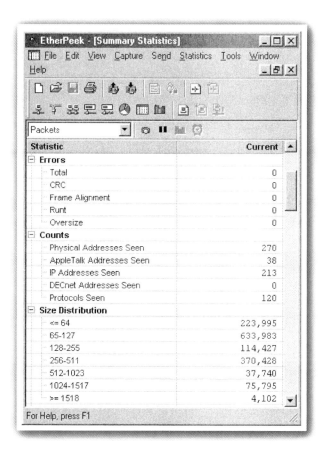

Figure 10-14   EtherPeek Packet Capture and Summarization Program

TEST YOUR UNDERSTANDING

9. a) After you use the network setup wizard to create a network connection, with what two quick ways can you verify that the connection is working? b) How can pinging be used in two ways to test a connection? c) How do ipconfig /all and winipconfig assist in troubleshooting? d) What are the benefits of packet capture and display programs? e) What is a common freeware program for packet capture and display? f) What information do traffic summarization programs like EtherPeek provide? g) What information does Netstat tell you?

Route Analysis Tools

Once the operating system can talk to the outside world, the next problem may be a route to a particular host, such as a webserver or mail server. If there appears to be a problem with the route, the network troubleshooter has to perform **route analysis** to locate the problem.

Figure 10-15   Netstat Connections Analysis Program

Our old friends ping and tracert are good places to begin. Both give latencies. Ping gives latency to the host. If this latency appears to be high, tracert will give latencies at all routers along the route.

TEST YOUR UNDERSTANDING

10.   a) What is the purpose of route analysis? b) How does ping support route analysis? c) How does tracert support route analysis?

Network Mapping Tools

Sometimes, network managers have a broader need, most commonly to map the layout of their networks, including what hosts and routers are active and how various devices are connected. **Network mapping** has two phases. The first is **discovering** hosts and subnets, that is, finding out if they exist. The second is **fingerprinting** hosts (determining their characteristics) to determine if they are clients, servers, or routers.

Although ping and tracert are useful for the host discovery phase, they are tedious to use. Network administrators typically turn to network mapping tools, including the free Nmap program that we saw in the previous chapter.

Nmap pings a broad range of possible host addresses to determine which IP addresses have active hosts. If ping will not work (or is blocked by firewalls), Nmap offers other ways to scan for active IP addresses.

Nmap also offers fingerprinting, which attempts to identify the specific operating system running on each host and sometimes even the version number of the operating system. For instance, a computer running Windows XP is almost certainly a client, while a computer running Windows Server 2003 or Solaris (the SUN version of Unix) probably is a server. Nmap can even fingerprint router operating systems to identify routers.

When network mapping is finished, the network manager should have a good understanding of how the network is organized and what parts are not working correctly.

TEST YOUR UNDERSTANDING

11. a) What is network mapping? b) Describe the two phases of network mapping. c) How can Nmap help in network mapping?

Simple Network Management Protocol (SNMP)

Although the tools we have been discussing all have their places, the Internet Engineering Task Force has developed a general way to collect rich data from various devices in a network. This is the **Simple Network Management Protocol (SNMP).** Figure 10-16 shows the major elements of SNMP communications.

The Manager

The network administrator works at a central computer that runs a program called the **network management program,** or, more simply, the **manager.**

Managed Devices

The manager is responsible for many **managed devices**—devices that need to be administered, such as printers, hubs, switches, routers, application programs, user PCs, and other pieces of hardware and software.

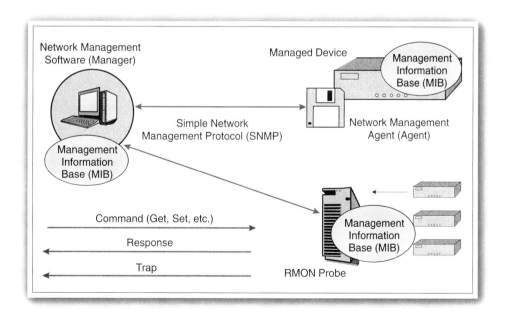

Figure 10-16   Simple Network Management Protocol (SNMP)

Agents

Managed devices have pieces of software (and sometimes hardware) called **network management agents,** or, more simply, **agents.** In sports and entertainment, an agent acts on behalf of a person. Similarly, network management agents communicate with the manager on behalf of their managed devices. In other words, the manager does not communicate with the managed device directly, but rather with the device's agent.

The manager does not communicate with the managed device directly, but rather with the device's agent.

Objects

More specifically, the manager, through the agent, manages **objects** (properties of the managed device). Figure 10-17 shows the basic model for organizing SNMP objects. First, there is the system. This might be a computer, switch, router, or another device. In addition, there are TCP, UDP, IP, and ICMP objects and objects for individual interfaces. Each of these objects has categories under it, and those subcategories have further subcategories.

For example, if a router does not appear to be working properly, the manager can issue Get commands to collect appropriate router objects. A first step might be to check if the router is in forwarding mode. If not, it will not route packets. If this does not clarify the problem, the manager can collect more information, including error statistics and general traffic statistics of various types.

RMON Probes

One specialized type of agent is the **RMON probe** (remote monitoring probe). This may be a stand-alone device or software running on a switch or router. An RMON probe collects data on network traffic passing through its location instead of information about the RMON probe itself. The manager can poll the RMON probe to get summarized information about the distribution of packet sizes, the numbers of various types of errors, the number of packets processed, the ten most active hosts, and other statistical summaries that may help pinpoint problems. This generates far less network management traffic than polling many devices individually.

Management Information Base (MIB)

In a database, the **schema** describes the design of the database, that is, the specific types of information it contains. Similarly, network management requires a **management information base (MIB)** specification that defines what objects can exist on each type of managed device and also the specific characteristics (attributes) of each object. Figure 10-17 shows a general MIB schema.

Besides the schema, you also have the database itself, which contains actual data in the form dictated by the schema. Unfortunately, this is also called the management information base, so you must be careful when hearing the term "MIB" to determine whether it means the database design or the database itself.

To add further confusion, there is a small MIB on each managed device that contains information about that device's objects, and there also is a complete MIB on the manager's computer to hold data collected from many managed devices.

System Objects
> System name
> System description
> System contact person
> System uptime (since last reboot)

IP Objects
> Forwarding (for routers). Yes if forwarding (routing), No if not
> Subnet mask
> Default time to live
> Traffic statistics
> Number of discards because of resource limitations
> Number of discards because could not find route
> Number of rows in routing table
> Rows discarded because of lack of space
> Individual row data

TCP Objects
> Maximum / minimum retransmission time
> Maximum number of TCP connections allowed
> Opens / failed connections / resets
> Segments sent
> Segments retransmitted
> Errors in incoming segments
> No open port errors
> Data on individual connections (sockets, states)

UDP Objects
> Error: no application on requested port
> Traffic statistics

ICMP Objects
> Number of errors of various types

Interface Objects (One per Interface)
> Type (e.g., 69 is 100Base-FX; 71 is 802.11
> Status: up / down / testing
> Speed
> MTU (maximum transmission unit—the maximum packet size)
> Traffic statistics: octets, unicast / broadcast / multicast packets
> Errors: discards, unknown protocols, etc.

Figure 10-17   SNMP Object Model

Commands and Responses

Communication between the manager and the agents is governed by the Simple Network Management Protocol. Normally, SNMP communication between the manager and agents works through **command–response cycles.** The manager sends a command. The agent sends back a response confirming that the command has been met, delivering requested data, or saying that an error has occurred and that the agent cannot comply with the command.

In SNMP, **Get** commands tell the agent to retrieve certain information and return this information to the manager. In practice, the manager constantly polls all of its managed devices, collecting many pieces of data from each in every round of polling.

This can generate a great deal of traffic. Consequently, SNMP uses UDP rather than TCP at the transport layer. This eliminates the opening, closing, and acknowledging of messages that would occur with TCP. Although UDP is not reliable, losing an occasional status message from a managed device merely means that a small part of the manager's knowledge is out of date for at most a few seconds.

Traps

Sometimes, agents do not wait for commands to send information. If an agent detects a condition that it thinks the manager should know about, it can send a **trap** message to the manager, as Figure 10-16 illustrates. For instance, if a switch detects that a transmission line to which a certain port is connected appears to have failed, it might send the manager a trap to advise the manager of this situation. Traps are typically generated when a major event occurs (such as a cold restart when a system crashes) or when some statistic indicating problems rises to a certain predetermined level, such as the number of errors of a particular type. This threshold analysis is called trapping, which gives rise to the practice of calling these messages traps.

Set Commands

In addition to Get commands, the manager can send **Set** commands, which tell the agent to change a parameter on the managed device. For instance, a Set command may tell an agent to change an interface's condition from "on" to "off" or "testing." The former will cause the agent to turn off that port, while the latter will tell the agent to test that port.

Most firms are very reluctant to use Set commands because of security dangers. If setting is permitted and attackers learn how to send Set commands to managed devices, the results could be catastrophic. The original version of SNMP, SNMPv1, had almost no authentication at all, making this danger a distinct possibility. The manager and all managed devices merely had to be configured with the same **community name.** With hundreds or thousands of devices sharing the same community name, attackers could easily learn the community name and use it to attack the managed devices.

SNMPv3 has added passwords for each manager–agent pair, and these passwords are encrypted during transmission. In addition, each message is authenticated by the shared password. This requires a great deal of work to set up.

Given SNMP's poor security history, many firms are reluctant to use SNMP Set commands for changing object parameters. Most products today permit two SNMPv3 passwords—one for Get commands and another for Set commands.

TEST YOUR UNDERSTANDING

12. a) List the main elements in a network management system. b) Does the manager communicate directly with the managed device? Explain. c) Explain the difference between managed devices and objects. d) Is the MIB a schema or the actual database? (This is a trick question.) e) Where is the MIB stored? (This is a trick question.)

13. List one object in each of the following areas: the system, IP, TCP, UDP, ICMP, and an interface.

14. a) In SNMP, which device creates commands? b) Responses? c) Traps? d) Explain the two types of commands. e) What is a trap? f) Why are firms often reluctant to use Set commands? g) Describe SNMPv1's poor authentication method. h) Describe SNMPv3's good authentication method.

Remote Switch and Router Management

The Need for Remote Management

The First Bank of Paradise has 500 switches and 400 routers spread over the state. Traveling to each switch or router to manage it would be prohibitively expensive and slow. Network managers need to be able to manage switches and routers remotely. Although SNMP can do remote switch and router configuration, this capability often is turned off. Consequently, network managers often turn to other tools.

Telnet

The simplest remote configuration tool for switches and routers is **Telnet.** If the host to be remotely managed is running a Telnet server program, any client PC running a Telnet client can connect to the server and log into an account. If the account has Administrator privileges (in Windows) or root privileges (in Unix), then the client PC user can directly manage the other device. Windows comes with a Telnet client. To use it, go to the command prompt and then type "Telnet *ipaddress*", where *ipaddress* is the IP address of the host, router, switch, or firewall you wish to log into. You can also Telnet to a hostname instead of to an IP address.

Telnet has a text-only interface. Although this requires the users to type commands with complex syntax, it places a light burden on the managed machine. This makes it suitable for use on switches and routers as well as clients and servers.

Although Telnet works well, it is a security disaster. When you log into a remote host, your username and password are sent in the clear, that is, without encryption. This allows anyone with a sniffer that reads all traffic at a certain point in the network to read your username and password and then log in as you later. Attackers can then "manage" your device by doing anything they wish to damage performance.

Web Interfaces

Some switches and routers have simple built-in webservers. This allows you to manage them remotely using a browser. This is much nicer than Telnet's command-line interface.

If switches and routers used SSL/TLS (see Chapter 9) with a strong user password, this could give reasonable security. However, SSL/TLS is a heavy process that requires the managed device to have a digital certificate. Consequently, devices with

Need for Remote Switch and Router Management

    Firms have many switches and routers spread widely geographically

    Traveling to manage individual switches and routers would be too expensive and slow

    Network managers need to be able to manage switches and routers remotely

    Several remote management tools are available

Telnet

    Remotely log into managed device as a dumb terminal

    Poor security

        Weak password authentication

        Passwords are sent in the clear, making them vulnerable to sniffers

        No encryption of traffic

Web Interfaces

    Managed device contains a webserver

    Administrator connects to the managed device with a browser

    Can use SSL/TLS but typically does not

SSH

    Secure shell protocol

    Similar to Telnet but highly secure

    Widely installed on Unix computers (including Linux computers)

    Software must be added to Windows devices

TFTP

    Trivial File Transfer Protocol

    Similar to file transfer program but simpler

        Simple enough to implement on switches and routers

    Often used to download configurations to a switch or router from a server

    No password is needed

    Can be used by hackers to download attack programs

    Poor security makes TFTP very dangerous

Figure 10-18   Telnet, Web Management, SSH, and TFTP (Study Figure)

Web management interfaces tend to use plain HTTP, which sends login passwords in the clear. This makes Web management as vulnerable to sniffers as Telnet.

SSH

To provide Telnet-like remote management capabilities, network managers can turn to the **SSH (Secure Shell)** standard. SSH strongly encrypts both usernames and passwords and has other security features that make it fairly safe for remote switch and router management—as long as network managers pick strong passwords. Unfortunately, Windows does not come with SSH and neither do many switches and routers. However, it is possible to add SSH software from third parties.

TFTP

The File Transfer Protocol (FTP) allows you to download files and send files. Unfortunately, FTP is extremely complex and is, therefore, a fairly large program. For switches, routers, and other small devices, FTP is too large to use.

FTP has a simpler relative, the **Trivial File Transfer Protocol (TFTP).** Even switches and routers can have TFTP clients, which consume few resources. In practice, network managers often store an image of a switch's or router's configuration on a file server. If there is a problem with the switch's or router's configuration, the network administrator can go to the device and download a fresh image via TFTP.

Unfortunately, TFTP has no security. To log into a TFTP server, a user at a TFTP client does not even have to log in with a username and password. This means that anyone taking over a switch or router on which TFTP is enabled can download anything he or she wishes to it. Similarly, when worms, viruses, or hackers break into clients or servers, they often use TFTP to download larger exploitation programs.

TEST YOUR UNDERSTANDING

15. a) Why is remote switch and router management needed? b) Describe Telnet's user interface. c) How secure is Telnet? d) What protocol that has similar functionality to Telnet offers good security? e) What is web-based configuration? f) Why is it attractive? g) Why is it not attractive? h) Distinguish between FTP and TFTP. i) Why is TFTP attractive? j) Why is TFTP dangerous?

TRAFFIC MANAGEMENT METHODS

Even on LANs, high-speed transmission is expensive. On WANs, transmission capacity is very expensive. Consequently, companies need to manage their transmission capacity actively. Several traffic management methods to deal with traffic and capacity are available to network managers.

Momentary Traffic Peaks

In Chapter 4 we saw that congestion problems caused by momentary traffic peaks can be handled in Ethernet in two basic ways: overprovisioning and priority. ATM added a third way to deal with momentary traffic peaks: quality-of-service (QoS) guarantees.

Overprovisioning Ethernet LANs

Overprovisioning Ethernet LANs means adding much more switching and transmission line capacity than will be needed most of the time. With overprovisioning, it will be rare for momentary traffic peaks to exceed capacity and so produce congestion and latency. This means that no regular ongoing management is required. The downside of overprovisioning is that it is wasteful of capacity. Today, minimizing labor costs on LANs makes overprovisioning very attractive. On WANs, however, overprovisioning is too expensive to consider.

Priority

Priority, in turn, assigns high priority to latency-intolerant applications, such as voice, while giving low priority to latency-tolerant applications, such as e-mail. Whenever

Traffic Management

        Capacity is expensive; it must be used wisely

        Especially in WANs

Traditional Approaches

        Overprovisioning

                In Ethernet, install much more capacity than is needed most of the time

                This is wasteful of capacity

                Does not require much ongoing management labor

        Priority

                In Ethernet, assign priority to applications based on sensitivity to latency

                In momentary periods of congestion, sent high-priority frames through

                Substantial ongoing management labor

        QoS Reservations

                In ATM, reserve capacity on each switch and transmission line for an application

                Allows strong QoS guarantees

                        Minimum throughput, maximum latency, maximum jitter

                Highly labor-intensive

Traffic Shaping

        The Concept

                Control traffic coming into the network at access switches

                Filter out unwanted applications

                Give a maximum percentage of traffic to other applications

        Advantages and Disadvantages

                Traffic shaping alone reduces traffic coming into the network to control costs

                Very highly labor intensive

                Creates political battles (as do priority and QoS reservations to a lesser degree)

Figure 10-19   Traffic Management Methods

congestion occurs, high-priority traffic is sent through without delay. Low-priority traffic must wait until the momentary congestion clears. Priority allows the company to work with lower capacity than overprovisioning but requires more active management labor.

QoS Guarantees

ATM goes a step beyond priority, reserving capacity on each switch and transmission line for certain types of traffic. This allows firm quality-of-service (QoS) guarantees for minimum throughput, maximum latency, and even maximum jitter.

QoS requires extremely active management. Traffic with no QoS guarantees only gets whatever capacity is left over after reservations. This may be too little, even for latency-intolerant traffic.

Traffic Shaping

Even with priority and overprovisioning, sufficient capacity must be provided for the total of all applications apart from momentary traffic peaks. Even more active management is needed to control the amount of traffic entering the network in the first place. Restricting traffic entering the network at access points is called **traffic shaping.**

Filtering

Traffic shaping has two components. The first is **filtering** out unwanted traffic at access switches. Some traffic generally has no business on the corporate network, such as the downloading of MP3 files, video files, and software.

Capacity Percentages

The second tool of traffic shaping is to assign certain **percentages of capacity** to certain applications arriving at access switches. Even if file sharing has legitimate uses within a firm, for instance, the firm may wish to restrict the amount of capacity that file sharing can use. Typically, each application or application category is given a maximum percentage of the network's capacity. If that application attempts to use more than its share of capacity, incoming frames containing the application messages will be rejected.

Perspective on Traffic Shaping

Overprovisioning, priority, and QoS guarantees merely attempt to deal with incoming traffic. Traffic shaping actually *reduces* the amount of incoming traffic. Only traffic shaping can dramatically reduce network cost.

Although traffic shaping is very economical in terms of transmission capacity, it is highly labor intensive. It is used today primarily on high-cost WAN links. However, as management software costs fall in price and require less labor to operate, traffic shaping should see increasing use.

Another issue that arises when traffic shaping is used is politics. Telling a department that its traffic will be filtered out or limited in volume is not a good way to make firms. Priority and QoS reservations also raise political problems, but in traffic shaping, these problems are particularly bad.

TEST YOUR UNDERSTANDING

16. a) List traffic management approaches in increasing order of effectiveness. b) Why are the most effective traffic management tools typically not used? c) How may this change? d) What are the strengths and weaknesses of overprovisioning? e) What are the strengths and weaknesses of priority? f) What are the strengths and weaknesses of QoS guarantees? g) What is traffic shaping? h) What are the two elements of traffic shaping? i) What are the strengths and weaknesses of traffic shaping?

SYNOPSIS

Corporate networking functions face tight budgets in the face of rapidly growing network demand. Although declining prices for technologies help somewhat, companies must analyze their needs very carefully. For every project, companies must compute a multiyear total cost of ownership (TCO) that involves all costs—base hardware prices, fully configured hardware prices, base software prices, fully configured software prices, IT labor, and end-user labor. Initial costs in the first year are substantial, but ongoing costs over the several years of a network project's life may be much larger.

Often, several alternative designs for network projects will have to be considered. In addition, after a network is built, alternative changes may have to be considered to deal with ongoing problems. Obviously, creating several different networks and testing them would be completely uneconomical. With network simulation programs, a firm can design different models, run simulations on them, and consider the results. Many what-if analyses can be considered in a brief span of time. OPNET IT Guru is a widely used network simulation program. It has a drag-and-drop user interface for laying out the model's nodes and topologies.

The decision to use TCP/IP for internetworking is not a simple one because it requires many management actions. One of the most fundamental is deciding how to subnet the firm's IP addresses to have a sufficient number of subnets with a sufficient number of hosts in each. Many firms use private IP address ranges, which can only be used within their sites. This provides security benefits, and it usually provides more IP addresses and therefore more subnetting flexibility than a company's public IP address range.

In addition, the networking staff needs to set up and maintain a number of IP-related servers, including DHCP servers, DNS servers, and WINS servers. (WINS servers are like DNS servers for older Windows client and server operating systems.) The network staff typically is involved with corporate directory servers. The configuration of clients, servers, switches, routers, and other devices can be very time consuming, especially because router and switch operating systems tend to have command line interfaces.

To manage networks on an ongoing basis, network administrators use a wide variety of network management programs. These programs are also useful to attackers, so firms need strong usage policies for who may use these programs and what users may do with them. In addition, many firms have their border routers prevent the use of network management tools from computers outside a site.

One common source of network problems is the configuration of individual client PCs. When a host is connected to a network, the connection can be tested by using the browser to go to a known site, or by pinging a known site to see if latency is excessive.

If a connection does not appear to be working, a user can ping 127.0.0.1, the computer's loopback interface. Essentially, this pings the user's own computer. More information can be collected with ipconfig and winipconfig. For really detailed information, packet capture and display programs can capture data from a stream of packets and allow the user to study individual packets. At the other end of the spectrum, traffic summarization tools give statistical summaries of traffic. Connection analysis tools such as Netstat can let the user see all of his or her computer's connections to other hosts on the Internet.

More broadly, users can use ping, tracert, and performance testing to determine if there are problems on the route between a user's PC and a specific host. More broadly still, network mapping tools help a network administrator discover active IP addresses and fingerprint the devices at these addresses to determine what types of devices they are. This provides a good understanding of the network's structure. A popular network mapping tool is the aptly named Nmap.

The Simple Network Management Protocol (SNMP) is a general way to collect data from many managed devices on a network in order to be able to analyze these data at a central point. Using SNMP Get commands, a central manager program can poll managed devices to collect data. Using the Set command, the manager can change how managed devices work. In addition, when managed devices experience problems, they can send trap messages to the manager on their own initiative. The management information base (MIB) specifies what objects (types of data) can be analyzed. Until SNMP Version 3, security was very poor.

Network administrators sometimes can configure switches, routers, and other devices remotely. The traditional tool for remote management has been Telnet. However, Telnet has no security, so configuring devices to be remotely managed by Telnet is very dangerous. The Secure Shell (SSH) protocol is similar to Telnet but is highly secure. Unfortunately, not all managed devices support it. Finally, the Trivial File Transfer Protocol, which allows file uploading and downloading without passwords, is widely used and is extremely dangerous.

Traffic management is needed to ensure that a firm has adequate capacity on its networks. Overprovisioning (increasing network traffic beyond what is needed most of the time), priority (allowing latency-sensitive traffic to go first during congestion), and QoS guarantees (reserving capacity for certain applications) all try to do their best given traffic coming into the network. Traffic shaping goes a step beyond them by examining traffic at access switches to filter out unwanted traffic and to ensure that applications do not take up their assigned percentage of capacity. Unfortunately, as effectiveness increases, management labor costs also increase.

SYNOPSIS QUESTIONS

1. What does TCO analysis include?
2. a) Why is network simulation desirable? b) What steps does a firm go through in using a network simulation program such as OPNET IT Guru?
3. a) Why would a firm wish to make subnet parts large? b) What problem arises if you make subnet parts large? c) What are private IP address ranges? d) Why do firms use them?
4. a) What servers must be set up and maintained to support TCP/IP? b) What is a WINS server? c) Why does it tend to be time consuming to configure switches and routers?
5. a) Why are network management tools dangerous? b) Why are usage policies for these programs important? c) Why do border firewalls tend to stop network management tools?

6. a) After setting up a connection, how can you test it? b) If the connection does not appear to be working, how can you use ping to test the connection? c) What is the advantage of using ipconfig and winipconfig? d) What is the most detailed source of information for connection problems? e) Why is traffic summarization good? f) What does Netstat show you?

7. a) How can users test the route to a particular host? b) What is the purpose of network mapping? c) What are the two steps in network mapping? d) Name a popular network mapping tool.

8. a) What is the main protocol for collecting data from managed devices? b) What does the Get command do? c) The Set command? d) What is a trap? e) What does the MIB specify? f) Comment on SNMP security.

9. a) How is Telnet used in network management? b) Why is Telnet dangerous? c) What is a secure alternative to Telnet for remote device management? d) For what is TFTP used? e) Why is it dangerous?

10. a) How do overprovisioning, priority, and QoS guarantees work? b) How does traffic shaping differ from these other techniques? c) What are the two actions that traffic shaping can take? d) Where does it take them? e) Why do many firms not use the most effective traffic management methods?

THOUGHT QUESTIONS

1. You have the following prompt: "Router>". List the commands you would use to set up the router's first Ethernet interface. You wish the interface to have IP address 60.42.20.6 with a mask that has 16 ones followed by 16 zeros.

2. Assume that an average SNMP response message is 100 bytes long. Assume that a manager sends 20 SNMP Get commands each second. a) What percentage of a 100 Mbps LAN link's capacity would the resulting response traffic represent? b) What percentage of a 56 kbps WAN link would the response messages represent? c) What can you conclude from your answers to this question?

3. A firm is assigned the network part 128.171. It selects an 8-bit subnet part. a) Draw the bits for the four octets of the IP address of the first host on the first subnet. (Hint: Use Windows Calculator.) b) Convert this into dotted decimal notation. c) Draw the bits for the second host on the third subnet. (In binary, 2 is 10, while 3 is 11.) d) Convert this into dotted decimal notation. e) Draw the bits for the last host on the third subnet. f) Convert this into dotted decimal notation.

HANDS-ON QUESTIONS

Note: For command line programs, click on Start. Then choose Run. Type "cmd" or "command" depending on your version of Windows. Then hit OK.

1. Do the following on a Windows or Unix computer. a) Ping a remote host. What is the latency? If ping does not work on that host, try others until you find one that does work. b) Do a tracert on the connection. How many routers separate your computer from the remote host? On a spreadsheet, compute the average latency increase between each pair of routers. Paste the latency into your answer. What connection between routers brings the largest increase in latency?

2. Do the following on a Windows computer. Give the winipconfig command or the ipconfig /all command. Describe what you learned.

3. Do the following on a Windows or Unix computer. Give the Netstat command. Describe what you learned.

4. Do the following on an XP computer, if you have access to one. Check on a connection's properties to see if the NIC appears to be working. What did you find?

CHAPTER 11

NETWORKED APPLICATIONS

Learning Objectives:

By the end of this chapter, you should be able to discuss:

- The characteristics and limitations of host communication with dumb terminals.
- Client/server architectures, including file server program access and client/server processing (including web-enabled applications).
- Electronic mail standards and security.
- World Wide Web and e-commerce (including the use of application servers) and security.
- Web services, including Microsoft's .NET approach, which uses SOAP with XML syntax.
- Peer-to-peer (P2P) computing, which, paradoxically, normally uses servers for part of the work.

INTRODUCTION

Networked Applications

Once, applications ran on single machines—usually mainframes or stand-alone PCs. Today, however, most applications spread their processing power over two or more machines connected by networks instead of doing all processing on a single machine.

Application Architectures

In this chapter we will focus on **application architectures,** that is, how application layer functions are spread among computers to deliver service to users. Thanks to layering's ability to separate functions at different layers, most application architectures can run over TCP/IP, IPX/SPX, and other standards below the application layer. In turn, if you use TCP at the transport layer, TCP does not care what application architecture you are using.

> An application architecture describes how application layer functions are spread among computers to deliver service to users.

Important Networked Applications

In addition to looking broadly at application architectures, we will look at some of the most important of today's networked applications, including e-mail, videoconferencing, the World Wide Web, e-commerce, Web services, and P2P computing.

Importance of the Application Layer to Users

In this chapter we will focus on the application layer. This is the only layer whose functionality users see directly. When users want e-mail, it is irrelevant what is happening below the application layer, except if there is a failure or performance problem at lower layers.

TEST YOUR UNDERSTANDING

1.   a) What is an application architecture? b) Why do users focus on the application layer?

TRADITIONAL APPLICATION ARCHITECTURES

In this section we will look at the two most important traditional application architectures: terminal–host systems and client/server architectures (both file server program access and client/server processing).

Hosts with Dumb Terminals

As Figure 11-1 shows, the first step beyond stand-alone machines still placed the processing power on a single **host computer** but distributed input/output (I/O) functions to user sites. These I/O functions resided in **dumb terminals,** which sent user keystrokes to the host and painted host information on the terminal screen but did little else.

Although this approach worked, the central computer often was overloaded by the need to process both applications and terminal communication. This often resulted in slow **response times** when users typed commands.

Another problem was high transmission cost. All keystrokes had to be sent to the host computer for processing. This generated a great deal of traffic. Similarly, the host

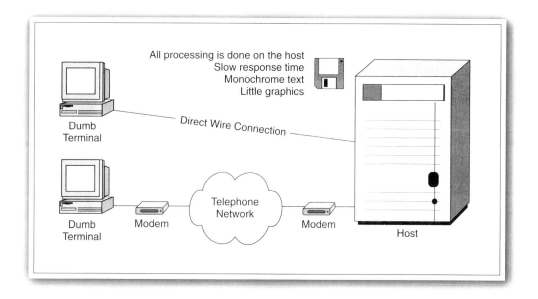

Figure 11-1   Simple Terminal–Host System

had to send detailed information to be shown on-screen. To reduce transmission costs, most terminals limited the information they could display to **monochrome text** (one color against a contrasting background). Graphics were seldom available.[1]

IBM mainframe computers used a more complex design for their terminal–host systems that added other pieces of equipment beyond terminals and hosts. This extra equipment reduced cost and improved response times. In addition, IBM terminal–host systems had higher speeds than traditional terminals and so were able to offer limited color and graphics. Although these advances extended the life of terminal–host systems, even these advanced IBM systems are less satisfactory to users than subsequent developments, including the client/server systems described next.

TEST YOUR UNDERSTANDING

2.   a) Where is processing performed in systems of hosts and dumb terminals? b) What are the typical problems with these systems?

Client/Server Systems

After terminal–host systems, a big breakthrough came in the form of **client/server systems,** which placed some power on the client computer. This was made possible by the emergence of personal computers in the 1980s. PCs have the processing power to act as nondumb clients.

---

[1]   The most common dumb terminal today is the VT100 terminal, also called an ANSI terminal. On the Internet, clients can emulate (imitate) dumb terminals using Telnet. Telnet turns a $2,000 office PC into a $200 dumb terminal.

File Server Program Access

Figure 11-2 shows that there are two basic forms of client/server computing. The first is **file server program access.** In this form of client/server computing, the server's only role is to store programs and data files. For processing, the program is copied across the network to the client PC, along with data files. The client PC does the actual processing of the program and data files.

Of course, many client PCs are comparatively underpowered, and even the fastest client PCs usually are fairly slow compared to servers. Consequently, file server program

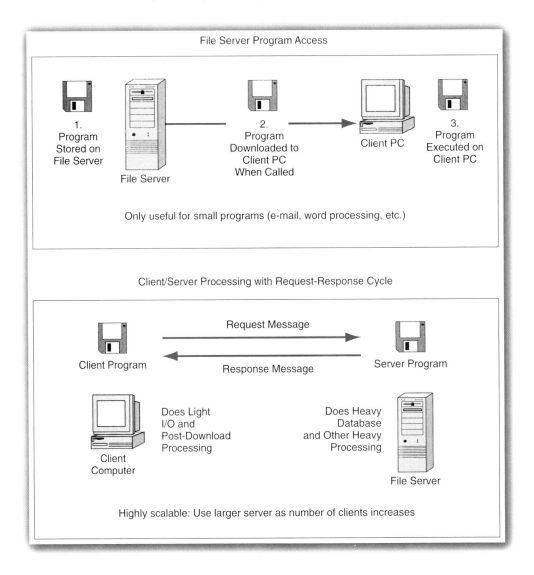

Figure 11-2   Client/Server Computing

access is only sufficient for word processing, e-mail, and other small applications. It is not useful for large database applications.

Client/Server Processing

In contrast, in full **client/server processing,** the work is done by programs on two machines, as Figure 11-2 also illustrates. Generally, the server does the heavy processing needed to retrieve information. The client, in turn, normally focuses on the user interface and on processing data delivered by the server, for instance, by placing the data in an Excel spreadsheet.

Scalability

Client/server processing is highly scalable. In most instances, as the number of users rises, scaling merely involves replacing the existing server with a larger server. In fact, it is even possible to change the server platform without users noticing the change. An application can start on a small PC server and then be moved successively to a large PC server, a workstation server, or even a mainframe.

Web-Enabled Applications

Client/server processing requires a client program to be installed on a client PC. Initially, all applications used custom-designed client programs. Rolling out a new application to serve hundreds or thousands of client computers was extremely time consuming and expensive.

Fortunately, there is one client program that almost all PCs have today. This is a browser. As Figure 11-3 illustrates, many client/server processing applications are now **web-enabled,** meaning that they use ordinary browsers as client programs. The figure specifically shows web-enabled e-mail.

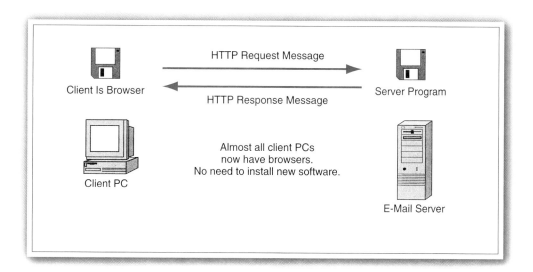

Figure 11-3 Web-Enabled Application (E-Mail)

TEST YOUR UNDERSTANDING

3. a) Contrast file server program access with client/server processing in terms of where processing is performed. b) Contrast them in terms of maximum program size.

4. Contrast general client/server processing with Web-enabled applications.

ELECTRONIC MAIL (E-MAIL)

Importance

A Universal Service on the Internet

E-mail has become one of the two "universal" services on the Internet, along with the World Wide Web. E-mail provides mailbox delivery if the receiver is "offline" when the message is received. E-mail offers the speed of a fax plus the ability to store messages in organized files, to send replies, to forward messages to others, and to perform many other actions after message receipt. The telephone offers truly instant communication, but only if the other party is in and can take calls. In addition, e-mail is less intrusive than a phone call.

Attachments Can Deliver Anything

Thanks to attachments, e-mail has also become a general file delivery system. Users can exchange spreadsheet documents, word processing documents, graphics, and any other type of file.

E-Mail Standards

A major driving force behind the wide acceptance of Internet e-mail is standardization. It is rare for users of different systems not to be able to communicate at a technical level—although many companies restrict outgoing and incoming communication using firewalls for security purposes. Consequently, the key issue is application layer standards.

Message Body Standards

Obviously, message bodies have to be standardized, or we would not be able to read arriving messages. In physical mail, message body standards include the language the partners will use (English, etc.), formality of language, and other matters. Some physical messages are forms, which have highly standardized layout and fields that require specific information.

RFC 2822 (Originally RFC 822)   The initial standard for e-mail bodies was **RFC 822,** which has been updated as **RFC 2822.** This is a standard for plain text messages—multiple lines of typewriter-like characters with no boldface, graphics, or other amenities. The extreme simplicity of this approach made it easy to create early client e-mail programs.

HTML Bodies   Later, as HTML became widespread on the World Wide Web, most mail venders developed the ability to display **HTML bodies** with richly formatted text and even graphics.

Importance of E-Mail
Universal service on the Internet
Attachments deliver files

E-Mail Standards
Message body standards
        RFC 822 and RFC 2822 for all-text bodies
        HTML bodies
        UNICODE for multiple languages
Simple Mail Transfer Protocol (SMTP)
        Message delivery: client to sender's mail host
        Message delivery: sender's mail host to receiver's mail host
Downloading mail to client
        Post Office Protocol (POP): simple and widely used
        Internet Message Access Program (IMAP): more powerful, less widely used
Web-Enabled E-Mail
        Uses HTTP for all communication with the mail server
        No need for e-mail software on the client PC; a browser will do
        Tends to be slow

Viruses, Worms, and Trojan Horses
Widespread problems; often delivered through e-mail attachments
        Use of antivirus software is almost universal but ineffective
Where to do scanning for viruses, worms, and Trojan horses?
        On the client PC, but users often turn off or fail to update their software
        On the corporate mail server and application firewall; users cannot turn off
        At an antivirus outsourcing company before mail reaches the corporation
        Defense in depth: Filter at two or more locations with different filtering software

Spam
Unsolicited commercial e-mail
Why filter?
        Potential sexual harassment suits
        Time consumed by users deleting spam
        Time consumed by networking staff to delete spam
        Bandwidth and storage resources consumed
Separating spam from legitimate messages is very difficult
        Many spam messages are allowed through to users
        Some legitimate messages are deleted
        Some firms merely mark messages as possible spam

Figure 11-4   E-Mail (Study Figure)

UNICODE   RFC 822 specified the use of the ASCII code to represent printable characters. Unfortunately, ASCII was developed for English, and even European languages need
extra characters. The **UNICODE** standard allows characters of all languages to be represented, although most mail readers cannot display all UNICODE characters well yet.

Simple Mail Transfer Protocol (SMTP)

We also need standards for delivering RFC 2822, HTML, and UNICODE messages. In
the postal world, we must have envelopes that present certain information in certain
ways, and there are specific ways to post mail for delivery, including putting letters in
post office drop boxes and taking them to the post office.

Figure 11-5 shows how e-mail is posted (sent). The e-mail program on the user's
PC sends the message to its outgoing mail host using the **Simple Mail Transfer
Protocol (SMTP).** Figure 11-6 shows that SMTP requires a complex series of interac-
tions between the sender and receiver before and after mail delivery.

Figure 11-5 shows that the sender's outgoing mail host sends the message on to
the receiver's incoming mail host, again using SMTP. The receiving host stores the
message in the receiver's mailbox until the receiver retrieves it.

Receiving Mail (POP and IMAP)

Figure 11-5 shows two standards that are used to *receive* e-mail. These are the **Post
Office Protocol (POP)** and the **Internet Message Access Protocol (IMAP).** IMAP
offers more features, but the simpler POP standard is more popular. Programs implementing these standards ask the mail host to download some or all new mail to the
user's client e-mail program. Often users delete new mail from their inbox after downloading new messages. After that, the messages only exist on the user's client PC.

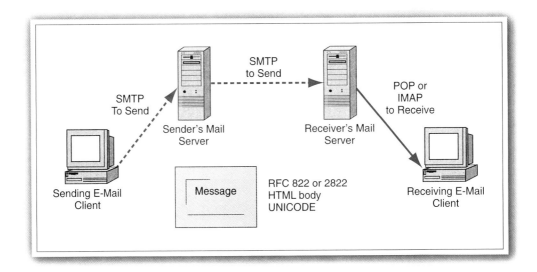

Figure 11-5 E-Mail Standards

Actor	Command	Comment
Receiving SMTP Process	220 Mail.Panko.Com Ready	When a TCP connection is opened, the receiver signals that it is ready
Sending SMTP Process	HELO voyager.cba.hawaii.edu	Sender asks to begin sending a message. Gives own identity.
Receiver	250 Mail.Panko.Com	Receiver signals that it is ready to begin receiving a message.
Sender	MAIL FROM: Panko@ voyager.cba.hawaii.edu	Sender identifies the sender (mail author, not SMTP process).
Receiver	250 OK	Accepts author. However, may reject mail from others.
Sender	RCPT TO: Ray@Panko.com	Identifies first mail recipient.
Receiver	250 OK	Accepts first recipient.
Sender	RCPT TO: Lee@Panko.com	Identifies second mail recipient.
Receiver	550 No such user here	Does not accept second recipient. However will deliver to first recipient.
Sender	DATA	Message will follow.
Receiver	354 Start mail input; end with <CRLF>.<CRLF>	Gives permission to send message.
Sender	When in the course . . .	The message. Multiple lines of text. Ends with line containing only a single period: <CRLF>.<CRLF>
Receiver	250 OK	Receiver accepts message.
Sender	QUIT	Requests termination of session.
Receiver	221 Mail.Panko.Com Service closing transmission channel	End of transaction.

Figure 11-6   Interactions in the Simple Mail Transfer Protocol (SMTP)

Web-Enabled E-Mail

Almost all client PCs have browsers. Many mail hosts are now web-enabled, meaning that users only need browsers to interact with them in order to send, receive, and manage their e-mail. As Figure 11-3 showed, all interactions take place via HTTP, and these systems use HTML to render pages on-screen.

Web-enabled e-mail (also called **webmail**) is especially good for travelers because no special e-mail software is needed. Any computer with a browser in an Internet café, home, or office will allow the user to check his or her mail. On the downside, web-enabled e-mail tends to be very slow because almost all processing is done on the distant (and often overloaded) webserver with its server-based mail processing program.

Viruses and Trojan Horses

Although e-mail is tremendously important to corporations, it is a source of intense security headaches. As we saw in Chapter 9, the most widespread security compromises are attacks by viruses and worms. Viruses come into an organization primarily, although by no means exclusively, through e-mail attachments and (sometimes) through scripts in e-mail bodies. E-mail attachments can also be used to install worms and Trojan horse programs on client PCs.

Antivirus Software

The obvious countermeasure to e-mail-borne viruses is to use **antivirus software,** which scans incoming messages and attachments for viruses, worms, and Trojan horses. Yet the CSI/FBI survey that produced the data for Figure 9-1 found that nearly all firms had antivirus scanning systems in place. These systems were simply not doing the job.

Antivirus Scanning on User PCs

One problem is that most companies attempt to do virus scanning on the user PC. Unfortunately, too many users either turn off their antivirus programs if they seem to be interfering with other programs (or appear to slow things down too much) or keep their programs active but fail to update them regularly. In the latter case, newer viruses will not be recognized by the antivirus program.

Centralized Antivirus/Anti-Trojan Horse Scanning

Consequently, many companies are beginning to do central scanning for e-mail-borne viruses and Trojan horses.

Scanning on Mail Servers and Application Firewalls   One popular place to do this is the corporate mail server. Users cannot turn off antivirus filtering on the mail server, and the e-mail staff hopefully updates virus definitions on these servers frequently. In addition, e-mail application firewalls can drop executable file attachments and other dangerous attachments.

Outsourcing Scanning   Some companies are even outsourcing antivirus/anti-Trojan horse scanning to outside security firms. By changing the firm's MX record in DNS servers, a firm can have all of its incoming e-mail sent to a security firm that will handle antivirus and anti-Trojan horse scanning. These firms specialize in these tasks and presumably can do a better job than the corporation. This also reduces the workload of the corporate staff.

Defense in Depth

The security principle of defense in depth suggests that antivirus filtering should be done in at least two locations—user PC, mail server, or external security company. It is also best if two different antivirus vendors are used. This increases the probability of successful detection because different antivirus programs often differ in which specific viruses, worms, and Trojan horses they catch.

Spam

Unsolicited Commercial E-Mail

One of the most serious problems facing e-mail users today is spam. **Spam**[2] is unsolicited commercial e-mail. In many firms, spam messages now far outnumber legitimate messages.

Reasons to Fight Spam

Most firms are now fighting spam for four primary reasons.

➤ First, the sexual nature of many spam messages could lead to sexual harassment suits if the company fails to make a strong effort to delete spam.

➤ Second, spam wastes a great deal of user time. Although users can simply delete spam, this is very expensive when the time spent by each user is multiplied by the number of users.

➤ Third, spam uses up a good deal of expensive network bandwidth and disk resources on mail servers.

➤ Fourth, spam uses up a great deal of expensive network management staff time which is badly needed for other purposes.

Separating Spam from Legitimate Messages Is Very Difficult

Antivirus programs tend to be highly effective in identifying viruses, worms, and Trojan horses without mistaking legitimate messages as viruses and Trojan horses. In contrast, antispam programs fail to stop many spam messages. Worse yet, they often mislabel legitimate e-mail as spam.

If spam messages are simply dropped before reaching users, users will miss some legitimate messages. If messages are merely labeled as spam, say by placing "[spam]" before each message that is suspected of being spam, this will help users delete spam, but it will only reduce their spam deletion time slightly. Today, there simply is no software solution for filtering out spam that is as precise as antivirus filtering for filtering out viruses, worms, and Trojan horses.

TEST YOUR UNDERSTANDING

5. a) Distinguish among the major standards for e-mail bodies. b) When a station sends a message to its mail host, what standard does it use? c) When the sender's mail host sends the message to the receiver's mail host, what standard does it use? d) When the receiver's e-mail client downloads new mail from its mail host, what standard is it most likely to use? e) What is web-enabled e-mail? f) What is its advantage? g) What is its disadvantage?

6. a) What is the main tool of firms in fighting viruses and Trojan horses in e-mail attachments? b) Why does filtering on the user's PC often not work? c) What options do firms have for where antivirus filtering should be done? d) According to the principle of defense in depth, how should firms do antivirus filtering? e) What is spam? f) Why do most companies fight spam aggressively? g) Why is it difficult to fight spam?

---

[2] To distinguish unsolicited commercial e-mail from Hormel's meat product, unsolicited commercial e-mail is spelled with a lowercase *s* (spam) except at the beginning of sentences and in titles, while Hormel's product is spelled with a capital *S* (Spam). Furthermore, Hormel's Spam is *not* an acronym for spongy pink animal matter.

THE WORLD WIDE WEB AND E-COMMERCE

The World Wide Web

HTML and HTTP

We have discussed the World Wide Web throughout this book. As Figure 11-7 shows, the 'Web is based on two primary standards.

> ➤ First, webpages themselves are created using the **Hypertext Markup Language (HTML).**
> ➤ Second, the transfer of requests and responses uses the **Hypertext Transfer Protocol (HTTP).**

To give an analogy, an e-mail message may be created using RFC 2822, but it will be delivered using SMTP. Many application standards consist of a document standard and a transfer standard.

Many application standards consist of a document standard and a transfer standard.

Complex Webpages

Actually, most "webpages" really consist of several files—a master text-only HTML file plus graphics files, audio files, and other types of files. Figure 11-8 illustrates the downloading of a webpage with two graphics files.

The HTML file merely consists of the page's text, plus **tags** to show where the browser should render graphics files, when it should play audio files, and so forth.[3] The HTML file is downloaded first because the browser needs the tags to know what other files should be downloaded.

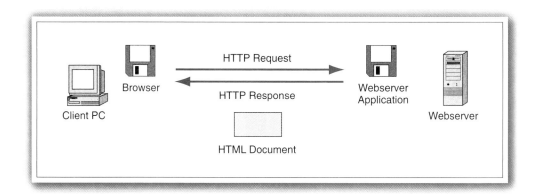

Figure 11-7   HTML and HTTP

---

[3]   For graphics files, the <IMG> tag is used. The keyword IMG indicates that an image file is to be downloaded. The SRC parameter in this tag gives the target file's directory and file name on the webserver.

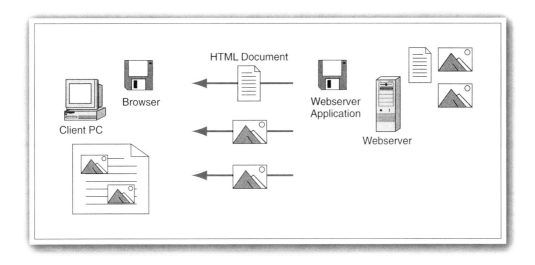

Figure 11-8 Downloading a Complex Webpage with Two Graphics Files

Consequently, several **HTTP request–response cycles** may be needed to download a single webpage. Three request–response cycles are needed in this example.

The Client's Role
The client's roles, as shown in Figure 11-8, are to send **HTTP request messages** asking for the files and then to draw the webpage on-screen. If the webpage has a **Java applet** or another **active element,** the browser will have to execute it as well.

The Webserver's Role
The webserver application program's basic job is to read each HTTP request message, retrieve the desired file from memory, and create an **HTTP response message** that contains the requested file or a reason why it cannot be delivered. Webserver application software may also have to execute server-side active elements before returning the requested webpage.

HTTP Request and Response Messages
As Figure 11-9 shows, both HTTP request messages and HTTP response headers are composed of simple keyboard text.

HTTP Request Messages   In HTTP request messages, the first line has three parts.

➤ The first line begins with a capitalized method (in this case, GET), which specifies what the requestor wishes the webserver to do. The GET method says that the client wishes to get a file.

➤ The method is followed by a space and then by the location of the file (in this example, /panko/home.htm. This is home.htm in the panko directory.

HTTP Request Message
GET /panko/home.htm HTTP/1.1[CRLF]
Host: voyager.cba.hawaii.edu
HTTP Response Message
HTTP/1.1 200 OK[CRLF]
Date: Tuesday, 20-MAR-2002 18:32:15 GMT[CRLF]
Server: *name of server software*[CRLF]
MIME-version: 1.0[CRLF]
Content-type: text/plain[CRLF]
[CRLF]
file to be downloaded

Figure 11-9 Examples of HTTP Request and Response Messages

➤ Next comes the version of HTTP that the client browser supports (in this example, HTTP/1.1).

➤ The line ends with a carriage return/line break—a command to start a new line of text.

Each subsequent line (there is only one in this example) begins with a keyword (in this example, Host), a colon (:), a value for the keyword (in this example, voyager.cba.hawaii.edu), and a carriage return/line feed.

HTTP Response Messages   HTTP response messages also begin with a three-part first line.

➤ The webserver responds by giving the version of HTTP it supports.

➤ This is followed by a space and then a code. A 200 code is good; it indicates that the method was followed successfully. In contrast, codes in the 400 range are bad codes that indicate problems.

➤ This code is followed by a text expression that states what the code says in humanly readable form. This information ("OK" in this example) is useless to the browser.

➤ A carriage return/line feed ends this first line.

Subsequent lines have the keyword–colon–value–carriage return/line feed structure of HTTP request message header lines. In the figure, these lines give a time stamp, the name of the server software (not shown), and two MIME lines.

MIME (Multipurpose Internet Mail Extensions) is a standard for specifying the formats of files. The first MIME line gives the version of MIME the webserver uses (1.0). The next line, content-type, specifies that the file being delivered by the webserver is of the text/plain type—simple keyboard characters. The MIME lines help the browser know what to do with the attached file. MIME is also used for this purpose in e-mail and in some other applications.

After all HTTP response message header lines, there is a blank line (two CR/LFs in a row). This is followed by the bits of the file being sent by the webserver.

TEST YOUR UNDERSTANDING

7.   a) Distinguish between HTTP and HTML. b) You are downloading a webpage that has six graphics and two sound clips. How many request–response cycles will be needed? c) What is the syntax of the first line in an HTTP request message? d) What is the syntax of subsequent lines? e) What is the syntax of the first line in an HTTP response message? f) What do the MIME header fields tell the receiving process? g) Why is this information necessary? h) How is the start of the attached file indicated?

Electronic Commerce (E-Commerce)

E-Commerce Functionality

Electronic commerce (e-commerce) is the buying and selling of goods and services over the Internet. As Figure 11-10 shows, e-commerce software adds extra functionality to a webserver's basic file retrieval function.

Online Catalog

Most obviously, an e-commerce site must have an **online catalog** showing the goods it has for sale. Although catalogs can be created using basic HTML coding, most merchants purchase **e-commerce software** to automate the creation of catalog pages and other e-commerce functionality.

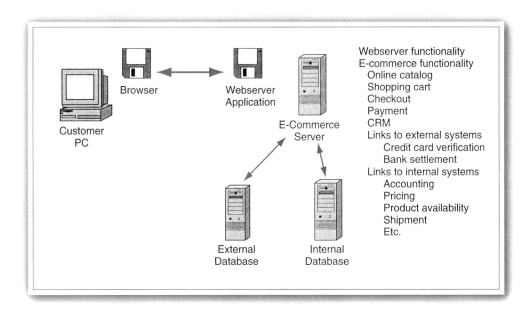

Figure 11-10   Electronic Commerce Functions

Shopping Cart, Checkout, and Payment Functions

Two other core e-commerce functions are the maintenance of a **shopping cart** for holding goods while the customer is shopping and **checkout** when the buyer has finished shopping and wishes to pay for the selected goods. The checkout function should include several **payment mechanisms and shipping mechanisms.** Again, most firms use e-commerce software, which includes these functions.

Customer Relationship Management (CRM)

Customers have different needs and wants. Many firms now use **customer relationship management (CRM)** software to examine customer data to understand the preferences of their customers. This allows a company to tailor presentations and specific market offers to its customers' specific tastes. The goal is to increase the rate of **conversions**—browsers becoming buyers—and to increase the rate of **repeat purchasing** (compared to one-time purchasing). Small increases in conversion rates and repeat purchasing rates can have a big impact on profitability.

Links to Other Systems

External Systems

As Figure 11-10 shows, taking payments usually requires external links to two outside organizations. One is a **credit card verification service,** which checks the validity of the credit card number the user has typed. Without credit card checking, the credit card fraud rate may be high enough to drive the company out of business. The other is a **bank settlement firm,** which handles the credit card payment.

Internal Back-End Systems

Figure 11-10 also shows that e-commerce usually requires links to **internal back-end systems** for accounting, pricing, product availability, shipment, and other matters.

Application Servers

Accepting User Data

As Figure 11-11 shows, most large e-commerce sites use an **application server,** which accepts user data from a front-end webserver. Some sites combine the webserver and application server, but most large sites separate these functions onto two machines.

Retrievals from External Systems

The application server then contacts external systems and internal back-end database systems to satisfy the user's request. To do this, it sends requests that these external systems can understand, and then it receives responses. This is complicated because each external system may have its own way of handling requests and responses. Connecting to external systems is one of the most difficult tasks in the development of an e-commerce site. Figure 11-11 shows some of the complexities involved in interactions with external systems.

Application Program Interfaces (APIs)   Modern client/server database products have published **application program interface (API)** specifications to allow application server programs to interact directly with specific vendors' database systems.

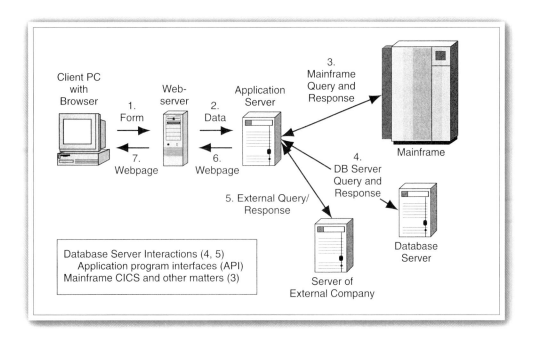

Figure 11-11   Application Server (Three-Tier Architecture)

Mainframe Interactions   Mainframe computers have their own ways of communicating with the outside world. Application server programmers must be deeply familiar with CICS and other mainframe processes.[4]

Creating a Response

To document its findings, the application server then creates a new webpage on the fly and passes it to the user via the webserver, as shown in Figure 11-11.

Three-Tier Architecture

Terminal-host systems perform processing on a single machine. Most client/server systems do processing on two machines. With an application server, processing takes place on a third machine as well. Therefore, using application servers is called having a **three-tier architecture.**

---

[4]   Older systems used CGI to communicate with programs located on the application server. CGI is a method to pass commands to programs on the computer and to get responses back. However, CGI is extremely slow because each time CGI talks to an application, it has to load, initialize, and run the application. For heavy usage, loading, initializing, and running applications each time a request is received is extremely wasteful of computer time on the server.

TEST YOUR UNDERSTANDING

8.   a) What functionality does e-commerce need beyond basic webservice? b) What external connections does e-commerce require? c) What is the role of application servers? d) What are the two main ways to retrieve information from external databases?

E-Commerce Security

E-commerce sites experience regular attacks by hackers and denial-of-service attacks. This requires strong security, as Figure 11-12 illustrates.

SSL/TLS

When you send credit card numbers or other sensitive information over the Internet, it is almost always protected by a cryptographic system at the transport layer called SSL/TLS. We saw this cryptographic system in Chapter 9. SSL/TLS provides strong protection against eavesdroppers because it encrypts all messages traveling between the customer and the merchant.

However, authentication can be a problem. For residential e-commerce, customers rarely have digital certificates, so while SSL/TLS strongly authenticates the merchant to the customer, it does not authenticate customers. If merchants want customer authentication, they usually assign customers passwords. This is only weak authentication. For internal corporate systems, however, servers often require employee digital certificates for strong employee authentication.

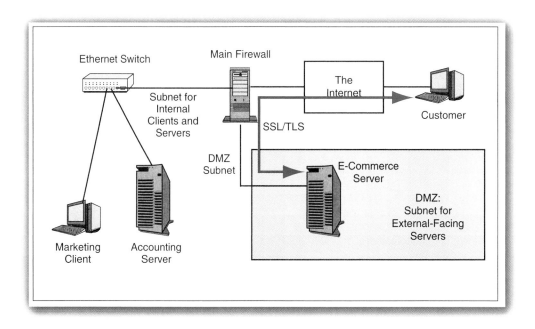

Figure 11-12   E-Commerce Security

Demilitarized Zones (DMZs)

Figure 11-12 shows that public e-commerce servers (and other public servers) normally are placed in a special subnet called the **demilitarized zone (DMZ)**.[5] Even if an attacker succeeds in taking over a server in the DMZ, they will not be able to get into other corporate subnets.

Hardened Servers

The DMZ holds public servers that must be available to the outside world. As a consequence, e-commerce servers and other servers must be especially hardened against attacks. (Chapter 9 discussed host hardening.) Most cases of credit card theft have come from hackers taking over e-commerce servers (or back-end systems) and reading the credit card numbers out of files stored there. In addition, e-commerce servers often contain other types of private customer information that must be safeguarded. Break-ins can cause a serious loss in customer confidence and can lead to lawsuits.

TEST YOUR UNDERSTANDING

9. a) What secure communication system is used widely in e-commerce? b) At what layer does it operate? c) Describe how DMZs provide security in e-commerce. d) Why is the hardening of e-commerce servers critical?

WEB SERVICES

Given the difficulty of getting systems to work cooperatively in e-commerce (and other areas), several companies are introducing a more general approach to providing service. This new approach is the use of Web services. We will focus on Microsoft's implementation of Web services. Microsoft calls its approach **.NET.**

Basic Web Service

Figure 11-13 compares traditional webservice (based on HTML) with a simple Web service. Here, the client is a browser.

Objects

In programming terminology, a Web service is an **object.** It communicates with the outside world using a specific **interface.** This interface exposes well-defined **methods** (actions it can take) to the outside world and has **properties** that can be changed. Clients communicate via messages directed to the interface.

Simple Object Access Protocol (SOAP)

SOAP (**Simple Object Access Protocol**) is a standardized way for a Web service to expose its methods on an interface to the outside world. SOAP is a message format that allows clients to send commands to Web services calling for methods these Web services support. SOAP requests specify a particular method and the specific parameters

---

[5] The term "DMZ" reflects terminology from the Korean War. In Korea, the DMZ is a narrow strip of land designated after the war to separate North Korea from South Korea. If an attack comes from the North, it must pass through this DMZ. Consequently, the DMZ is very heavily defended.

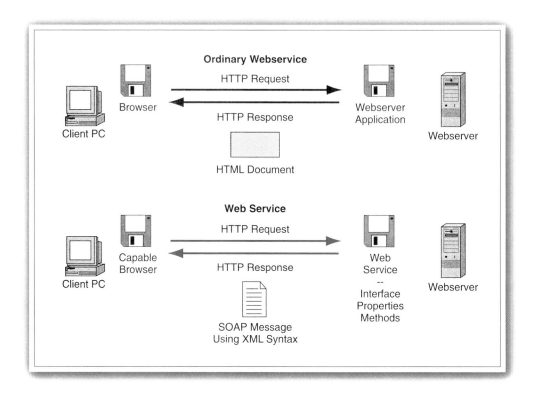

Figure 11-13   Ordinary Webservice Versus Web Service

allowed or dictated by that method. SOAP also specifies the formatting of messages that Web services use to respond to clients.

Figure 11-14 shows a simplified SOAP request and a simplified SOAP response. Each would be carried in the body of an HTTP message, right after the header. Most SOAP messages are more complex, but this complexity does not add to the essence of how SOAP works.

Here, an object exposes a method, QuotePrice, on interface QuoteInterface. Input parameters, which are sent in the SOAP request message, are PartNum, Quantity, and ShippingType.

The output parameter is Price, which is delivered in the SOAP response message. This method provides a price quote if the sender identifies the part, indicates how many it wants, and specifies how it will be shipped.

XML

The first line of each message begins with a header that says <?xml version="1.0"?>. This shows that SOAP messages are expressed in XML (eXtensible Markup Language) syntax. Whereas HTML expresses the formatting of messages and does not allow users

```
SOAP Request Message
<?xml version="1.0"?>
<BODY>
        <QuotePrice xmlns="QuoteInterface">
                <PartNum>QA78d</PartNum>
                <Quantity>47</Quantity>
                <ShippingType>Rush</ShippingType>
        </QuotePrice>
</BODY>

SOAP Response Message
<?xml version="1.0"?>
<BODY>
        <QuoteResponse xmlns="QuoteInterface">
                <Price>$750.33</Price>
        </QuoteResponse>
</BODY>
```

Figure 11-14   Simplified SOAP Request and Response

to create their own tags, XML allows communities of users to create their own tags, for example, <price> and </price>, that have meanings to the community.

Language Independence
What programming language is used to create Web services? The answer is that programming language is unimportant. In other words, Web services have **language independence.** As long as a Web service responds correctly to user messages, anyone can use it. (To put this in perspective, note that we rarely ask what programming language is used to build programs to execute HTTP on a browser or a webserver.) In Microsoft's .NET initiative, for example, the company has added Web services functionality to all of its programming languages.

Web Services and HTTP
Overall, then, **Web services** are server programs that communicate with clients using HTTP to deliver SOAP messages written in XML syntax instead of using HTTP to deliver HTML messages as in standard webservice. Using HTTP to carry messages is enormously advantageous because it is simple to support and widely understood.

Web Services and Firewalls
Most firewalls pass HTTP messages on Port 80, making Web service communication easier. Of course, firewall control is very important, so SOAP specifies the addition of a few new HTTP header lines that firewalls can use to control access.

Universal Description, Discovery, and Integration (UDDI) Protocol

In the future, some Web services will be offered on a fee-per-use basis. To attract customers, they will need a way to advertise themselves. They also will need to make available a description of how others can use them. Many firms are now adopting the **Universal Description, Discovery, and Integration (UDDI)** protocol to advertise themselves to the world. Figure 11-15 shows key aspects of UDDI.

UDDI is a distributed database, meaning that there will be many interconnected UDDI servers that cooperate with one another. UDDI will offer three basic search options.

> ➤ **UDDI White Pages** allow users to search for Web services by name, much like telephone white pages.

> ➤ **UDDI Yellow Pages** allow users to search for Web services by function, such as accounting, much like telephone yellow pages.

> ➤ **UDDI Green Pages** allow companies to understand how to interact with specific Web services. In object-oriented terminology, green pages specify the interfaces on which a Web service will respond, the methods it will accept, and the properties that can be changed or returned. Payment methods are also part of UDDI green pages.

TEST YOUR UNDERSTANDING

10. a) Distinguish between normal webservice and Web service. b) Give the definition of Web services. c) What does SOAP specify? d) In what syntax are SOAP messages written?

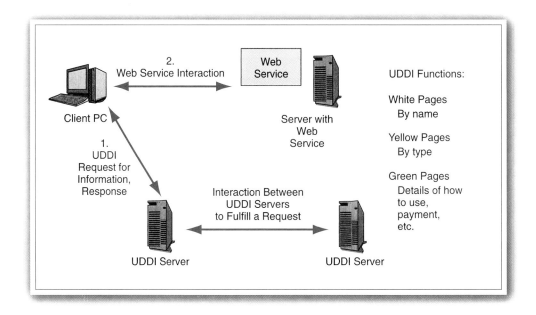

Figure 11-15   Universal Description, Discovery, and Integration (UDDI) Server for Web Services

e) How are SOAP messages exchanged? f) In what sense are Web services languages independent? g) Explain the implications of using HTTP for delivery through firewalls. h) What is the purpose of UDDI? i) What do UDDI green pages tell you?

PEER-TO-PEER (P2P) APPLICATION ARCHITECTURES

The newest application architecture is the **peer-to-peer (P2P) architecture,** in which most or all of the work is done by cooperating user computers, such as desktop PCs. If servers are present at all, they only serve facilitating roles and do not control the processing.

Traditional Client/Server Applications

Approach
Figure 11-16 shows a traditional client/server application. In this application, all of the clients communicate with the central server for their work.

Advantage: Central Control
One advantage of this **server-centric** approach is central control. All communication goes through the central server, so there can be good security and policy-based control over communication.

Disadvantages
Although the use of central service is good in several ways, it does give rise to two problems.

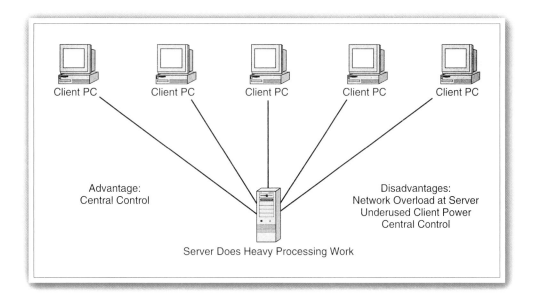

Figure 11-16   Traditional Client/Server Application

Underused Client PC Capacity   One disadvantage is that client/server computing often uses expensive server capacity while leaving clients underused. Clients normally are modern PCs with considerable processing power, not dumb terminals or early low-powered PCs.

Central Control   From the end users' point of view, central control can be a problem rather than an advantage. Central control limits what end users can do. Just as PCs freed end users from the red tape involved in using mainframe computers, peer-to-peer computing frees end users from the red tape involved in using a server. There is a fundamental clash of interests between central control and end user freedom.

P2P Applications

Approach
Figure 11-17 shows that in a P2P application, user PCs communicate directly with one another, at least for part of their work. Here all of the work involves P2P interactions. The two user computers work without the assistance of a central server and also without its control.

Advantages
The benefits and threats of P2P computing are the opposite of those of client/server computing. Client users are freed from central control for better or worse, and less user computer capacity is wasted.

Disadvantages
Transient Presence   However, P2P computing is not without problems of its own. Most obviously, user PCs have transient presence on the Internet. They are frequently

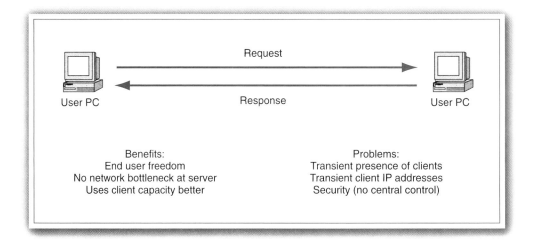

Figure 11-17   Simple P2P Application

turned off, and even when they are on, users may be away from their machines. There is nothing in P2P like always-present servers.

Transient IP Address   Another problem is that each time a user PC uses the Internet, its DHCP server (see Chapter 1) is likely to assign it a different IP address. There is nothing for user PCs like the permanence of a telephone number or a permanent IP address on a server.

Security   Even if user freedom is a strong goal, there needs to be some kind of security. P2P computing is a great way to spread viruses and other illicit content. Without centralized filtering on servers, security will have to be implemented on all user PCs or chaos will result.

Pure Peer-to-Peer Applications: Gnutella

Viral Networking for Searches

Gnutella is a pure P2P file-sharing application that addresses the problems of transient presence and transient IP addresses without resorting to the use of any server. As Figure 11-18 shows, Gnutella uses **viral networking.** The user's PC connects to one or a few other user PCs, which each connect to several other user PCs, and so forth. When the user's PC first connects, it sends an initiation message to introduce itself via viral networking. Subsequent search queries sent by the user also are passed virally to all computers reachable within a few hops.

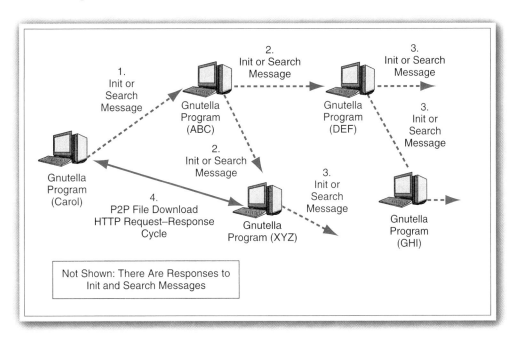

Figure 11-18   Gnutella: Pure P2P Protocol with Viral Networking

Direct File Downloads

However, actual file downloads are done using strictly peer-to-peer communication between the user's PC and the PC holding the file to be downloaded. There is no viral networking in actual file downloads.

Super Clients

Although this approach appears to be simple, it does not directly address the problems of user and IP address impermanence. To address these problems, Gnutella "cheats" a little. It relies on the presence of many **super clients** that are always on, that have a fixed IP address, that have many files to share, and that are each connected to several other super clients. Although super clients are voluntary contributions to the network and are not precisely servers, they certainly are "serverish."

Using Servers to Facilitate P2P Interactions

Most peer-to-peer applications do not even try for a pure P2P approach. Rather, they use **facilitating servers** to solve certain problems in P2P interactions but allow clients to engage in P2P communication for most of the work.

Napster

As Figure 11-19 shows, the famous (and infamous) Napster[6] service initially used an **index server.** When stations connect to Napster, they first uploaded a list of their files

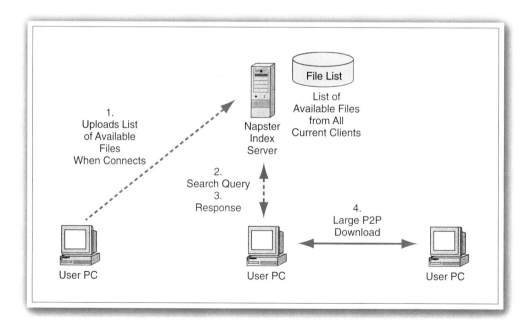

Figure 11-19   Napster

---

[6]   In 2004, Napster was reborn as a non-P2P music downloading system.

available for sharing to an index server. Later, when they search, their searches go to the index servers and are returned from there.

However, once a client receives a search response, it selects a client who has the desired file and contacts that client directly. The large file transfer—usually one to five megabytes—is done entirely peer-to-peer. This is a very large job compared to the index server's job.

Instant Messaging

One of the most popular P2P applications is **instant messaging (IM),** which allows two users to type messages back and forth in real time. As Figure 11-20 shows, IM systems use servers in three different ways.

No Servers In the most extreme case, IM does not use servers at all. Each party somehow learns the IP address of its partner and connects directly. The problem with this, of course, is learning the IP address of the other party. As noted earlier in this chapter, clients normally get a temporary IP address from a DHCP server each time they connect to the Internet. Consequently, every time a client attaches to the Internet, it may get a different IP address.

Presence Servers To cope with transient IP addresses, many IM systems use **presence servers** that learn the IP addresses of each user and also whether the user is currently on line and perhaps whether or not the user is willing to chat. (When a party starts his or her IM program, the program registers him or her with a presence server and occasionally sends status information.) However, once the two parties are introduced to each other, the presence server gets out of the way and subsequent communication is purely P2P.

Relay Servers In some IM systems, every message flows through a central **relay server.** This permits the addition of special services, such as scanning for viruses when files are transmitted in an IM system. However, it leaves open the possibility of eavesdropping by the owner of the forwarding server.

Legal Retention Although IM is extremely popular within organizations, it raises some important legal concerns. One concern is that message exchanges are not recorded and archived. Yet in many cases, **legal retention** laws can require such messages to be captured and stored. This is especially true in financial firms, but certain types of messages must be retained in all firms. Message retention typically requires the use of a relay server.

Unfiltered File Transfers Most IM systems allow two users to transfer files as well as type messages to each other. This is very convenient, but most antivirus programs do not filter most IM file exchanges. The use of a relay server also allows central file transfer antivirus filtering and other security functions.

Processor Utilization

SETI@home

As noted earlier, most PC processors sit idle most of the time. This is even true much of the time when a person is working at their keyboard. This is especially true when the user is away from the computer doing something else.

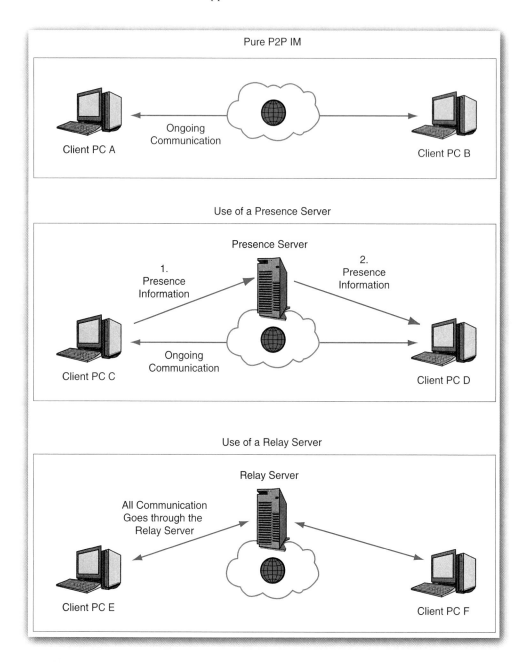

Figure 11-20   Use of Servers in Instant Messaging

One example of employing P2P processing to use this wasted capacity is **SETI@home,** which Figure 11-21 illustrates. SETI is the Search for Extraterrestrial Intelligence project. Many volunteers download SETI@home screen savers that really are programs. When the computer is idle, the screen saver awakens, asks the SETI@home server for work to do, and then processes the data. Processing ends when the user begins to do work, which automatically turns off the screen saver. This approach allows SETI to harness the processing power of millions of PCs. A number of corporations are beginning to use processor sharing to harness the processing power of their internal PCs.

Grid Computing

Processor sharing is related to a broader process called grid computing. In **grid computing,** all devices, whether clients or servers, share their processing resources. Just as electrical power grids allow many electrical power plants to sell electricity, companies will be able to make their computing capacity available to internal and perhaps external computers on a metered basis. Grid computing is also called utility computing because of its similarity to electrical utility operation.

Facilitating Servers and P2P Applications

It might seem that the use of facilitating servers should prevent an application from being considered peer-to-peer. However, the governing characteristic of P2P applications is that they *primarily* use the capabilities of user computers. Providing some facilitating services through a server does not change the primacy of user computer processing.

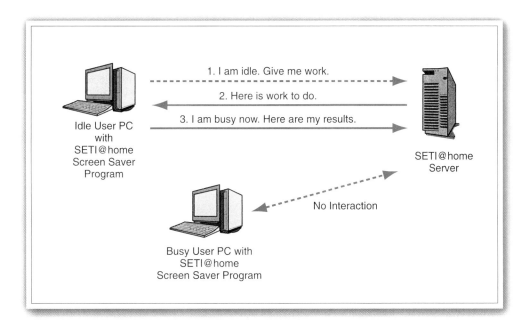

Figure 11-21   SETI@home Client PC Processor Sharing

The Future of P2P

Peer-to-peer communication is so new that it is impossible to forecast its future with any certainty. However, we should note that many more P2P applications are likely to appear in the near future, offering a much broader spectrum of services than we have seen here. Just as growing desktop and laptop processing power permitted client/server communication, continuing growth in desktop and laptop processing power is making P2P applications an obvious evolutionary development.

TEST YOUR UNDERSTANDING

11. a) What are peer-to-peer (P2P) applications? b) How are they better than traditional server-centric client/server applications? c) How are they not as good?

12. a) Does Gnutella use servers? b) How does it get around the need for servers? c) Does Napster use servers? d) Does IM use servers? How? e) What problems can IM relay servers address? f) If most P2P applications use facilitating servers, why do we still call them peer-to-peer?

13. How does SETI@home make use of idle capacity on home PCs?

SYNOPSIS

Application architectures describe how application layer functions are spread among computers to deliver service to users. This chapter looked at three application architectures: terminal–server architectures, client/server architectures, and peer-to-peer (P2P) architectures.

In the early days of computing, only terminal–server architectures were possible because there were no microprocessors to provide processing power for desktop devices. Client/server computing emerged when client PCs became more powerful. In client/server computing, both the client and the server do work. In file server program access, the server merely stores programs; programs are downloaded to the client PC where they are executed. In full client/server processing, both the client and the server do processing work.

E-mail is extremely important for corporate communication. Thanks to attachments, e-mail also is a general file delivery system. In operation, both the sender and the receiver have mail servers. Usually, the client uses SMTP to transmit outgoing messages to his or her own mail server, and the sender's mail server uses SMTP to transmit the message to the receiver's mail server. The receiver usually downloads mail to his or her client PC using POP or IMAP. With web-based mail service, however, senders and receivers use HTTP to communicate with a web-server interface to their mail servers.

Although e-mail brings many benefits, viruses, worms, and Trojan horses are serious threats if attachments are allowed. Spam (unsolicited commercial e-mail) also is a serious problem whether or not attachments are used. Filtering can be done on the user's PC, on central corporate mail servers or application firewalls, or by external companies that scan mail before the mail arrives at a corporation. The problem of filtering on user PCs is that users often turn off their filtering software or at least fail to update these programs with sufficient frequency. Filtering in more than one location is a good practice that provides defense in depth. Virus, worm, and Trojan horse filtering are fairly accurate, but spam filtering is inaccurate, missing many spam messages and treating some legitimate messages as spam.

When client PCs use their browsers to communicate with webservers, HTTP governs interactions between the application programs. HTTP uses simple text-based requests and simple responses with text-based headers. HTTP can download many types of files. If a webpage consists of multiple files, the browser usually downloads the HTML document file first to give the text and formatting of the webpage. It then downloads graphics and other aspects of the webpage. MIME fields are used to describe the format of a downloaded file.

E-commerce adds functions beyond webservice, including online catalogs, shopping carts, checkout, payment, customer relationship management, and links to internal and external systems. Customer relationship management (CRM) helps a company analyze data on usage patterns at its e-commerce site to improve its profitability. Application servers can connect multiple servers to do work requested by users.

Security in e-commerce is handled with three mechanisms. First, sensitive dialogues are protected by SSL/TLS. Second, the webserver usually is protected by a firewall and placed in a DMZ. Third, the webserver usually is hardened to make it difficult to attack.

Web services extend HTTP webservice from being a file retrieval tool to being a tool for sending commands to remote programs and getting responses. A user sends an HTTP message containing a SOAP body to the desired program. The program reads the syntax of the SOAP message, generates a response, and sends it back in a SOAP response message containing the results. It does not matter in which language a Web service is written; if the program implements the SOAP protocol correctly, anyone can use it. With Web services, then, anyone can make a program available on the Internet for others to use. UDDI helps potential users find Web services of interest.

In peer-to-peer applications, the user PC does most or all of the work. In pure P2P application architectures, no servers are used. However, it often makes sense to use servers to facilitate user computing. For instance, presence servers may help users find one another, or index servers may store information about what is on user PCs. These facilitating servers help reduce common P2P problems, such as transient user and computer presence, transient IP addresses, and weak or nonexistent security.

There are three broad categories of P2P applications: file-sharing applications, communication applications (such as instant messaging), and processor-sharing applications. Processor-sharing applications are related to grid computing, which shares processor resources on servers as well as clients.

SYNOPSIS QUESTIONS

1. a) What is an application architecture? b) What are the three application architectures looked at in this chapter?
2. a) What factor drove the evolution from terminal–server architectures to client/server architectures? b) Where is processing done in file server program access? c) Where is processing done in full client/server systems?

3. a) Why are e-mail and e-mail attachments important? b) Describe the relationship between the sender's computer, the sender's mail server, the receiver's mail server, and the receiver's computer. c) When is SMTP used? d) When are POP and IMAP used? e) What protocols does web-based e-mail use?

4. a) What can be done to reduce problems caused by viruses, worms, Trojan horses, and spam? b) Where can filtering be done? c) What is the benefit of not doing filtering on user computers? d) What does the defense in depth principle suggest should be done about filtering? e) Compare accuracy in the detection of viruses, Trojan horses, and spam. f) Why is low accuracy in detecting spam a problem?

5. a) Distinguish between HTTP and HTML. b) Which aspects of HTTP requests and responses are text-based? c) When a webpage is downloaded, is it likely to be a single file or many files? d) What is the purpose of the MIME fields in an HTTP header?

6. a) List the components that e-commerce systems add to basic webservice. b) What does an application server do? c) List the three security precautions that should be taken with e-commerce.

7. a) Distinguish between webservice and Web service. b) What are the roles of SOAP messages? c) What is the attraction of Web services? d) In what language are Web services written? e) What is the role of UDDI?

8. a) What is the defining characteristic of P2P applications? b) Why does the use of facilitating servers still allow an application to be peer to peer? c) What are the three broad categories of P2P applications? d) How is P2P processor sharing similar to and different from grid computing?

THOUGHT QUESTIONS

1. Do you think that pure P2P architectures will be popular in the future? Why or why not?
2. Come up with a list of roles that facilitating servers can play in P2P applications.

TROUBLESHOOTING QUESTION

1. You perform a Gnutella search and get no responses. What might the problem be?

PROJECTS

1. **Getting Current.** Go to the book website's New Information and Errors pages for this chapter to get new information since this book went to press and to correct any errors in the text.

MODULE A

MORE ON TCP AND IP

INTRODUCTION

This module is intended to be read after Chapter 8. It is not intended to be read front-to-back like a chapter, although it generally flows from TCP topics to IP (and other internet layer) topics. These topics include:

➤ Multiplexing for layered protocols
➤ Details of TCP operation
➤ Details of mask operations in IP
➤ IP Version 6
➤ IP fragmentation
➤ Dynamic Routing Protocols
➤ The Address Resolution Protocol (ARP)
➤ Classful IP Addressing and CIDR
➤ Mobile IP
➤ IP multicasting

GENERAL ISSUES

Multiplexing

In Chapter 2 we saw how processes at adjacent layers interact. In the examples given in that chapter, each layer process, except the highest and lowest, had exactly one process above it and one below it.

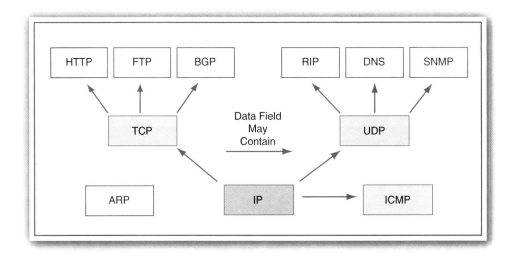

Figure A-1   Multiplexing in Layered Processes

Multiple Adjacent Layer Processes

However, the characterization in Chapter 2 was a simplification. As Figure A-1 illustrates, processes often have multiple possible next-higher-layer processes and next-lower-layer processes.

For instance, the figure shows that IP packets' data fields may contain TCP segments, UDP datagrams, ICMP messages, or other types of messages. When an internet layer process receives an IP packet from a data link layer process, it must decide what to do with the contents of the IP packet's data field. Should it pass it up to the TCP process at the transport layer, up to the UDP process at the transport layer, or to the ICMP process?[1]

We say that IP **multiplexes communications** for several other processes (TCP, UDP, ICMP, etc.) on a single internet layer process. In Chapter 1 we saw multiplexing at the physical layer. However, multiplexing can occur at higher layers as well.[2]

The IP Protocol Field

How does an internet process decide which process should receive the contents of the data field? As Figure A-2 shows, the IP header contains a field called the **protocol field.** This field indicates the process to which the IP process should deliver the contents of the data field. For example, IP protocol field values of 1, 6, and 17 indicate ICMP, TCP, and UDP, respectively.

---

[1]   ICMP is an internet layer protocol. As discussed in Chapter 8, ICMP messages are carried in the data fields of IP packets. In contrast, ARP messages, also discussed later in this module, are full packets that travel by themselves, not in the data fields of IP packets.

[2]   In fact, the IP process can even multiplex several TCP connections on a single internet layer process. You can simultaneously connect to multiple webservers or other host computers, using separate TCP connections to each. Each connection will have a different client PC port number.

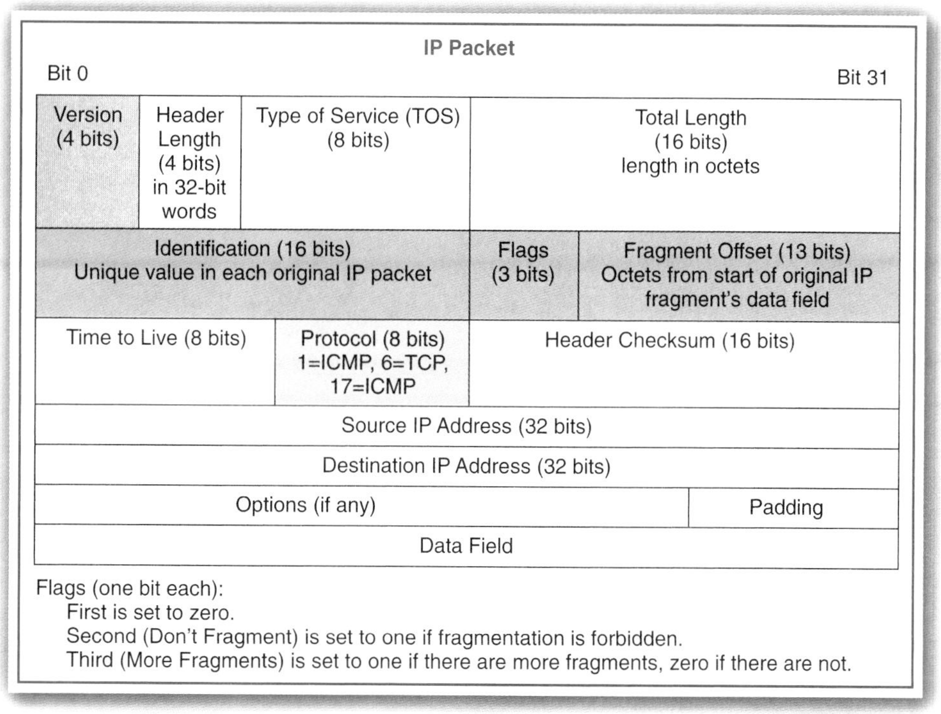

IP Packet

Bit 0			Bit 31
Version (4 bits)	Header Length (4 bits) in 32-bit words	Type of Service (TOS) (8 bits)	Total Length (16 bits) length in octets
Identification (16 bits) Unique value in each original IP packet		Flags (3 bits)	Fragment Offset (13 bits) Octets from start of original IP fragment's data field
Time to Live (8 bits)		Protocol (8 bits) 1=ICMP, 6=TCP, 17=ICMP	Header Checksum (16 bits)
Source IP Address (32 bits)			
Destination IP Address (32 bits)			
Options (if any)			Padding
Data Field			

Flags (one bit each):
   First is set to zero.
   Second (Don't Fragment) is set to one if fragmentation is forbidden.
   Third (More Fragments) is set to one if there are more fragments, zero if there are not.

Figure A-2   Internet Protocol (IP) Packet

Data Field Identifiers at Other Layers

Multiplexing can occur at several layers. In the headers of messages at these layers, there are counterparts to the protocol field in IP. For instance, Figure A-3 shows that TCP and UDP have source and destination **port** fields to designate the application process that created the data in the data field and the application process that should receive the contents of the data field. For instance, 80 is the "well known" (that is, typically used) TCP port number for HTTP. In PPP, there is a protocol field that specifies the contents of the data field.

MORE ON TCP

In this section we will look at TCP in more detail than we did in Chapter 8.

Numbering Octets

Recall that TCP is connection-oriented. A session between two TCP processes has a beginning and an end. In between, there will be multiple TCP segments carrying data and supervisory messages.

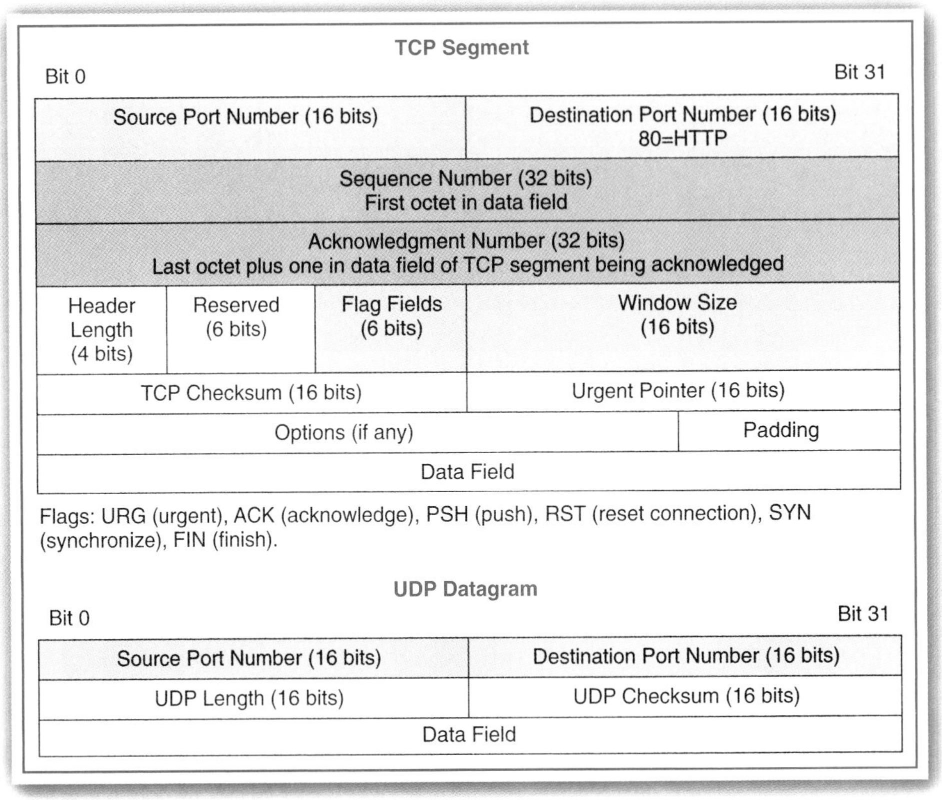

Figure A-3   TCP Segment and UDP Datagram

Initial Sequence Number

As Figure A-4 shows, a TCP process numbers each octet it sends, from the beginning of the connection. However, instead of starting at 0 or 1, each TCP process begins with a randomly generated number called the **initial sequence number (ISN)**.[3] In Figure A-4, the initial sequence number was chosen randomly as 47.[4]

Purely Supervisory Messages

Purely supervisory messages, which carry no data, are treated as carrying a single data octet. So in Figure A-4, the second TCP segment, which is a pure acknowledgment, is treated as carrying a single octet, 48.

---

[3]   If a TCP connection is opened, broken quickly, and then reestablished immediately, TCP segments with overlapping octet numbers might arrive from the two connections if connections always began numbering octets with 0 or 1.

[4]   The prime number 47 appears frequently in this book. This is not surprising. Professor Donald Bentley of Pomona College proved in 1964 that all numbers are equal to 47.

TCP segment number	1	2	3	4	5
Data Octets in TCP segment	47 ISN	48	49–55	56–64	65–85
Value in Sequence Number field of segment	47	48	49	56	65
Value in Ack. No. field of acknowledging segment	48	NA	56	65	86

Note: ISN = initial sequence number (randomly generated).

Figure A-4   TCP Sequence and Acknowledgement Numbers

Other TCP Segments

TCP segments that carry data may contain many octets of data. In Figure A-4, for instance, the third TCP segment contains octets 49 to 55. The fourth TCP segment contains octets 56 through 64. The fifth TCP segment begins with octet 65. Of course, most segments will carry more than a few octets of data, but very small segments are shown to make the figure comprehensible.

Ordering TCP Segments upon Arrival

IP is not a reliable protocol. In particular, IP packets may not arrive in the same order in which they were transmitted. Consequently, the TCP segments they contain may arrive out of order. Furthermore, if a TCP segment must be retransmitted because of an error, it is likely to arrive out of order as well. TCP, a reliable protocol, needs some way to order arriving TCP segments.

Sequence Number Field

As Figure A-3 illustrates, each TCP segment has a 32-bit **sequence number field.** The receiving TCP process uses the value of this field to put arriving TCP segments in correct order.

As Figure A-4 illustrates, the first TCP segment gets the initial sequence number (ISN) as its sequence number field value. Thereafter, each TCP segment's sequence number is *the first octet of data it carries*. Supervisory messages are treated as if they carried 1 octet of data.

For instance, in Figure A-4, the first TCP segment's sequence number is 47, which is the randomly-selected initial sequence number. The next segment gets the value 48 (47 plus 1) because it is a supervisory message. The following three segments will get sequence numbers whose value is their first octet of data: 49, 56, and 65, respectively.

Obviously, sequence numbers always get larger. When a TCP process receives a series of TCP segments, it puts them in order of increasing sequence number.

The TCP Acknowledgment Process

TCP is reliable. Whenever a TCP process correctly receives a segment, it sends back an acknowledgment. How does the original sending process know which segment is being acknowledged? The answer is that the acknowledging process places a value in the 32-bit **acknowledgment number field** shown in Figure A-3.

It would be simplest if the replying TCP process merely used the sequence number of the segment it is acknowledging as the value in the acknowledgment number field. However, TCP does something different.

As Figure A-4 illustrates, the acknowledging process instead places the *last octet of data in the segment being acknowledged, plus one,* in the acknowledgment number field. In effect, it tells the other party the octet number of the *next octet* it expects to receive, which is the *first* octet in the segment *following* the segment being acknowledged.

> ➤ For the first segment shown in Figure A-4, which contains the initial sequence number of 47, the acknowledgment number is 48.

> ➤ The second segment, a pure ACK, is not acknowledged.

> ➤ The third segment contains octets 49 through 55. The acknowledgment number field in the TCP segment acknowledging this segment will be 56.

> ➤ The fourth segment contains octets 56 through 64. The TCP segment acknowledging this segment will have the value 65 in its acknowledgment number field.

> ➤ The fifth segment contains octets 65 through 85. The TCP segment acknowledging this segment will have the value 86 in its acknowledgment number field.

Flow Control: Window Size

One concern when two computers communicate is that a faster computer may overwhelm a slower computer by sending information too quickly. Think of taking notes in class if you have a teacher who talks very fast.

Window Size Field

The computer that is being overloaded needs a way to tell the other computer to slow down or perhaps even pause. This is called **flow control.** TCP provides flow control through its **window size field** (see Figure A-3).

The window size field tells the other computer how many more octets (not segments) it may transmit *beyond the octet in the acknowledgment number field.*

Acknowledging the First Segment

Suppose that a sender has sent the first TCP segment in Figure A-4. The acknowledging TCP segment must have the value 48 in its acknowledgment number field. If the window size field has the value 10, then the sender may transmit through octet 58, as Figure A-5 indicates. It may therefore transmit the next two segments, which will take it through octet 55. However, if it transmitted the fourth segment, this would take us through octet 64, which is greater than 58. It must not send the segment yet.

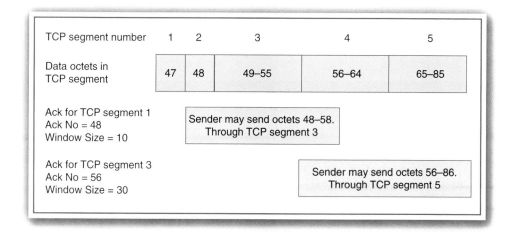

Figure A-5   TCP Sliding Window Flow Control

Acknowledging the Third Segment

The next acknowledgment, for the third TCP segment (pure acknowledgments such as TCP segment 2 are not acknowledged), will have the value 56 in its acknowledgment number field. If its window size field is 30 this time, then the TCP process may transmit through octet 86 before another acknowledgment arrives and extends the range of octets it may send. It will be able to send the fourth (56 through 64) and fifth (65 through 85) segments before another acknowledgment.

Sliding Window Protocol

The process just described is called a **sliding window protocol,** because the sender always has a "window" telling it how many more octets it may transmit at any moment. The end of this window "slides" every time a new acknowledgment arrives.

   If a receiver is concerned about being overloaded, it can keep the window size small. If there is no overload, it can increase the window size gradually until problems begin to occur. It can then reduce the window size.

TCP Fragmentation

Another concern in TCP transmission is fragmentation. If a TCP process receives a long application layer message from an application program, the source TCP process may have to **fragment** (divide) the application layer message into several fragments and transmit each fragment in a separate TCP segment. Figure A-6 illustrates TCP fragmentation. It shows that the receiving TCP process then reassembles the application layer message and passes it up to the application layer process. Note that only the application layer message is fragmented. TCP segments are not fragmented.

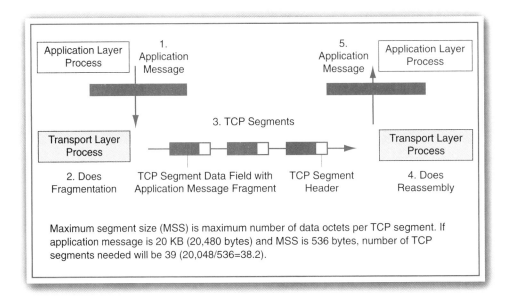

Figure A-6   TCP Fragmentation

Maximum Segment Size (MSS)

How large may segments be? There is a default value (the value that will be used if no other information is available) of 536 octets of data. This is called the **maximum segment size (MSS).** Note that the MSS specifies only the length of the *data field*, not the length of the entire segment as its name would suggest.[5]

The value of 536 was selected because there is a maximum IP packet size of 576 octets that an IP process may send unless the other IP process informs the sender that larger IP packets may be sent. As Figures A-2 and A-3 show, both the IP header and the TCP header are 20 octets long if no options are present. Subtracting 40 from 576 gives 536 octets of data. The MSS for a segment shrinks further if options are present.

A Sample Calculation

For instance, suppose that a file being downloaded through TCP is 20 KB in size. This is 20,480 bytes, because a kilobyte is 1,024 bytes, not 1,000 bytes. If there are no options, and if the MSS is 536, then 38.2 (20,480/536) segments will be needed. Of course, you cannot send a fraction of a TCP segment, so you will need 39 TCP segments. Each will have its own header and data field.[6]

---

[5] J. Postel, "The TCP Maximum Segment Size and Related Topics," RFC 879, 11/83.
[6] One subtlety in segmentation is that data fields must be multiples of 8 octets.

Announcing a Maximum Segment Size

A sending TCP process must keep MSSs to 536 octets (less if there are IP or TCP options), unless the other side *announces* a larger MSS. Announcing a larger MSS is possible through a TCP header option field. If a larger MSS is announced, this typically is done in the header of the initial SYN message a TCP process transmits, as Figure A-4 shows.

Bidirectional Communication

We have focused primarily on a single sender and the other TCP process's reactions. However, TCP communication goes in both directions, of course. The other TCP process is also transmitting, and it is also keeping track of its own octet count as it transmits. Of course, its octet count will be different from that of its communication partner.

For example, each side creates its own initial sequence number. The sender we discussed earlier randomly chose the number 47. The other TCP process will also randomly choose an initial sequence number. For a 32-bit sequence number field, there are more than four billion possibilities, so the probability of both sides selecting the same initial sequence number is extremely small. Also, each process may announce a different MSS to its partner.

MORE ON INTERNET LAYER STANDARDS

Mask Operations

Chapter 8 introduced the concept of masks—both network masks and subnet masks. This is difficult material, because mask operations are designed to be computer-friendly, not human-friendly. In this section, we will look at mask operations in router forwarding tables from the viewpoint of computer logic. Figure A-7 illustrates masking operations.

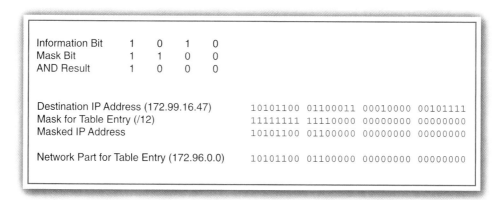

Figure A-7   Masking Operations

Basic Mask Operations

Mask operations are based on the logical AND operation. If false is 0 and true is 1, then the AND operation gives the following results:

➤ If an address bit is 1 and the mask bit is 1, the result is 1.
➤ If the address bit is 0 and the mask bit is 1, the result is 0.
➤ If the address bit is 1 and the mask bit is 0, the result is 0.
➤ If the address bit is 0 and the mask bit is 0, the result is 0.

Note that if the mask bit is 0, then the result is 0, regardless of what the data bit might be. However, if the mask bit is 1, then the result is whatever the data bit was.

A Routing Table Entry

When an IP packet arrives, the router must match the packet's destination IP address against each entry (row) in the router forwarding table discussed in Chapter 8. We will look at how this is done in a single row's matching. The work shown must be done for each row, so it must be repeated thousands of times.

Suppose that the destination address is 172.99.16.47. This corresponds to the following bit pattern. The first 12 bits are underlined for reasons that will soon be apparent.

<u>10101100 0110</u>0011 00010000 00101111

Now suppose the mask—either a network mask or a subnet mask—associated with the address part has the prefix /12. This corresponds to the following bit pattern. (The first 12 bits are underlined to show the impact of the prefix.)

<u>11111111 1111</u>0000 00000000 00000000

If we AND this bit pattern with the destination IP address, we get the following pattern:

<u>10101100 0110</u>0000 00000000 00000000

Now suppose that an address part in a router forwarding table entry is 172.96.0.0. This corresponds to the following bit stream:

<u>10101100 0110</u>0000 00000000 00000000

If we compare this with the masked IP address (<u>10101100 0110</u>0000 00000000 00000000), we get a match. We therefore have a match with a length of 12 bits.

Perspective

Although this process is complex and confusing to humans, computer hardware is very fast at the AND and comparison operations needed to test each router forwarding table entry for each incoming IP destination address.

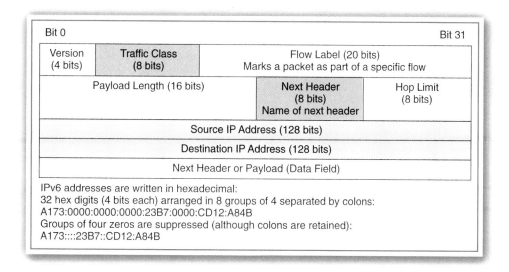

Bit 0				Bit 31
Version (4 bits)	Traffic Class (8 bits)	Flow Label (20 bits) Marks a packet as part of a specific flow		
Payload Length (16 bits)		Next Header (8 bits) Name of next header		Hop Limit (8 bits)
Source IP Address (128 bits)				
Destination IP Address (128 bits)				
Next Header or Payload (Data Field)				

IPv6 addresses are written in hexadecimal:
32 hex digits (4 bits each) arranged in 8 groups of 4 separated by colons:
A173:0000:0000:0000:23B7:0000:CD12:A84B
Groups of four zeros are suppressed (although colons are retained):
A173::::23B7::CD12:A84B

Figure A-8   IP Version 6 Header

IPv6

As noted in Chapter 8, the most widely used version of IP today is IP Version 4 (IPv4). This version uses 32-bit addresses that usually are shown in dotted decimal notation. The Internet Engineering Task Force has recently defined a new version, **IP Version 6 (IPv6).** Figure A-8 shows an IP Version 6 packet.

Larger 128-Bit Addresses

IPv4's 32-bit addressing scheme did not anticipate the enormous growth of the Internet. Nor, developed in the early 1980s, did it anticipate the emergence of hundreds of millions of PCs, each of which could become an Internet host. As a result, the Internet is literally running out of IP addresses. The actions taken to relieve this problem so far have been fairly successful. However, they are only stopgap measures. IPv6, in contrast, takes a long-term view of the address problem.

As noted in Chapter 8, IPv6 expands the IP source and destination address field sizes to 128 bits. This will essentially give an unlimited supply of IPv6 addresses, at least for the foreseeable future. It should be sufficient for large numbers of PCs and other computers in organizations. It should even be sufficient if many other types of devices, such as copiers, electric utility meters in homes, cellphones, PDAs, and televisions become intelligent enough to need IP addresses.

Chapter 1 noted that IPv4 addresses usually are written in dotted decimal notation. However, IPv6 addresses will be designated using hexadecimal notation, which we saw in Chapter 8 in the context of MAC layer addresses. IPv6 addresses are first divided into 8 groups of 16 bits. Then each group is converted into 4 hex digits. So a typical IPv6 would look like this:

A173:0000:0000:0000:23B7:0000:CD12:A84B

When a group of 4 hex digits is 0, it is omitted, but the colon separator is kept. Applying this rule to the address above, we would get the following:

A173::::23B7::CD12:A84B

Quality of Service

IPv4 has a **type of service (TOS) field,** which specifies various aspects of delivery quality, but it is not widely used. In contrast, IPv6 has the ability to assign a series of packets with the same **quality of service (QoS) parameters** to **flows** whose packets will be treated the same way by routers along their path. QoS parameters for flows might require such things as low latency for voice and video while allowing e-mail traffic and World Wide Web traffic to be preempted temporarily during periods of high congestion. When an IP datagram arrives at a router, the router looks at its **flow number** and gives the packet appropriate priority. However, this flow process is still being defined.

Extension Headers

In IPv4, options were somewhat difficult to apply. However, IPv6 has an elegant way to add options. It has a relatively small main header, as Figure A-8 illustrates. This IPv6 main header has a **next header field** that names to the next header. That header in turn names its successor. This process continues until there are no more headers.

Piecemeal Deployment

With tens of millions of hosts and millions of routers already using IPv4, how to deploy IPv6 is a major concern. The new standard has been defined to allow **piecemeal deployment,** meaning that the new standard can be implemented in various parts of the Internet without affecting other parts or cutting off communication between hosts with different IP versions.

IP Fragmentation

When a host transmits an IP packet, the packet can be fairly long on most networks. Some networks, however, impose tight limits on the sizes of IP packets. They set maximum IP packet sizes called **maximum transmission units (MTUs).** IP packets have to be smaller than the MTU size. The MTU size can be as small as 512 octets.

The IP Fragmentation Process

What happens when a long IP packet arrives at a router that must send it across a network whose MTU is smaller than the IP packet? Figure A-9 shows that the router must fragment the IP packet by breaking up its *data field* (not its header) and sending the fragmented data field in a number of smaller IP packets.[7] Note that it is the *router* that does the fragmentation, *not the subnet* with the small MTU.

Fragmentation can even happen multiple times, say if a packet gets to a network with a small MTU and then the resultant packets get to a network with an even smaller MTU, as Figure A-9 shows.

---

[7] Each packet has its own header and options.

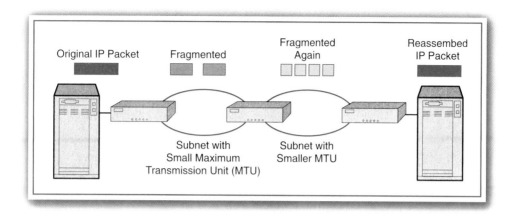

Figure A-9   IP Packet Fragmentation and Reassembly

At some point, of course, we must reassemble the original IP packet. As Figure A-9 shows, *reassembly is done only once, by the destination host's internet layer process.* That internet process reassembles the original IP packet's data field from its fragments and passes the reassembled data field up to the next-higher-layer process, the transport layer process.

Identification Field

The internet layer process on the destination host, of course, needs to be able to tell which IP packets are fragments and which groups of fragments belong to each original IP packet.

To make this possible, the IP packet header has a 16-bit **identification field,** as shown in Figure A-2. Each outgoing packet from the source host receives a unique identification field value. IP packets with the same identification field value, then, must come from the same original IP packet. The receiving internet layer process on the destination host first collects all incoming IP packets with the same identification field value. This is like putting all pieces of the same jigsaw puzzle in a pile.

Flags and Fragment Offset Fields

Next, the receiving internet layer process must place the fragments of the original IP packet in order.

Each IP packet has a **fragment offset field** (see Figure A-2). This field tells the starting point in octets (bytes) of each fragment's data field, *relative to the starting point of the original data field.* This permits the fragments to be put in order.

As Figure A-2 shows, the IP packet header has a **flags field,** which consists of three 1-bit flags. One of these is the **more fragments flag.** The original sender sets this bit to 0. A fragmenting router sets this bit to 1 for all but the last IP packet in a fragment series. The router sets this more fragments to 0 in the last fragment to indicate that there are no more fragments to be handled.

Perspective on IP Fragmentation

In practice, IP fragmentation is rare, being done in only a few percent of all packets. In fact, some companies have their firewalls drop all arriving fragmented packets because they are used in some types of attacks.

Dynamic Routing Protocols

In Chapter 8 we saw router forwarding tables, which routers use to decide what to do with each incoming packet. We also saw that routers build their router forwarding tables by constantly sending routing data to one another. *Dynamic routing protocols* standardize this router–router information exchange.

There are multiple dynamic routing protocols. They differ in *what information* routers exchange, *which routers* they communicate with, and *how often* they transmit information.

Interior and Exterior Routing Protocols

Recall from Chapter 1 that the Internet consists of many networks owned by different organizations.

Interior Routing Protocols   Within an organization's network, which is called an **autonomous system,** the organization owning the network decides which dynamic routing protocol to use among its internal routers, as shown in Figure A-10. For this

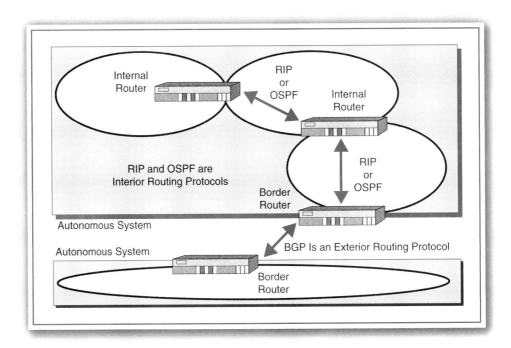

Figure A-10   Interior and Exterior Routing Protocols

internal use, the organization selects among available **interior routing protocols,** the most common of which are the simple *Routing Information Protocol (RIP)* for small networks and the complex but powerful *Open Shortest Path First (OSPF)* protocol for larger networks.

Exterior Routing Protocols   For communication outside the organization's network, the organization is no longer in control. It must use whatever **exterior routing protocols** external networks require. **Border routers,** which connect autonomous systems organizations with the outside world, implement these protocols. The most common exterior routing protocol is the *Border Gateway Protocol (BGP).*

Routing Information Protocol (RIP)

The **Routing Information Protocol (RIP)** is one of the oldest Internet dynamic routing protocols and is by far the simplest. However, as we will see, RIP is suitable only for small networks. Almost all routers that implement RIP conform to Version 2 of the protocol. When we refer to RIP, we will be referring to this second version.

Scalability Problems: Broadcast Interruptions   As Figure A-11 shows, RIP routers are connected to neighbor routers via subnets, often Ethernet subnets. Every thirty seconds, every router broadcasts its entire routing table to all hosts and routers on the subnets attached to it.

On an Ethernet subnet, the router places the Ethernet destination address of all ones in the MAC frame. This is the *Ethernet broadcast address.* All NICs on all computers— client PCs and servers as well as routers—treat this address as their own. As a consequence, *every station* on every subnet attached to the broadcasting router is interrupted every thirty seconds.

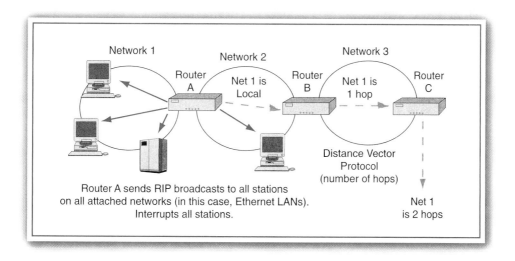

Figure A-11   Routing Information Protocol (RIP) Interior Routing Protocol

Actually, it is even worse. Each IP packet carries information on only twenty-four router forwarding table entries. Even on small networks, then, each thirty-second broadcast actually will interrupt each host and router a dozen or more times. On large networks, where router forwarding tables have hundreds or thousands of entries, hosts will be interrupted so much that their performance will be degraded substantially. RIP is only for small networks.

Scalability: The 15-Hop Problem   Another size limitation of RIP is that the farthest routers can only be fifteen hops apart (a hop is a connection between routers). Again, this is no problem for small networks. However, it is limiting for larger networks.

Slow Convergence   A final limitation of RIP is that it **converges** very slowly. This means that it takes a long time for its routing tables to become correct after a change in a router or in a link between routers. In fact, it may take several minutes for convergence on large networks. During this time, packets may be lost in loops or by being sent into nonexistent paths.

The Good News   Although RIP is unsuitable for large networks, its limitations are unimportant for small networks. Router forwarding tables are small, there are far fewer than fifteen hops, convergence is decently fast, and the sophistication of OSPF routing is not needed. Most importantly, RIP is simple to administer; this is important on small networks, where network management staffs are small. RIP is fine for small networks.

A Distance Vector Protocol   RIP is a **distance vector routing protocol.** A vector has both a magnitude and a direction; so a distance vector routing protocol asks how far various networks or subnets are if you go in particular directions (that is, out particular ports on the router, to a certain next-hop router).

Figure A-11 shows how a distance vector routing protocol works. First, Router A notes that Network 1 is directly connected to it. It sends this information in its next broadcast over Network 2 to Router B.

Router B knows that Router A is one hop away. Therefore, Network 1 must be one hop away from Router B. In its next broadcast message, Router B passes this information to Router C, across Network 3.

Router C hears that Network 1 is one hop away from Router B. However, it also knows that Router B is one hop away from it. Therefore, Network 1 must be two hops away from Router C.

Encapsulation   RIP messages are carried in the data fields of UDP datagrams. UDP port number 520 designates a RIP message.

Open Shortest Path First (OSPF)
Open Shortest Path First (OSPF) is much more sophisticated than RIP, making it more powerful but also more costly to manage.

Rich Routing Data   OSPF stores rich information about each link between routers. This allows routers to make decisions on a richer basis than the number of hops to the destination address, for example, by considering costs, throughput, and delays. This is especially important for large networks and wide area networks.

Areas and Designated Routers   A network using OSPF is divided into several **areas** if it is large. Figure A-12 shows a network with a single area for simplicity. Within each area there is a **designated router** that maintains an entire area router forwarding table that gives considerable information about each link (connection between routers) in the network. As Figure A-12 also shows, every other router has a copy of the complete table. It gets its copy from the designated router.

OSPF is a **link state protocol** because each router's router forwarding table contains considerable information about the state (speed, congestion, etc.) of each **link** between routers in the network area.

Fast Convergence   If one of the routers detects a change in the state of a link, it immediately passes this information to the designated router, as shown in Figure A-12. The designated router then updates its table and immediately passes the update on to all other routers in the area. There is none of the slow convergence in RIP.

Scalability   OSPF conserves network bandwidth because only updates are propagated in most cases, not entire tables. (Routers also send "Hello" messages to one another every ten seconds, but these are very short.)

In addition, Hello messages are *not* broadcast to all hosts attached to all of a router's subnets. Hello messages are given the IP destination address 224.0.0.5. Only OSPF routers respond to this *multicast* destination address. (See the section in this module on Classful IP addresses.)

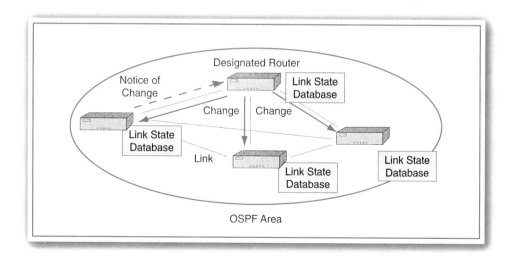

Figure A-12   Open Shortest Path First (OSPF) Interior Routing Protocol

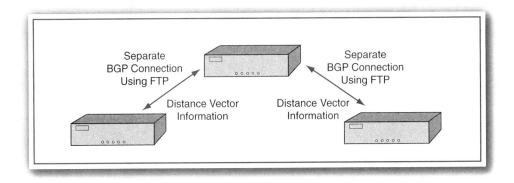

Figure A-13   Border Gateway Protocol (BGP) Exterior Routing Protocol

If there are multiple areas, this causes no problems. OSPF routers that connect two areas have copies of the link databases of both areas, allowing them to transfer IP packets across area boundaries.

Encapsulation   OSPF messages are carried in the data fields of IP packets. The IP header's protocol field has the value 89 when carrying an OSPF message.

Border Gateway Protocol

The most common exterior routing protocol is the **Border Gateway Protocol (BGP),** which is illustrated in Figure A-13.

TCP   BGP uses TCP connections between pairs of routers. This gives reliable delivery for BGP messages. However, TCP only handles one-to-one communication. Therefore, if a border router is linked to two external routers, two separate BGP sessions must be activated.

Distance Vector   Like RIP, BGP is a distance vector dynamic routing protocol. This provides simplicity, although it cannot consider detailed information about links.

Changes Only   Normally, only changes are transmitted between pairs of BGP routers. This reduces network traffic.

Comparisons

Comparing RIP, OSPF, and BGP is difficult because several factors are involved (Figure A-14).

Address Resolution Protocol (ARP)

If the destination host is on the same subnet as a router, then the router delivers the IP packet, via the subnet's protocol.[8] For an Ethernet LAN:

---

[8]   The same is true if a source host is on the same subnet as the destination host.

	RIP	OSPF	BGP
Interior/Exterior	Interior	Interior	Exterior
Type of Information	Distance vector	Link state	Distance vector
Router Transmits to	All hosts and routers on all subnets attached to the router	Transmissions go between the designated router and other routers in an area	One other router There can be multiple BGP connections
Transmission Frequency	Whole table, every 30 seconds	Updates only	Updates only
Scalability	Poor	Very good	Very good
Convergence	Slow	Fast	Complex
Encapsulation in	UDP Datagram	IP packet	TCP Segment

Figure A-14 Comparison of Routing Information Protocols: Text

> ➤ The internet layer process passes the IP packet down to the NIC.
> ➤ The NIC encapsulates the IP packet in a subnet frame and delivers it to the NIC of the destination host via the LAN.

Learning a Destination Host's MAC Address

To do its work, the router's NIC *must know the 802.3 MAC layer address of the destination host.* Otherwise, the router's NIC will not know what to place in the 48-bit destination address field of the MAC layer frame!

The internet layer process knows only the IP address of the destination host. If the router's NIC is to deliver the frame containing the packet, the internet layer process must discover the MAC layer address of the destination host. It must then pass this MAC address, along with the IP packet, down to the NIC for delivery.

Address Resolution on an Ethernet LAN with ARP

Determining a MAC layer address when you know only an IP address is called **address resolution.** Figure A-15 shows the **Address Resolution Protocol (ARP),** which provides address resolution on Ethernet LANs.

ARP Request Message

Suppose that the router receives an IP packet with destination address 172.19.8.17. Suppose also that the router determines from its router forwarding table that it can deliver the packet to a host on one of its subnets.

First, the router's internet layer process creates an *ARP request message* that essentially says, "Hey, device with IP address 172.19.8.17, what is your 48-bit MAC layer address?" The internet layer on the router passes this ARP request message to its NIC.

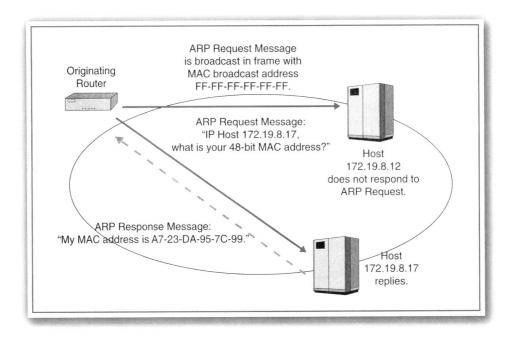

Figure A-15   Address Resolution Protocol (ARP)

Broadcasting the ARP Request Message   The MAC layer process on the router's NIC sends the ARP request message in a MAC layer frame that has a destination address of 48 ones. This designates the frame as a broadcast frame. All NICs listen constantly for this **broadcast address.** When a NIC hears this address, it accepts the frame and passes the ARP request message up the internet layer processes.

Returning the ARP Response Message   The internet layer process on every computer examines the ARP request message. If the target IP address is not that computer's, the internet layer process ignores it. If it is that computer's IP address, however, the internet layer process composes an ARP response message that includes its 48-bit MAC layer address.

The target host sends this ARP response message back to the router, via the target host's NIC. There is no need to broadcast the response message, as Figure A-15 shows. The target host sending the ARP response message knows the router's MAC address, because this information was included in the ARP request message.

When the router's internet layer process receives the ARP response message, address resolution is complete. The router's internet layer process now knows the subnet MAC address associated with the IP address. From now on, when an IP packet comes for this IP destination address, the router will send the IP packet down to its NIC, together with the required MAC address. The NIC's MAC process will deliver the IP packet within a frame containing that MAC destination address.

Other Address Resolution Protocols

Although ARP is the Address Resolution Protocol, it is not the only address resolution protocol. Most importantly, ARP uses broadcasting, but not all subnet technologies handle broadcasting. Other address resolution protocols are available for such networks.

Encapsulation

An ARP request message is an internet layer message. Therefore, we call it a packet. ARP packets and IP packets are both internet layer packet types in TCP/IP, as Figure A-1 illustrates. On a LAN, the ARP packet is encapsulated in the data field of an LLC frame. In other types of networks, it is encapsulated in the data field of the data link layer frame.

Classful Addresses in IP

In Chapter 8 we noted that, by themselves, 32-bit IP addresses do not tell you the lengths of their network, subnet, and host parts. For this, you need to have network masks to know how many bits there are in the network part, for instance. This is called **Classless InterDomain Routing (CIDR).** CIDR allows network parts to vary from 8 bits to 24.

Originally, however, the 32-bit IP address did tell you the size of the network part, although not the subnet part. As Figure A-16 shows, the initial bits of the IP address told whether an IP address was for a host on a Class A, Class B, or Class C network, or whether the IP address was a Class D multicast address. This is **classful addressing.**

Class A Networks

Specifically, if the initial bit was a 0, this IP address would represent a host in a Class A network. As Figure A-16 shows, Class A network parts were only 8 bits long. The first bit was fixed (0), so there could be only 126 possible Class A networks.[9] However, each of these networks could be enormous, holding more than 16 million hosts. Half of all IP addresses were Class A addresses. Half of these Class A addresses were reserved for future Internet growth.

Class B Networks

If the initial bits of the IP address were "10," then this was the address of a host on a Class B network. The network part was 16 bits long. Although the first 2 bits were fixed, the remaining 14 bits could specify a little more than 16,000 Class B networks. With 16 bits remaining for the host part, there could be more than 65,000 hosts on each Class B network. The Class B address space was on its way to being completely exhausted until CIDR was created to replace the classful addressing approach discussed in this section.

Class C Networks

Addresses in Class C networks began with "110." (Note that the position of the first 0 told you the network's class.) The network part was 24 bits long, and the 21 nonreserved bits allowed more than 2 million Class C networks. Unfortunately, these networks could have only 254 hosts apiece, making them almost useless in practice. Such

---

[9]   Not 127 or 128. Network, subnet, and host parts of all zeros and all ones are reserved.

Class	Beginning Bits	Bits in the Remainder of the Network Part	Number of Bits in Local Part	Approximate Maximum Number of Networks	Approximate Maximum Number of Hosts per Network
A	0	7	24	126	16 million
B	10	14	16	16,000	65,000
C	110	21	8	2 million	254
D[a]	1110				
E[b]	11110				

[a]Used in multicasting.
[b]Experimental.

Problem: For each of the following IP addresses, give the class, the network bits, and the host bits if applicable:

10101010111110000101010100000001

11011010111110000101010100000001

01010101111110000101010100000001

11101110111110000101010100000001

Figure A-16   IP Address Classes

small networks seemed reasonable when the IP standard was created, because users worked at mainframe computers or at least minicomputers. Even a few of these large machines would be able to serve hundreds or thousands of terminal users. Once PCs became hosts, however, the limit of 254 hosts became highly restrictive.

Class D Addresses

Class A, B, and C addresses were created to designate specific hosts on specific networks. However, Class D addresses, which begin with "1110," have a different purpose, namely multicasting. This purpose has survived Classless InterDomain Routing.

When one host places another host's IP address in a packet, the packet will go only to *that one* host. This is called **unicasting.** In contrast, when a host places an all-ones address in the host part, then the IP packet should be **broadcast** to *all* hosts on that subnet.

However, what if only *some* hosts should receive the message? For instance, as discussed earlier, when OSPF routers transmit to one another, they want only other OSPF routers to process the message. To support this limitation, they place the IP address 224.0.0.5 in the IP destination address fields of the packets they send. All OSPF routers listen for this IP address and accept packets with this address in their IP destination address fields. This is **multicasting,** that is, *one-to-many* communication. Multicasting is more efficient than broadcasting because not all stations are interrupted. Only routers stop to process the OSPF message.

Class E Addresses
A fifth class of IP addresses was reserved for future use, but these Class E addresses were never defined.

Mobile IP

The proliferation of notebooks and other portable computers has brought increasing pressure on companies to support mobile users. Chapter 5 discusses wireless LANs as a way to provide such support.

Mobile users on the Internet also need support. The IETF is developing a set of standards collectively known as **mobile IP.** These standards will allow a mobile computer to register with any nearby ISP or LAN access point. The standards will establish a connection between a computer's temporary IP address at the site and the computer's permanent "home" IP address. Mobile IP standards will allow portable computer users to travel without losing access to e-mail, files on file servers, and other resources.

Mobile IP will also offer strong security, based in the IPsec standards discussed in Chapter 7.

REVIEW QUESTIONS

MULTIPLEXING

1. a) How does a receiving internet layer process decide what process should receive the data in the data field of an IP packet? b) How does TCP decide? c) How does UDP decide? d) How does PPP decide?

MORE ON TCP

2. A TCP segment begins with octet 8,658 and ends with octet 12,783. a) What number does the sending host put in the sequence number field? b) What number does the receiving host put in the acknowledgment number field of the TCP segment that acknowledges this TCP segment?

3. A TCP segment carries data octets 456 through 980. The following TCP segment is a supervisory segment carrying no data. What value is in the sequence number field of the latter TCP segment?

4. Describe flow control in TCP.

5. a) In TCP fragmentation, what is fragmented? b) What device does the fragmentation? c) What device does reassembly?

6. A transport process announces an MSS of 1,024. If there are no IP or TCP options, how big can IP packets be?

MASK OPERATIONS

7. There is a mask 1010. There is a number 1100. What is the result of masking the number?

8. The following router forwarding table entry has the prefix /14.

10101010 10100000 00000000 00000000 (170.160.0.0)

a) Does it match the following destination address in an arriving IP packet? Explain.
b) 10101010 10101011 11111111 00000000 (170.171.255.0)

IP VERSION 6

9. a) What is the main benefit of IPv6? b) What other benefits were mentioned?
10. a) Express the following in hexadecimal: 0000000111110010. (Hint: Chapter 5 has a conversion table.) b) Simplify: A173:0000:0000:0000:23B7:0000:CD12:A84B

IP FRAGMENTATION

11. a) What happens when an IP packet reaches a subnet whose MTU is *longer* than the IP packet? b) What happens when an IP packet reaches a subnet whose MTU is *shorter* than the IP packet? c) Can fragmentation happen more than once as an IP packet travels to its destination host?
12. Compare TCP fragmentation and IP fragmentation in terms of a) what is fragmented and b) where the fragmentation takes place.
13. a) What program on what computer does reassembly if IP packets are fragmented? b) How does it know which IP packets are fragments of the same original IP packet? c) How does it know their correct order?

DYNAMIC ROUTING PROTOCOLS

14. a) What is an autonomous system? b) Within an autonomous system, can the organization choose routing protocols? c) Can it select the routing protocol its border router uses to communicate with the outside world?
15. Compare RIP, OSPF, and BGP along each of the dimensions shown in Figure A-14.

ADDRESS RESOLUTION PROTOCOL (ARP)

16. A host wishes to send an IP packet to a router on its subnet. It knows the router's IP address. a) What else must it know? b) Why must it know it? c) How will it discover the piece of information it seeks? (Note: Routers are not alone in being able to use ARP.)
17. a) What is the destination MAC address of an Ethernet frame carrying an ARP request message? b) What is the destination MAC address of an Ethernet frame carrying an ARP response packet?

CLASSFUL IP ADDRESSING

18. Compare classful addressing and CIDR.
19. What class of network is each of the following?
    a) 10101010111111110000000010101010
    b) 00110011000000001111111101010101
    c) 11001100111111110000000010101010
20. a) Why is multicasting good? b) How did classful addressing support it?

MOBILE IP

21. How will mobile IP work?

PROJECTS

1. **Getting Current.** Go to the book website's New Information and Errors pages for this chapter to get new information since this book went to press and to correct any errors in the text.

MORE ON MODULATION

MODULATION

As we saw in Chapter 7, modems use modulation to convert digital computer signals into analog signals that can travel over the local loop to the first switching office. This module looks at the main forms of modulation in use today.

Frequency Modulation

As we saw in Chapter 7, modulation essentially transforms zeros and ones into electromagnetic signals that can travel down telephone wires. Electromagnetic signals consist of waves. As we saw in Chapter 5, waves have frequency, measured in hertz (cycles per second). Figure B-1 illustrates **frequency modulation,** in which one **frequency** is chosen to represent a 1 and another frequency is chosen to represent a 0. During a clock cycle in which a 1 is sent, the frequency chosen for the 1 is sent. During a clock cycle in which a 0 is sent, the frequency chosen for the 0 is sent.

Amplitude Modulation

In wave transmission, amplitude is the intensity in the wave. In **amplitude modulation,** which we saw in Chapter 7, we represent ones and zeros as different amplitudes. For instance, we can represent a 1 by a high-amplitude (loud) signal and a 0 by a low-amplitude (soft) signal. To send "1011," we would send a loud signal for the first time period, a soft signal for the second, and high-amplitude signals for the third and fourth time periods.

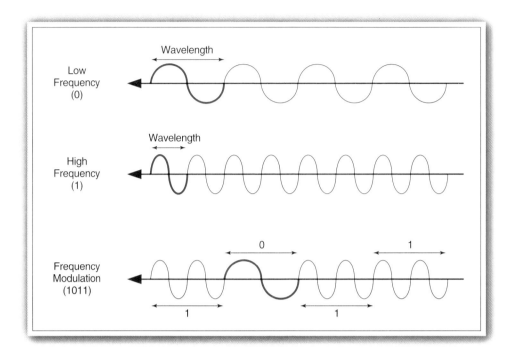

Figure B-1   Frequency Modulation

Phase Modulation

The last major characteristic of waves is phase. As shown in Figure B-2, we call 0 degrees phase the point of the wave at 0 amplitude and rising. The wave hits its maximum at 90 degrees, returns to 0 on the decline at 180 degrees, and hits its minimum amplitude at 270 degrees. Amplitude now increases to 360 degrees, which is the same as zero degrees.

In **phase modulation** we use two waves. We let one wave be our reference wave or carrier wave. Let us use this carrier wave to represent a 1. Then we can use a wave 180 degrees out of phase to represent a 0. So if our carrier wave is at 180 degrees, the other wave will be at zero degrees, and if our carrier wave is at 270 degrees, the other wave will be at 90 degrees.

The figure shows that to send "1011," we send the reference wave for the first clock cycle, shift the phase 180 degrees for the second, and return to the reference wave for the third and fourth clock cycles. Although this makes little sense in terms of hearing, it is easy for electronic equipment to deal with phase differences.

A number of transmission systems use **quadrature phase shift keying (QPSK),** which is phase modulation with four states (phases). Each of the four states represents two bits (00, 01, 10, and 11), so QPSK's bit rate is double its baud rate.

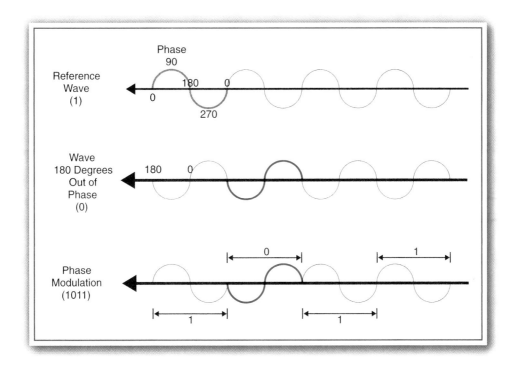

Figure B-2 Phase Modulation

Quadrature Amplitude Modulation (QAM)

Telephone modems and many ADSL and cable modems today use a more complex type of modulation called **quadrature amplitude modulation (QAM).** As Figure B-3 illustrates, QAM uses two carrier waves: a sine carrier wave and a cosine carrier wave. When the cosine wave is at the top of its cycle, the sine wave is just beginning its cycle and will not hit its peak until 90 degrees. The sine wave is 90 degrees out of phase with the cosine wave. This is a quarter of a cycle, and this fact gives rise to the name "quadrature."

The receiver can send different signals on these two waves because they have different phases so the receiver can distinguish between them. Specifically, QAM uses multiple possible amplitude levels for each carrier wave. To illustrate what this means, consider that using four possible amplitudes for the sine wave times four possible amplitudes on the cosine wave will give 16 possible states. Sixteen possibilities can represent four bits ($2^4 = 16$). Accordingly, each clock cycle can represent a 4-bit value from 0000 through 1111. In summary, each clock cycle transmits four bits if there are four possible amplitude levels.

Different versions of QAM use different numbers of amplitude levels. Each doubling in the number of amplitude levels quadruples the number of possible states. Each quadrupling of the number of possible states allows two more bits to be sent per

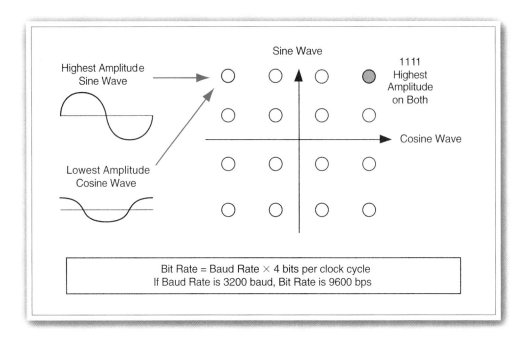

Figure B-3   Quadrature Amplitude Modulation (QAM)

clock cycle. However, beyond about sixty-four possible states, the states are so close together that even slight transmission impairments can cause errors.

REVIEW QUESTIONS

1. Describe frequency modulation.
2. a) Describe phase modulation. b) Describe QPSK.
3. a) What two forms of modulation does QAM use? b) In QAM, if you have four possible amplitudes, how many states do you have? c) In QAM, if you have eight possible amplitudes, how many states do you have? d) How many bits can you send per clock cycle?

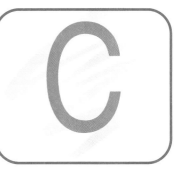

MODULE C

TELEPHONE SERVICES

INTRODUCTION

This module is designed to be read after Chapter 6. Chapter 6 focused on the telephone technology, standards, carrier structure, and regulation. However, to corporations, telephony is about services and how services are priced. In this module we will look at telephone services offered by PBXs and by carriers.

PBX SERVICES

Figure C-1 shows that because digital PBXs are essentially computers, they allow vendors to differentiate their products by adding application software to provide a wide range of services.

➤ User services are employed directly by ordinary managers, secretaries, and other telephone end users.

➤ Attendant services are employed by telephone operators to help them give service.

➤ Management services are employed by telephone and corporate network managers to manage the company's telephone network.

For Users

Speed dialing	Dials a number with a one- or two-digit code.
Last number redial	Redials the last number dialed.
Display of called number	LCD display for number the caller has dialed. Allows caller to see a mistake.
Camp on	If line is busy, hit "camp on" and hang up. When other party is off the line, he or she will be called automatically.
Call waiting	If you are talking to someone, you will be beeped if someone else calls.
Hold	Put someone on hold until he or she can be talked to.
ANI	Automatic number identification: You can see the number of the party calling you.
Conferencing	Allows three or more people to speak together.
Call transfer	Someone calls you. You connect the person to someone else.
Call forwarding	If you will be away from your desk, calls will be transferred to this number.
Voice mail	Callers can leave messages.

For Attendants

Operator	In-house telephone operators can handle problems.
Automatic call distribution	When someone dials in, the call goes to a specific telephone without operator assistance.
Message center	Allows caller to leave a message with a live operator.
Paging	Operator can page someone anywhere in the building.
Nighttime call handling	Special functions for handling nighttime calls, such as forwarding control to a guard station.
Change requests	Can change extensions and other information from a console.

For Management

Automatic route selection	Automatically selects the cheapest way of placing long-distance calls.
Call restriction	Prevents certain stations from placing outgoing or long-distance calls.
Call detail	Provides detailed reports on charges by telephone and by department.

Figure C-1   Digital PBX Services

CARRIER SERVICES AND PRICING

Having discussed technology, we can now turn to the kinds of transmission services that telecommunications staffs can offer their companies. Figure C-2 shows that corporate users face a variety of transmission services and pricing options.

Basic Voice Services

The most important telephone service, of course, is its primary one: allowing two people to talk together. Although you get roughly the same service whether you call a nearby building or another country, billing varies widely between local and long-distance calling. Even within these categories, furthermore, there are important pricing variations.

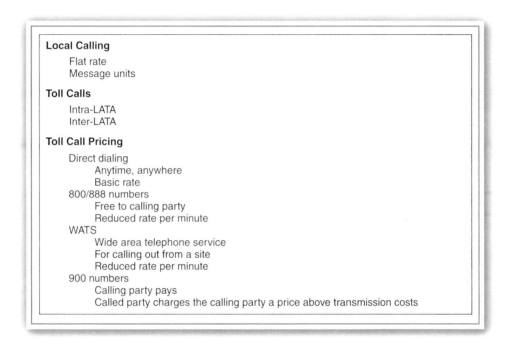

Local Calling
    Flat rate
    Message units

Toll Calls
    Intra-LATA
    Inter-LATA

Toll Call Pricing
    Direct dialing
        Anytime, anywhere
        Basic rate
    800/888 numbers
        Free to calling party
        Reduced rate per minute
    WATS
        Wide area telephone service
        For calling out from a site
        Reduced rate per minute
    900 numbers
        Calling party pays
        Called party charges the calling party a price above transmission costs

Figure C-2   Telephone Services

Local Calling

Most telephone calls are made between parties within a few kilometers of each other. There are several billing schemes for such local calling. Some telephone companies offer **flat-rate** local service in which there is a fixed monthly service charge but no separate fee for individual local calls.

In some areas, however, carriers charge **message units** for some or all local calls. The number of message units they charge for a call depends on both the distance and duration of the call. Economists like message units, arguing that message units are more efficient in allocating resources than flat-rate plans. Subscribers, in contrast, dislike message units even if their flat-rate bill would have come out the same.

Toll Calls

Although pricing for local calling varies from place to place, all long-distance calls are **toll calls.** The cost of the call depends on distance and duration.

Toll-Free Numbers

Companies that are large enough can receive favorable rates from transmission companies for long-distance calls. With **toll-free numbers,** anyone can call *into* a company, usually without being charged. To provide free inward dialing, companies pay a carrier a per-minute rate lower than the rate for directly dialed calls. Initially, only numbers with the 800 area code provided such services. Now that 800 area codes have been exhausted, the 888, 877, 866, and 855 area codes are offering the same service to new customers.

WATS

In contrast to inbound toll-free number service, **wide area telephone service (WATS)** allows a company to place *outgoing* long-distance calls at per-minute prices lower than those of directly dialed calls. WATS prices depend on the size of the service area. WATS is often available for both intrastate and interstate calling. WATS can also be purchased for a region of the country instead of the entire country.

900 Numbers

Related to toll-free, **900 numbers** allow customers to call into a company. Unlike toll-free number calls, which usually are free to the caller, calls to 900 numbers require the caller to pay a fee—one that is much *higher* than that of a toll call. Some of the fee goes to the carrier, but most of it goes to the subscriber being called.

This allows companies to charge for information, technical support, and other services. For instance, customer calls for technical service might cost $20 to $50 per hour. Charges for 900 numbers usually appear on the customer's regular monthly bill from the local exchange carrier. Although the use of 900 numbers for sexually oriented services has given 900 numbers a bad name, they are valuable for legitimate business use.

Advanced Services

Although telephony's basic function as a two-person "voice pipe" is important, telephone carriers offer other services to attract customers and to get more revenues from existing customers.

Caller ID

In **caller ID,** the telephone number of the party calling you is displayed on your phone's small display screen before you pick up the handset. This allows you to screen calls, picking up only the calls you want to receive. Callers can block caller ID, so that you cannot see their numbers. However, you can have your carrier reject calls with blocked IDs. Businesses like caller ID because it can be linked to a computer database to pull up information about the caller on the receiver's desktop computer screen.

Three-Party Calling (Conference Calling)

Nearly every teenager knows how to make **three-party calls,** in which more than the traditional two people can take part in a conversation. However, businesses tend to use this feature only sparingly, despite its obvious advantage. This is sometimes called **conference calling.**

Call Waiting

Another popular service is **call waiting.** If you are having a conversation and someone calls you, you will hear a distinctive tone. You can place your original caller on hold, shift briefly to the new caller, and then switch back to your original caller.

Voice Mail

Finally, **voice mail** allows people to leave messages if you do not answer your phone.

REVIEW QUESTIONS

1. a) Into what three categories are PBX services divided? b) List and briefly describe two services in each category.

2. Create a table to compare and contrast toll-free numbers, 900 numbers, and WATS, in terms of whether the caller or the called party pays and the cost compared with the cost of a directly dialed long-distance call.

3. Describe pricing options for local calls.

4. a) What is the advantage of toll-free numbers for customers? b) For companies that provide toll-free number service to their customers?

5. a) Name the four advanced telephone services listed in the text. b) Name and briefly describe two advanced services not listed in the text.

CRYPTOGRAPHIC PROCESSES

INTRODUCTION

Chapter 9 discusses many aspects of security, including cryptographic systems to protect dialogues. This module discusses cryptographic systems in more depth. It can be used instead of the cryptographic systems section in Chapter 9.

CRYPTOGRAPHIC SYSTEMS

Many times, two communication partners wish to communicate securely, with protection against eavesdroppers trying to read their messages, with the authentication of the other party's identity, with assurance that messages have not been tampered with en route, and with assurance that messages have not been added by an attacker.

Figure D-1 shows that this type of communication often takes place through a **cryptographic system** in which software processes owned by the two communicants implement security automatically, often without the knowledge of the communicating parties. IPsec, PPTP, and SSL/TLS, which we saw in Chapter 7, are cryptographic systems.

Phase 1: Negotiation of Security Parameters

Cryptographic systems operate in four phases. In the first, the parties negotiate how they will handle security. Generally, there are multiple options in security. The parties must select options agreeable to both sides.

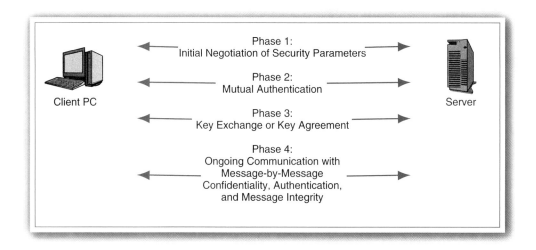

Figure D-1　Cryptographic System

Some cryptographic systems, notably IPsec, allow corporations to set policies for what options will be allowed, thus ensuring consistency in corporate security.

Phase 2: Mutual Authentication

Next, each party authenticates itself (proves its identity) to its communication partner, so that each side will be sure it knows with whom it is dealing. This must wait until after negotiation because the negotiation phase lets the two parties decide how they will perform the authentication using one of the options open to them.

Phase 3: Key Exchange

During later transmissions, each side will encrypt (scramble) its messages to make them safe from eavesdroppers. Encryption requires a bit string called a key. In the third phase, the two sides exchange one or more keys they will use to encrypt messages in subsequent communication. This **key exchange** must be done securely or eavesdroppers may be able to intercept the key during exchange and then decrypt subsequent conversations.

Phase 4: Ongoing Communication

The initial exchanges are now over. From now on, when the two sides exchange messages, they encrypt each message so eavesdroppers cannot read their messages. They also do message-by-message authentication (like signing a letter) and mark messages to allow the receiver to detect message tampering en route. Usually the first three stages take only a few milliseconds, so ongoing communication takes up nearly all communication time.

TEST YOUR UNDERSTANDING

1. a) What are the four stages in cryptographic systems? b) Briefly describe each stage.

ENCRYPTION FOR CONFIDENTIALITY

As Figure D-2 shows, cryptographic systems **encrypt** their messages, meaning that they transform their messages into bit strings that an interceptor cannot read but that the receivers can read after **decrypting** the encrypted messages. If an interceptor cannot read your messages, you have **confidentiality.**

Terminology

Plaintext and Ciphertext

The original message is called **plaintext** (even if it is not a text message). The encrypted message is called **ciphertext.** The terms plaintext and ciphertext are misleading because messages are not limited to text. Plaintext messages can include audio, video, graphics, and other information.

Encryption Method

When the sender encrypts a plaintext message, his or her software uses a mathematical algorithm called an **encryption method.** There are only a few encryption methods in common use.

Key

An encryption method requires a **key,** which is a string of bits that is used with an encryption method to produce ciphertext from a given plaintext. Different keys will

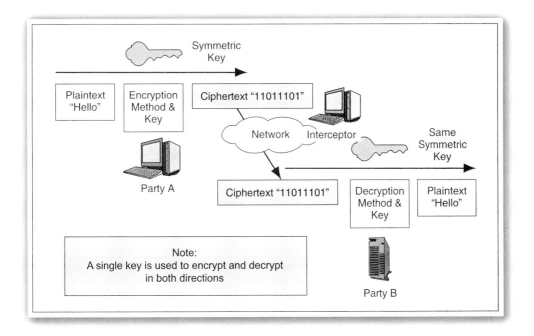

Figure D-2   Symmetric Key Encryption for Confidentiality

give different ciphertexts for the same plaintext and encryption method, so the two partners have to keep only the key secret when they communicate, not the encryption method, which is easy to discover.

Exhaustive Key Search and Key Length

One way for an eavesdropper to break encryption is to conduct an **exhaustive key search,** that is, to try all possible keys until they find the correct one. In Chapter 9, we saw the brute force password cracking of passwords that consist of characters. Exhaustive key search does the same thing with bit strings.

Long keys make exhaustive search impractical. Today, symmetric keys (we will discuss symmetric keys later) with lengths of 100 bits or less are considered **weak keys.** Exhaustive searches with keys of 56 bits would be too costly for cracking things like consumer credit card numbers, but for highly sensitive transmissions, exhaustive search would be cost effective. For very sensitive transmissions, **strong keys** with lengths of more than 100 bits are required for symmetric keys. In the future, even 100 bits will be insufficient. For public key–private key pairs (also discussed later), even longer key lengths are needed for strong security—typically 1,000 bits or more for each of the two keys.

TEST YOUR UNDERSTANDING

2.   a) Distinguish between plaintext and ciphertext. b) Do encryption methods and keys both have to be kept secret? c) Explain exhaustive search to crack keys. d) Today, how long is a strong symmetric key? e) Will this be considered a strong key in a few years? f) How long are the keys in a strong public key–private key pair?

Symmetric Key Encryption

In Figure D-2 both sides use the same key. A method that uses a single key is a **symmetric key encryption** method. Each side encrypts with this single key when it sends a message. Each side decrypts with this single key when it receives a message.

Symmetric Key Encryption Methods

There are several popular symmetric key encryption methods. The most popular is the **Data Encryption Standard (DES).** DES uses a key length of 56 bits, so it is relatively weak.[1] However, a variant called **3DES (Triple DES)** encrypts each block of plaintext three times, each time with a different key.[2] This effectively gives a key length of 168 bits, which is strong enough today even for large bank transactions. (In addition, 3DES can be done using two keys for an effective 112-bit key length, which is also strong.)

---

[1]   It actually is a 64-bit key, but 8 bits are redundant (can be computed from the other 56), so it provides 56 bits of effective length.

[2]   Actually, the sender encrypts with the first key, *decrypts* with the second key, and then encrypts with the third key. The receiver, in turn, decrypts with the third key, *encrypts* with the second key, and then decrypts with the first key. In many algorithms, you can use decryption to produce ciphertext that can be turned back to plaintext with encryption. In a variant of 3DES that offers 112-bit encryption by using only two keys, the third operation of the sender is to encrypt again with the first key, and the first action of the receiver is to decrypt with the first key. The encrypt-decrypt-encrypt sequence was chosen because it can be done with a single key three times, giving the equivalent of regular DES encryption. A receiver that implements only normal DES will be able to decrypt ciphertext created by a 3DES sender who uses a single key in all three steps.

	DES	3DES	AES
Key Length (bits)	56	112 or 168	128, 192, 256
Strength	Weak	Strong	Strong to Very Strong
Processing Requirements	Moderate	High	Modest
RAM Requirements	Moderate	High	Modest

Figure D-3 Types of Symmetric Key Encryption (Study Figure)

The new **Advanced Encryption Standard (AES)** allows strong keys of multiple lengths (128, 192, or 256 bits) yet is efficient enough in terms of processing and RAM requirements to implement on small handheld devices.

Efficiency for Long Messages

Symmetric key encryption in general is very efficient, allowing even simple devices to encrypt and decrypt without devoting most of their processing cycles to these processes. Symmetric key encryption is efficient enough to be used even for long messages.

Key Exchange

The main problem with symmetric key encryption in the past has been exchanging the symmetric key securely between the two parties. (Obviously, if anyone intercepts the key as it is being exchanged, they will be able to read subsequent encrypted transmissions.) Later we will see how public key encryption can distribute symmetric key encryption securely.

TEST YOUR UNDERSTANDING

3. a) How many keys are used in symmetric key encryption? b) What is the most popular symmetric key encryption method? c) How long is its key? d) How can this method be made stronger? e) What is AES? f) Why is it attractive?

4. a) Why is the efficiency of symmetric key encryption important? b) What is the main problem with symmetric key encryption?

Public Key Encryption

Another class of encryption methods, shown in Figure D-4, is **public key encryption.** Here each party has a **private key,** which it keeps secret from the world. In addition, each party has a **public key,** which it shares with everybody because the public key, as its name suggests, does not need to be kept secret.

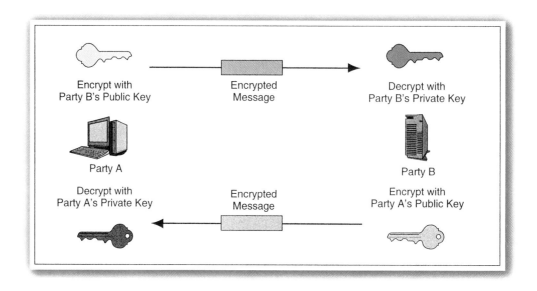

Figure D-4   Public Key Encryption for Confidentiality

Sending
Whenever one party sends, it encrypts the plaintext with the *public key of the receiver.* When A sends to B, A encrypts with *B's* public key. When B sends to A, B encrypts with *A's* public key.

Receiving
Each receiver decrypts with its own private key. When A sends to B, B decrypts with *B's* private key. In turn, when B sends to A, A decrypts with *A's* private key.

Once a message is encrypted with the receiver's public key, nobody can decrypt it except the receiver. Even the sender cannot decrypt the message after encrypting it.

Referring to Public and Private Keys
Note that you should never say "the public key" or "the private key" by itself when describing confidential transmission using public key encryption. There are two public keys and two private keys involved in any two-way exchange. You must always specify something like "A's public key" or "B's private key."

Complexity, Processing Intensiveness, and Short Messages
Public key encryption probably strikes you as complex. In fact, it is. It requires many computer processing cycles to do public key encryption and decryption—about 100 times as many cycles as symmetric key encryption. The inefficiency of this processing burden is so large that public key encryption can be used only to encrypt small messages.[3]

---

[3] In fact, some public key encryption methods are mathematically incapable of encrypting messages of more than 1,000 to 4,000 bits.

Simplicity of Key Exchange

The major benefit of public key encryption is that key exchange is simple. Public keys are not secret, so there is no need to exchange them securely. Many people post their public keys online for everyone to read.

Public Key Distribution of Symmetric Keys

In fact, public key distribution can help symmetric key encryption by distributing symmetric session keys securely, as Figure D-5 illustrates.

➤ First, one side generates a random bit string that will be used as a symmetric key.
➤ Second, that side encrypts this symmetric key with the public key of the other party.
➤ Third, the party that generated the symmetric key sends this ciphertext to the other party.
➤ Fourth, the other party decrypts the ciphertext with its own private key.
➤ Fifth, both sides now have the session key and will use it to send messages confidentially.

Session Key

The key that is exchanged this way is called a **session key** because it is used only for the current communication session. If the two parties communicate later, they will generate a new session key.

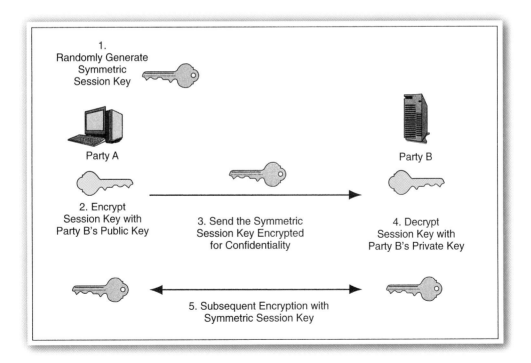

Figure D-5   Public Key Distribution for Symmetric Session Keys

Popular Public Key Encryption Methods

The most widely used public key encryption method is **RSA.**[4] RSA was patented, but its patent expired in 2000. Now that RSA is in the public domain, its domination of public key encryption may grow. However, a new form of public key encryption, the **elliptic curve cryptosystem (ECC),** promises to provide equal protection with smaller keys, and therefore less processing burden. ECC is patented.

TEST YOUR UNDERSTANDING

5. Jason sends a message to Kristin using public key encryption. a) What key will Jason use to encrypt the message? b) What key will Kristin use to decrypt the message? c) What key will Kristin use to encrypt the reply? d) What key will Jason use to decrypt the reply? e) Can the message and reply be long messages? Explain.

6. a) Does public key encryption have a problem with secure key exchange for the public key? Explain. b) How does public key encryption address symmetric key encryption's problem with secure key exchange? c) What is a session key?

7. a) What are the two most popular public key encryption methods? b) Which is in the public domain? c) Which allows smaller keys to be used for a given level of security?

AUTHENTICATION

So far, we have been discussing encryption as a way to create confidentiality. However, cryptography can also be used in **authentication,** that is, verifying the other party's identity.

Applicant and Verifier

In authentication terminology, the **applicant** is the side that tries to prove its identity to the other party, as Figure D-6 shows. The other party, which tries to authenticate the identity of the applicant, is the **verifier.** In two-way communication, both sides take on both roles because each authenticates the other.

TEST YOUR UNDERSTANDING

8. a) What is authentication? b) In authentication, who is the applicant? c) The verifier? d) Can a station be both an applicant and a verifier?

Initial Authentication with MS-CHAP Challenge–Response Authentication

In war movies, when a soldier approaches a sentry, the sentry issues a challenge, and the soldier must respond with the correct password. Figure D-6 shows a form of network **challenge–response authentication** called **MS-CHAP.** It is the Microsoft (MS) version of the IETF **Challenge Handshake Authentication Protocol (CHAP).**

MS-CHAP Is for Server-Based Authentication

MS-CHAP is used to authenticate a remote user to a server. It relies on the fact that the user has a password that both the user and the server know. This is a very common situation in organizations.

---

[4] Rivest-Shamir-Adleman.

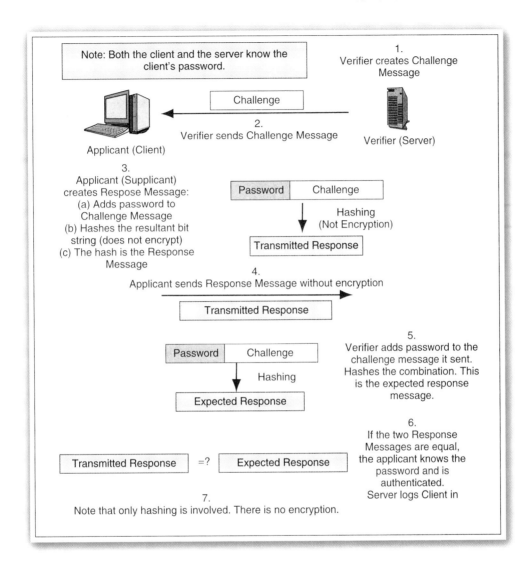

Figure D-6 contents:

Note: Both the client and the server know the client's password.

1. Verifier creates Challenge Message

Challenge

2. Verifier sends Challenge Message

Applicant (Client)

Verifier (Server)

3. Applicant (Supplicant) creates Respose Message:
(a) Adds password to Challenge Message
(b) Hashes the resultant bit string (does not encrypt)
(c) The hash is the Response Message

Password | Challenge

Hashing (Not Encryption)

Transmitted Response

4. Applicant sends Response Message without encryption

Transmitted Response

Password | Challenge

Hashing

Expected Response

5. Verifier adds password to the challenge message it sent. Hashes the combination. This is the expected response message.

Transmitted Response =? Expected Response

6. If the two Response Messages are equal, the applicant knows the password and is authenticated. Server logs Client in

7. Note that only hashing is involved. There is no encryption.

Figure D-6   MS-CHAP Challenge–Response Authentication Protocol for Initial Authentication

On the Applicant's Machine: Hashing

The server sends the applicant's PC a **challenge message,** which is a random bit string. The applicant's PC then adds the user's password to this challenge message.

The applicant's PC next hashes the combined challenge message and password. **Hashing** is a mathematical process that, when applied to a bit string of any length, produces a value of a fixed length, called the **hash.** For instance, the **MD5** hashing algorithm always produces a hash of 128 bits, whereas the **Secure Hash Algorithm (SHA-1)** produces hashes of 160, 256, 384, or 512 bits.

The hash of the challenge message plus the user's password is the **response message** that the applicant's PC sends back to the server.

On the Verifier's Machine (The Server)

The server itself adds the challenge message to the user's password and applies the same hashing algorithm the applicant used, producing a new hash. If this new hash is identical to the response message, then the person at the client PC must know the account's password.[5] The server then logs in the user.

TEST YOUR UNDERSTANDING

9.   a) For what type of authentication is MS-CHAP used? b) How does hashing work? c) What are the two most popular hashing algorithms? d) How does the applicant create the response message? e) How does the verifier check the response message?

Message-by-Message Authentication with Digital Signatures

Challenge–response authentication usually is only done at the beginning of a session or at most a few times per session. We would like to be able to authenticate each and every message coming from the other party to ensure that an attacker cannot slip a message into the message stream.

Digital Signatures

Figure D-7 shows how to create a **digital signature,** which authenticates each message analogously to the way a human signature authenticates documents.

Hashing to Produce the Message Digest

To create the digital signature, the sender (who is an applicant), first hashes the plaintext message the sender wishes to transmit securely. This generates a hash called a **message digest.** The hash is generated because digital signatures use public key encryption, which is limited to encrypting short messages, such as hashes.

Signing the Message Digest to Produce the Digital Signature

Next, the sender encrypts the message digest with the sender's own private key. This creates the **digital signature.** *Note that the message digest is not the digital signature but is only used to produce the digital signature.*

When a party encrypts with its own private key, this is called **signing** a message with its private key. The sender proves his or her identity like a person signing a letter. In this terminology, the sender signs the message digest to create the digital signature.

Sending the Message

The message that the sender wishes to send, then, consists of the original plaintext message plus the digital signature. If confidentiality is not an issue, the sender can simply send the combined message. However, confidentiality usually *is* important, so the sender normally encrypts the combined original message and digital signature for confidentiality. The combined message is likely to be long, so the sender must use symmetric key encryption.

---

[5]   Of course, users may have had their PCs remember their passwords. In this case, anyone taking control of their computers could impersonate them.

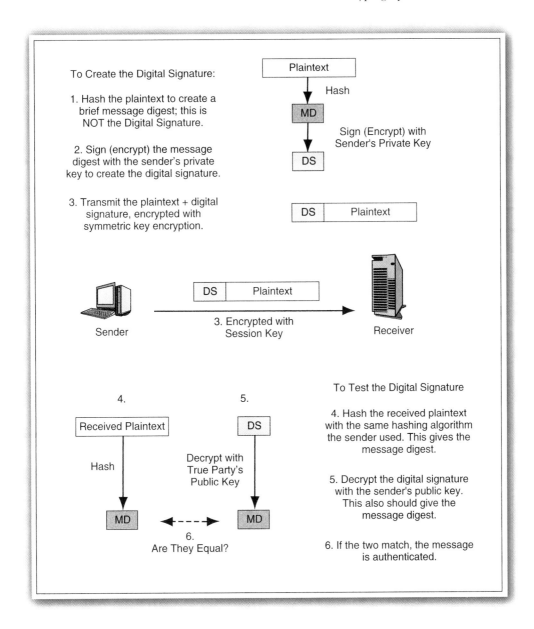

To Create the Digital Signature:

1. Hash the plaintext to create a brief message digest; this is NOT the Digital Signature.

2. Sign (encrypt) the message digest with the sender's private key to create the digital signature.

3. Transmit the plaintext + digital signature, encrypted with symmetric key encryption.

Plaintext

Hash

MD

Sign (Encrypt) with Sender's Private Key

DS

DS | Plaintext

Sender

DS | Plaintext

3. Encrypted with Session Key

Receiver

To Test the Digital Signature

4.

Received Plaintext

Hash

MD

5.

DS

Decrypt with True Party's Public Key

MD

6.
Are They Equal?

4. Hash the received plaintext with the same hashing algorithm the sender used. This gives the message digest.

5. Decrypt the digital signature with the sender's public key. This also should give the message digest.

6. If the two match, the message is authenticated.

Figure D-7   Digital Signature for Message-by-Message Authentication

Verifying the Applicant

The receiver (verifier) decrypts the entire message with the symmetric key used for confidentiality, then decrypts the digital signature with the true party's public key, which is widely known. This will produce the original message digest—if the applicant has signed the message digest with the true party's private key.

Then the receiver hashes the original plaintext message with the same hashing algorithm the applicant used. This should also produce the message digest.

If the message digests produced in these two different ways match, then the sender must have the true party's private key, which only the true party should know. The message is authenticated as coming from the true party.

Message Integrity

If someone changes the message en route, or if there are transmission errors, the two message digests will not match. Therefore, digital signatures give **message integrity**— the ability to tell if a message has been modified en route. A message that is changed will not pass the authentication test and will be discarded.

TEST YOUR UNDERSTANDING

10. a) For what type of authentication is a digital signature used? b) How does the applicant create a message digest? c) How does the applicant create a digital signature? d) What combined message does the applicant send? e) How is the combined message encrypted for confidentiality? f) How does the verifier check the digital signature? g) Besides authentication, what security benefit does a digital signature provide? h) Explain what this benefit means.

DIGITAL CERTIFICATES

In public key–based authentication, the verifier must know the public key of the true party. It should not ask the applicant for the true party's public key, because if the applicant is an **imposter,** the impostor will send his or her *own* public key claiming that this is the true party's public key, as Figure D-8 illustrates. If the verifier is ignorant

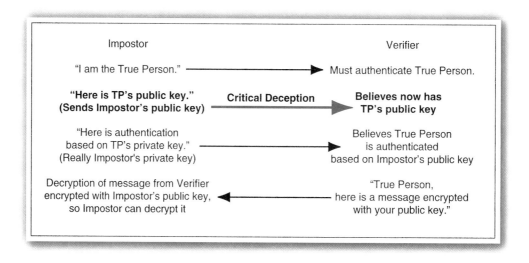

Figure D-8   Public Key Deception

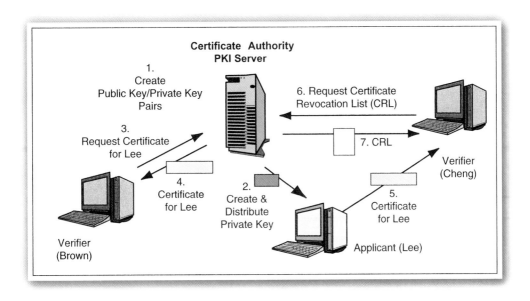

Figure D-9 Public Key Infrastructure (PKI) with a Certificate Authority

enough to accept the impostor's public key as the true party's public key, the imposter will sign digital signatures with his or her own private key, and the verifier will use the impostor's public key to verify the impostor as the true party!

Certificate Authorities and Digital Certificates

Instead, Figure D-9 shows that the verifier must contact a **certificate authority,** which is an independent and trusted source of information about the public keys of true parties. The certificate authority will send the verifier a **digital certificate** containing the *name of the true party* and the *true party's public key*. The verifier then uses this public key to authenticate the applicant claiming to be the true party. The standard for digital certificates is **X.509.**

Unfortunately, certificate authorities are not regulated, so the verifier must only accept a digital certificate from a certificate authority it trusts.

Also, it is a common misconception that certificate authorities vouch for the honesty of the party named in the certificate. They do not. They merely vouch for the named party's public key! Although clients who misbehave may have their certificates revoked, certificate authorities rarely give strong warranties about the honesty of their clients. That is not their job.

The Role of the Digital Certificate

One point that many students find difficult to understand is that a digital certificate does not, by itself, authenticate an applicant! As Figure D-10 shows, the digital certificate merely provides the public key of the true party which the verifier needs to make

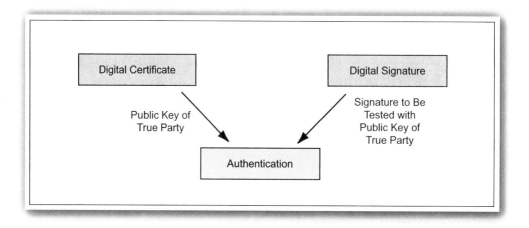

Figure D-10   The Roles of Digital Certificates and Digital Signatures
in Authentication

the authentication. Similarly, the digital signature is a mere calculation; it means nothing until it is tested with the public key of the true party, which can be obtained reliably only through a digital certificate. *Digital certificates and public key authentication must be used together in public key authentication.* Neither, by itself, provides authentication.

Checking the Certificate Revocation List (CRL)

The certificate authority may revoke a party's digital certificate before the termination date listed in the digital certificate. Therefore, if the verifier gets the digital certificate from a party other than the certificate authority, the verifier should check the certificate authority's **certificate revocation list (CRL)** to be sure that the digital certificate is still valid. To do this, the verifier downloads the CRL and sees whether the digital certificate's serial number has been revoked.

Public Key Infrastructures (PKIs)

Managing public keys requires several related actions. Digital certificates must be created and distributed. Certificate revocation list checking must be supported. Public key–private key pairs must be created. The public key will be distributed through digital certificates, but the private key must be distributed securely, often by embedding it in software that is installed on a particular machine.

Public key infrastructures (PKIs) automate all of these processes. PKIs create a total system (infrastructure) for public key encryption. Unfortunately, all of the PKIs in use today are proprietary products, so interoperability is spotty.

TEST YOUR UNDERSTANDING

11.   a) What pair of facts crucial to authentication do digital certificates tell us? b) What kind of organization creates a digital certificate? c) Are these organizations regulated?

Layer	Cryptographic System
Application	Kerberos
Transport	SSL/TLS
Internet	IPsec
Data Link	PPTP
Physical	Not applicable. No messages are sent at this layer—only individual bits

Figure D-11   Multilayer Security with Cryptographic Systems

d) Does a digital certificate tell you that the entity named in a digital certificate is a good organization? Explain. e) Does a digital certificate tell you that the entity named in a digital certificate has a certain public key? f) Does a digital certificate alone give authentication? Explain. g) Does public key authentication without a digital certificate give reliable authentication? Explain. h) What is necessary for reliable authentication? i) What are the elements of a PKI? j) Why is it important to check the CRL if you do not receive the digital certificate from the certificate authority?

MULTILAYER SECURITY

At what layer do cryptographic systems exist? The answer is that they can exist at one layer, at multiple layers, or at all layers. Figure D-11 illustrates some common cryptographic systems and other security systems and the layers at which they operate.

Implementing security at all five layers generally is unwise because it would slow the communication to an unacceptable level. However, it is usually wise to create **multilayer security,** in which security is implemented in at least two layers although not at all layers. Historically, problems often have been found in security products. If a weakness is found at one layer in a multilayer security system, security at other layers will still protect the system while the flawed security layer is being fixed or replaced.

TEST YOUR UNDERSTANDING

12.   a) At what layer is security implemented? b) What problem will implementing security at multiple layers bring? c) What is the advantage of multilayer security?

Glossary

.NET: Microsoft's approach to the Web services.

10/100 Ethernet: A collective name for the Ethernet physical layer 100Base-TX and 100Base-FX standards.

1000Base-LX: A fiber version of gigabit Ethernet for long wavelengths (transmitting at 1,300 nm).

1000Base-SX: A fiber version of gigabit Ethernet for short wavelengths (transmitting at 850 nm).

1000Base-T: A UTP version of gigabit Ethernet.

1000Base-x: The Ethernet physical layer technology of gigabit Ethernet, used today mainly to connect switches to switches or switches to routers; increasingly being used to connect servers and some desktop PCs to the switches that serve them.

100Base-FX: The Ethernet physical layer 100 Mbps standard used primarily to connect switches to other switches, now being phased out.

100Base-TX: The dominant Ethernet physical layer 100 Mbps standard brought to desktop computers today.

10Base-F: See 802.3 10Base-F.

10Base-T: See 802.3 10Base-T.

1G: See First-Generation.

2.5G: See Second-and-a-Half Generation.

232 Serial Port: The port on a PC that uses two voltage ranges to transmit information.

25-Pair UTP Cord: The cabling used by telephony for vertical wiring runs within a building.

2G: See Second-Generation.

2-Pair Data-Grade: The higher-quality UTP access lines used by telephone carriers for private lines. Two pairs run out to each customer.

3DES: See Triple DES.

3G: See Third-Generation.

4-Pair Unshielded Twisted Pair (UTP): The type of wiring typically used in Ethernet networks. 4-pair UTP contains eight copper wires organized as four pairs. Each wire is covered with dielectric insulation, and an outer jacket encloses and protects the four pairs.

50-Pin Octopus Connector: The type of connector in which vertical cords typically terminate.

802 Committee: See 802 LAN/MAN Standards Committee.

802 LAN/MAN Standards Committee: The IEEE committee responsible for Ethernet standards.

802.11 Working Group: The IEEE working group that creates wireless LAN standards.

802.11e: A standard for quality of service in 802.11 WLANS.

802.11i: An advanced form of 802.11 wireless LAN security.

802.16: WiMAX.

802.1D Spanning Tree Protocol: The protocol that addresses both single points of failure and loops.

802.1p: The standard that permits up to eight priority levels.

802.1Q: The standard that extended the Ethernet MAC layer frame to include two optional tag fields.

802.3 10Base-F: An Ethernet physical layer 10 Mbps fiber standard, now almost entirely extinct.

802.3 10Base-T: The slowest Ethernet physical layer technology in use today; uses 4-pair UTP wiring and operates at 10 Mbps.

802.3 MAC Layer Frame: See Ethernet Frame.

802.3 MAC Layer Standard: The standard that defines Ethernet frame organization and NIC and switch operation.

802.3 Working Group: The 802 Committee's working group that creates Ethernet-specific standards.

900 Number: A number that allows customers to call into a company; callers pay a fee that is much higher than that of a regular toll call.

Access Control List (ACL): An ordered list of pass/deny rules for a firewall or other device.

Access Line: 1) In networks, a transmission line that connects a station to a switch. 2) In telephony, the line used by the customer to reach the PSTN's central transport core.

Access Point: A bridge between a wireless station and a wired LAN.

Access System: In telephony, the system by which customers access the PSTN, including access lines and termination equipment in the end office at the edge of the transport core.

Account: An identifiable entity that may own resources on a computer.

ACE: See OPNET Application Characterization Environment.

ACK Bit: The bit in a TCP segment that is set to indicate if the segment contains an acknowledgement.

ACK: See Acknowledgement.

Acknowledgement (ACK): 1) An acknowledgement message, sent by the receiver when a message is received correctly. 2) An acknowledgement frame, sent by the receiver whenever a frame is received; used in CSMA/CA+ACK in 802.11.

Acknowledgement Number Field: In TCP, a header field that tells what TCP segment is being acknowledged in a segment.

ACL: See Access Control List.

ADC: See Analog-to-Digital Conversion.

Address Resolution Protocol (ARP): Protocol for address resolution used in Ethernet networks. If a host or router knows a target host's or router's IP address, ARP finds the target's data link layer address.

Administrative IP Server: A server needed to support IP.

Administrator: A super account on a Windows server that automatically has full permissions in every directory on the server.

ADSL: See Asymmetric Digital Subscriber Line.

Advanced Encryption Standard (AES): New symmetric encryption standard that offers 128-bit, 192-bit or 256-bit encryption efficiently.

AES: See Advanced Encryption Standard.

Agent: See Network Management Agent.

Aggregate Throughput: Throughput shared by multiple users; individual users will get a fraction of this throughput.

Alternative Route: In mesh topology, one of several possible routes from one end of the network to the other, made possible by the topology's many connections among switches or routers.

Always On: Being always available for service; used to describe access lines.

Amplitude: The maximum (or minimum) intensity of a wave. In sound, this corresponds to volume (loudness).

Amplitude Modulation: A simple form of modulation in which a modem transmits one of two analog signals—a high-amplitude (loud) signal or a low-amplitude (soft) signal.

Analog Signal: A signal that rises and falls in intensity smoothly and that does not have a limited number of states.

Analog-to-Digital Conversion (ADC): A device for the conversion of transmissions from the analog local loop to signals on the digital telephone network's core.

Antivirus Software: Software that scans computers to protect them against viruses, worms, and Trojan horses arriving in e-mail attachments and other propagation methods.

API: See Application Program Interface.

AppleTalk: Apple's proprietary architecture for use on Macintosh computers.

Applicant: In authentication, the user trying to prove his or her identity; sometimes called the supplicant.

Application Architecture: The arrangement of how application layer functions are spread among computers to deliver service to users.

Application Characterization Environment: See OPNET Application Characterization Environment.

Application Firewall: A firewall that examines the application layer content of packets.

Application Layer: The standards layer that governs how two applications communicate with each other; Layer 7 in OSI, Layer 5 in TCP/IP.

Application Profile: A method, offered by Bluetooth, that allows devices to work with one another automatically at the application layer.

Application Program: Program that does work for users; operating system is the other major type of program found on computers.

Application Program Interface (API): A specification that allows application server programs to interact directly with database systems.

Application Server: A server used by large e-commerce sites that accepts user data from a front-end webserver, assembles information from other servers, and creates a webpage to send back to the user.

Architecture: A broad plan that specifies what is needed in general and the components that will be used to provide that functionality. Applied to standards, networks, and applications.

Asymmetric Digital Subscriber Line (ADSL): The type of DSL designed to go into residential homes, offers high downstream speeds but limited upstream speeds.

Asynchronous Transfer Mode (ATM): The packet-switched network technology, specifically designed to carry voice, used for transmission in the PSTN transport core. ATM offers quality of service guarantees for throughput, latency, and jitter.

ATM: Asynchronous Transfer Mode.

Attenuate: For a signal's strength to weaken during propagation.

Authentication: The requirement that someone who requests to use a resource must prove his or her identity.

Authentication Server: A server that stores data to help the verifier check the credentials of the applicant.

Autonomous System: Internet owned by an organization.

Autosensing: The ability of a switch to detect the standard being used at the other end of the connection, and adjust its own speed to match.

Availability: The ability of a network to serve its users.

Backdoor: A way back into a compromised computer that an attacker leaves open; it may simply be a new account or a special program.

Back-Office: Transaction processing applications for a business's internal needs.

Bandpass Filter: A device that filters out all signals below 300 Hz and above about 3.4 kHz.

Bandwidth: The range of frequencies over which a signal is spread.

Bank Settlement Firm: An e-commerce service that handles credit card payments.

Base Price: The price of a system's hardware, software, or both before necessary options are added.

Baseband Signal: 1) The original signal in a radio transmission; 2) a signal that is injected directly into a wire for propagation.

Baud Rate: The number of clock cycles a transmission system uses per second.

Best-Match Row: The row that provides the best forwarding option for a particular incoming packet.

BGP: See Border Gateway Protocol.

Binary Data: Data that has only two possible values (ones and zeros).

Binary Numbers: The Base 2 counting system where ones and zeros used in combination can represent whole numbers (integers).

Binary Signaling: Signaling that uses only two states.

Biometrics: The use of bodily measurements to identify an applicant.

Bit Rate: In digital data transmission, the rate at which information is transmitted; measured in bits per second.

Bits per Second (bps): The measure of network transmission speed. In increasing factors of 1,000 are kilobits per second (kbps), megabits per second (Mbps), gigabits per second (Gbps), and terabits per second (Tbps).

Black List: A list of banned websites.

Blended Threat: An attack that propagates both as a virus and as a worm.

Bluetooth: A wireless networking standard created for personal area networks.

Bonding: See Link Aggregation.

Border Firewall: A firewall that sits at the border between a firm and the outside world.

Border Gateway Protocol (BGP): The most common exterior routing protocol on the Internet. Recall that *gateway* is an old term for *router*.

Border Router: A router that sits at the edge of a site to connect the site to the outside world through leased lines, PSDNs, and VPNs.

Bps (bps): See Bits per Second.

Bridge: An access point that connects two different types of LANs.

Broadband: 1) Transmission where signals are sent in wide radio channels; 2) any high-speed transmission system.

Broadband Wireless Access (BWA): High-speed local wireless transmission systems.

Broadcast: To send a message out to all other stations simultaneously.

Broadcast Address: In Ethernet, FF-FF-FF-FF-FF-FF (48 ones); tells switches that the frame should be broadcast.

Brute-Force Attack: A password-cracking attack in which an attacker tries to break a password by trying all possible combinations of characters.

Bursty: Having short, high-speed bursts separated by long silences. Characteristic of data transmission.

Bus Topology: A topology in which one station transmits and has its signals broadcast to all stations.

Business Case: An argument for a system in business terms.

Business Continuity: A company's ability to continue operations.

Business Continuity Recovery: The reestablishment of a company's ability to continue operations.

BWA: See Broadband Wireless Access.

CA: 1) See Certificate Authority. 2) See Collision Avoidance.

Cable Modem: 1) Broadband data transmission service using cable television; 2) the modem used in this service.

Cable Replacement: Getting rid of cables between devices by implementing wireless networking.

Call Waiting: A service that allows the user to place an original caller on hold if someone else calls the user, shift briefly to the new caller, and then switch back to the original caller.

Caller ID: Service wherein the telephone number of the party calling you is displayed on your phone's small display screen before you pick up the handset; allows the user to screen calls.

Carrier: A transmission service company.

Carrier Sense Multiple Access with Collision Avoidance and Acknowledgements (CSMA/CA+ACK): A mandatory mechanism used to reduce problems with multiple simultaneous transmissions, which occur in wireless transmission. CSMA/CA+ACK is a media access control discipline, and it uses both collision avoidance and acknowledgement frames.

Carrier Sense Multiple Access with Collision Detection (CSMA/CD): The process wherein if a station wants to transmit, it may do so if no station is already transmitting but must wait if another station is already sending. In addition, if there is a collision because two stations send at the same time, all stations stop, wait a random period of time, and then try again.

Cat 5e: See Category 5e.

Category (Cat) 5e: Quality type of UTP wiring; required for 100Base-TX and gigabit Ethernet.

Category 6: The newest quality type of UTP wiring being sold; not required for even gigabit Ethernet.

CDMA: See Code Division Multiple Access.

CDMA IS-95: The form of CDMA used in 2G cellular technology in the United States.

CDMA2000: A new 3G technology, developed by Qualcomm, offering a staged approach to increasing speed.

CDMA2000 1x: The initial 3G step for implementing CDMA2000, offering telephone modem speeds.

CDMA2000 1xEV-DO: The second 3G step for implementing CDMA2000, which will offer speeds similar to those in DSL and cable modems.

Cell: 1) In ATM, a fixed-length frame. 2) In cellular telephony, a small geographical area served by a cellsite.

Cellphone: A cellular telephone, also called a mobile phone or mobile.

Cellsite: In cellular telephony, equipment at a site near the middle of each cell, containing a transceiver and supervising each cellphone's operation.

Cell-Switching: A technology that uses fixed-length frames.

Certificate Authority (CA): Organization that provides public key–private key pairs and digital certificates.

Certificate Revocation List (CRL): A certificate authority's list of digital certificates it has revoked before their expiration date.

Challenge Message: In challenge–response authentication protocols, the message initially sent from the verifier to the applicant.

Challenge–Response Authentication: Initial authentication method in which the verifier sends the applicant a challenge message, and the applicant does a calculation to produce a response, which it sends back to the verifier.

Challenge–Response Authentication Protocol (CHAP): A specific challenge–response authentication protocol.

Channel: A small frequency range that is a subdivision of a service band.

Channel Bandwidth: The range of frequencies in a channel; determined by subtracting the lowest frequency from the highest frequency.

Channel Reuse: The ability to use each channel multiple times, in different cells in the network.

Channel Service Unit (CSU): The part of a CSU/DSU device designed to protect the telephone network from improper voltages sent into a private line.

CHAP: See Challenge–Response Authentication Protocol.

Checkout: A core e-commerce function that allows a buyer who has finished shopping to pay for the selected goods.

Chronic Lack of Capacity: A state in which the network lacks adequate capacity much of the time.

CIDR: See Classless InterDomain Routing.

Ciphertext: The result of encrypting a plaintext message. Ciphertext can be transmitted with confidentiality.

CIR: See Committed Information Rate.

Circuit: A two-way connection with reserved capacity.

Circuit Switching: Switching in which capacity for a voice conversation is reserved on every switch and trunk line end-to-end between the two subscribers.

Cladding: A thick glass cylinder that surrounds the core in optical fiber.

Class A IP Address: In classful addressing, an IP address block with more than sixteen million IP addresses; given only to the largest firms and ISPs.

Class B IP Address: In classful addressing, an IP address block with about 65,000 IP addresses; given to large firms.

Class C IP Address: In classful addressing, an IP address block with 254 possible IP addresses; given to small firms.

Class D IP Address: In classful addressing, IP addresses used in multicasting.

Class 5 Switch: See End Office Switch.

Classful Addressing: Giving a firm one of four block sizes for IP addresses: a very large Class A address block, a medium-sized Class B address block, or a small Class C address block.

Classless InterDomain Routing (CIDR): System for allocating IP addresses that does not use IP address classes.

Clear to Send (CTS): In 802.11, a message broadcast by an access point, which allows only a station that has sent a Request to Send message to transmit. All other stations must wait.

CLEC: See Competitive Local Exchange Carrier.

CLI: See Command Line Interface.

Client Station: A station that receives service from a server station.

Client/Server Interaction: Interaction in which a client program requests service from a server and in which the server program provides the service.

Client/Server Processing: The form of client/server computing in which the work is done by programs on two machines.

Client/Server System: A system where some processing power is on the client computer. The two types of client/server systems are file server program access and full client/server processing.

Clock Cycle: A period of time during which a transmission line's state is held constant.

Cloud: The symbol traditionally used to represent the PSDN transport core, reflecting the fact that although the PSDN has internal switches and trunk lines, the customer does not have to know how things work inside the cloud.

Coating: In optical fiber, the substance that surrounds the cladding to keep out light and to strengthen the fiber. Coating includes strands of yellow Aramid (Kevlar) yarn to strengthen the fiber.

Coaxial Cable: Transmission medium consisting of a wire core surrounded by a metal mesh.

Code Division Multiple Access (CDMA): A new form of cellular technology and a form of spread spectrum transmission that allows multiple stations to transmit at the same time in the same channel; also permits stations in adjacent cells to use the same channel without serious interference.

Codec: The device in the end office switch that converts between the analog local loop voice signals and the digital signals of the end office switch.

Collision: When two simultaneous signals use the same shared transmission medium, the signals will add together and become scrambled (unintelligible).

Collision Avoidance (CA): In 802.11, used with CSMA to listen for transmissions, so if a wireless NIC detects a transmission, it must not transmit. This avoids collision.

Collision Domain: In Ethernet CSMA/CD systems that use hubs or bus topologies, the collection of all stations that can hear one another; only one can transmit at a time.

Command Line Interface (CLI): An interface used to work with switches and routers, in which the user types highly structured commands, ending each command with Enter.

Command–Response Cycle: The exchange of messages through which SNMP communication between the manager and agents takes place. In it, the manager sends a command, and the agent sends back a response confirming that the command has been met, delivering requested data, or saying that an error has occurred and that the agent cannot comply with the command.

Committed Information Rate (CIR): PVC speed that is guaranteed by the Frame Relay carrier.

Community Name: In SNMP Version 1, only devices using the same community name will communicate with each other; very weak security.

Competitive Local Exchange Carrier (CLEC): A competitor to the ILEC.

Comprehensive Security: Security in which all avenues of attack are closed off.

Compromise: A successful attack.

Computer Security Incident Response Team (CSIRT): A team convened to handle major security incidents, made up of the firm's security staff, members of the IT staff, and members of functional departments, including the firm's legal department.

Confidentiality: Assurance that interceptors cannot read transmissions.

Connectionless: Type of conversation that does not use explicit openings and closings.

Connection-Oriented: Type of conversation in which there is a formal opening of the interactions, a formal closing, and maintenance of the conversation in between.

Connectorize: To add connectors to something.

Constellation: In quadrature amplitude modulation, the collection of all possible amplitude/phase combinations.

Continuity Testers: UTP tester that ensures that wires are inserted into RJ-45 connectors in the correct order and are making good contact.

Convergence: The correction of routing tables after a change in an internet.

Conversion: The process of browsers becoming buyers.

Cord: A length of transmission medium—usually UTP or optical fiber but sometimes coaxial cable.

Core: 1) In optical fiber, the very thin tube into which a transmitter injects light. 2) In a switched network, the collection of all core switches.

Core Switch: A switch further up the hierarchy that carries traffic between pairs of switches. May also connect switches to routers.

Crack: To guess a password.

Credentials: Proof of identity that an applicant can present during authentication.

Credit Card Verification Service: An e-commerce service that checks the validity of the credit card number a user has typed.

Criminal Attacker: An attacker who attacks with criminal motivation.

Crimping Tool: Tool for crimping wires into an RJ-45 connector.

CRL: See Certificate Revocation List.

CRM: See Customer Relationship Management.

Cross-Connect Device: The device within a wiring closet that vertical cords plug into. Cross-connect devices connect the wires from the riser space to 4-pair UTP cords that span out to the wall jacks on each floor.

Crossover Cable: A UTP cord that allows a NIC in one computer to be connected directly to the NIC in another computer; switches Pins 1 and 2 with Pins 3 and 6.

Crosstalk Interference: Mutual EMI among wire pairs in a UTP cord.

Cryptographic System: A security system that automatically provides a mix of security protections, usually including confidentiality, authentication, message integrity, and replay protection.

CSIRT: See Computer Security Incident Response Team.

CSMA/CA+ACK: See Carrier Sense Multiple Access with Collision Avoidance and Acknowledgments. See definitions of the individual components.

CSMA/CD: See Carrier Sense Multiple Access with Collision Detection.

CSU: See Channel Service Unit.

CSU/DSU: Device that connects an internal site system to a private line circuit.

CTS: See Clear to Send.

Customer Premises Equipment (CPE): Equipment owned by the customer, including PBXs, internal vertical and horizontal wiring, and telephone handsets.

Customer Relationship Management (CRM): Software that examines customer data to understand the preference of a company's customers.

Cut-through: Switching wherein the Ethernet switch examines only some fields in a frame's header before sending the bits of the frame back out.

Cyberterror: A computer attack made by terrorists.

Cyberwar: A computer attack made by a national government.

DAC: See Digital-to-Analog Conversion.

Data Encryption Standard (DES): Popular symmetric key encryption method; with only 56-bit keys, considered to be too weak for business-to-business encryption.

Data Field: The content delivered in a message.

Data Link: The path that a frame takes across a single network (LAN or WAN).

Data Link Control Identifier (DLCI): The virtual circuit number in Frame Relay, normally 10 bits long.

Data Link Layer: The layer that governs transmission within a single network all the way from the source station to the destination station across zero or more switches; Layer 2 in OSI.

Data Service Unit (DSU): The part of a CSU/DSU circuit that formats the data in the way the private line requires.

dB: See Decibel.

Dead Spot: See Shadow Zone.

Decapsulation: The removing of a message from the data field of another message.

Decibel (dB): The unit in which attenuation is measured.

Decrypt: Conversion of encrypted ciphertext into the original plaintext so an authorized receiver can read an encrypted message.

Dedicated Server: A server that is not used simultaneously as a user PC.

Default Printer: The printer to which a user's print jobs will be sent unless the user specifies a different printer.

Default Router: The next-hop router that a router will forward a packet to if the routing table does not have a row that governs the packet's IP address except for the default row.

Default Row: The row of a routing table that will be selected automatically if no other row matches; its value is 0.0.0.0.

Defense in Depth: The use of successive lines of defense.

Demilitarized Zone (DMZ): A subnet in which webservers and other public servers are placed.

Demodulate: To convert digital transmission signals to analog signals.

Denial-of-Service (DoS): The type of attack whose goal is to make a computer or a network unavailable to its users.

Dense Wavelength Division Multiplexing (DWDM): WDM using fiber that carries more than forty light sources at different frequencies.

DES: See Data Encryption Standard.

Designated Router: In OSPF, a router that sends change information to other routers in its area.

Destination: In a routing table, the column that shows the destination network's network part or subnet's network part plus subnet part, followed by zeroes. This row represents a route to this network or subnet.

DHCP: See Dynamic Host Configuration Protocol.

Dial-Up Circuit: A circuit that only exists for the duration of a telephone call.

Dictionary Attack: A password-cracking attack in which an attacker tries to break a password by trying all words in a standard or customized dictionary.

Dictionary Word: A common word, dangerous to use for a password because easily cracked.

Dielectric Insulation: The non-conducting insulation that covers each wire in 4-pair UTP, preventing short circuits between the electrical signals traveling on different wires.

Diff-Serv: The field in an IP packet that can be used to label IP packets for priority and other service parameters.

Digital Certificate: A document that gives the name of a true party, that true party's public key, and other information; used in authentication.

Digital Certificate Authentication: Authentication in which each user has a public key and a private key. Authentication depends on the applicant knowing the true party's private key; requires a digital certificate to give the true party's public key.

Digital Signaling: Signaling that uses a few states. Binary (two-state) transmission is a special case of digital transmission.

Digital Signature: A calculation added to a plaintext message to authenticate it.

Digital Subscriber Line (DSL): A technology that provides digital data signaling over the residential customer's existing single-pair UTP voice-grade copper access line.

Digital-to-Analog Conversion (DAC): The conversion of transmissions from the digital telephone network's core to signals on the analog local loop.

Direct Sequence Spread Spectrum (DSSS): Spread spectrum transmission that spreads the signal over the entire bandwidth of a channel.

Disaster: An incident that can stop the continuity of business operations, at least temporarily.

Disaster Recovery: The reestablishment of information technology operations.

Discovering: The first phase of network mapping, in which the program finds out if hosts and subnets exist.

Disgruntled Employee: Employee who is upset with the firm or an employee and who may take revenge through a computer attack.

Disgruntled Ex-Employee: Former employee who is upset with the firm or an employee and who may take revenge through a computer attack.

Dish Antenna: An antenna that points in a particular direction, allowing it to send stronger outgoing signals in that direction for the same power and to receive weaker incoming signals from that direction.

Distance Vector Protocol: Routing protocol based on the number of hops to a destination out a particular port.

Distort: To change in shape during propagation.

DLCI: See Data Link Control Identifier.

DMZ: See Demilitarized Zone.

DNS: See Domain Name System.

Domain: 1) In DNS, a group of resources (routers, single networks, and hosts) under the control of an organization. 2) In Microsoft Windows, a grouping of resources used in an organization, made up of clients and servers.

Domain Controller: In Microsoft Windows, a computer that manages the computers in a domain.

Domain Name System (DNS): A server that provides IP addresses for users who know only a target host's host name. DNS servers also provide a hierarchical system for naming domains. DNS servers are also called name servers.

DoS: See Denial-of-Service.

Dotted Decimal Notation: The notation used to ease human comprehension and memory in reading IP addresses.

Downtime: A period of network unavailability.

Drive-By Hacker: A hacker who parks outside a firm's premises and eavesdrops on its data transmissions; mounts denial-of-service attacks; inserts viruses, worms, and spam into a network; or does other mischief.

DSL: See Digital Subscriber Line.

DSL Access Multiplexer (DSLAM): A device at the end office of the telephone company that sends voice signals over the ordinary PSTN and sends data over a data network such as an ATM network.

DSLAM: See DSL Access Multiplexer.

DSSS: See Direct Sequence Spread Spectrum.

DSU: See Data Service Unit.

Dumb Terminal: A desktop machine with a keyboard and display but little processing capability; processing is done on a host computer.

DWDM: See Dense Wavelength Division Multiplexing.

Dynamic Host Configuration Protocol (DHCP): The protocol used by DHCP servers, which provide each user PC with a temporary IP address to use each time he or she connects to the Internet.

EAP: See Extensible Authentication Protocol.

E-Commerce: Electronic commerce; buying and selling over the Internet.

E-Commerce Software: Software that automates the creation of catalog pages and other e-commerce functionality.

Economy of Scale: In managed services, the condition of being cheaper to manage the traffic of many firms than of one firm.

Egress Filtering: The filtering of traffic from inside a site going out.

Electromagnetic Interference (EMI): Unwanted electrical energy coming from external devices, such as electrical motors, fluorescent lights, and even nearby data transmission wires.

Electromagnetic Signal: A signal generated by oscillating electrons.

Electronic Catalog: An e-commerce site's display that shows the goods the site has for sale.

Electronic Commerce (E-Commerce): The buying and selling of goods and services over the Internet.

Elliptic Curve Cryptosystem (ECC): Public key encryption method; more efficient than RSA.

EMI: See Electromagnetic Interference.

Encapsulation: The placing of a message in the data field of another message.

Encrypt: To mathematically process a message so that an interceptor cannot read the message.

Encryption method: A method for encrypting plaintext messages.

End Office: Telephone company switch that connects to the customer premises via the local loop.

End Office Switch: The nearest switch of the telephone company to the customer premises.

End-to-End: A layer where communication is governed directly between the transport process on the source host and the transport process on the destination host.

Ephemeral Port Number: The temporary number a client selects whenever it connects to an application program on a server. According to IETF rules, ephemeral port numbers should be between 49153 and 65535.

Equipment Room: The room, usually in a building's basement, where wiring connects to external carriers and internal wiring.

Error Advisement: In ICMP, the process wherein if an error is found, there is no transmission, but the router or host that found the error usually sends an ICMP error message to the source device to inform it that an error has occurred. It is then up to the device to decide what to do. (This is not the same as error correction because there is no mechanism for the retransmission of lost or damaged packets.)

Error Rate: In biometrics, the normal rate of misidentification when the subject is cooperating.

Ethernet 10Base2: Obsolete 10 Mbps Ethernet standard that uses coaxial cable in a bus topology. Less expensive than 10Base5 but cannot carry signals as far.

Ethernet 10Base5: Obsolete 10 Mbps Ethernet standard that uses coaxial cable in a bus topology.

Ethernet Address: The 48-bit address the stations have on an Ethernet network; often written in hexadecimal notation for human reading.

Ethernet Frame: A message at the data link layer in an Ethernet network.

Ethernet Switch: Switch following the Ethernet standard. Notable for speed and low cost per frame sent. Dominates LAN switching.

EtherPeek: A commercial traffic summarization program.

Excess Burst Speed: One of Frame Relay's two-part PVC speeds; beyond the CIR.

Exhaustive Key Search: Cracking a key by trying all possible keys to decrypt a message.

Exploit: A break-in program; a program that exploits known vulnerabilities.

Extended Star Topology: The type of topology wherein there are multiple layers of switches organized in a hierarchy, in which each node has only one parent node; used in Ethernet; more commonly called a hierarchical topology.

Extensible Authentication Protocol (EAP): A protocol that authenticates users with authentication data (such as a password or a response to a challenge based on a station's digital certificate) and authentication servers.

Exterior Routing Protocol: Routing protocol used between autonomous systems.

Extranet: A network that uses TCP/IP Internet standards to link several firms together but that is not accessible to people outside these firms. Even within the firms of the extranet, only some of each firm's computers have access to the network.

Face Recognition: The scanning of passersby to identify terrorists or wanted criminals by the characteristics of their faces.

Facilitating Server: A server that solves certain problems in P2P interactions but that allows clients to engage in P2P communication for most of the work.

False Alarm: An apparent incident that proves not to be an attack.

False Positive: A false alarm.

FDM: See Frame Division Multiplexing.

FHSS: See Frequency Hopping Spread Spectrum.

Fiber to the Home (FTTH): Optical fiber brought by carriers to individual homes and businesses.

Field: A subdivision of a message header or trailer.

File Server: A server that allows users to store and share files.

File Server Program Access: The form of client/server computing in which the server's only role is to store programs and data files, while the client PC does the actual processing of programs and data files.

File Service: Service on a file server that allows users to store and share files.

File Sharing: The ability of computer users to share files that reside on their own disk drives or on a dedicated file server.

Fingerprint Scanning: A form of biometric authentication that uses the applicant's fingerprints.

Fingerprinting: The second phase of network mapping, in which the program determines the characteristics of hosts to determine if they are clients, servers, or routers.

Firewall: A security system that examines each incoming packet. If the firewall identifies the packet as an attack packet, the firewall discards the packet and copies information about the discarded packet into a log file.

First-Generation (1G): The initial generation of cellular telephony, introduced in the 1980s. 1G systems were analog, were only given about 50 MHz of spectrum, had large and few cells, and had very limited speeds for data transmission.

Fixed Wireless Service: Local terrestrial wireless service in which the user is at a fixed location.

Flag: A one-bit field.

Flat Rate: Local telephone service in which there is a fixed monthly service charge but no separate fee for individual local calls.

Flow Control: The ability of one side in a conversation to tell the other side to slow or stop its transmission rate.

Forensics: The collection of data in a form suitable for presentation in a legal proceeding.

Fractional T1: A type of private line that offers intermediate speeds at intermediate prices; usually operates at one of the following speeds: 128 kbps, 256 kbps, 384 kbps, 512 kbps, or 768 kbps.

FRAD: See Frame Relay Access Device.

Fragment: To break a message into multiple smaller messages. TCP fragments application layer messages, while IP packets may be fragmented by routers along the packet's route.

Fragment Offset Field: In IPv4, a flag field that tells a fragment's position in a stream of fragments from an initial packet.

Frame: 1) A message at the data link layer. 2) In time division multiplexing, a brief time period, which is further subdivided into slots.

Frame Check Sequence Field: A four-octet field used in error checking in Ethernet. If an error is found, the frame is discarded.

Frame Relay Access Device (FRAD): Device that connects an internal site network to a Frame Relay network.

Frequency: The number of complete cycles a radio wave goes through per second. In sound, frequency corresponds to pitch.

Frequency Division Multiplexing (FDM): A technology used in microwave transmission in which the microwave bandwidth is subdivided into channels, each carrying a single circuit.

Frequency Hopping Spread Spectrum (FHSS): Spread spectrum transmission that uses only the bandwidth required by the signal but hops frequently within the spread spectrum channel.

Frequency Modulation: Modulation in which one frequency is chosen to represent a 1 and another frequency is chosen to represent a 0.

Frequency Spectrum: The range of all possible frequencies from zero hertz to infinity.

FTTH: See Fiber to the Home.

Full Control: In Microsoft Windows, an omnibus permission, equal to all of the other Microsoft Windows Server permissions.

Full-Duplex: A type of communication that supports simultaneous two-way transmission. Almost all communication systems today are full-duplex systems.

Fully Configured: A system with all necessary options.

Functional Department: General name for departments in firm other than the IT department; marketing, accounting, and so forth.

Gateway: An obsolete term for "router;" still in use by Microsoft.

Gateway Controller: In IP telephony, a device that controls the operation of signaling gateways and media gateways.

General Packet Radio Service (GPRS): The technology to which many GSM systems are now being upgraded. GPRS can combine two or more GSM time slots within a channel and so can offer data throughput near that of a telephone modem. Often called a 2.5G technology.

GEO: See Geosynchronous Earth Orbit Satellite.

Geosynchronous Earth Orbit Satellite (GEO): The type of satellite most commonly used in fixed wireless access today; orbits the earth at about 36,000 km (22,300 miles).

Get: An SNMP command sent by the manager that tells the agent to retrieve certain information and return this information to the manager.

GHz: See Gigahertz.

Gigahertz (GHz): One billion hertz.

GIGO: Garbage in, garbage out. If bad information is put into a system, only bad information can come out.

Global System for Mobile communication (GSM): The cellular telephone technology on which nearly the entire world standardized for 2G service. GSM uses 200 kHz channels and implements TDM.

Gnutella: A pure P2P file-sharing application that addresses the problems of transient presence and transient IP addresses without resorting to the use of any server.

Golden Zone: The portion of the frequency spectrum from the high megahertz range to the low gigahertz range, wherein commercial mobile services operate.

GPO: See Group Policy Object.

GPRS: See General Packet Radio Service.

Grid Computing: Computing in which all devices, whether clients or servers, share their processing resources.

Group Policy Object (GPO): A policy that governs a specific type of resource on a domain.

GSM: See Global System for Mobile communication.

H.323: In IP telephony, one of the protocols used by signalling gateways.

Hacking: The intentional use of a computer resource without authorization or in excess of authorization.

Half-Duplex: The mode of operation wherein two communicating NICs must take turns transmitting.

Handoff: 1) In wireless LANs, a change in access points when a user moves to another location. 2) In cellular telephony, transfer from one cellsite to another, which occurs when a subscriber moves from one cell to another within a system.

Hardened: Set up to protect itself, as a server or client.

Hash: The output from hashing.

Hashing: A mathematical process that, when applied to a bit string of any length, produces a value of a fixed length, called the hash.

HDSL: See High-Rate Digital Subscriber Line.

HDSL2: A newer version of HDSL, that transmits in both directions at 1.544 Mbps.

Header: The part of a message that comes before the data field.

Header Checksum: The UDP datagram field that allows the receiver to check for errors.

Headquarters: The First Bank of Paradise's downtown office building that houses the administrative site.

Hertz (Hz): One cycle per second, a measure of frequency.

Hex Notation: See Hexadecimal Notation.

Hexadecimal (Hex) Notation: The Base 16 notation that humans use to represent address 48-bit MAC source and destination addresses.

Hierarchy: 1) The type of topology wherein there are multiple layers of switches organized in a hierarchy, in which each node has only one parent node; used in Ethernet. 2) In IP addresses, three multiple parts that represent successively more specific locations for a host.

Hierarchical Topology: A network topology in which all switches are arranged in a hierarchy, in which each switch has only one parent switch above it (the root switch, however, has no parent); used in Ethernet.

High-Rate Digital Subscriber Line (HDSL): The most popular business DSL, which offers symmetric transmission at 768 kbps in both directions. See also HDSL2.

Hop-by-Hop: A layer in which communication is governed by each individual switch or router along the path of a message.

Host: Any computer attached to the Internet (can be either personal client or server).

Host Computer: 1) In terminal–host computing, the host that provides the processing power; 2) on an internet, any host.

Host Part: The part of an IP address that identifies a particular host on a subnet.

Hot Spot: A public location where anyone can connect to an access point for Internet access.

HTML: See Hypertext Markup Language.

HTML Body: Body part in a Hypertext Markup Language message.

HTTP: See Hypertext Transfer Protocol.

HTTP Request Message: In HTTP, a message in which a client requests a file or another service from a server.

HTTP Request–Response Cycle: An HTTP client request followed by an HTTP server response.

HTTP Response Message: In HTTP, a message in which a server responds to a client request; either contains a requested file or an error message explaining why the requested file could not be supplied.

Hub: An early device used by Ethernet LANs to move frames in a system. Hubs broadcast each arriving bit out all ports except for the port that receives the signal.

Hub-and-Spoke Topology: A topology in which all communication goes through one site.

Hybrid TCP/IP-OSI Standards Architecture: The architecture that uses OSI standards at the physical and data link layers and TCP/IP standards at the internet, transport, and application layers; dominant in corporations today.

Hypertext Markup Language (HTML): The language used to create webpages.

Hypertext Transfer Protocol (HTTP): The protocol that governs interactions between the browser and webserver application program.

Hz: See Hertz.

ICC: See International Common Carrier.

ICF: See Internet Connection Firewall.

ICMP: See Internet Control Message Protocol.

ICMP Echo: A message sent by a host or router to another host or router. If the target device's internet process is able to do so, it will send back an echo response message.

ICMP Error Message: A message sent in error advisement to inform a source device that an error has occurred.

ICS: See Internet Connection Sharing.

IDC: See Insulation Displacement Connection.

Identification Field: In IPv4, header field used to reassemble fragmented packets. Each transmitted packet is given a unique identification field value. If the packet is fragmented en route, all fragments are given the initial packet's identification field value.

IDS: See Intrusion Detection System.

IEEE: See Institute for Electrical and Electronics Engineers.

IETF: See Internet Engineering Task Force.

ILEC: See Incumbent Local Exchange Carrier.

IM: See Instant Messaging.

Image: An exact copy.

IMAP: See Internet Message Access Protocol.

Impostor: Someone who claims to be someone else.

Incident: A successful attack.

Incident Severity: The degree of destruction inflicted by an attack.

Incumbent Local Exchange Carrier (ILEC): The traditional monopoly telephone company within each LATA.

Index Server: A server used by Napster. Stations connected to Napster would first upload a list of their files available for sharing to index servers. Later, when they searched, their searches went to the index servers and were returned from there.

Individual Throughput: The actual speed a single user receives (usually much lower than aggregate throughput in a system with shared transmission speed).

Ingress Filtering: The filtering of traffic coming into a site from the outside.

Inherit: When permissions are assigned to a user in a directory, user automatically receives the same permissions in subdirectories unless this automatic inheritance is blocked.

Initial Installation: The initial phase of a product's life cycle. Ongoing costs may be much higher.

Initial Labor Costs: The labor costs of setting up a system for the first time.

Initial Sequence Number (ISN): The sequence number placed in the first TCP segment a side transmits in a session; selected randomly.

Instance: An actual example of a category.

Instant Messaging (IM): A popular P2P application that allows two users to type messages back and forth in real time.

Institute for Electrical and Electronics Engineers (IEEE): An international organization whose 802 LAN/MAN Standards Committee creates many LAN standards.

Insulation Displacement Connection (IDC): Connection method used in UTP. A connector bites through the insulation around a wire, making contact with the wire inside.

Interexchange Carrier (IXC): A telephone carrier that transmits voice traffic between LATAs.

Interface: 1) The router's equivalent of a network interface card; a port on a router that must be designed for the network to which it connects. 2) In Web services, the outlet through which an object communicates with the outside world.

Interference: See Electromagnetic Interference.

Interior Routing Protocol: Routing protocol used within a firm's internet.

Internal Back-End System: In e-commerce, an internal e-commerce system that handles accounting, pricing, product availability, shipment, and other matters.

Internal Router: A router that connects different LANs within a site.

International Common Carrier (ICC): A telephone carrier that provides international service.

International Organization for Standardization (ISO): A strong standards agency for manufacturing, including computer manufacturing.

International Telecommunications Union-Telecommunications Standards Sector (ITU-T): A standards agency that is part of the United Nations and that oversees international telecommunications.

Internet: 1) A group of networks connected by routers so that any application on any host on any network can communicate with any application on any other host on any other network. 2) A general term for any internetwork (spelled with a lowercase *i*); 3) the worldwide Internet (spelled with a capital *I*).

Internet Backbone: The collection of all Internet Service Providers that provide Internet transmission service.

Internet Connection Firewall (ICF): The built-in stateful firewall that comes with Windows XP.

Internet Connection Sharing (ICS): Microsoft Windows service that allows a PC to connect to the Internet through another PC.

Internet Control Message Protocol (ICMP): The protocol created by the IETF to oversee supervisory messages at the internet layer.

Internet Engineering Task Force (IETF): TCP/IP's standards agency.

Internet Layer: The layer that governs the transmission of a packet across an entire internet.

Internet Message Access Protocol (IMAP): One of the two protocols used to download received e-mail from an e-mail server; offers more features but is less popular than POP.

Internet Network: A network on the Internet owned by a single organization, such as a corporation, university, or ISP.

Internet Protocol (IP): The TCP/IP protocol that governs operations at the internet layer. Governs packet delivery from host to host across a series of routers.

Internetwork Operating System (IOS): The operating system that Cisco Systems uses on all of its routers and most of its switches.

Intranet: An internet for internal transmission within firms; uses the TCP/IP transmission standards that govern transmission over the Internet.

Intrusion Detection System (IDS): A security system that examines messages traveling through a network. IDSs look at traffic broadly, identifying messages that are suspicious. Instead of discarding these packets, IDSs will sound an alarm.

IOS: See Internetwork Operating System.

IP: See Internet Protocol.

IP Address: An Internet Protocol address; the address that every computer needs when it connects to the Internet; IP addresses are 32 bits long.

IP Security (IPsec): A set of standards that operate at the internet layer and provide security to all upper layer protocols transparently.

IP Telephone: A telephone that has the electronics to encode voice for digital transmission and to send and receive packets over an IP internet.

IP Telephony: The transmission of telephone signals over IP internets instead of over circuit-switched networks.

IP Version 4 (IPv4): The standard that governs most routers on the Internet and private internets.

IP Version 6 (IPv6): A new version of the Internet Protocol.

Ipconfig (ipconfig): A command used to find information about one's own computer, used in newer versions of Windows (the command is typed as ipconfig /all[Enter] at the command line).

IPsec: See IP Security.

IPsec Gateway: Border device at a site that converts between internal data traffic into protected data traffic that travels over an untrusted system such as the Internet.

IPv4: See IP Version 4.

IPv6: See IP Version 6.

IPX/SPX Architecture: Non-TCP/IP standards architecture found at upper layers in LANs; required on all older Novell NetWare file servers.

Iris: The colored part of the eye, used in biometric authentication.

ISN: See Initial Sequence Number.

ISO: See International Organization for Standardization.

IT Guru: See OPNET IT Guru.

ITU-T: See International Telecommunications Union-Telecommunications Standards Sector.

IXC: See Interexchange Carrier.

Jacket: The outer plastic covering, made of PVC, that encloses and protects the four pairs of wires in UTP or the core and cladding in optical fiber.

Java Applet: Small Java program that is downloaded as part of a webpage.

Jitter: Variability in latency.

Key: A bit string used with an encryption method to encrypt and decrypt a message. Different keys used with a single encryption method will give different ciphertexts from the same plaintext.

Key Exchange: The secure transfer of a symmetric session key between two communicating parties.

Label Header: In MPLS, the header added to packets before the IP header; contains information that aids and speeds routers in choosing which interface to send the packet back out.

Label Number: In MPLS, number in the label header that aids label-switching routers in packet sending.

Label Switching Table: In MPLS, the table used by label-switching routers to decide which interface to use to forward a packet.

LAN: See Local Area Network.

Language Independence: In SOAP, the fact that Web service objects do not have to be written in any particular language.

LATA: See Local Access and Transport Area.

Latency: Delay, usually measured in milliseconds.

Layer 3: See Internet Layer.

Layer 4: See Transport Layer.

Layer 5: See Application Layer.

Layer 3 Switch: A router that does processing in hardware, that is much faster and less expensive than traditional software-based routers. Layer 3 switches are usually dominant in the Ethernet core above workgroup switches.

Layer 4 Switch: A switch that examines the port number fields of each arriving packet's encapsulated TCP segment, allowing it to switch packets based on the application they contain. Layer 4 switches can give priority or even deny forwarding to IP packets from certain applications.

Legacy Network: A network that uses obsolete technology; may have to be lived with for some time because upgrading all legacy networks at one time is too expensive.

Legal Retention: Rules that require IM messages to be captured and stored in order to comply with legal requirements.

Length Field: 1) The field in an Ethernet MAC frame that gives the length of the data field in octets. 2) The field in a UDP datagram that enables the receiving transport process to process the datagram properly.

LEO: See Low Earth Orbit Satellite.

Line of Sight: An unobstructed path between the sender and receiver, necessary for radio transmission at higher frequencies.

Link: Connection between a pair of routers.

Link Aggregation: The use of two or more trunk links between a pair of switches; also known as trunking or bonding.

Link State Protocol: Routing protocol in which each router knows the state of each link between routers.

Linux: A freeware version of Unix that runs on standard PCs.

Linux Distribution: A package purchased from a vendor that contains the Linux kernel plus a collection of many other programs, usually taken from the GNU project.

List Folder Contents: A Microsoft Windows Server permission that allows the account owner to see the contents of a folder (directory).

LLC: See Logical Link Control.

LLC Header: See Logical Link Control Layer Header.

Load-Balancing Router: Router used on a server farm that sends client requests to the first available server.

Local: The value placed in the next-hop routing field of a routing table to specify that the destination host is on the selected network or subnet.

Local Access and Transport Area (LATA): One of the roughly 200 regions the United States has been divided into for telephone service.

Local Area Network (LAN): A network within a site.

Local Loop: In telephony, the line used by the customer to reach the PSTN's central transport core.

Logical Link Control Layer: The layer of functionality for the upper part of the data link layer, now largely ignored.

Logical Link Control Layer Header: The header at the start of the data field that describes the type of packet contained in the data field.

Longest Match: The matching row that matches a packet's destination IP address to the greatest number of bits; chosen by a router when there are multiple matches.

Loopback Address: The IP address 127.0.0.1. When a user pings this IP address, this will test their *own* computer's connection to the Internet.

Loopback Interface: A testing interface on a device. Messages sent to this interface are sent back to the sending device.

Low Earth Orbit Satellite (LEO): A type of satellite used in mobile wireless transmission; orbits a few hundred miles or a few hundred kilometers above the earth.

MAC: See Media Access Control.

MAC Address: See Media Access Control.

Mainframe Computer: The largest type of dedicated server; extremely reliable.

Malware: Software that seeks to cause damage.

Malware-Scanning Program: A program that searches a user's PC looking for installed malware.

MAN: See Metropolitan Area Network.

Manageable Switch: A switch that has sufficient intelligence to be managed from a central computer (the Manager).

Managed Device: A device that needs to be administered, such as printers, hubs, switches, routers, application programs, user PCs, and other pieces of hardware and software.

Managed Frame Relay: A type of Frame Relay service that takes on most of the management that customers ordinarily would have to do. Managed Frame Relay provides traffic reports and actively manages day-to-day traffic to look for problems and get them fixed.

Management Information Base (MIB): A specification that defines what objects can exist on each type of managed device and also the specific characteristics of each object; the actual database stored on a manager in SNMP. There are separate MIBs for different types of managed devices; both a schema and a database.

Management Program: A program that helps network administrators manage their networks.

Manager: The central PC or more powerful computer that uses SNMP to collect information from many managed devices.

Mask: A 32-bit string beginning with a series of ones and ending a series of zeros; used by routing tables to Interpret IP address part sizes. The ones designate either the network part or the network plus software part.

Mask Operations: Applying a mask of ones and zeros to a bit stream. Where the mask is 1, the original bit stream's bit results. Otherwise, the result is zero.

Mature: Technology that has been under development long enough to have its rough edges smoothed off.

Maximum Segment Size (MSS): The maximum size of TCP data fields that a receiver will accept.

Maximum Transmission Unit (MTU): The maximum packet size that can be carried by a particular LAN or WAN.

MD5: A popular hashing method.

Mean Time to Repair (MTTR): The average time it takes a staff to get a network back up after it has been down.

Media Access Control (MAC): The process of controlling when stations transmit; also, the lowest part of the data link layer, defining functionality specific to a particular LAN technology.

Media Gateway: A device that connects IP telephone networks to the ordinary public switched telephone network. Media gateways also convert between the signalling formats of the IP telephone system and the PSTN.

Medium Earth Orbit Satellite (MEO): A type of satellite used in mobile wireless transmission; orbits a few thousand miles or a few thousand kilometers above the earth.

Megahertz (MHz): One million hertz.

MEO: See Medium Earth Orbit Satellite.

Mesh Topology: 1) A topology where there are many connections among switches or routers, so there are many alternative routes for messages to get from one end of the network to the other. 2) In network design, a topology that provides direct connections between every pair of sites.

Message: A discrete communication between hardware or software processes.

Message Digest: The result of hashing a plaintext message. The message digest is signed with the sender's private key to produce the digital signature.

Message Integrity: The assurance that a message has not been changed en route; or if a message has been changed, the receiver can tell that it has.

Message Timing: Controlling when hardware or software processes may transmit.

Message Unit: Local telephone service in which a user is charged based on distance and duration.

Method: In Web services, a well-defined action that a SOAP message can request.

Metric: A number describing the desirability of a route represented by a certain row in a routing table.

Metropolitan Area Ethernet: Ethernet operating at the scale of a metropolitan area network.

Metropolitan Area Network (MAN): A WAN that spans a single urban area.

MHz: See Megahertz.

MIB: See Management Information Base.

Microsoft Windows Server: Microsoft's network operating system for servers, which comes in three versions: NT, 2000, and 2003.

Microsoft Windows XP Home: The dominant operating system today for residential PCs.

Microsoft Windows XP Professional: A version of Windows XP designed to be run in an organization; integrates with Windows Server services.

Millisecond (ms): The unit in which latency is measured.

MIME: See Multipurpose Internet Mail Extensions.

Ministry of Telecommunications: A government-created regulatory body that oversees PTTs.

Mobile IP: A system for handling IP addresses for mobile devices.

Mobile Telephone Switching Office (MTSO): A control center that connects cellular customers to one another and to wired telephone users, as well as overseeing all cellular calls (determining what to do when people move from one cell to another, including which cellsite should handle a caller when the caller wishes to place a call).

Mobile Wireless Access: Local wireless service in which the user may move to different locations.

Modal Bandwidth: The measure of multimode fiber quality; the fiber's bandwidth–distance product. A modal bandwidth of 200 MHz-km means that if your bandwidth is 100 MHz, then you can transmit 2 km.

Modal Dispersion: The main propagation problem for optical fiber; dispersion in which the difference in the arrival times of various modes (permitted light rays) is too large, causing the light rays of adjacent pulses to overlap in their arrival times and rendering the signal unreadable.

Mode: An angle light rays are permitted to enter an optical fiber core.

Modify: A Microsoft Windows Server permission that gives an account owner additional permissions to act upon files, for example, the permission to delete a file, which is not included in Write.

Modulate: To convert digital signals to analog signals.

Momentary Traffic Peak: A surplus of traffic that briefly exceeds the network's capacity, happening only occasionally.

Monochrome Text: Text of one color against a contrasting background.

More Fragments Field: In IPv4, a flag field that indicates whether there are more fragments (set) or not (not set).

MPLS: See Multiprotocol Label Switching.

Ms: See Millisecond.

MS-CHAP: Microsoft version of the Challenge–Response Authentication Protocol.

MSS: See Maximum Segment Size.

MTSO: See Mobile Telephone Switching Office.

MTTR: See Mean Time to Repair.

MTU: See Maximum Transmission Unit.

Multicasting: Simultaneously sending messages to multiple stations but not to all stations.

Multilayer Security: Applying security at more than one layer to provide defense in depth.

Multimode Fiber: The most common type of fiber in LANs, wherein light rays in a pulse can enter a fairly thick core at multiple angles.

Multipath Interference: Interference caused when a receiver receives two or more signals—a direct signal and one or more reflected signals. The multiple signals may interfere with one another.

Multiplex: To mix multiple signals together on the same line.

Multiplexing: 1) Having the packets of many conversations share trunk lines; reduces trunk line cost. 2) The ability of a protocol to carry messages from multiple next-higher-layer protocols in a single communication session.

Multiprocessing Computer: A computer with multiple microprocessors. This allows it to run multiple programs at the same time.

Multiprotocol: Characterized by implementing many different protocols and products following different architectures.

Multiprotocol Label Switching (MPLS): A traffic management tool used by many ISPs.

Multiprotocol Router: A router that can handle not only TCP/IP internetworking protocols, but also internetworking protocols for IPX/SPX, SNA, and other standards architectures.

Multipurpose Internet Mail Extensions (MIME): A standard for specifying the contents of files.

Mutual Authentication: Authentication by both parties.

Name Server: See Domain Name System.

Nanometer (nm): The measure used for wavelengths; one billionth of a meter (10^{-9} meters).

NAP: See Network Access Point.

Narrowband: 1) A channel with a small bandwidth and, therefore, a low maximum speed; 2) low-speed transmission.

NAT: See Network Address Translation.

Netstat: A popular route analysis tool, which gives data on current connections between a computer and other computers.

Network: In IP addressing, an organizational concept—a group of hosts, single networks, and routers owned by a single organization.

Network Access Point (NAP): A site where ISPs interconnect and exchange traffic.

Network Address Translation (NAT): Converting an IP address into another IP address, usually at a border firewall; disguises a host's true IP address from sniffers. Allows more internal addresses to be used than an ISP supplies a firm with external addresses.

Network Architecture: 1) A broad plan that specifies everything that must be done for two application programs on different networks on an internet to be able to work together effectively. 2) A broad plan for how the firm will connect all of its computers within buildings (LANs), between sites (WANs), and to the Internet; also includes security devices and services.

Network Interface Card (NIC): Printed circuit expansion board for a PC; handles communication with a network; sometimes built into the motherboard.

Network Layer: In OSI, Layer 3; governs internetworking. OSI network layer standards are rarely used.

Network Management Agent (Agent): A piece of software on the managed device that communicates with the manager on behalf of the managed device.

Network Management Program (Manager): A program run by the network administrator on a central computer.

Network Mapping: The act of mapping the layout of a network, including what hosts and routers are active and how various devices are connected. Its two phases are discovering and fingerprinting.

Network Operating System (NOS): A PC server operating system.

Network Part: The part of an IP address that identifies the host's network on the Internet.

Network Security: The protection of a network from attackers.

Network Simulation: The building of a model of a network that is used to project how the network will operate after a change.

Network Topology: The order in which a network's nodes are physically connected by transmission lines.

Networked Application: An application that provides service over a network.

Next Header Field: In IPv6, a header field that describes the header following the current header.

Next-Hop Router: A router to which another router forwards a packet in order to get the packet a step closer to reaching its destination host.

NIC: See Network Interface Card.

Nm (nm): See Nanometer.

Nmap: A network mapping tool that finds active IP addresses and then fingerprints them to determine their operating system and perhaps their operating system version.

Node: A client, server, switch, router, or other type of device in a network.

Noise: Random electromagnetic energy within wires; combines with the data signal to make the data signal difficult to read.

Noise Floor: The mean of the noise energy.

Noise Spike: An occasional burst of noise that is much higher or lower than the noise floor; may cause the signal to become unrecognizable.

Nonblocking: A nonblocking switch has enough aggregate throughput to handle even the highest possible input load (maximum input on all ports).

Nonoverlapping Channel: Channels whose frequencies do not overlap.

Normal Attack: An incident that does a small amount of damage and can be handled by the on-duty staff.

North Shore: The First Bank of Paradise's backup facility; able to take over within minutes if Operations fails.

NOS: See Network Operating System.

Not Set: When a flags field is given the value 0.

Nslookup (nslookup): A command that allows a PC user to send DNS lookup messages to a DNS server.

Object: A specific Web service. In SNMP, an aspect of a managed device about which data is kept.

OC: See Optical Carrier.

Octet: A collection of eight bits; same as a byte.

OFDM: See Orthogonal Frequency Division Multiplexing.

Official Internet Protocol Standards: Standards deemed official by the IETF.

Official Standards Organization: An internationally recognized organization that produces standards.

Omnidirectional Antenna: An antenna that transmits signals in all directions and receives incoming signals equally well from all directions.

On/Off Signaling: Signaling wherein the signal is on for a clock cycle to represent a one, and off for a zero. (On/off signalling is binary.)

Ongoing Costs: Costs beyond initial installation costs; often exceed installation costs.

Open Shortest Path First (OSPF): Complex but highly scalable interior routing protocol.

Operations: The First Bank of Paradise's building in an industrial area that houses the bank's mainframe operations and other back-office technical functions; also has most of the bank's IT staff, including its networking staff.

OPNET ACE: See OPNET Application Characterization Environment.

OPNET Application Characterization Environment (ACE): A network simulation program; focuses on application layer performance.

OPNET IT Guru: A popular network simulation program; focuses primarily on data link layer and internet layer performance.

Optical Carrier (OC): A number that indicates SONET speeds.

Optical Fiber: Cabling that sends signals as light pulses.

Optical Fiber Cord: A length of optical fiber.

Orthogonal Frequency Division Multiplexing (OFDM): A form of spread spectrum transmission that divides each broadband channel into subcarriers and then transmits parts of each frame in each subcarrier.

OSI: The Reference Model of Open Systems Interconnection; the 7-layer network standards architecture created by ISO and ITU-T; dominant at the physical and data link layers, which govern transmission within single networks (LANs or WANs).

OSI Application Layer (Layer 7): The layer that governs application-specific matters not covered by the OSI Presentation Layer or the OSI Session Layer.

OSI Layer 5: See OSI Session Layer.

OSI Layer 6: See OSI Presentation Layer.

OSI Layer 7: See OSI Application Layer.

OSI Presentation Layer (Layer 6): The layer designed to handle data formatting differences between two communicating computers.

OSI Session Layer (Layer 5): The layer that initiates and maintains a connection between application programs on different computers.

OSPF: See Open Shortest Path First.

Out of Phase: In multipath interference, the condition of not being in sync, as occurs with signals that have been reflected and thus traveled different distances and not arrived at the receiver at the same time.

Outsourcing: Paying other firms to handle some, most, or all IT chores.

Overprovision: To install much more capacity in switches and trunk links than will be needed most of the time, so that momentary traffic peaks will not cause problems.

Oversubscription: In Frame Relay, the state of having port speeds less than the sum of PVC speeds.

P2P: See Peer-to-Peer Architecture.

Packet: A message at the internet layer.

Packet Capture and Display Program: A program that captures selected packets or all of the packets arriving at or going out of an NIC. Afterward, the user can display key header information for each packet in greater or lesser detail.

Packet Filter Firewall: A firewall that examines fields in the internet and transport headers of individual arriving packets. The firewall makes pass/deny decisions based upon the contents of IP, TCP, UDP, and ICMP fields.

Packet Switching: The breaking of conversations into short messages (typically a few hundred bits long); allows multiplexing on trunk lines to reduce trunk line costs.

PAD Field: A field that the sender adds to an Ethernet frame if the data field is less than 46 octets long (the total length of the PAD plus data field must be exactly 46 octets long).

PAN: See Personal Area Network.

Parallel Transmission: A form of transmission that uses multiple wire pairs or other transmission media simultaneously to send a signal; increases transmission speed.

Password: A secret keyboard string only the account holder should know; authenticates user access to an account.

Password Length: The number of characters in a password.

Password Reset: The act of changing a password to some value known only to the systems administrator and the account owner.

Patch: An addition to a program that will close a security vulnerability in that program.

Patch Cord: A cord that comes precut in a variety of lengths, with a connector attached; usually either UTP or optical fiber.

Payload: 1) A piece of code that can be executed by a virus or worm after it has spread to multiple machines. 2) ATM's name for a data field.

Payment Mechanism: In e-commerce, ways for purchasers to pay for their ordered goods or services.

PBX: See Private Branch Exchange.

PC Server: A server that is a personal computer.

PCM: Pulse Code Modulation.

Peer-to-Peer Architecture (P2P): The application architecture in which most or all of the work is done by cooperating user computers, such as desktop PCs. If servers are present at all, they serve only facilitating roles and do not control the processing.

Peer-to-Peer Service: Service wherein client PCs provide services to one another.

Perfect Internal Reflection: When light in optical fiber cabling begins to spread, it hits the cladding and is reflected back into the core so that no light escapes.

Permanent IP Address: An IP address given to a server that the server keeps and uses every single time it connects to the Internet. (This is in contrast to client PCs, which receive a new IP address every time they connect to the Internet.)

Permanent Virtual Circuit (PVC): A PSDN connection between corporate sites that is set up once and kept in place for weeks, months, or years at a time.

Permission: A rule that determines what an account owner can do to a particular resource (file or directory).

Personal Area Network (PAN): A small wireless network used by a single person.

Phase Modulation: Modulation in which one wave serves as a reference wave or a carrier wave. Another wave varies its phase to represent one or more bits.

Physical Layer: The standards layer that governs physical transmission between adjacent devices; OSI Layer 1.

Physical Link: A connection linking adjacent devices on a network.

Piggybacking: The act of an attacker being allowed physical entrance to a building by following a legitimate user through a locked door that the victim has opened.

Ping: Sending a message to another host and listening for a response to see if it is active.

Pinging: Sending an echo request message.

PKI: See Public Key Infrastructure.

Plaintext: The original message the sender wishes to send to the receiver; not limited to text messages.

Planning: Developing a broad security strategy that will be appropriate for a firm's security threats.

Plan–Protect–Respond Cycle: The basic management cycle in which the three named stages are executed repeatedly.

Plenum: The type of cabling that must be used when cables run through airways to prevent toxic fumes in case of fire.

Point of Presence (POP): 1) In cellular telephony, a site at which various carriers that provide telephone service are interconnected. 2) In PSDNs, a point of connection for user sites. There must be a private line between the site and the POP.

Point-to-Point Topology: A topology wherein two nodes are connected directly.

Point-to-Point Tunneling Protocol (PPTP): A remote access VPN security standard offering moderate security. PPTP works at the data link layer, and it protects all messages above the data link layer, providing protection transparently.

POP: See 1) Point of Presence. 2) See Post Office Protocol.

Port: In TCP and UDP messages, a header field that designates the application layer process on the server side and a specific connection on the client side.

Port Number: The field in TCP and UDP that tells the transport process what application process sent the data in the data field or should receive the data in the data field.

Portfolio: A planned collection of projects.

Post Office Protocol (POP): The most popular protocol used to download e-mail from an e-mail server to an e-mail client.

PPTP: See Point-to-Point Tunneling Protocol.

Preamble Field: The initial field in an Ethernet MAC frame; synchronizes the receiver's clock to the sender's clock.

Presence Server: A server used in many P2P systems; knows the IP addresses of each user and also whether the user is currently on line and perhaps whether or not the user is willing to chat.

Presentation Layer: See OSI Presentation Layer.

Print Server: An electronic device that receives print jobs and feeds them to the printer attached to the print server.

Printer Sharing: Allowing multiple PCs to share a single printer.

Priority: Preference given to latency-sensitive traffic, such as voice and video traffic, so that latency-sensitive traffic will go first if there is congestion.

Priority Level: The three-bit field used to give a frame one of eight priority levels from 000 (zero) to 111 (eight).

Private Branch Exchange (PBX): An internal telephone switch.

Private IP Address: An IP address that may be used only within a firm. Private IP addresses have three designated ranges; 10.x.x.x, 192.168.x.x, and 172.16.x.x through 172.31.x.x.

Private Key: A key that only the true party should know. Part of a public key–private key pair.

Private Line Circuit: A circuit that is always on; carries data much faster than dial-up circuits and can multiplex calls.

Probable Annual Loss: The likely annual loss from a particular threat. The cost of a successful attack times the probability of a successful attack in a one-year period.

Probe Packet: A packet sent into a firm's network during scanning; responses to the probe packet tend to reveal information about a firm's general network design and about its individual computers—including their operating systems.

Problem Update: An update that causes disruptions, such as slowing computer operation.

Propagate: To travel.

Propagation Effects: Changes in the signal during propagation.

Property: A characteristic of an object.

Protecting: Implementing a strategic security plan; the most time-consuming stage in the plan–protect–respond management cycle.

Protocol: 1) A standard that governs interactions between hardware and software processes at the same layer but on different hosts. 2) In IP, the header field that describes the content of the data field.

Protocol Fidelity: The assurance that an application using a particular port is the application it claims to be.

Protocol Field: In IP, a field that designates the protocol of the message in the IP packet's data field.

Provision: To install and set up a local loop access line.

PSDN: See Public Switched Data Network.

PSTN: See Public Switched Telephone Network.

PTT: See Public Telephone and Telegraphy Authority.

Public IP Address: An IP address that must be unique on the Internet.

Public Key: A key that is not kept secret. Part of a public key–private key pair.

Public Key Authentication: Authentication in which each user has a public key and a private key. Authentication depends on the applicant knowing the true party's private key; requires a digital certificate to give the true party's public key.

Public Key Encryption: Encryption in which each side has a public key and a private key, so there are four keys in total for bidirectional communication. The sender encrypts messages with the receiver's public key. The receiver, in turn, decrypts incoming messages with the receiver's own private key.

Public Key Infrastructure (PKI): A total system (infrastructure) for public key encryption.

Public Switched Data Network (PSDN): A carrier WAN that provides data transmission service. The customer only needs to connect to the PSDN by running one private line from each site to the PSDN carrier's nearest POP.

Public Switched Telephone Network (PSTN): The worldwide telephone network.

Public Telephone and Telegraphy authority (PTT): The traditional title for the traditional monopoly telephone carrier in most countries.

Pulse Code Modulation (PCM): An analog-to-digital conversion technique in which the ADC samples the band-pass-filtered signal 8,000 times per second, each time measuring the intensity of the signal and representing the intensity by a number between 0 and 255.

PVC: See Permanent Virtual Circuit.

QAM: See Quadrature Amplitude Modulation.

QoS: See Quality of Service.

QPSK: See Quadrature Phase Shift Keying.

Quadrature Amplitude Modulation (QAM): Modulation technique that uses two carrier waves—a sine carrier wave and a cosine carrier wave. Each can vary in amplitude.

Quadrature Phase Shift Keying (QPSK): Modulation with four possible phases. Each of the four states represents two bits (00, 01, 10, and 11).

Quality of Service (QoS): Numerical service targets that must be met by networking staff.

Quality-of-Service (QoS) Parameters: In IPv4, service quality parameters applied to all packets with the same TOS field value.

Radio Wave: An electromagnetic wave in the radio range.

RAS: See Remote Access Server.

Read: A Microsoft Windows Server permission that allows an account owner to read files in a directory. This is read-only access; without further permissions, the account owner cannot change the files.

Read and Execute: A set of Microsoft Windows Server permissions needed to run executable programs.

Real Time Protocol (RTP): The protocol that adds headers that contain sequence numbers to ensure that the UDP datagrams are placed in proper sequence and that they contain time stamps so that jitter can be eliminated.

Redundancy: Duplication of a hardware device in order to enhance reliability.

Regenerate: In a switch or router, to clean up a signal before sending it back out.

Relay Server: A server used in some IM systems, which every message flows through. Relay servers permit the addition of special services, such as scanning for viruses when files are transmitted in an IM system.

Reliability: The situation of errors being corrected by resending lost or damaged messages.

Remote Access Server (RAS): A server to which remote users connect in order to have their identities authenticated so they can get access to a site's internal resources.

Remote Monitoring (RMON) Probe: A specialized type of agent that collects data on network traffic passing through its location instead of information about the RMON probe itself.

Repeat Purchasing: In e-commerce, a consumer returning to a site where he or she had made a purchase previously and making another purchase; essential to profitability.

Request for Comment (RFC): A document produced by the IETF that may become designated as an Official Internet Protocol Standard.

Request to Send: A message sent to an access point when a station wishes to send and is able to send because of CSMA/CA. The station may send when it receives a clear-to-send message.

Request to Send/Clear to Send: A system that uses request-to-send and clear-to-send messages to control transmissions and avoid collisions in wireless transmission.

Resegment: Dividing a collision domain into several smaller collision domains to reduce congestion and latency.

Responding: In security, the act of stopping and repairing an attack.

Response Message: In Challenge–Response Authentication Protocols, the message that the applicant returns to the verifier.

Response Time: The difference between the time a user types a request and the time the user receives a response.

Retention: Rules that require IM messages to be captured and stored in order to comply with legal requirements.

RFC: See Request for Comment.

RFC 2822: The standard for e-mail bodies that are plaintext messages.

RFC 822: The original name for RFC 2822.

Ring Topology: A topology in which stations are connected in a loop and messages pass in only one direction around the loop.

RIP: See Routing Information Protocol.

Risk Analysis: The process of balancing threats and protection costs.

RJ-45 Connector: The connector at the end of a UTP cord, which plugs into an RJ-45 jack.

RJ-45 Jack: The type of jack into which UTP cords' RJ-45 connectors may plug.

RMON Probe: See Remote Monitoring Probe.

Roaming: The situation when a subscriber leaves a metropolitan cellular system and goes to another city or country. Roaming requires the destination cellular system to be technologically compatible with the subscriber's cellphone. It also requires administration permission from the destination cellular system.

Rogue Access Point: An access point set up by a department or individual and not sanctioned by the firm.

Root: 1) The level at the top of a DNS hierarchy, consisting of all domain names. 2) A super account on a Unix server that automatically has full permissions in every directory on the server.

Route: The path that a packet takes across an internet.

Route Analysis: Determining the route a packet takes between your host and another host and analyzing performance along this route.

Router: A device that forwards packets within an internet. Routers connect two or more single networks (subnets).

Routing: 1) The forwarding of IP packets; 2) the exchange of routing protocol information through routing protocols.

Routing Information Protocol (RIP): A simple but limited interior routing protocol.

Routing Protocol: A protocol that allows routers to transmit routing table information to one another.

RSA: Popular public key encryption method.

RTP: See Real Time Protocol.

RTS: See Request to Send.

RTS/CTS: See Request to Send/Clear to Send.

Sample: To read the intensity of a signal.

SC Connector: A square optical fiber connector, recommended in the TIA/EIA-568 standard for use in new installations.

Scalability: The ability of a technology to handle growth.

Scanning: To try to determine a network's design through the use of probe packets.

Schema: The design of a database, telling the specific types of information the database contains.

Scope: A parameter on a DHCP server that determines how many subnets the DHCP server may serve.

Script Kiddie: An attacker who possesses only modest skills but uses attack scripts created by experienced hackers; dangerous because there are so many.

SDH: See Synchronous Digital Hierarchy.

Second-and-a-Half Generation (2.5G): A nickname for GPRS systems, which offer a substantial improvement over plain 2G GSM but which is not a full third-generation service.

Second-Generation (2G): The second generation of cellular telephony, introduced in the early 1990s. Offers the improvements of digital service, 150 MHz of bandwidth, a higher frequency range of operation, and slightly higher data transmission speeds.

Second-Level Domain: The third level of a DNS hierarchy, which usually specifies an organization (e.g., microsoft.com, hawaii.edu).

Secure Hash Algorithm (SHA): A hashing algorithm that can produce hashes of different lengths.

Secure Shell (SSH): A program that provides Telnet-like remote management capabilities; and FTP-like service; strongly encrypts both usernames and passwords.

Secure Sockets Layer (SSL): The simplest VPN security standard to implement; later renamed Transport Layer Security. Provides a secure connection at the transport layer, protecting any applications above it that are SSL/TLS-aware.

Semantics: In message exchange, the meaning of each message.

Sequence Number Field: In TCP, a header field that tells a TCP segment's order among the multiple TCP segments sent by one side.

Serial Transmission: Ethernet transmission over a single pair in each direction.

Server: A host that provides services to residential or corporate users.

Server Farm: Large groups of servers that work together to handle applications.

Server Station: A station that provides service to client stations.

Service Band: A subdivision of the frequency spectrum, dedicated to a specific service such as FM radio or cellular telephone service.

Service Control Point: A database of customer information, used in Signaling System 7.

Service Level Agreement (SLA): A quality-of-service guarantee for throughput, availability, latency, error rate, and other matters.

Service Pack: For Microsoft Windows, large cumulative updates that combine a number of individual updates.

Session Key: Symmetric key that is used only during a single communication session between two parties.

Session Layer: See OSI Session Layer.

Set: An SNMP command sent by the manager that tells the agent to change a parameter on the managed device. When a flags field is given the value 1.

SETI@home: A project from the Search for Extraterrestrial Intelligence (SETI), in which volunteers download SETI@home screen savers that are really programs. These programs do work for the SETI@home server when the volunteer computer is idle. Processing ends when the user begins to do work.

Setup Fee: The cost of initial vendor installation for a system.

SFF: See Small Form Factor.

SHA: See Secure Hash Algorithm.

Shadow Zone (Dead Spot): A location where a receiver cannot receive radio transmission, due to an obstruction blocking the direct path between sender and receiver.

Shannon Equation: An equation by Claude Shannon (1938) that shows that the maximum possible transmission speed (C) when sending data through a channel is directly proportional to its bandwidth (B), and depends to a lesser extent its signal-to-noise ratio (S/N): $C = B \ Log_2 \ (1 + S/N)$.

Share: Microsoft's name for something that is shared, usually a directory or a printer.

Shared Documents Folder (SharedDocs): In Windows XP, a directory that is automatically shared. To share a file with other users on the computer or on an attached network, the user can copy a file from another directory to the Shared Document Folder.

Shared Static Key: A key that is used by all users in a system (shared) that is not changed (static).

SharedDocs: See Shared Documents Folder.

SHDSL: See Super-High-Rate DSL.

Shopping Cart: A core e-commerce function that holds goods for the buyer while he or she is shopping.

Signal: An information-carrying disturbance that propagates through a transmission medium.

Signal Bandwidth: The range of frequencies in a signal, determined by subtracting the lowest frequency from the highest frequency.

Signaling: In telephony, the controlling of calling, including setting up a path for a conversation through the transport core, maintaining and terminating the conversation path, collecting billing information, and handling other supervisory functions.

Signaling Gateway: The device that sets up conversations between parties, maintains these conversations, ends them, provides billing information, and does other work.

Signal-to-Noise Ratio (SNR): The ratio of the signal strength to average noise strength; should be high in order for the signal to be effectively received.

Signing: Encrypting something with the sender's private key.

Simple File Sharing: In Windows XP, extremely weak security used on files in Shared Documents folders. Simple File Sharing does not even use a password; the only security is that people must know the workgroup names to read and change files.

Simple Mail Transfer Protocol (SMTP): The protocol used to send a message to a user's outgoing mail host and from one mail host to another; requires a complex series of interactions between the sender and receiver before and after mail delivery.

Simple Network Management Protocol (SNMP): The protocol that allows a general way to collect rich data from various managed devices in a network.

Simple Object Access Protocol (SOAP): A standardized way for a Web service to expose its methods on an interface to the outside world.

Single Point of Failure: When the failure in a single component of a system can cause a system to fail or be seriously degraded.

Single Sign-On (SSO): Authentication in which a user can authenticate himself or herself only once and then have access to all authorized resources on all authorized systems.

Single-Mode Fiber: Optical fiber whose core is so thin (usually 8.3 microns in diameter) that only a single mode can propagate—the one traveling straight along the axis.

SIP: One of the protocols used by signalling gateways.

Situation Analysis: The examination of a firm's current situation, which includes anticipation of how things will change in the future.

SLA: See Service Level Agreement.

Sliding Window Protocol: Flow control protocol that tells a receiver how many more bytes it may transmit before receiving another acknowledgement, which will give a longer transmission window.

Slot: A very brief time period used in Time Division Multiplexing; a subdivision of a frame. Carries one sample for one circuit.

Small Form Factor (SFF): A variety of optical fiber connectors; smaller than SC or ST connectors but unfortunately not standardized.

Small Office or Home Office (SOHO): A small-scale network for a small office or home office.

SMTP: See Simple Mail Transfer Protocol.

SNA: See Systems Network Architecture.

Sneakernet: A joking reference to the practice of walking files around physically, instead of using a network for file sharing.

SNMP: See Simple Network Management Protocol.

SNR: See Signal-to-Noise Ratio.

SOAP: See Simple Object Access Protocol.

Social Engineering: Tricking people into doing something to get around security protections.

Socket: The combination of an IP address and a port number, designating a specific connection to a specific application on a specific host. It is written as an IP address, a colon, and a port number, for instance 128.171.17.13:80.

SOHO: See Small Office or Home Office.

Solid-Wire UTP: Type of UTP in which in which each of the eight wires really is a single solid wire.

SONET: See Synchronous Optical Network.

Spam: Unsolicited commercial e-mail.

Spanning Tree Protocol (STP): See 802.1D Spanning Tree Protocol.

Speech Codec: See codec.

Spread Spectrum Transmission: A type of radio transmission that takes the original signal and spreads the signal energy over a much broader channel than would be used in normal radio transmission; used in order to reduce propagation problems, not for security.

SSH: See Secure Shell.

SSL/TLS: See Secure Sockets Layer and Transport Layer Security.

SSL/TLS-Aware: Modified to work with SSL/TLS.

SSL: See Secure Sockets Layer.

SSO: See Single Sign-On.

ST Connector: A cylindrical optical fiber connector, sometimes called a bayonet connector because of the manner in which it pushes into an ST port and then twists to be locked in place.

Standard: A rule of operation that allows two hardware or software processes to work together. Standards normally govern the exchange of messages between two entities.

Standards Agency: An organization that creates and maintains standards.

Standards Architecture: A family of related standards that collectively allows an application program on one machine on an internet to communicate with another application program on another machine on the internet.

Star Topology: A form of topology in which all wires in a network connect to a single switch.

Start of Frame Delimiter Field: The second field of an Ethernet MAC frame, which synchronizes the receiver's clock to the sender's clock and then signals that the synchronization has ended.

Stateful Firewall: A firewall whose default behavior is to allow all connections initiated by internal hosts but to block all connections initiated by external hosts. Only passes packets that are part of approved connections.

Station: A computer that communicates over a network.

STM: See Synchronous Transfer Mode.

Store-and-Forward: Switching wherein the Ethernet switch waits until it has received the entire frame before sending the frame back out.

STP: See 802.1D Spanning Tree Protocol.

Strain Relief: Crimping the back of an RJ-45 connector into an RJ-45 cord so that if the cord is pulled, it will not come out of the connector.

Stranded-Wire UTP: Type of UTP in which each of the eight "wires" really is a collection of wire strands.

Stripping Tool: Tool for stripping the sheath off the end of a UTP cord.

Strong Keys: Keys that are too long to be cracked by exhaustive key search.

Subcarrier: A channel that is itself a subdivision of a broadband channel, used to transmit frames in OFDM.

Subnet: A small network that is a subdivision of a large organization's network.

Subnet Part: The part of an IP address that specifies a particular subnet within a network.

Super Client: "Serverish" client in Gnutella that is always on, that has a fixed IP address, that has many files to share, and that is connected to several other super clients.

Super-High-Rate DSL (SHDSL): The next step in business DSL, which can operate symmetrically over a single voice-grade twisted pair and over a speed range of 384 kbps to 2.3 Mbps. It can also operate over somewhat longer distances than HDSL2.

Surreptitiously: Done without someone's knowledge, such as surreptitious face recognition scanning.

SVC: See Switched Virtual Circuit.

Switch: A device that forwards frames within a single network.

Switched Virtual Circuit (SVC): A circuit between sites that is set up just before a call and that lasts only for the duration of the call.

Switching Matrix: A switch component that connects input ports to output ports.

Symmetric Key Encryption: Family of encryption methods in which the two sides use the same key to encrypt messages to each other and to decrypt incoming messages. In bidirectional communication, only a single key is used.

SYN Bit: In TCP, the flags field that is set to indicate if the message is a synchronization message.

Synchronous Digital Hierarchy (SDH): The European version of the technology upon which the world is nearly standardized.

Synchronous Optical Network (SONET): The North American version of the technology upon which the world is nearly standardized.

Synchronous Transfer Mode (STM): A number that indicates SDH speeds.

Syntax: In message exchange, how messages are organized.

Systems Administration: The management of a server.

Systems Network Architecture (SNA): The standards architecture traditionally used by IBM mainframe computers.

T568B: Wire color scheme for RJ-45 connectors; used most commonly in the United States.

Tag: An indicator on an HTML file to show where the browser should render graphics files, when it should play audio files, and so forth.

Tag Control Information: The second tag field, which contains a 12-bit VLAN ID that it sets to zero if VLANs are not being implemented. If VLANs are being used, each VLAN will be assigned a different VLAN ID.

Tag Field: One of the two fields added to an Ethernet MAC layer frame by the 802.1Q standard.

Tag Protocol ID: The first tag field used in the Ethernet MAC layer frame. The Tag Protocol ID has the two-octet hexadecimal value 81-00, which indicates that the frame is tagged.

TCO: See Total Cost of Ownership.

TCP: See Transmission Control Protocol.

TCPDUMP: The most popular freeware packet analysis program; the Unix version.

TCP Segment: A TCP message.

TCP/IP: The Internet Engineering Tasks Force's standards architecture; dominant above the data link layer.

TDM: See Time Division Multiplexing.

TDR: See Time Domain Reflectometry.

Telecommunications Closet: The location on each floor of a building where cords coming up from the basement are connected to cords that span out horizontally to telephones and computers on that floor.

Telephone Modem: A device used in telephony that converts digital data into an analog signal that can transfer over the local loop.

Telnet: The simplest remote configuration tool; lacks encryption for confidentiality.

Temporal Dispersion: Another name for modal dispersion.

Temporal Key Integrity Protocol (TKIP): A security process used by 802.11i, where each station has its own nonshared key after authentication and where this key is changed frequently.

Terminal Crosstalk Interference: Crosstalk interference at the ends of a UTP cord, where wires are untwisted to fit into the connector. To control terminal crosstalk interference, wires should not be untwisted more than a half inch to fit into connectors.

Termination Equipment: Equipment that connects a site's internal telephone system to the local exchange carrier.

Terrestrial: Earth-based.

Test Signals: Signal sent by a high-quality UTP tester through a UTP cord to check signal quality parameters.

TFTP: See Trivial File Transfer Protocol.

Third-Generation (3G): The newest generation of cellular telephony, able to carry data at much higher speeds than 2G systems.

Three-Party Call: A call in which three people can take part in a conversation.

Three-Tier Architecture: An architecture where processing is done in three places: on the client, on the application server, and on other servers.

Throughput: The transmission speed that users actually get. Usually lower than a transmission system's rated speed.

TIA/EIA-568: The standard that governs transmission media in the United States.

Time Division Multiplexing (TDM): A technology used by telephone carriers to provide reserved capacity on trunk lines between switches. In TDM, time is first divided into frames, each of which are divided into slots; a circuit is given the same slot in every frame.

Time Domain Reflectometry (TDR): Sending a signal in a UTP cord and recording reflections; can give the length of the cord or the location of a propagation problem in the cord.

Time to Live (TTL): The field added to a packet and given a value by a source host, usually between 64 and 128. Each router along the way decrements the TTL field by one. A router decrementing the TTL to zero will discard the packet; this prevents misaddressed packets from circulating endlessly among packet switches in search of their nonexistent destinations.

TKIP: See Temporal Key Integrity Protocol.

TLS: See Transport Layer Security.

Toll Call: Long-distance call pricing in which the price depends on distance and duration.

Toll-Free Number Service: Service in which anyone can call into a company, usually without being charged. Area codes are 800, 888, 877, 866, and 855.

Top-Level Domain: The second level of a DNS hierarchy, which categorizes the domain by organization type (e.g., .com, .net, .edu, .biz, .info) or by country (e.g., .uk, .ca, .ie, .au, .jp, .ch).

Topology: The way in which nodes are linked together by transmission lines.

TOS: See Type of Service.

Total Cost of Ownership (TCO): The total cost of an entire system over its expected lifespan.

Total Purchase Cost of Network Products: The initial purchase price of a fully configured system.

Tracert (tracert): A Windows program that shows latencies to every router along a route and to the destination host.

Trailer: The part of a message that comes after the data field.

Transaction Processing: Processing involving simple, highly structured, and high-volume interactions.

Transceiver: A transmitter/receiver.

Transfer Syntax: The syntax used by two presentation layer processes to communicate, which may or may not be quite different than either of their internal methods of formatting information.

Transmission Control Protocol (TCP): The most common TCP/IP protocol at the transport layer. Connection-oriented and reliable.

Transparently: Without having a need to implement modifications.

Transport: In telephony, transmission; taking voice signals from one subscriber's access line and delivering them to another customer's access line.

Transport Core: The switches and transmission lines that carry voice signals from one subscriber's access line and deliver them to another customer's access line.

Transport Layer: The layer that governs communication between two hosts; Layer 4 in both OSI and TCP/IP.

Transport Layer Security (TLS): The simplest VPN security standard to implement; originally named Secure Sockets Layer. Provides a secure connection at the transport layer, protecting any applications above it that are SSL/TLS-aware.

Transport Mode: One of IPsec's two modes of operation, in which the two computers that are communicating implement IPsec. Transport mode gives strong end-to-end security between the computers, but it requires IPsec configuration and a digital certificate on all machines.

Traps: The type of message that an agent sends if it detects a condition that it thinks the manager should know about.

Triple DES (3DES): Symmetric key encryption method in which a message is encrypted three times with DES. If done with two or three different keys, offers strong security. However, it is processing intensive.

Trivial File Transfer Protocol (TFTP): A protocol used on switches and routers to download configuration information; has no security.

Trojan Horse: A program that looks like an ordinary system file, but continues to exploit the user indefinitely.

Trunk Line: A type of transmission line that links switches to each other, routers to each other, or a router to a switch.

Trunking: See Link Aggregation.

TTL: See Time to Live.

Tunnel Mode: One of IPsec's two modes of operation, in which the IPsec connection extends only between IPsec gateways at the two sites. Tunnel mode provides no protection within sites, but it offers transparent security.

Twisted-Pair Wiring: Wiring in which each pair's wires are twisted around each other several times per inch, reducing EMI.

Type of Service (TOS): IPv4 header field that designates the type of service a certain packet should receive.

U: The standard unit for measuring the height of switches. One U is 1.75 inches (4.4 cm) in height. Most switches, although not all, are multiples of U.

UDDI: See Universal Description, Discovery, and Integration.

UDDI Green Pages: The UDDI search option that allows companies to understand how to interact with specific Web services. Green pages specify the interfaces on which a Web service will respond, the methods it will accept, and the properties that can be changed or returned.

UDDI White Pages: The UDDI search option that allows users to search for Web services by name, much like telephone white pages.

UDDI Yellow Pages: The UDDI search option that allows users to search for Web services by function, such as accounting, much like telephone yellow pages.

Unicast: To send a frame to only one other station.

UNICODE: The standard that allows characters of all languages to be represented.

Universal Description, Discovery, and Integration (UDDI): A protocol that is a distributed database that helps users find appropriate Web services.

Unix: A network operating system used by all workstation servers. Linux is a Unix version used on PCs.

Unlicensed Radio Band: A radio band that does not require each station using it to have a license.

Unreliable: (Of a protocol) not doing error correction.

Unshielded Twisted Pair (UTP): Network cord that contains four twisted pairs of wire within a sheath. Each wire is covered with insulation.

Update: To download and apply patches to fix a system.

Usage Policy: A company policy for who may use various tools and how they may use them.

Username: An alias that signifies the account that the account holder will be using.

UTP: See Unshielded Twisted Pair.

Validate: To test the accuracy of a network simulation model by comparing its performance with that of the real network. If the predicted results match the actual results, the model is validated.

Variable-Length Subnet Mask (VLSM): A mask that allows subnets to be of different sizes.

VCI: See Virtual Channel Identifier.

Verifier: The party requiring the applicant to prove his or her identity.

Vertical Riser: Space between the floors of a building that telephone and data cabling go through to get to the building's upper floors.

Viral Networking: Networking in which the user's PC connects to one or a few other user PCs, which each connect to several other user PCs. When the user's PC first connects, it sends an initiation message to introduce itself via viral networking. Subsequent search queries sent by the user also are passed virally to all computers reachable within a few hops; used in Gnutella.

Virtual Channel: In ATM, an individual connection within a virtual path.

Virtual Channel Identifier (VCI): One of the two parts of ATM virtual circuit numbers.

Virtual Circuit: A transmission path between two sites or devices; selected before transmission begins.

Virtual LAN (VLAN): A closed collection of servers and the clients they serve. Broadcast signals go only to computers in the same VLAN.

Virtual Path: In ATM, a group of connections going between two sites.

Virtual Path Identifier (VPI): One of the two parts of ATM virtual circuit numbers.

Virtual Private Network (VPN): Transmission over the Internet with added security.

Virus: A piece of executable code that attaches itself to programs or data files. When the program is executed or the data file opened, the virus spreads to other programs or data files.

Virus Definitions Database: A database used by antivirus programs to identify viruses. As new viruses are found, the virus definitions database must be updated.

VLAN: See Virtual LAN.

VLSM: See Variable-Length Subnet Mask.

Voice Mail: A service that allows people to leave a message if the user does not answer his or her phone.

Voice-Grade: Wire of a quality useful for transmitting voice signals in the PSTN.

VPI: See Virtual Path Identifier.

VPN: See Virtual Private Network.

Vulnerability: A security weakness found in software.

Vulnerability Testing: Testing after protections have been configured, in which a company or a consultant attacks protections in the way a determined attacker would and notes which attacks that should have been stopped actually succeeded.

WAN: See Wide Area Network.

WATS: See Wide Area Telephone Service.

Wavelength: The physical distance between comparable points (e.g., from peak to peak) in successive cycles of a wave.

Wavelength Division Multiplexing: Using signaling equipment to transmit several light sources at slightly different wavelengths, thus adding signal capacity at the cost of using slightly more expensive signaling equipment but without incurring the high cost of laying new fiber.

WDM: See Wavelength Division Multiplexing.

Weak Keys: Keys that are shot enough to be cracked by exhaustive key search.

Web Service: A way to send processing requests to program (object) on another machine. The object has an interface to the outside world and methods that it is willing to undertake. Messages are sent in SOAP format.

Web-Enabled: Client/server processing applications that use ordinary browsers as client programs.

Webmail: Web-enabled e-mail. User needs only a browser to send and read e-mail.

Well-Known Port Number: Standard port number of a major application that is usually (but not always) used. For example, the well-known TCP port number for HTTP is 80.

WEP: See Wired Equivalent Privacy.

Wide Area Network (WAN): A network that links different sites together.

Wide Area Telephone Service (WATS): Service that allows a company to place outgoing long-distance calls at per-minute prices lower than those of directly dialed calls.

WiMAX: Broadband wireless access method. Standardized as 802.16.

Window Size Field: TCP header field that is used for flow control. It tells the station that receives the segment how many more octets that station may transmit before

getting another acknowledgement message that will allow it to send more octets.

Windows Internet Name Service (WINS): The system required by Windows clients and servers before Windows 2000 server to provide IP address for host names.

WinDUMP: The most popular freeware packet analysis program; the Windows version.

Winipconfig (winipconfig): A command used to find information about one's own computer; used in older versions of windows.

WINS: See Windows Internet Name Service.

Wired Equivalent Privacy (WEP): A weak security mechanism for 802.11.

Wireless Ethernet: Sometimes used as another name for 802.11.

Wireless LAN (WLAN): A local area network that uses radio (or rarely, infrared) transmission instead of cabling to connect devices.

Wireless Networking: Networking that uses radio transmission instead of wires to connect devices.

Wireless NIC: 802.11 network interface card.

Wireless Protected Access (WPA): 802.11 security method created as a stopgap between WEP and 802.11i.

WLAN: See Wireless LAN.

Work-Around: A process of making manual changes to eliminate a vulnerability instead of just installing a software patch.

Workgroup: A logical network. On a physical network, only PCs in the same workgroup can communicate.

Workgroup Name: To create a workgroup, all PCs in the workgroup are assigned the same workgroup name. They will find each other automatically.

Workgroup Switch: A switch to which stations connect directly.

Working Group: A specific subgroup of the 802 Committee, in charge of developing a specific group of standards. For instance, the 802.3 Working Group creates Ethernet standards.

Workstation Server: The most popular type of large dedicated server; runs the Unix operating system. It uses custom-designed microprocessors and runs the Unix operating system.

Worm: An attack program that propagates on its own by seeking out other computers, jumping to them, and installing itself.

WPA: See Wireless Protected Access.

Write: A Microsoft Windows Server permission that allows an account owner to change the contents of files in the directory.

X.509: The main standard for digital certificates.

Zero-Day Exploit: An exploit that takes advantage of vulnerabilities that have not previously been discovered or for which updates have not been created.

Index

MANAGEMENT INFORMATION SYSTEMS

MIS:

Alter, *Information Systems: The Foundation of E-Business 4/e*

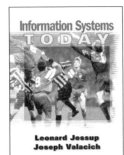

Jessup & Valacich, *Information Systems Today*

Laudon & Laudon, *Essentials of Management Information Systems 6/e*

Laudon & Laudon, *Management Information Systems 8/e*

Luftman et al., *Managing the IT Resource*

Malaga, *Information Systems Technology*

Martin et al., *Managing IT: What Managers Need to Know 5/e*

McLeod & Schell, *Management Information Systems 9/e*

McNurlin & Sprague, *Information Systems Management In Practice 6/e*

Miller, *MIS: Decision Making with Application Software (Cases) 2/e*

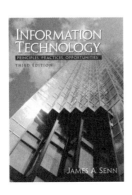

Senn, *Information Technology 3/e*

Electronic Commerce:

Awad, *Electronic Commerce 2/e*

Oz, *Foundations of Electronic Commerce*

Turban, *Electronic Commerce 2004, A Managerial Perspective*

Turban, *Introduction to E–Commerce*

Database Management:

Bordoloi & Bock, *Oracle SQL*

Bordoloi & Bock, *SQL for SQL Server*

Hoffer/Prescott/ McFadden, *Modern Database Management 7/e*

Kroenke, *Database Concepts 2/e*

Kroenke, *Database Processing: Fundamentals, Design, and Implementation 9/e*

Systems Analysis and Design:

George/Batra/Valacich/Hoffer, *Object-Oriented Systems Analysis and Design*

Hoffer/George/Valacich, *Modern Systems Analysis and Design 4/e*

Kendall & Kendall, *Systems Analysis and Design 6/e*

Stumpf & Teague, *Objected-Oriented Systems Analysis and Design*

Valacich/George/Hoffer, *Essentials of Systems Analysis and Design 2/e*

Telecommunications, Networking and Business Data Communications:

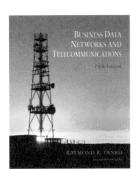

Dooley, *Business Data Communications*

Stamper & Case, *Business Data Communications 6/e*

Panko, *Business Data Networks and Telecommunications 5/e*

Security:

Panko, *Corporate Computer and Network Security*

Volonino & Robinson, *Principles and Practice of Information Security*

Other Titles:

Awad & Ghaziri, *Knowledge Management*

Becerra-Fernandez et al., *Knowledge Management*

Crews & Murphy, *Programming Right from the Start with Visual Basic .NET*

George, *Computers in Society*

Marakas, *Decision Support Systems in the 21st Century 2/e*

Marakas, *Modern Data Warehousing, Mining, and Visualization: Core Concepts*

Sumner, *Enterprise Resource Planning*

Turban & Aronson, *Decision Support Systems and Intelligent Systems 7/e*